00724

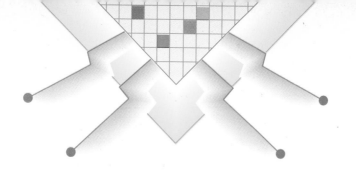

New Progress in
Mathematics

with Pre-Algebra Readiness

Rose Anita McDonnell

Catherine D. LeTourneau

Anne Veronica Burrows

with

Dr. Elinor R. Ford

Sadlier-Oxford
A Division of William H. Sadlier, Inc.

Table of Contents

Chapter 10

Plane Geometry

Chapter 11

Polygons and Circles

Chapter 12

Ratio and Proportion

Chapter 13

Percent

Chapter 14

Consumer Mathematics

Chapter 15

Surface Area and Volume

Chapter 16

Probability, Statistics, and Graphing

OPTIONAL: Algebra Readiness

Chapter 17

Algebra and Rational Number Topics

End of Book Material

Photo Credits:
ALLSTOCK: Raymond Gendreau 235.
Jeffrey Aranita 23, 45, 73, 313, 337, 385, 419.
Van Bucher 121.
FPG International: Michael Simpson 363.
THE IMAGE BANK: Jacques Cochin 151; Neal Farris 1; Nicholas Foster 287; Steven Hunt 99; Eric Meola 179.
THE STOCK MARKET: Masahiro Sano 257.
TSW/ Chicago: Franz Edson 211.

Design: Grace Kao, Kelly Kao.

Home Office: 9 Pine Street, New York, NY 10005
ISBN: 0-8215-1707-4
    6789/99

Dear Student,

Mastery in problem solving is dependent on critical thinking.
To think critically, it is essential for you to organize your thoughts.
Here are a set of steps that will help you do just that.

IMAGINE	NAME	THINK	COMPUTE	CHECK
Create a mental picture.	List the facts and the questions.	Choose and outline a plan.	Work the plan.	Test that the solution is reasonable.

When working to solve a problem, these steps help you to form a plan that will lead you to choose one or more of these problem-solving strategies:

USE THESE STRATEGIES:
Use Simpler Numbers
Working Backwards
Use/Make a Table
Find a Pattern
Write a Question
Guess and Test

USE THESE STRATEGIES:
Missing/Extra Information
Write an Equation
Combining Strategies
Hidden Information
Logical Reasoning
Interpret the Remainder

USE THESE STRATEGIES:
Use a Map/Chart
Make an Organized List
Use a Model/Drawing
Multi-Step Problem
Use a Formula
More Than One Equation

1 IMAGINE

Create a mental picture.

As you read a problem, create a picture in your mind. Make believe you are there in the problem. This will help you think about:

- what facts you will need;
- what the problem is asking;
- how you will solve the problem.

After reading the problem, draw and label a picture of what you imagine the problem is all about.

2 NAME

List the facts and the questions.

Name or list all the facts given in the problem. Be aware of *extra* information not needed to solve the problem. Look for *hidden* information to help solve the problem. Name the question or questions the problem is asking.

3 THINK

Choose and outline a plan.

Think about how to solve the problem by:

- looking at the picture you drew;
- thinking about what you did when you solved similar problems;
- choosing a strategy or strategies for solving the problem.

4 COMPUTE

Work the plan.

Work with the listed facts and the strategy to find the solution. Sometimes a problem will require you to add, subtract, multiply, or divide. Two-step problems require more than one choice of operation or strategy. It is good to *estimate* the answer before you compute.

5 CHECK

Test that the solution is reasonable.

Ask yourself:

- "Have you answered the question?"
- "Is the answer reasonable?"

Check the answer by comparing it to the estimate. If the answer is not reasonable, check your computation. You may use a calculator.

Problem-Solving Strategy: Substituting Simpler Numbers

Problem: To win first place in a figure-skating competition, Marilou needs an average of at least 5.65. If three of the four judges award her scores of 5.4, 5.8, and 5.5, what is the smallest score she can receive from the fourth judge and win first place?

1 IMAGINE Draw and label a scoreboard with Marilou's scores and the average score she needs to win.

Judge 1	Judge 2	Judge 3	Judge 4
5.4	5.8	5.5	?

needs an average of 5.65

2 NAME

Facts: 3 Scores — 5.4; 5.8; 5.5
　　　　　Average — 5.65
Question: 4th Score = __?__ to get 5.65 average

3 THINK To get an average of 4 scores,
add the four scores and divide by 4.
But the average is given and one of the scores is missing.
Substitute simpler numbers for the given decimal numbers so you can figure out mentally what to do.

Let 6 be substituted for average of 5.65 in this problem and 5 for the other 3 scores that are given.
Let S represent the missing score. This means:

$$\frac{(5 + 5 + 5 + S)}{4} = 6 \quad OR \quad (5 + 5 + 5 + S) = 6 \times 4 = 24$$

The sum of the 4 scores must, therefore, equal 24.

$(5 + 5 + 5 + S) = 24 \longrightarrow 15 + S = 24 \longrightarrow S = 24 - 15 \longrightarrow S = 9$
The fourth score here is 9.

What did you do?
1. Multiplied the average by 4.
2. Subtracted the sum of the 3 given scores from the product. The result is the fourth score.

4 COMPUTE Using the given numbers, do these two steps.

$5.65 \times 4 = 22.6$
$S = 22.6 - (5.4 + 5.8 + 5.5)$
$S = 22.6 - 16.7$
$S = 5.9$　This is the smallest 4th score needed to win.

5 CHECK Find the average of the 4 scores and see if you get back the given average, 5.65.

$$\frac{(5.4 + 5.8 + 5.5 + 5.9)}{4} = \text{Average} = \frac{22.6}{4} = 5.65 \text{ Answer checks.}$$

Problem-Solving Strategy: Working Backwards

Problem: The sum of four consecutive numbers is 174. What are the numbers?

1 IMAGINE Draw and label a picture of the 4 numbers whose sum is 174.

2 NAME *Facts:* 4 consecutive numbers
      sum of the 4 numbers: 174

Question: What are the 4 numbers?

3 THINK Usually we are asked to find a sum given the numbers. Here we have to work backwards from the sum to find the numbers.

$$n, (n + 1), (n + 2), (n + 3)$$

If we divide 174 by 4, we get the *average* of the numbers, 43.5. We know the 4 numbers are around the "40's." To find them exactly, look at your *Imagine* picture.

  Let n = 1st number $\longrightarrow$ $(n + 1)$ = 2nd number
  and $(n + 1 + 1)$ or $(n + 2)$ = 3rd number
  and $(n + 2 + 1)$ or $(n + 3)$ = 4th number

Write a word sentence showing the sum.

$$\underline{n + (n + 1) + (n + 2) + (n + 3)} = 174$$
   the sum of 4 consecutive numbers

Now simplify the above by grouping all the n's and all the whole numbers together. (The commutative and associative properties let us do this.)

$$(n + n + n + n) + (1 + 2 + 3) = 174$$
$$4n + 6 = 174$$
$$4n + 6 - 6 = 174 - 6 \quad \longleftarrow \text{Subtract 6 from both sides.}$$
$$\frac{4n}{4} = \frac{168}{4} \quad \longleftarrow \text{Divide both sides by 4.}$$
$$n = 42 \longrightarrow (n+1) = 43; (n+2) = 44; \text{ and } (n+3) = 45$$

4 CHECK Add: $(42 + 43 + 44 + 45) \overset{?}{=} 174 \longrightarrow 174 = 174$

The answers are reasonable since the averaging done in the *THINK* step showed us the numbers would be in the "40's."

Problem-Solving Strategy: Make a Table/Find a Pattern

Problem: Garim's has 100 special edition baseball cards for sale. The first customer buys 5 baseball cards but returns 1 card. Each of the next 5 customers buys 5 more cards than the preceding customer, and returns 1 more card than the preceding customer. After the sixth customer, how many of these baseball cards has Garim's sold?

1 IMAGINE Draw and label a diagram or table showing the purchases and returns of the 6 customers.

	Six Customers (C_n)						
	C_1	C_2	C_3	C_4	C_5	C_6	
Purchased	5	5 + 5 = 10	10 + 5 = 15	15 + 5 = 20	20 + 5 = 25	25 + 5 = 30	Grand
Returned	1	1 + 1 = 2	2 + 1 = 3	3 + 1 = 4	4 + 1 = 5	5 + 1 = 6	Total
Total Bought	4	8	12	16	?	?	?

2 NAME

Facts: C_1 (1st Customer) — 5 cards bought
1 card returned

C_2, C_3, C_4, C_5, C_6 — 5 more cards bought than preceding customer

1 more card returned than preceding customer

Total customers — 6

Total number of cards — 100

Question: After returns, how many cards were sold altogether?

3 THINK Complete the table made in the *Imagine* step.

Do you see a pattern in the table for finding:

- the number of cards purchased?
  (The number of cards purchased are multiples of 5.)
- the number of cards returned?
  (The number of cards returned are the counting numbers.)
- the total number of cards bought?

The total sold is computed by adding the difference between the cards bought and returned to the preceding number of total cards bought:

4 + 8 = 12 and 12 + 12 = 24 and 24 + 16 = 40 . . .

4 COMPUTE Finish the table by completing the calculations.

5 CHECK Put the answers from your computations into the table.
Do the answers fit the patterns you have observed?
Check your computations. Is your answer reasonable?

Problem-Solving Strategy: Writing the Question

Problem: Two space explorers from planet Sevenoid visited Earth to collect mineral samples. XYZ-0 collected twice as much gold and half the amount of silver as CDE-3. CDE-3 collected 40 samples of gold and 30 samples of silver. Each sample of gold weighed 2 kg and each sample of silver weighed 4 kg. The two explorers' spaceship can carry up to 550 kg of mineral samples.

	Gold		Silver	
	Samples	Wt	Samples	Wt
CDE-3	40	2 kg	30	4 kg
XYZ-0	twice CDE-3	2 kg	half CDE-3	4 kg
Spaceship — 500 kg of metal (capacity)				

1 IMAGINE

Draw and label a picture of the minerals collected by the explorers.

2 NAME

Facts:

	CDE-3	40 samples gold	2 kg each
	(c)	30 samples silver	4 kg each

XYZ-0 $x = 2$ *c*'s gold

(x) $x = \frac{1}{2}$ *c*'s silver

Spaceship   500 kg of metal (capacity)

Question: Write questions that can be solved using the facts given above. More than one operation should be used to answer each question.

3 THINK

How many kilograms of minerals did each explorer collect?

CDE-3 (c) collected

40 samples of gold each weighing 2 kg	+	30 samples of silver each weighing 4 kg

(40×2) + $(30 \times 4) = \underline{?}$

XYZ-0 (x) collected

twice the amount of *c*'s gold   +   half the amount of *c*'s silver

$2 \times (40 \times 2)$ + $\dfrac{(30 \times 4)}{2} = \underline{?}$

4 COMPUTE

(40×2) $+ (30 \times 4)$ $= \underline{?}$ kg of metal collected by CDE-3
80   +   120   = 200 kg of metal collected by CDE-3

$2 \times (40 \times 2) + \dfrac{(30 \times 4)}{2} = \underline{?}$ kg of metal collected by XYZ-0

160   +   60   = 220 kg of metal collected by XYZ-0

Use the facts in the problem to make up more problems.

5 CHECK

Check your computations. Did you answer the questions you wrote?

Problem-Solving Strategy: Using a Frequency Table and Graph

Problem: In a class of twenty-six students Ross and Lisa collected data on the eye color of the boys and girls in the class. They listed the data on a frequency table. Next they drew a double bar graph to help them analyze the data. Last they wrote several conclusions based on their discovery.

Which of Ross and Lisa's conclusions are true based on the data they collected?

1 IMAGINE You collected the data, drew the graph and came to Ross and Lisa's conclusions as shown here.

2 NAME *Facts:* Eye color of class as shown on the frequency table.

Question: Which of their conclusions are true based on this data?

3 THINK Study the frequency table and the graph that was drawn.

Examine the graph to see if it matches the data in the frequency table.

Read each of the conclusions.

Which conclusions can be drawn from the actual given data?

Which conclusions are based on incorrect data?

4 COMPUTE Compute the sums in the frequency table.

Recheck the graph to see if it is accurate.

Correct any false conclusion by rewriting it as a true one.

5 CHECK Compare your new conclusions with others who worked on this problem.

> **MAKE UP YOUR OWN...**

Survey your class and construct a frequency table on a topic like: critical ecological issues, career choices, and so on. Graph your results. Try to draw conclusions from your data. Base each conclusion on the data gathered.

Frequency Table

Eye Color	Boy	Girl
Brown	︀卌	卌 l
Blue	卌	l
Green	l	ll
Hazel	ll	llll

Graph

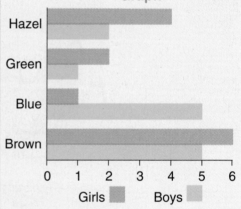

Conclusions
(Write a T next to each True one.)

1. In the class most boys have brown eyes.

2. In the class girls outnumber boys.

3. No girls have blue eyes.

4. Fewer boys than girls have green eyes.

5. Most students have hazel eyes.

6. The number of boys having hazel eyes is half the number of boys having brown eyes.

7. The number of girls having brown eyes is three times the number of girls having green eyes.

1 Numbers and Numeration

In this chapter you will:

- Read and write numerals for whole numbers, and decimals
- Compare, order, and round whole numbers and decimals
- Solve problems: guess and test
- Use technology: software tools

Do you remember?

Standard numeral is the name for a number as it is written or read aloud.

Any number can be expressed as a standard numeral with just these ten digits: 0, 1, 2, 3, 4, 5, 6, 7, 8, 9.

Period is the name for every group of three places in a numeral. Either commas or spaces separate the digits in a numeral into periods. The periods are *ones, thousands, millions, billions, trillions,* etc.

RESEARCHING TOGETHER

Money in Base 8

You are the wealthiest person on an imaginary planet! On Earth, you would have $5,654,095. But on your planet you do not have a base 10 system, you have a base 8 system.

Create a chart for your system using up to 12 places. Give names to your place values. Now show what $5,654,095 looks like in your money.

1-1 Working with Whole Numbers

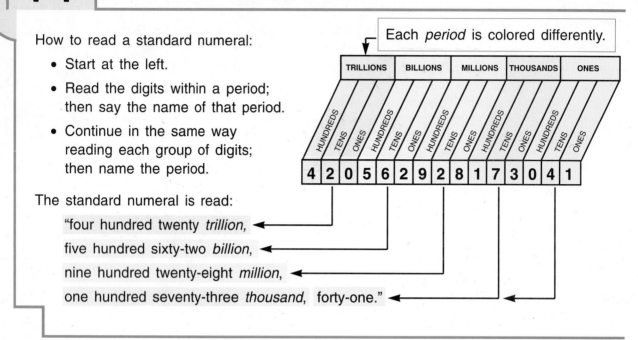

How to read a standard numeral:

- Start at the left.
- Read the digits within a period; then say the name of that period.
- Continue in the same way reading each group of digits; then name the period.

Each *period* is colored differently.

| TRILLIONS | BILLIONS | MILLIONS | THOUSANDS | ONES |

4 2 0 5 6 2 9 2 8 1 7 3 0 4 1

The standard numeral is read:

"four hundred twenty *trillion,*

five hundred sixty-two *billion,*

nine hundred twenty-eight *million,*

one hundred seventy-three *thousand,* forty-one."

Name the period of the underlined digits.

1. 68,<u>437</u>,219
2. 43,<u>739</u>,918
3. 306,221,<u>048</u>
4. 173,158,<u>470</u>
5. 492,<u>783</u>,007,512
6. 506,<u>010</u>,999,099

Read the standard numeral.

7. 27,250
8. 91,506
9. 4,857,914
10. 2,272,120
11. 101,000,101,000
12. 800,456,001,112

Write the standard numeral.*

13. four hundred five million, three thousand, twenty
14. two hundred seventy-three thousand, fifty-three
15. twelve billion, ninety-six million, fifty-three thousand, one hundred one
16. three trillion, eighteen million, eleven thousand, five hundred eleven
17. one hundred forty-two billion, two hundred thousand, twenty-six
18. eight hundred billion, five hundred thousand, fifty-five

> MAKE UP YOUR OWN... **Write the greatest possible number that you can express:**

19. using 10 digits.
20. using each digit 0 through 9 just once.

*See page 447 for more practice.

1-2 Place Value

The value of a digit depends on its position, or place, in a standard numeral.

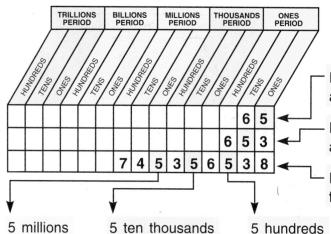

5 millions   5 ten thousands   5 hundreds

The *place-value chart* helps you see that:

In 65 the five is in the *ones* place and means 5 *ones*, or 5.

In 653 the five is in the *tens* place and means 5 *tens*, or 50.

In 745,356,538 see why each five has a different value.

In what place is the underlined digit? What is its value?*

1. 7498
2. 9073
3. 40,006
4. 13,118
5. 556,394,221
6. 449,000,000
7. 286,439,658
8. 747,680,200

Use these numbers: a. 1,947,836   b. 345,897

9. The 7 in **a** is how many times greater than the 7 in **b**?
10. The 9 in **b** is how many times less than the 9 in **a**?
11. Is the 4 in **a** greater than the 4 in **b**? Why or why not?
12. The 3 in **a** is how many times less than the 3 in **b**?

What place has a place value of:

13. 1 hundred times 10?
14. 1 hundred times 100?
15. one tenth of 100?
16. one tenth of 1000?
17. one tenth of 100,000?
18. one tenth of 1,000,000?
19. ten times 10,000?
20. ten times 1,000,000?
21. 1 one-hundredth of 100?

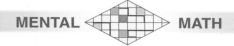

MENTAL ◁▨▷ MATH

Use the number 3,257,815. What number is:

22. 1000 greater?
23. 1,000,000 greater?
24. 100 less?
25. 2000 less?

How much less than 3,257,815 is:

26. 3,257,810?
27. 2,257,815?
28. 3,257,000?
29. 2,257,915?

*See page 447 for more practice.   3

1-3 Working with Decimals

The place-value chart is extended to include numerals for numbers less than 1. Such numbers are called *decimals*.

How to read a standard numeral for a decimal:

- Read the whole-number part first (if there is one).
- Read the decimal point as "and."
- Read the decimal part as you would read a whole number, ending with the name of the place of the last digit.

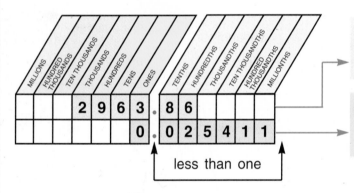

Read 2963.86 as:
"two thousand, nine hundred sixty-three and eighty-six *hundredths*."

Read 0.025411 as:
"twenty-five thousand, four hundred eleven *millionths*."

less than one

Every numeral that has only a decimal part always has a value *less than one*. The value of each decimal place is *one tenth* the value of the *next* place to the *left*.

Write the letter of the correct answer.

1. The decimal *one hundred forty-two thousandths* is written as:

    a. 142   **b.** 0.142   **c.** 0.0142   **d.** 1420

2. The decimal *six thousandths* is written as:

    a. 0.6000   **b.** 0.6   **c.** 0.006   **d.** 6000

3. The decimal *four thousand, two hundred fifty-seven millionths* is written as:

    a. 0.4200057   **b.** 0.004257   **c.** 4,200,057   **d.** 0.4257

4. The value of the hundredths place is what part of the value of the tenths place?

5. The value of the thousandths place is what part of the value of the hundredths place?

6. The value of the thousandths place is what part of the value of the tenths place?

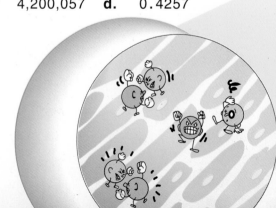

Use these numbers: **a.** 0.9456   **b.** 0.0549

7. The 9 in **a** is (ten, one hundred, one thousand) times greater than the 9 in **b**.

8. The 4 in **b** is (one tenth, one hundredth, one thousandth) of the 4 in **a**.

Read the numeral.

9. 24.6   10. 86.9   11. 0.095   12. 0.216

13. 0.0008   14. 0.5609   15. 0.00364   16. 207.09092

17. 68.0075   18. 100.00102   19. 39.08   20. 603.605057

Write the numeral.

21. four hundred six and ninety-five hundredths

22. forty-five and eighty-seven hundredths

23. twenty-seven and five hundred thirty-six ten thousandths

24. six hundred three and five thousand, three hundred forty-one ten thousandths

25. ninety-five millionths

26. one thousand and one millionths

CALCULATOR ACTIVITY

While examining species, EZ 2 DO often gets results like these.

Show these numbers on a calculator.
(Remember to press the decimal key for "and.")

27. thirty-two and ten hundredths

28. nine hundred seventy-six and nine tenths

29. six and three hundred sixty-four thousandths

30. four hundred six and ninety-five ten thousandths

31. nine hundred forty-five thousand, one hundred fifteen millionths

32. ten and eleven thousandths

33. seven hundred three and ten thousand, two hundred nine millionths

Show the number 0.367 on a calculator. Then find what number is:

34. 1000 greater   35. 100,000 greater   36. 0.1 less   37. 0.001 less

On a calculator find how much less than 4.6103 is:

38. 2.98   39. 0.9999   40. 3.006   41. 4.5

5

1-4 Whole Numbers and Expanded Form

The value of each digit in a standard numeral is shown by writing the numeral in *expanded form*.

		MILLIONS			THOUSANDS			ONES	
		HUNDRED MILLIONS 100,000,000	TEN MILLIONS 10,000,000	MILLIONS 1,000,000	HUNDRED THOUSANDS 100,000	TEN THOUSANDS 10,000	THOUSANDS 1000	HUNDREDS 100	TENS 10	ONES 1

6 4 9 0 7 3 5

$5 \times 1 = 5$
$3 \times 10 = 30$
$7 \times 100 = 700$
$0 \times 1000 = 0$
$9 \times 10,000 = 90,000$
$4 \times 100,000 = 400,000$
$6 \times 1,000,000 = 6,000,000$

In this numeral the 6 is in the millions place. Its value is $6 \times 1,000,000$ or 6,000,000.

Standard Numeral: 6,490,735

Expanded Form:

$(6 \times 1,000,000) + (4 \times 100,000) + (9 \times 10,000) + (0 \times 1000) + (7 \times 100) + (3 \times 10) + (5 \times 1)$

OR

$(6 \times 1,000,000) + (4 \times 100,000) + (9 \times 10,000) + (7 \times 100) + (3 \times 10) + (5 \times 1) = 6,490,735$

Write each standard numeral.

1. $(3 \times 100,000) + (7 \times 10,000) + (1 \times 1000) + (3 \times 100) + (2 \times 10) + (4 \times 1)$
2. $(6 \times 10,000,000) + (4 \times 100,000) + (2 \times 1000) + (1 \times 100) + (9 \times 10) + (1 \times 1)$
3. $(9 \times 1,000,000) + (7 \times 1000) + (3 \times 100)$
4. $(8 \times 10,000) + (8 \times 1000) + (8 \times 1)$

Copy and complete.

5. $37,685 = (3 \times ?) + (7 \times ?) + (? \times 100) + (8 \times ?) + (5 \times ?)$
6. $5,980,099 = (? \times 1,000,000) + (9 \times ?) + (? \times 10,000) + (9 \times ?) + (? \times 1)$
7. $444,044 = (? \times 100,000) + (4 \times ?) + (4 \times ?) + (? \times 10) + (? \times 1)$

Write each standard numeral in expanded form.*

8. 163,258   9. 6,420,058   10. 3007   11. 50,005

12. 87,005,400   13. 209,109,000   14. 68,047,007   15. 780,000,645

1-5 Decimals and Expanded Form

For decimals, the expanded form follows the same pattern as for whole numbers.

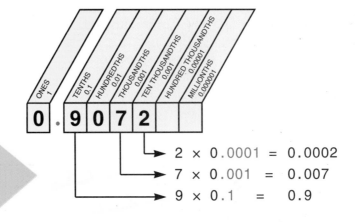

$$2 \times 0.0001 = 0.0002$$
$$7 \times 0.001 = 0.007$$
$$9 \times 0.1 = 0.9$$

In this decimal the 2 is in the ten-thousandths place. Its value is 2×0.0001 or 0.0002.

Standard Numeral: 0.9072

Expanded Form: $(9 \times 0.1) + (7 \times 0.001) + (2 \times 0.0001) = 0.9072$

In what place is the underlined digit? What is its value?

1. 0.7<u>8</u>
2. 0.9<u>7</u>8
3. 0.097<u>8</u>
4. 0.0097<u>8</u>
5. 2.643<u>9</u>

Write the standard numeral.

6. $(7 \times 0.1) + (2 \times 0.001) + (9 \times 0.0001)$
7. $(6 \times 10) + (8 \times 0.001) + (5 \times 0.00001) + (3 \times 0.000001)$
8. $(4 \times 1000) + (3 \times 1) + (1 \times 0.1) + (5 \times 0.01)$

Write the numeral in expanded form.*

9. 0.2115
10. 0.00095
11. 41.7
12. 4.404
13. 3.037
14. 0.00314
15. 63.4009
16. 700.8804

What place has a place value of:

17. ten times 0.001?
18. one tenth of 0.001?
19. ten times 0.1?

MENTAL MATH   **Use the number 0.74158. What number is:**

20. one tenth greater?
21. one hundredth less?
22. nine and one tenth greater?

CHALLENGE   **How much greater than 0.620528 is:**

23. 0.720528?
24. 0.620628?
25. 0.621528?
26. 0.820528?

*See page 447 for more practice.   7

1-6 Comparing Whole Numbers and Decimals

Compare 75,100 and 72,927.

To compare whole numbers:

- Write one number under the other so their place values line up.

- Compare the digits in the greatest place-value position.

- If these are the same, compare the next digits to the right. Do this until the digits are NOT the same.

- Compare the value of these digits to find which number is greater.

< means "is less than."
> means "is greater than."

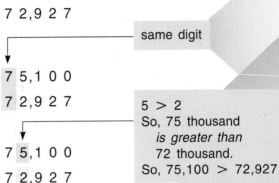

7 5,1 0 0
7 2,9 2 7

same digit

7 5,1 0 0
7 2,9 2 7

5 > 2
So, 75 thousand
is greater than
72 thousand.
So, 75,100 > 72,927

7 5,1 0 0
7 2,9 2 7

Compare 671.04 and 671.4.

To compare decimals:

- Write one number under the other so their place values line up.

- If necessary, insert zeros so the numbers have the SAME number of places.

- Compare the whole-number parts. If the whole-number parts are the same, then compare each number or digit in the decimal part until the digits are NOT the same.

- Compare the value of these digits to find which number is greater.

Whole Number Part | Decimal Part

6 7 1 . 0 4
6 7 1 . 4 0 ← Insert a zero.

6 7 1 = 6 7 1

$0.0\,4 \overset{?}{=} 0.4\,0$

because 4 hundredths
is less than
40 hundredths
So, 671.04 < 671.4

$0.0\,4 < 0.4\,0$

Compare these numbers. Write <, =, or >.*

1. 306,243 __?__ 299,798

2. 4,216,302 __?__ 4,216,301

3. 939,878,795 __?__ 946,221,584

4. 3,876,209,462 __?__ 3,881,100,005

5. 0.43 __?__ 0.444

6. 21.3541 __?__ 21.3544

7. 0.235 __?__ 0.246

8. 1.081 __?__ 1.0810

9. 1.4 __?__ 1.3948

10. 19.06 __?__ 19.0610

*See page 447 for more practice.

1-7 Ordering Whole Numbers and Decimals

To order whole numbers and decimals:

I. Line up the digits in each number. (Write any needed zeros in decimal numbers.)

II. Compare the digits in each place-value position, starting with the greatest.

III. According to the values of these digits, order the numbers either from *least* to *greatest* or from *greatest* to *least*.

Order these numbers **a, b, c, d** from least to greatest.

a. 4 3,6 9 5
b. 4 3,6 9 1
c. 4 3,5 6 9
d. 4 3,6 6 3

"43" is the same in each number.
Look at the next place.
  The "5" in **c** < "6" in **a, b,** and **d.**
  So, **c** or 43,569 is the least, or smallest number.
Now look at the tens place for **a, b,** and **d.**
  The "6" in **d** < "9" in **a** and **b.**
  So, **d** or 43,663 is the next smallest number.
Now look at the ones place for **a** and **b.**
  The "1" in **b** < "5" in **a.**
  So, **b** or 43,691 is the third smallest number.

So, the order from *least* to *greatest* is: 43,569 < 43,663 < 43,691 < 43,695

Order these numbers **a, b, c, d** from greatest to least.

a. 0.5 3 4
b. 0.5 2 9
c. 0.5 3 5
d. 0.5 3 0

> Write a zero to line the numbers up.

"5" is the same in each number.
Look at the hundredths place.
  The "2" in **b** < "3" in **a, c,** and **d.**
  So, **a, c, d** > **b**
  So, 0.529 is the smallest number.
Now look at the thousandths place.
  The "5" in **c** > "4" in **a** and "0" in **d.**
  So, **c** > **a** > **d.**
Each of these is greater than **b**, the smallest number.

So, the order from *greatest* to *least* is: 0.535 > 0.534 > 0.530 > 0.529

Order these numbers from least to greatest.

1. 14,683; 14,638; 14,688; 14,863
2. 35,514; 35,114; 35,541; 35,551
3. 2.453; 2.345; 2.435; 2.445
4. 0.6118; 0.61; 0.618; 0.6128
5. 0.109; 0.9; 0.091; 0.190
6. 0.0362; 0.602; 0.263; 0.006

7. Which is closest to 1?   **a.** 0.8969   **b.** 0.9001   **c.** 0.9498
8. Which is closest to 0.1?   **a.** 0.0899   **b.** 0.09   **c.** 0.099

1-8 Rounding Whole Numbers

The NASA computer recorded that a space probe flew a distance of 396,532 miles. A TV reporter said it flew about 400,000 miles.

To round 396,532 to the *greatest place-value position*:

1. Find the digit in the greatest place-value position.

   3 9 6,5 3 2

   3 is in the greatest place-value position.

2. Look at the next digit.
   - If this digit is **5 or more**, round the first digit UP.

   9 > 5

   3⑨6,5 3 2

   9 > 5, so round 3 up to 4.

   - If this digit is **less than 5**, leave the first digit alone.

   9 > 5

3. Replace all the digits after the greatest place-value digit with zeros.

   3 9 6,5 3 2
   ↓ ↓ ↓ ↓ ↓ ↓
   4 0 0,0 0 0

396,532 rounded to the greatest place-value position is 400,000.

To round 346,215 to the greatest place-value position:

4 < 5, so 3 remains the same.

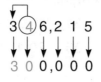

3④6,2 1 5
↓ ↓ ↓ ↓ ↓ ↓
3 0 0,0 0 0

Think
5 or more: Round *up*.
Less than 5: Leave alone.

346,215 rounded to the greatest place-value position is 300,000.

Do the same to round to the other place-value positions.

To round 49,721 to the nearest *thousand*:

7 > 5

4 9,⑦2 1
+   1
‾‾‾‾‾‾‾‾
5 0,0 0 0

9 is in the thousands place. So, look at the next digit, 7.

7 > 5, so round 9 thousand to 10 thousand or 49 thousand to 50 thousand.

Replace the other digits with zeros.

49,721 rounded to the nearest thousand is 50,000.

10

Round each number to its greatest place-value position.

1. 7359
2. 5459
3. 89,024
4. 851,492
5. 3,086,562
6. 67,876,678
7. 888,888
8. 999,999

Round to the nearest hundred.

9. 156
10. 8089
11. 271
12. 125,391

Round to the nearest thousand.

13. 35,429
14. 6802
15. 11,909
16. 401,500

Round to the nearest million.

17. 1,380,672
18. 4,627,111
19. 3,076,584
20. 8,552,999

Round to the nearest hundred million.

21. 760,841,296
22. 107,572,009
23. 295,008,721
24. 818,326,945
25. Which of these is almost 40,000?

    a. 32,988
    b. 39,501
    c. 34,999
    d. none of these

26. Which of these is almost 1 million?

    a. 982,307,000
    b. 1,750,000
    c. 986,240
    d. none of these

27. A number is rounded to the next greater thousand if the digit in which place is 5 or greater?

    a. ten thousands
    b. thousands
    c. hundreds
    d. none of these

Round each number to the place represented by the underlined digit.

28. 5867
29. 21,682
30. 67,207
31. 545,296
32. 64,987,789
33. 39,075
34. 163,596
35. 20,110,976

CRITICAL THINKING   In each THINK BOX name what does NOT belong.

Design your own THINK BOXES.

1-9 | Rounding Decimals

To round decimals, use the same rules as for whole numbers.

▶ **To round 0.38709 to the *nearest tenth*:**

$$8 > 5$$

0.3⑧7 0 9

$$8 > 5$$

So, round 3 up to 4.

Drop the remaining digits because these become zeros.

$$8 > 5$$

0.3⑧7 0 9
↓ ↓ ↓ ↓ ↓ ↓
0.4 0̸ 0̸ 0̸ 0̸ = 0.4

0.3 8 7 0 9 rounded to the nearest tenth is 0.4.

▶ **To round 0.3851 to the *nearest hundredth*:**

$$5 = 5$$

0.3 8 ⑤1

$$5 = 5$$

So, round 8 up to 9.

Drop the remaining digits after the hundredths place.

$$5 = 5$$

0.3 8 ⑤1
↓ ↓ ↓ ↓ ↓
0.3 9 0̸ 0̸ = 0.3 9

0.3 8 5 1 rounded to the nearest hundredth is 0.39.

To round 23.869 to the *nearest whole number*:

$$8 > 5, \text{ so round 3 up to 4.}$$

2 3.⑧6 9
↓ ↓ ↓ ↓ ↓
2 4.0̸ 0̸ 0̸ = 24

23.869 rounded to the nearest whole number is 24.

▶ Round 39.51 to the *nearest whole number*.

$$5 = 5, \text{ so round 9 up to 10 or 39 to 40.}$$

  3 9.⑤1
+ 1 ↓ ↓
‾‾‾‾‾‾‾‾‾‾
  4 0.0̸ 0̸ = 4 0

39.51 rounded to the nearest whole number is 40.

▶ Round 73.42 to the *nearest whole number*.

$$4 < 5, \text{ so leave 3 alone.}$$

7 3.④2
↓ ↓ ↓ ↓
7 3.0̸ 0̸ = 7 3

73.42 rounded to the nearest whole number is 73.

Round to the nearest tenth.

1. 0.32   **2.** 0.672   **3.** 0.063   **4.** 39.48   **5.** 7.082

Round to the nearest hundredth.

6. 0.126   **7.** 0.512   **8.** 0.005   **9.** 8.0508   **10.** 75.099

Round to the nearest whole number.

11. 0.888   **12.** 25.456   **13.** 7.0078   **14.** 29.821   **15.** 63.3786

To what place would you round? Why? What is the rounded answer?

16. 21.562 miles per gallon   **17.** 2.043 packages   **18.** 0.37468 batting average

Place a decimal point in the number to make each statement reasonable.
Then round the decimal to the nearest whole number.

19. 2 0 3 3 3 might be the average number of school days per month
in the regular school year.

20. If you were looking at a typical dictionary page, you might find that 6 1 4 8 5 words
were defined on that page.

21. The number 2 1 2 3 7 could be the average number of petals on a daisy.

Choose the correct answer.

22. A number is rounded up to the next greater hundredth
if the digit in which place is 5 or greater?

    a. tenths   **b.** hundredths   **c.** thousandths   **d.** none of these

23. A number has been rounded to the nearest thousandth. The rounded number is
0.493. Which of these could be the original number?

    a. 0.49362   **b.** 0.49247   **c.** both of these   **d.** neither of these

---

Rounding Money

Money is often rounded to the nearest:

Dollar	Dime	Cent
$1.98 = $2	$.67 = $.70	$17.293 = $17.29
↑ one	↑ tenth	↑ hundredth

Round to the nearest:

	Dollar	Dime	Cent
24.	$ 6.32	$ .97	$1.499
25.	$17.67	$1.22	$ .624

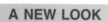 **A NEW LOOK**   **SUPPOSE THAT...**

26. Suppose the "Gas and Go" is selling hi-test gasoline for
$1.629 a gallon and regular unleaded for $1.479.
To the nearest cent, what is the cost per gallon for each gasoline?

Discover the World of Software Tools

Did you know that you can talk to computers? Indeed you can. And if you do it correctly, they may even answer you! But you need *software* to do it. Computers can be programmed, or given instructions, to perform many different types of jobs.

Kim is using a *word-processing program* to prepare a social studies report. She can easily correct her errors, insert or delete information, arrange the report in different ways, and even check her spelling. When she is satisfied with the report, she will print the final copy.

This skill will help Kim if she pursues a business, publishing, or other professional career.

Kevin is a library aide and is good at using the computerized card catalog. This is a specialized *data base file* of information that can list all books on a specific topic or by a certain author. Another data base management program keeps a record of the library circulation. Now Kevin can list all the books due on a certain date.

Kevin's skills could help him as a reservation clerk, as a medical-records technician, as a sales inventory clerk, or in many areas of business administration.

Juan is conducting a science experiment on the effects of different fertilizers on the growth of plants. He uses an *electronic spreadsheet* to plan and keep a record of his different mixtures and to relate the rate of growth to the various chemicals used.

This skill will help Juan if he pursues a career as a research scientist, investment counselor, or accountant.

What can you discover?

1. Name some of the capabilities of a word-processing program.

2. Name some of the capabilities of a data base management program.

3. Name some of the capabilities of an electronic spreadsheet.

4. Find out how computers are used in each area of employment.
   - medical profession
   - law enforcement
   - military service
   - space program
   - architecture and design
   - engineering
   - government
   - manufacturing

5. Find out what kind of work is done in each job in the computer field.
   - computer operator
   - computer engineer
   - systems analyst
   - computer programmer

6. Explain the use of each type of software.
   - CAD (Computer-Assisted Design)
   - CAI (Computer-Assisted Instruction)
   - EFT (Electronic Fund Transfer)
   - Natural Language Processing
   - Integrated Software
   - Telecommunications
   - Artificial Intelligence
   - Simulation

7. Make a collage of newspaper ads for computer-related jobs. In each ad underline the qualifications required for the job.

8. Write a report on a newspaper or magazine article about the positive and the negative impact of computers on society.

9. Interview someone who works with computers. Find out exactly how that person makes use of the computer. Ask about any difficulties that may arise from computer use. Report your findings to the group.

Problem: Guy filled these boxes with 490 peanuts. If Box 2 held twice as many peanuts as Box 1 and Box 3 held twice as many as Box 2, how many peanuts did he put in each box?

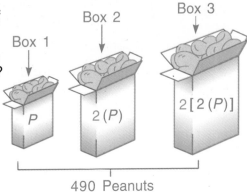

Box 1   Box 2   Box 3

490 Peanuts

1 IMAGINE   Draw and label a picture of what is happening.

2 NAME

Facts: Box 2 — twice Box 1
        Box 3 — twice Box 2
          490 = Total peanuts

Question: How many peanuts are in each box?

3 THINK

Solve this problem by making an "educated guess" from the clues given in the problem. Then test your guess. This is called the "guess and test" strategy.

To make a good "educated guess," first put 50 peanuts in the first box. Then try the problem with 100 peanuts. See if 490 comes closer to the "50" guess or the "100" guess.

4 COMPUTE

> Guess 1:   Box 1, $P =$ 50
>   Box 2, $2P = 2 \times 50$ $= 100$
>   Box 3, $2(2P)$
>   or $4P = 4 \times 50$ $= \underline{200}$
>   350 Sum

> Guess 2:   Box 1, $P =$ 100
>   Box 2, $2P = 2 \times 100$ $= 200$
>   Box 3, $2(2P)$
> $4P = 4 \times 100$ $= \underline{400}$
>   700 Sum

The sum, 490, is between 350 and 700. A "70" guess and an "80" guess is between 50 and 100. Since 490 is closer to 350 than 700, try the smaller, "70" guess.

> Guess 3:   Box 1, $P =$ 70
>   Box 2, $2P = 2 \times 70$ $= 140$
>   Box 3, $2(2P)$
> $4P = 4 \times 70$ $= \underline{280}$
>   490 Sum Needed

Since sum 490 is correct, make sure the relationships among the boxes are correct.

5 CHECK

Box 1: 70 P
Box 2: 140 $\overset{?}{=}$ $2P$ ⟶ $140 = 2 \times 70$ (Correct Relationship)
Box 3: 280 $\overset{?}{=}$ $4P$ ⟶ $280 = 4 \times 70$ (Correct Relationship)

Solve these problems using the guess-and-test strategy.
(You may find a calculator helpful.)

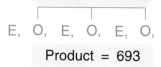

3 consecutive odds

1. The product of three consecutive odd numbers is 693. Find the three numbers. (Hint: 1, 3, 5, ... are consecutive odd numbers.)

E, O, E, O, E, O,

Product = 693

1 IMAGINE	Draw and label a picture.	2 NAME	→ Facts
			→ Question

3 THINK Try three consecutive odd numbers that when multiplied equal 693:

$3 \times 5 \times 7 = 15 \times 7 = 105$ ← Too small

$9 \times 11 \times 13 = 99 \times 13 = 1287$ ← Too large

Answer is between these.

4 COMPUTE ——→ **5 CHECK**

2. Juan, Joan, and Noah earned $105. If Joan earned $10 more than Juan and Noah earned $10 more than Joan, how much did each earn?

3. The seventh grade has 35 students. For every two students who like art there are three students who like music. How many of the students like art, and how many like music?

Exercise 4

4. When three consecutive even numbers are added, the sum is 30. Two of the even numbers are two-digit numbers. What are the three numbers?

5. The sum of Jan's and her mother's age is 51 years. The difference in their ages is 25. How old are Jan and her mother?

6. The product of a number and its square is 125. Find the number. (Hint: The square of a number is the product of the number and itself.)

7. Jess is twice his brother's age. The sum of their ages is 21. How old are Jess and his brother?

8. The Corner Store sold boxes of fruit at $5.00 a box and at $4.00 a box. How many boxes were sold at each price if 100 boxes were sold for a total of $460?

9. For every dollar the student council collected during a fund raiser, it donated $.75 to charity. How much money did the student council collect if it donated a total of $480 to the charity?

30

MAKE UP YOUR OWN...

10. Write a problem modeled after these that you can solve using the guess-and-test strategy. Solve your problem. Give your problem to a classmate to solve.

1-12 Problem Solving: Applications

Solve.

USE THESE STRATEGIES:
Use Simpler Numbers
More Than One Solution
Use a Graph
Write an Equation

1. According to the *Big Bang* theory, the universe began 15 billion years ago. Express this number as a standard numeral.

2. The mass of Pluto is 17 ten thousandths of the mass of Earth. Express this number as a standard numeral.

3. The sun changes 4.4 million tons of its mass into light every second. Write 4.4 million as a standard numeral.

4. The northern and southern lights occur in the atmosphere at distances ranging from about 50 to about 500 miles from Earth. How many times larger is 500 than 50?

5. The moon rotates on its axis about once every 27.3 days. If this number has been rounded to the nearest tenth, what might the original four-digit number have been?

6. Light from the sun reaches the earth in 499.12 seconds. Write this decimal in expanded form.

7. The surface temperature of the sun is 9940 °F. Its interior temperature is about 35 million °F. Sunspots have temperatures of about 7210 °F. List these temperatures in increasing order.

8. Six hundred twenty-five thousandths of the surface of Mars is covered by reddish rocks, sands, and soils. Write this number as a standard numeral.

9. A light year is the distance light travels in a year and is equal to 5 trillion, 900 billion miles. Round this number to the nearest trillion.

10. The chart at the right shows the distance in astronomical units (AU) from the sun to each planet. Round each distance to the nearest hundredth.

Planet	Distance in AU
Mercury	0.387099
Venus	0.723327
Earth	0.999999
Mars	1.523652
Jupiter	5.20316
Saturn	9.52355
Uranus	19.1690
Neptune	30.0468
Pluto	39.3395

11. The masses of three stony meteorites are 16.315 kg, 16.309 kg, and 16.319 kg. List the masses of these meteorites from greatest to least.

Use this graph to answer exercises 12–15.

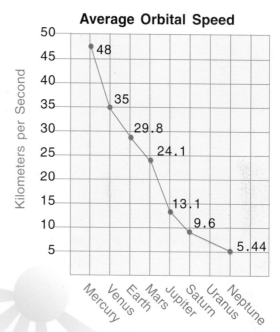

Average Orbital Speed

12. Round the average orbital speed of the first four planets on the graph to the nearest ten.

13. As the distance between the sun and a planet increases:
   a. average orbital speed increases.
   b. average orbital speed decreases.
   c. average orbital speed remains the same.

14. The average orbital speed of Uranus is *not* on the graph. Which of the following might it be?
   a. 16 km per sec
   b. 6.9 km per sec
   c. 3.7 km per sec

15. Which planet has an average orbital speed that is nearly twice the average orbital speed of Mars?

   a. Jupiter   **b.** Mercury   **c.** Venus

16. The fuel mixture used in the Saturn rocket that launched the Apollo 11 moon shot provided 7.5 million tons of thrust. Write this amount of thrust in expanded form.

17. An aircraft breaks the sound barrier at 740 mph. If a certain aircraft was traveling at a speed that was 80 mph faster than 5 times the speed of sound, how fast was it going?

18. During a solar eclipse the moon's shadow is 232,000 miles long. During a lunar eclipse the earth's shadow is 28,000 miles less than four times the moon's shadow during a solar eclipse. How long is the earth's shadow during a lunar eclipse?

19. The Eagle III, a radio-controlled glider, weighs 4.5 lb more than a 10-lb bag of potatoes. How much does the glider weigh?

> **MAKE UP YOUR OWN...**

20. Write some word problems using short word names for numbers. Then express the numbers in standard form and in expanded form.

More Practice

Write the standard numeral.

1. four million, two hundred sixty thousand, fifteen

2. one hundred nine thousand, eight hundred six

3. thirty-two thousand five and six hundred seven millionths

In the numeral 8,714,923.805 what digit is in:

4. the millions place?   5. the hundredths place?

6. the tens place?   7. the ten-thousands place?

Write the numeral in expanded form.

8. 503,979   9. 3,070,531   10. 33.002   11. 1.00302

Compare. Write <, =, or >.

12. 42.0640 __?__ 42.064   13. 6.844 __?__ 5.844   14. 707.025 __?__ 707.125

Name the greatest place-value position. Then round to that position.

15. 478,183   16. 20,888   17. 9777   18. 208

To what place has the decimal 0.6837 been rounded if the rounded decimal is:

19. 0.684   20. 0.7?   21. 0.68?   22. 1.0?

Use this chart to answer exercises 23–27.

23. Which is the most dense substance?

24. Which of these substances is the least dense?

25. List all the substances that are not as dense as sodium.

26. Which substances are more dense than potassium?

27. List the substances in order of their density, beginning with the most dense.

Substance	Density (g/cm^3)
lithium	0.534
sodium	0.97
zinc	7.14
lead	11.35
water	1.00
gold	19.32
helium	0.17847
calcium	1.54
potassium	0.862
hydrogen	0.07099
acetone	0.80
mercury	13.546

See *Still More Practice,* p. 447.

Math Probe

ANCIENT NUMERALS

Here are some ways numerals were once written.

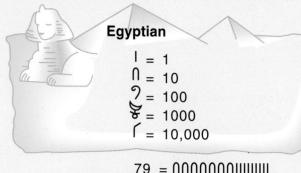

Egyptian

I = 1
∩ = 10
૭ = 100
⚡ = 1000
𝌂 = 10,000

79 = ∩∩∩∩∩∩∩IIIIIIIII
231 = ૭૭∩∩∩I

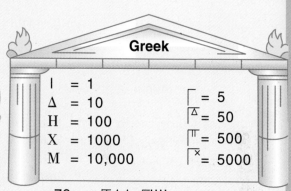

Greek

I = 1
Δ = 10
H = 100
X = 1000
M = 10,000

Γ = 5
Γᐃ = 50
Γᴨ = 500
Γˣ = 5000

79 = Γᐃ ΔΔ ΓIIII
231 = HHΔΔΔI

Roman

I = 1	V = 5	V̄ = 5000
X = 10	L = 50	X̄ = 10,000
C = 100	D = 500	C̄ = 100,000
M = 1000		D̄ = 500,000

Roman **Think:**

4 = IV ⟶ 5 − 1
40 = XL ⟶ 50 − 10
90 = XC ⟶ 100 − 10

Greek **Think:**

4 = IIII ⟶ 1 + 1 + 1 + 1
40 = ΔΔΔΔ ⟶ 10 + 10 + 10 + 10
90 = ΔΔΔΔΔΔΔΔΔ ⟶ 10 + 10 + 10 + 10 + 10 + 10 + 10 + 10 + 10

1. Explain the similarities and differences among these numerals.

Write each in Egyptian, Greek, and Roman numerals.

2. the year
3. your height in inches
4. 3600 seconds in an hour
5. your house number
6. your age in months
7. your zip code

Write each Roman numeral as a standard numeral.

8. bicentennial of the Bill of Rights, MCMXCI

9. average depth of the Pacific Ocean, X̄MMCMXXV feet

10. area of the Sahara, M̄M̄M̄D square miles

11. duration of space flight of first woman in space, ĪV̄CCI

12. Name several places in which Roman numerals are still used today.

Check Your Mastery

Write the numeral for:

See pp. 2–5

1. three hundred five billion, four hundred sixty-two million, two hundred thirty-two

2. six hundred eight million, seventy-five thousand and forty-one hundredths

Write each number in words. Name the value of each underlined digit.

3. <u>4</u>0,343

4. 5,7<u>38</u>,052

5. <u>6</u>4,000,630,477

6. 3,000,<u>2</u>04

7. 0.8<u>73</u>

8. 0.0905<u>4</u>

9. 9221.300<u>4</u>

10. 301.40<u>8</u>7

Write the standard numeral.

See pp. 6–7

11. $(7 \times 100,000) + (5 \times 10,000) + (8 \times 100) + (2 \times 10)$

12. $(4 \times 0.1) + (6 \times 0.001) + (2 \times 0.0001)$

13. $(3 \times 100) + (5 \times 1) + (9 \times 0.001) + (8 \times 0.0001)$

Write the numeral in expanded form.

14. 6,980,080

15. 0.0406

16. 30.70068

Order these numbers from greatest to least.

See pp. 8–9

17. 4,763,156; 4,673,651; 4,687,165; 4,637,186

18. 0.0981; 0.89; 0.881; 0.0881

19. 7.0819; 7.0891; 7.0981; 7.08

Round to the nearest million.

See pp. 10–11

20. 6,834,791

21. 1,476,980

22. 5,393,400

23. 8,506,633

Round to the nearest thousandth.

See pp. 10–13

24. 7.1195

25. 0.5443

26. 0.63219

27. 14.7666

28. In 1940 the land area of the United States was 2,977,128 square miles. Round this number to the nearest ten thousand square miles.

29. The atomic weight of phosphorus is 30.9738. Round this number to the nearest hundredth.

30. Tom boiled the frozen vegetables for 5.39 minutes. Round this number to the nearest tenth of a minute.

2 Addition and Subtraction of Whole Numbers and Decimals

In this chapter you will:

- Add and subtract whole numbers and decimals
- Use rounding strategies to estimate sums and differences for whole numbers and decimals
- Use patterns for mental computation
- Solve problems: finding missing information

Do you remember?

For regrouping, remember:

When Adding
10 thousandths	= 1 hundredth
10 hundredths	= 1 tenth
10 tenths	= 1 one
10 ones	= 1 ten
10 tens	= 1 hundred
10 hundreds	= 1 thousand

When Subtracting
1 thousand	= 10 hundreds
1 hundred	= 10 tens
1 ten	= 10 ones
1 one	= 10 tenths
1 tenth	= 10 hundredths
1 hundredth	= 10 thousandths

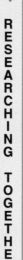

RESEARCHING TOGETHER

Mathspeak: Old and New

Can you speak Arabic? Greek? Latin? You will be surprised at how many math words come from other languages. Look at these. What do they mean?

zero, digit, cipher, geometry, abacus, radical, arithmetic, calculate

Create stickers. On each sticker write a different math term and make a design that hints at the meaning of the term or how the term began. Ask others to guess the meanings. Then explain all you have discovered about each word.

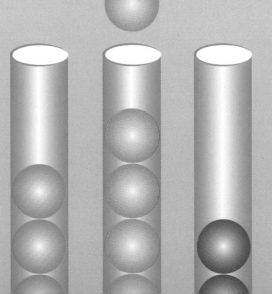

2-1 Adding Whole Numbers

To add two numbers:

- Start with the ones column.
- Add the numbers in each column, working from right to left.

The monthly computer costs for the ACZ Engineering Company during the first half of the year are listed. Find the sum of the costs for the months of March and June.

ACZ ENGINEERING Computer Costs

Month	Amount
January	$42,860
February	54,109
March	58,374
April	33,214
May	27,506
June	17,669

First estimate:

$50,000 + $10,000 + $10,000 = $70,000

```
   ①                ① 1              ① 1 1            ① 1  1 1         1 1  1 1
 $5 8, 3│7 4      $5 8,│3 7 4       $5│8, 3 7 4      $5 8, 3 7 4     $5 8, 3 7 4
 + 1 7, 6│6 9     + 1 7,│6 6 9      + 1│7, 6 6 9     + 1 7, 6 6 9    + 1 7, 6 6 9
        ③               ④3             ⓪4 3            ⑥0 4 3       $7 6, 0 4 3
```

13 ones = 1 ten 3 ones	14 tens = 1 hundred 4 tens	10 hundreds = 1 thousand	16 thousands = 1 ten thousand 6 thousands	**Sum**

Remember: Addend + Addend = Sum

Estimate. Then find the sum.

1.	435 +318	2.	780 +295	3.	809 +111	4.	2493 +5704	5.	789 +123

6.	1352 +3708	7.	4926 +8431	8.	25,098 +53,412	9.	30,279 +85,460	10.	827,600 + 17,899

Add.

11. 99 + 3977

12. 64,975 + 328

13. 5029 + 71

14. 4,728 + 44,069

15. 6473 + 3582

16. 7,283 + 88,475

17. 904,409 + 173,855

18. 2,040,671 + 53,949

19. 8,668,377 + 44,927

20. $37,914 + $584,037

21. $384,210 + $10,082

22. $877,240 + $64,040

23. $208,111 + $942,503

24. $457,019 + $75,396

25. $107,007 + $71,007

Solve. Use the chart on page 24.

26. How much were the computer costs for January and February together?

27. What were the total computer costs during the six months listed on the chart?

Adding Mentally

Look for tens or multiples of 5 to help you add a column of numbers mentally.

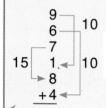

Think: 10 + 10 = 20
20 + 15 = 35

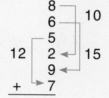

Think: 10 + 15 = 25
25 + 12 = 25 + (10 + 2)
= (25 + 10) + 2
= 37

Look for tens. Then add.

28.	**29.**	**30.**	**31.**	**32.**
4	4	36	4985	624
3	9	43	628	951
7	8	95	95	385
8	9	58	372	209
9	7	17	+ 83	+576
5	2	+44		
+7	+7			

33.	**34.**	**35.**	**36.**
490,768	437,963	$3,040,760	$51,736,222
5,830	75,807	560,409	17,191
607	97,458	834,572	421,040
+ 2,079	+ 99,672	+ 640,732	+ 567

 CRITICAL THINKING

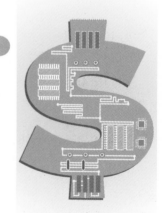

Use these clues to determine the amount of money spent by ACZ in July.

37. The ones digit is one less than the hundreds digit. The tens digit is the sum of the ones digit and the hundreds digit. The thousands digit is two more than the hundreds digit. Eight, the ten thousands digit, is two times the hundreds digit.

2-2 Subtracting Whole Numbers

Thinking in tens can help sometimes when regrouping in subtraction.

Subtract: $85 - 9 = \underline{\ ?\ }$

$(80 + 5) = (70 + 10) + 5 = 70 + (10 + 5) = $

$$
\begin{array}{r} 70 + 15 \\ -\ \ \ \ 9 \\ \hline 70 + 6 \end{array}
\qquad
\begin{array}{r} {}^{7\ 15}\!\!\not{8}\,\not{5} \\ -\ 9 \\ \hline 7\ 6 \end{array}
\text{ Difference}
$$

Subtract: $10{,}341 - 4{,}876 = \underline{\ ?\ }$

- First estimate: $10{,}000 - 4{,}000 = 6{,}000$
- Start with the ones column.
- Subtract the numbers in each column, working from right to left.

$$
\begin{array}{r} {}^{3\ 11}\\ 1\,0,3\,\not{4}\,\not{1} \\ -\ 4,8\,7\,6 \\ \hline 5 \end{array}
\qquad
\begin{array}{r} {}^{\ \ 13}\\ {}^{2\ 3\ 11}\\ 1\,0,\not{3}\,\not{4}\,\not{1} \\ -\ 4,8\,7\,6 \\ \hline 6\ 5 \end{array}
\qquad
\begin{array}{r} {}^{12\ 13}\\ {}^{9\ 2\ 3\ 11}\\ 1\,0,\not{3}\,\not{4}\,\not{1} \\ -\ \not{4},8\,7\,6 \\ \hline 4\ 6\ 5 \end{array}
\qquad
\begin{array}{r} {}^{12\ 13}\\ {}^{9\ 2\ 3\ 11}\\ 1\,0,\not{3}\,\not{4}\,\not{1} \\ -\ \not{4},8\,7\,6 \\ \hline 5\ 4\ 6\ 5 \end{array}
$$

41 =
3 tens
11 ones

330 =
2 hundreds
13 tens

10,200 =
9000 thousands
12 hundreds

Remember: Minuend – Subtrahend = Difference

Estimate. Then find the difference.

1. $\begin{array}{r} 96 \\ -18 \end{array}$
2. $\begin{array}{r} 68 \\ -49 \end{array}$
3. $\begin{array}{r} 568 \\ -237 \end{array}$
4. $\begin{array}{r} 878 \\ -794 \end{array}$
5. $\begin{array}{r} 3958 \\ -\ 499 \end{array}$

6. $\begin{array}{r} 8943 \\ -5698 \end{array}$
7. $\begin{array}{r} 6419 \\ -2840 \end{array}$
8. $\begin{array}{r} 57{,}224 \\ -56{,}148 \end{array}$
9. $\begin{array}{r} 728{,}965 \\ -\ \ \ \ \ 388 \end{array}$
10. $\begin{array}{r} 641{,}957 \\ -280{,}009 \end{array}$

Subtract.

11. $24{,}518 - 9{,}354$
12. $15{,}839 - 4{,}573$
13. $33{,}451 - 1{,}092$
14. $\$45{,}469 - \$3{,}528$
15. $\$194{,}671 - \$82{,}103$
16. $\$724{,}183 - \$71{,}489$
17. $\$27{,}381 - \$502$
18. $\$240{,}063 - \$148{,}284$
19. $\$5{,}010{,}901 - \$622{,}900$

Complete each subtraction.

20. $\begin{array}{r} {}^{\ \ 10\ 9}\\ {}^{3\ 0\ 10\ 18}\\ \not{4}\,\not{1}\,\not{0}\,\not{8} \\ -2\,7\,6\,9 \end{array}$

21. $\begin{array}{r} {}^{\ \ 9}\\ {}^{2\ 10\ 10}\\ \not{3}\,\not{0}\,\not{0}\,2 \\ -1\,9\,2\,2 \end{array}$

22. $\begin{array}{r} {}^{9\ \ 9\ 9}\\ {}^{4\ 10\ 10\ 10\ 10}\\ \not{5}\not{0},\not{0}\,\not{0}\,\not{0} \\ -2\,7,9\,6\,8 \end{array}$

Use these examples in solving exercises 23–26.

| a. | 47,820
− 28,351 | b. | 47,820
− 1,320 | c. | 86,400
− 69,185 | d. | 69,185
− 38,454 |

23. In which example is **NO** regrouping needed?

24. In which example must you regroup a hundred as tens *before* you can regroup a ten as ones?

25. In which example do you regroup only once?

26. In which example do you regroup in the tens place *and* the hundreds place *and* the ten-thousands place?

Subtract.

| 27. | 7083
−1298 | 28. | 6004
− 985 | 29. | 64,308
− 52,109 | 30. | 43,100
− 9,870 | 31. | 70,854
− 69,907 |

| 32. | 98,882
− 8,999 | 33. | 70,832
− 17,981 | 34. | 241,608
− 19,453 | 35. | 85,135
− 47,989 | 36. | 250,345
− 141,876 |

| 37. | 60,001
− 9,888 | 38. | 41,234
− 31,009 | 39. | 901,010
− 899,999 | 40. | 600,001
− 555,555 | 41. | 1,234,567
− 999,999 |

Find the sum or difference.

42. 17,390 − 8,598

43. 47,099 + 866

44. 30,004 − 3,769

45. $321,065 − $197

46. $258,603 + $70,989

47. $800,502 + $8,912

48. $754,916 + $84

49. $900,001 − $899,999

50. $444,000 + $56,000

51. On sale, a package of paper for the microcomputer printer costs $79.95 and a box of diskettes costs $39.95. The computer budget has $317.42 in it. What is the total cost of buying the printer paper and diskettes? After making these purchases, how much money is left in the budget?

This graph shows the computer hardware unit sales of the Chip Computer Company.

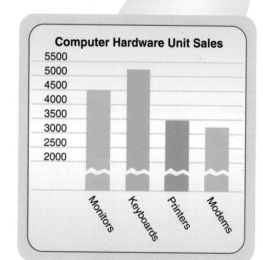

52. How many more keyboards were sold than printers?

53. How many monitors and printers were sold?

54. How many pieces of hardware were sold all together?

To estimate a sum, choose the rounding strategy that best fits the situation.
The sum 39,456 + 15,614 can be estimated in these four different ways:

Rounding Strategies

I. Round *both addends* to the greatest place-value position and add mentally.

$$3\,\textcircled{9},456 \approx 40,000$$
$$+1\,\textcircled{5},614 \approx +20,000$$
Estimated Sum $60,000$

Both addends were rounded *up*, so the exact sum, 55,070, is *less than* 60,000.

Round here to the ten-thousands place.

II. Round only *one addend* to the greatest place-value position and add mentally.

$$39,456 = 39,456$$
$$+1\,\textcircled{5},614 \approx +20,000$$
Estimated Sum $59,456$

One addend was rounded *up*, so the exact sum, 55,070, is *less than* 59,456.

III. Round *both addends* to the greatest period and add mentally.

$$39,\textcircled{4}56 \approx 39,000$$
$$+15,\textcircled{6}14 \approx +16,000$$
Estimated Sum $55,000$

The first addend was rounded *down*, and the other addend was rounded *up*, so the exact sum, 55,070, is *greater than* 55,000.

IV. Round only *one addend* to the greatest period and add mentally.

$$39,456 = 39,456$$
$$+15,\textcircled{6}14 \approx +16,000$$
Estimated Sum $55,456$

One addend was rounded *up*, so the exact sum, 55,070, is *less than* 55,456.

Which of the four estimated answers above is closest to the exact sum? Why?

Use this addition example: 47,505 + 28,709.
Explain how these addends were rounded to produce each of the following sums.

1. 80,000 2. 77,000 3. 78,709 4. 76,505

Use < or > to compare. Explain your rounding strategy choice.

5. 47,505 + 28,709 __?__ 80,000 6. 47,505 + 28,709 __?__ 78,709

7. 47,505 + 28,709 __?__ 77,000 8. 47,505 + 28,709 __?__ 76,505

Which rounding strategy gives the closest estimate for each of these? Why?*

9. 23,321 + 88,432 10. 36,847 + 45,231 11. 78,462 + 19,104

12. 22,408 + 19,326 13. 893,107 + 52,933 14. 652,999 + 180,321

2-4 | Estimating Differences for Whole Numbers

The rounding strategies that are used in estimating differences are like those in estimating sums. The difference 49,275 − 16,836 can be estimated in these four ways:

Rounding Strategies		
I. Round *both numbers* to the greatest place-value position and subtract mentally.	4⑨, 2 7 5 ≈ 5 0, 0 0 0 −1⑥, 8 3 6 ≈ −2 0, 0 0 0 Estimated Difference 3 0, 0 0 0	In both examples the actual subtrahend is less than 20,000, so the exact difference, 32,439, is *greater than* the estimated difference.
II. Round *one number* to the greatest place-value position and subtract mentally.	4 9, 2 7 5 = 4 9, 2 7 5 −1⑥, 8 3 6 ≈ −2 0, 0 0 0 Estimated Difference 2 9, 2 7 5	
III. Round *both numbers* to the greatest period and subtract mentally.	4 9,②7 5 ≈ 4 9, 0 0 0 −1 6,⑧3 6 ≈ −1 7, 0 0 0 Estimated Difference 3 2, 0 0 0	In both examples the actual subtrahend is less than 17,000, so the exact difference, 32,439, is *greater than* the estimated difference.
IV. Round just *one number* to the greatest period and subtract mentally.	4 9, 2 7 5 = 4 9, 2 7 5 −1 6,⑧3 6 ≈ −1 7, 0 0 0 Estimated Difference 3 2, 2 7 5	

In subtraction, when rounding only one number, it is usually best to round the subtrahend. Why?

Use this subtraction example: 76,453 − 38,676.
Explain the rounding strategy used for each of the following estimated differences:

1. 40,000   2. 36,453   3. 37,453   4. 37,000

Use < or > to compare. Explain your choice.

5. 76,453 − 38,676 __?__ 40,000   6. 76,453 − 38,676 __?__ 37,453

7. 76,453 − 38,676 __?__ 36,453   8. 76,453 − 38,676 __?__ 37,000

Which rounding strategy gives the closest estimate for each of these? Why?*

9. 37,643 − 18,267   10. 62,808 − 39,799   11. 88,516 − 21,555

12. 77,411 − 22,219   13. 217,079 − 12,689   14. 677,211 − 333,721

2-5 Mental Math Computation: Compensation

Thinking up or down to a ten, hundred, or thousand can help in mentally computing sums and differences. Try these without paper and pencil.

Add Mentally

Add 1 to make a 10.	$69 + 36 = \underline{?}$ $70 + 36 = 106$
Now, subtract the 1 that was added.	$106 - 1 = 105$ So, $69 + 36 = 105$
Add 22 to make 500.	$\underline{478} + 47 = \underline{?}$ $500 + 47 = 547$
Now, subtract the 22 that was added.	$547 - 22 = 525$ Think: $(547 - 20) - 2$ So, $478 + 47 = 525$

Subtract Mentally

Subtract 12 to make 100.	$112 - 25 = \underline{?}$ $100 - 25 = 75$
Now, add the 12 that was subtracted.	$75 + 12 = 87$ So, $112 - 25 = 87$
Add 20 to make 300.	$473 - \underline{280} = \underline{?}$ $473 - 300 = 173$
Now, 20 too much was subtracted; so, 20 must be added.	$473 + 20 = 193$ So, $473 - 280 = 193$

To make 100 out of a small number like 21:

$21 + \underline{?} = 100$ Think: 21 is close to 20.

$20 + 80 = 100$

Add 1 to 20. → $+ 1$ $- 1$ ← Subtract 1 from 80.

$21 + 79 = 100$

So, $21 + 79 = 100$

MENTAL MATH   Do all these exercises mentally.

How much must be added to each to make 100?

1. 81   2. 67   3. 52   4. 33

How much must be added to each to make 500?

5. 479   6. 456   7. 461   8. 488

How much must be added to each to make 800?

9. 747   10. 786   11. 751   12. 739

How much must be added to each to make 1000?

13. 946   14. 731   15. 654   16. 807

Now try these. Look for shortcuts.

17. $78 + 49$ 18. $57 + 65$ 19. $685 - 490$ 20. $725 - 155$

21. $472 + 87$ 22. $576 - 87$ 23. $3725 - 418$ 24. $2749 - 950$

25. $898 - 40$ 26. $1018 + 88$ 27. $2705 + 295$ 28. $3485 - 1799$

30

2-6 | Adding Decimals

A farm manager keeps a record of the number of kilograms of feed used for the chickens and the cows.

    chickens: 152.97 kg
    cows: 387.25 kg

Find the total amount of feed used.

To add decimals:

- Align the decimal points in each addend.
- Add as you would whole numbers.

Find the sum: 152.97 + 387.25.

First estimate:
100 + 300 + 100 = 500

The actual sum should be greater than 500.

```
  1  1  1  1
  1  5  2 . 9  7
+ 3  8  7 . 2  5
  5  4  0 . 2  2
```

Remember to align this decimal point in the answer with the ones above it.

Estimate. Then find the sum.

1. 0.621 +0.342	**2.** 43.86 + 2.07	**3.** 9.52 +16.50	**4.** 12.079 + 1.283	**5.** 3.296 +5.098
6. $62.93 + 9.07	**7.** $8.12 + 4.19	**8.** $ .66 + 1.35	**9.** $102.80 + 299.51	**10.** $944.08 + 7.93

Align and add.*

```
47.310    Insert zeros where needed
38.200    to line up the columns.
+ 6.123
```

11. 47.31 + 38.2 + 6.123 = _?_

12. 5.9 + 63

13. 0.7 + 2 + 4.08

14. 4.035 + 17 + 12.3

15. 36.981 + 127.24 + 0.6

16. 4 + 4.61 + 21.7 + 0.99

17. 30.09 + 8.2 + 0.111 + 6

18. 10 + 5 + 3 + 42.9081 + 0.44

19. 28.72 + 9.63 + 16.9 + 1.0001

20. Which has the same value as 104.4?

    a. 14.400   **b.** 104.40   **c.** 104.04   **d.** none of these

21. Which has the same value as 67.10?

    a. 671   **b.** 6.71   **c.** 67.01   **d.** none of these

2-7 Subtracting Decimals

Lyle Park's swim team made 592.13 points. Ross Valley's made 387.75 points. What is the difference in their scores?

To subtract decimals:

- Align the decimal points.
- Write any necessary zeros in the decimal part of the minuend.
- Subtract as you would whole numbers.

Find the difference: 592.13 − 387.75.

First estimate:
500 − 300 = 200

$$\begin{array}{r} 5\ 9\ 2\ .\ 1\ 3 \\ -\ 3\ 8\ 7\ .\ 7\ 5 \\ \hline 2\ 0\ 4\ .\ 3\ 8 \end{array}$$

Align and subtract: 49.6 − 44.574.

$$\begin{array}{r} 4\ 9\ .\ 6\ 0\ 0 \\ -\ 4\ 4\ .\ 5\ 7\ 4 \\ \hline 5\ .\ 0\ 2\ 6 \end{array}$$

Write two zeros in the minuend since the subtrahend has three decimal places.

Estimate. Then find the difference.

1. 46.75 −15.27	**2.** 589.48 − 24.67	**3.** 67.801 − 4.640

4. 62.871 − 0.999 **5.** 53.100 − 2.589

6. $32.01 − 13.64 **7.** $154.72 − 60.88 **8.** $408.61 − .88 **9.** $307.69 − 28.79 **10.** $111.11 − 99.99

Align and subtract. (Remember: if necessary, write zeros in the minuend.)

11. 59.7 − 18.12

12. 4.783 − 2.5

13. 39 − 4.62

14. 728 − 1.926

15. 819 − 81.927

16. 0.782 − 0.359

17. 1.205 − 0.426

18. 20.1 − 5.121

19. 7 − 6.403

20. $607.91 − $148.12

21. $50.25 − $19.46

22. $300.00 − $249.76

23. 8.050 − 0.1112

24. 0.35 − 0.19

25. 6.4001 − 0.9377

26. 806.1 − 479.35

27. 200.1 − 59.668

28. 3.407 − 2.0098

29. 0.04 − 0.006

30. 5.901 − 2.7

31. 0.369 − 0.0863

Compute. Remember to compute within the parentheses first.

32. $(35.6 + 65.187) - (26.08 + 11.09)$ **33.** $(0.93 - 0.5) + (3.91 - 1.006)$

34. $(421.1 + 0.99) - (6.843 + 3.009)$ **35.** $(4 - 1.601) + (0.8 - 0.009)$

36. $(3.801 - 0.13) + (25.6 - 17.01)$ **37.** $(8.01 - 0.2) + (17 - 3.06)$

Solve.

38. From 6.1 take 0.6895. (Hint: Write 6.1 as 6.1000. Then subtract.)

39. From 14.032 take 7.109.   **40.** Take 7.894 from 10.

41. How much money did Mrs. Anders spend when she bought a sweater for $32.95, a scarf for $9.99, and boots for $68.29?

42. Theo saved $6.19 when he bought three computer programs listed at $29.95 each. What did he spend?

43. The science class mixed 18.35 g of chemical X with 109.6 g of chemical Y. What was the mass of the mixture? How many grams more of chemical Y were in the mixture?

44. Brad measures the length and width of a stamp from his collection. If the stamp measures 0.43 dm by 0.29 dm, what is its perimeter?

45. Which calculator display shows the correct difference for $3.079 - 1.32$?

a. `2.047`   b. `1.759`   c. `2.47`

46. What numbers might have been entered to cause each of the incorrect answers?

CALCULATOR ACTIVITY   **Use your calculator to find four addends for each sum given below. Study the example.**

$2.4 = 0.6 + 0.7 + \underline{\ ?\ } + \underline{\ ?\ }$ Enter: `2.4` `−` `0.6` `−` `0.7` `=` `1.1`

Then enter: `−` `0.5` `=` `0.6`   These are the missing addends.

47. $0.75 = 0.18 + 0.15 + \underline{\ ?\ } + \underline{\ ?\ }$ **48.** $0.09 = 0.009 + 0.032 + \underline{\ ?\ } + \underline{\ ?\ }$

49. Use your calculator to check your answers to the exercises on pages 32 and 33.

CHALLENGE   **Find the missing addends.**

50. $\underline{\ ?\ } + 3.1 = 7.2$ **51.** $\underline{\ ?\ } + 10.09 = 15$ **52.** $7 + \underline{\ ?\ } = 11.6$

53. $10.8 + \underline{\ ?\ } = 11.98$ **54.** $\underline{\ ?\ } + 1.43 = 6.09$ **55.** $0.07 + \underline{\ ?\ } = 2.5$

56. $\underline{\ ?\ } + \underline{\ ?\ } = 5.5$ **57.** $\underline{\ ?\ } + \underline{\ ?\ } = 1.25$ **58.** $\underline{\ ?\ } + \underline{\ ?\ } = 7.013$

2-8 Estimating Sums and Differences for Decimals

To estimate with decimals, use the rounding strategies for whole numbers on pages 28–29. Round to the nearest ten, whole number, or tenth.

Estimate this sum three ways:

36.71 + 2.055 + 6.4 = _?_

```
  3 6 . 7 1
    2 . 0 5 5
+   6 . 4
_____
```

Estimated Sums

I Round to Nearest Ten	II Round to Nearest Whole Number	III Round to Nearest Tenth
40	37	36.7
0	2	2.1
10	6	6.4
50	45	45.2

Estimate this difference three ways:

18.236 – 7.65 = _?_

```
  1 8 . 2 3 6
–     7 . 6 5
_____
```

Estimated Differences

I Round to Nearest Ten	II Round to Nearest Whole Number	III Round to Nearest Tenth
20	18	18.2
10	8	7.7
10	10	10.5

Estimate the sum or the difference to the nearest whole number.

1.   436.8
    +298.4

2.   508.7
    − 356.1

3.   499.06
    +813.55

4.   602.88
    − 127.13

Use these addends: 17.1 + 88 + 39.78 + 6.14.
Estimate the sum to the nearest:

5. ten.

6. whole number.

7. tenth.

Use: 17.375 – 9.89.
Estimate the difference to the nearest:

8. ten.

9. whole number.

10. tenth.

Use < or > to compare. Explain.

11. 36.128 + 11.461 _?_ 50   12. 47.712 – 21.61 _?_ 30

13. 19.61 – 4.92 _?_ 15   14. 9.36 + 8.56 _?_ 17.5

15. 11.05 – 10.95 _?_ 1.1   16. 6.6 – 1.59 _?_ 5.99

Clustering

Clustering is another way to estimate.
Each of these four addends clusters around the number 40.

$$\boxed{37} + \boxed{43} + \boxed{39} + \boxed{42} \approx 40 + 40 + 40 + 40$$

Since each of the four addends is about 40, multiply 40 by 4.
$$4 \times 40 = 160 \text{ Estimated Sum}$$

Solve by estimation. Check, using a calculator.

17. 76 + 81 + 79   18. 18 + 21 + 22 + 19

19. 108 + 114 + 107 + 112   20. 9.6 + 9.9 + 11 + 10.5

21. A picture frame has a perimeter of 1.2 m. Leon paints 0.7 m of the edge of the frame with gold. How much more does he need to paint?

22. Andrea bought a compact disc player for $179.50 and four discs for $12.95 each. About how much did Andrea spend?

23. About how much change should Mr. Wilcox expect from a $50 bill if he purchased painting supplies for $29.75, $6.25, $10.99, and $1.09?

Finding Sums of 10 to Add Mentally

Sums of 10 are useful numbers for adding decimals quickly. Study this exercise.

$1.00 so, $2.70 + $1.30 = $4

$1.90 + $2.70 + $3.10 + $1.30 = $5 + $4 = $9

$1.00 so, $1.90 + $3.10 = $5

MENTAL MATH   **Solve.**

24. $4.06 + $6.05 + $1.05 + $1.04 = _?_

25. 0.002 + 0.006 + 0.001 + 0.004 + 1.008 = _?_

26. (1.4 + 1.5 + 1.6) – 0.5 = _?_

27. Use four different decimals as addends to make up sums of: $10, $1, and $.10.

STRATEGY
2-9 Problem Solving: Missing Information

Problem: Valley Junior High cheerleaders won the cheerleading competition with a total of 338.5 points won for 4 events. They earned 72.4 points for their common cheer and 92.6 points for their dance routine. How many points did they score in the other two events, the team cheer and mounts?

1 IMAGINE Draw and label a picture of the information given.

Common Cheer	Dance	Team Cheer	Mounts	Total Points
72.4	92.6	?	?	338.5

2 NAME *Facts:* 72.4 — Points for common cheer
92.6 — Points for dance routine
338.5 — Total points from the four parts of the competition

Question: How many points did they receive for team cheer and mounts?

3 THINK By examining your *Imagine* picture, you get this number sentence:

Common Cheer		Dance				Total Points
72.4	+	92.6	+ Team Cheer	+ Mounts	=	338.5

Since two addends (team cheer and mounts) are missing, there is not enough information to solve the problem.

4 COMPUTE By subtracting the points given for two of the categories, you can suggest some possible solutions to the problem.

Total Points – (Common Cheer + Dance) = (Team Cheer + Mounts)
 338.5   –   (72.4   + 92.6) =   ?

 338.5   –   165   =   173.5

Suppose that the team scored 90 points in the team cheer.
Find the points received for mounts by subtracting:
  173.5 – 90 = 83.5 Mounts Points

5 CHECK Since there is information missing from the problem, check your computation by substituting the numbers you supplied for the missing information.

Common Cheer	Dance	Team Cheer	Mounts	Total Points
72.4 +	92.6 +	90 +	83.5	= 338.5

36

Identify the missing information in each problem. Then make up some information and solve.

1. Mr. Todd's Deli is sponsoring a Little League softball team. Uniforms cost $40. How many uniforms did the sponsor buy for the team?

| **1 IMAGINE** | Draw and label a picture. ➔ | **2 NAME** | ➔ *Facts* |
| | | | ➔ *Question* |

3 THINK   Missing information — The amount of money that the sponsor spent on the uniforms.

Added information — A total of $600 was spent on uniforms.

To solve — Divide to find the number of uniforms that the sponsor bought.

4 COMPUTE ➔ **5 CHECK**

2. The batting averages of the top three batters, Bob, José, and Eddie, add up to 1.000. The top batter, Bob, has an average of 0.337. What are the averages of José and Eddie?

3. Sam and Eileen earned $75 for raking leaves around the neighborhood. How much money did they earn for each job?

4. Mr. Hobbs spent 1.5 hours lining the soccer field on Friday. He finished the job on Saturday. How much longer did he work on Saturday than on Friday?

5. During the basketball season, Rhonda scored a total of 170 points. What was her average points per game?

6. Tomi ran the first leg of the mile relay in 1.2 minutes. Ted ran the second leg in 1.35 minutes, Juan ran the third leg in 1.25 minutes, and Al ran the last leg. The first-place team finished the mile race in 4.8 minutes. How many seconds separated Tomi, Ted, Juan, and Al from the first-place team?

MAKE UP YOUR OWN...

7. Suppose that the Athletic Association collected $800 for a spring dance. Make up a budget for its use.

8. The physical education department has a budget of $1200. Make up a budget for buying new floor mats, two exercise bicycles, and ten new baseball gloves. Show how much money, if any, was left in the budget after buying this equipment.

9. Use this frequency table to write a word problem. Substitute a reasonable number for the missing information. Solve the problem.

Frequency Table

hits	|||| ||||	
home runs	||||
walks	|||
outs	||
times at bat	

2-10 | Problem Solving: Applications

USE THESE STRATEGIES:
Write an Equation
Use a Graph
Use Simpler Numbers
Guess and Test
Multi-Step Problem

1. The ancient Roman circus ended when Rome fell in 476 A.D. The modern circus began in England in 1768. How many years elapsed before the circus began again?

2. A certain traveling circus has one large ring with a standard diameter of 13 m and two smaller rings each 10.8 m in diameter. Placed side by side, how long will the rings be?

3. A tiger cub weighs 1.6 kg. A baby elephant weighs 88 kg. How many kilograms less does the tiger cub weigh than the baby elephant?

4. The elephants parade around the circus ring at 5.6 km per hour. In the wild an elephant can run at a speed of 34.4 km per hour faster than this. How fast can an elephant run?

The animal tamer records the food a tiger eats in a 5-day period. Use this graph to answer exercises 5–8.

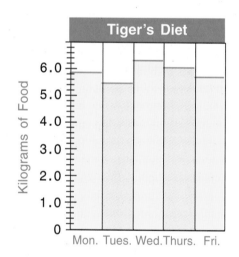

5. How many more kilograms did the tiger eat on Friday than on Tuesday?

6. How many fewer kilograms did the tiger eat on Monday than on Wednesday?

7. How many kilograms of food did the tiger eat during the 5-day period?

8. During a previous 5-day period, the tiger ate 30.5 kg of food. Has the tiger eaten more or less food during the 5 days graphed? Explain.

9. A trapeze act once took place 16,420 feet above the ground, while the artists were suspended from a hot-air balloon. Normally this act takes place thirty feet above the ground. How much higher was the record-breaking act than the normal trapeze act?

10. The bareback trick riders practiced 4.5 hours on Monday, 3.75 hours on Tuesday, 5.25 hours on Wednesday, and 4 hours on Thursday.

    a. How many hours have the riders practiced in all?

    b. If the riders practice at least 20 hours in a 5-day period, how many hours must they practice on Friday?

11. Over a 3-day period the circus ticket office sold 9857, 10,103, and 11,040 tickets. During the next 3-day period the ticket office hopes to sell a total of 32,000 tickets. How many more tickets do they hope to sell?

12. The open-air circus closed its tour with a fireworks display. The display included 32.5 lb of shooting stars, 60.5 lb of rainbow rockets, and 30.5 lb of flaming wheels. How many pounds of fireworks were used in the display?

13. Wild animals became part of the circus in 1831. Since then, performers have used as many as 50 different animals at once in an act. This is 10 more than the largest number of tigers performing at once in an act and 20 less than the largest number of horses performing at once in an act. What is the largest number of tigers ever to perform at once? of horses?

14. The circus tent has 12,000 seats available for each performance. If there were 103 empty seats on opening night, how many people attended the circus?

15. The admission price for the "Big Top" show is $15.95. The side-show admission price is $7.25. Mrs. Celeste spent $62.35 on 5 tickets. How many of each type of ticket did she buy?

16. Janet polled the class to determine which circus acts were their favorites. She tallied the results. Use the tally to answer these questions:

    a. How many more students like clowns than bareback riders?

    b. How many fewer students like jugglers than lion tamers?

    c. If each student listed one favorite circus act, how many students did Janet survey?

Name	Tally
trapeze	𝍦 |||
clown	𝍦 𝍦 ||||
high-wire	𝍦 𝍦	
lion tamer	𝍦 𝍦 |||
juggler	𝍦
barebacker rider	𝍦 ||||

MAKE UP YOUR OWN...

17. Use the tally that Janet prepared to make up problems that can be solved by adding and subtracting.

18. Make up a word problem involving decimals.

More Practice

Align and compute.

1. 83,205 + 67,825
2. 755 + 495
3. 895,340 + 138,265
4. 57,047 + 14,443
5. 391 − 235
6. 23,603 − 8,972
7. 98,243 − 61,659
8. 105,873 − 82,045

Align and compute.

9. 87.58 + 2.8045 + 0.88 + 9
10. 2.5 + 3.66 + 6.159
11. 3.48 + 60 + 8.0056 + 17.908
12. 912.3 + 0.054 + 5.76 + 11
13. 500.18 − 277.493
14. 10.762 − 0.716
15. 9.403 − 8.452
16. 212.3 − 18.521
17. 14 − 4.1807
18. 765.4 − 656.9201

Identify which of these rounding strategies was used in each example in exercises 19–22. Then complete the example and check your answer.

   a. Round both numbers to the greatest place-value position.
   b. Round one number to the greatest place-value position.
   c. Round both numbers to the greatest period.
   d. Round one number to the greatest period.

19.
$$855,321 \longrightarrow 855,000$$
$$+620,421 \longrightarrow 620,000$$

20.
$$6852 \longrightarrow 7000$$
$$+7095 \longrightarrow 7000$$

21.
$$53,942 \longrightarrow 54,000$$
$$-11,860 \longrightarrow 12,000$$

22.
$$26,480 \longrightarrow 26,480$$
$$-17,951 \longrightarrow 18,000$$

Solve mentally.

23. 7 + 5 + 1 + 2 + 8 + 9
24. 510 − 98
25. 181 − 120
26. 2.2 + 2.7 + 2.8 + 2.35
27. 583 + 119
28. 957 − 188

Estimate. Use < or > to compare.

29. 3694 + 1407 + 896 = __?__ 5000
30. 746 − 391 = __?__ 275
31. 0.86 + 0.109 + 0.7 = __?__ 1.8
32. 0.27 + 0.30 + 0.24 = __?__ 0.61

Math Probe

SHORT-WORD FORM

The chart at the right gives the estimated number of telephones in various countries. The numbers on this chart are expressed in short-word form.

Short-word form is commonly used when writing larger numbers.

For example: 3.2 million — 3,200,000

3.2 million is read: 3 point 2 million.

You can add or subtract larger numbers in short-word form when the *period* name is the same.

4.8 million + 2.72 million = 7.52 million

When the period name is *not* the same, first express the numbers in standard form. Then add or subtract.

1.5 million − 850 thousand = ?

1,500,000 − 850,000 = 650,000

Number of Telephones	
Country	Number
Canada	18.0 million
Mexico	6.4 million
U.S.	182.6 million
Peru	544 thousand
Argentina	2.65 million
S. Africa	4.5 million
Egypt	600 thousand
Philippines	873 thousand
Japan	64.0 million
France	39.2 million
Germany	40.3 million
Italy	25.6 million

Solve. Use the chart above when necessary.

1. Which country has the fewest telephones? the most?

2. How many telephones are there in France, Germany, and Italy?

3. How many more telephones are there in Canada than in Mexico?

4. How many telephones are there in Peru and Argentina combined?

5. How many more telephones are there in Japan than in the Philippines?

6. How many more telephones are needed in the Philippines to reach a total of 1.0 million?

7. Use a newspaper or other periodicals to search for large numbers in billions or trillions written in short-word form. Express these numbers in standard form.

8. Use an almanac to find examples of large numbers expressed in short-word form. Write original problems involving addition and subtraction like the ones above.

Check Your Mastery

Align and add.
See pp. 24–27

1. 58 + 27 + 125 + 55

2. 133,085 + 53,216

3. 91 + 107 + 892,765 + 98,643 + 91

4. 563 + 4396 + 56

Align and subtract.

5. 458,511 – 196,558

6. 619,917 – 114,606

7. 728,114 – 367,906

Estimate. Use < or > to compare.
See pp. 28–30

8. 5 + 16 + 9 + 21 + 4 + 8 _?_ 60

9. 26 + 34 + 37 + 32 + 13 _?_ 150

10. 616 + 388 _?_ 1100

11. 504 – 429 _?_ 100

How much should be added to each to make 2000?

12. 1698

13. 1110

14. 749

15. 502

Align and add.
See pp. 31–33

16. 0.42 + 1.7 + 17.05 + 16.258

17. 12.6 + 1.9 + 3.071 + 0.1 + 41

Align and subtract.

18. 215,138.53 – 24,072.9

19. 18.998 – 9.8

20. 35.42 – 2.601

21. 514.42 – 219.8

22. 72 – 0.346

23. 8.1 – 7.253

Choose the correct answer.
See pp. 32–33

24. From 7.2 take 0.384.
  a. 7.816
  b. 6.816
  c. 7.584
  d. 0.312

25. Take 6.075 from 20.
  a. 14.925
  b. 26.075
  c. 14.075
  d. 13.925

Estimate. Use < or > to compare.
See pp. 34–35

26. 14.709 + 6.09 + 7.2 _?_ 30

27. 242.398 – 9.249 _?_ 230

Solve.

28. The student council spent $187.55 for their dance. Advance tickets
brought in $202.75, and door sales totaled $116.
About how much profit did the student council realize?

Cumulative Review

Choose the correct answer.

1. The standard numeral for three billion, four hundred million, four thousand, four is:

   a. 3444 **b.** 3,400,404

   c. 3,400,004,004 **d.** 3440

2. What number is 10,000 greater than 2,325,624?

   a. 2,325,624 **b.** 2,425,724

   c. 2,335,624 **d.** 2,326,624

3. What number is closest to one?

   a. 0.8969 **b.** 0.9001

   c. 0.9498 **d.** 0.8899

4. Which has the same value as 85.03?

   a. 85.030 **b.** 8.503

   c. 85.30 **d.** 850.3

5. What is the standard numeral for $(6 \times 100) + (8 \times 0.1) + (4 \times 0.01) + (2 \times 0.001)$?

   a. 680.042 **b.** 608.042

   c. 600.842 **d.** 608.041

6. Which statement is true?
   a. $5,007,219 > 5,070,219$
   b. $4630.03 < 4603.30$
   c. $1.3 - 0.07 = 0.6$
   d. $0.07 + 0.03 = 0.1$

7. Which of these is another way to write 100.00102?
   a. $100 + 0.01 + 0.02$
   b. $100 + 0.001 + 0.00002$
   c. $100 + 0.01 + 0.00002$
   d. $100 + 0.1 + 0.02$

8. Which is the best estimate for the sum of 7.082 and 14.856?
   a. 20
   b. 21
   c. 23
   d. 22

9. Subtract 10,000 from the sum of 18,004 and 3.45. The difference will be
   a. greater than 10,000
   b. less than 10,000
   c. equal to 10,000
   d. less than 8,000

10. Using the digits 0, 1, 3, 5, and 9, what is the largest number you can write?
    a. 95,310
    b. 59,031
    c. 95,301
    d. 95,013

11. 0.034 is greater than 0.000034 by how much?
    a. 0.003966
    b. 0.000396
    c. 0.039966
    d. 0.033966

12. How much greater than 0.42018 is 2.62013?
    a. 2.29985
    b. 2.19995
    c. 2.19895
    d. 2.18995

13. Add 1 thousandth to the difference between 1.372 and 0.8691.
    a. 0.5029
    b. 0.5129
    c. 0.5019
    d. 0.5039

14. The estimated difference between two numbers is 20,000. What are these numbers?
    a. $37,643 - 18,267$
    b. $34,225 - 18,267$
    c. $41,265 - 26,450$
    d. $31,745 - 19,745$

Compute.

15.	16.	17.	18.	19.
65,202 + 914	59,488 + 811	19,765 − 2,183	887,029 −575,034	672.74 +271.18

20.	21.	22.	23.	24.
294.02 +100.98	770.94 −161.89	871.05 −361.36	268.057 +943.976	94.0164 + 2.8172

Use < or > to compare.

25. 36,805 + 2921 ___?___ 40,000

26. 65,322 − 36,215 ___?___ 40,000

Align and add or subtract.

27. $13.41 + $26.95 + $1.20 + $425.05 + $75.28 = ___?___

28. $39,440 + $259.07 − $63.55 = ___?___

29. 37,000 − 2843 = ___?___

30. Write the numerals. Then find the difference between *a* and *b*.
    a. six thousand five hundred and eight hundred seventeen thousandths
    b. one hundred and four hundred sixty-seven ten thousandths

Solve.

31. Thomas has a $25 gift certificate. He wants to use it to help pay for a camera on sale for $59.99. How much cash will he need to buy the camera?

32. The seventh-grade class collected 12,206 box tops. The sixth grade collected 19,842. What is the total number of box tops collected by both classes?

33. On January 1, Ms. Temple noted that her car's odometer read 26,456.2 mi. On May 1 she noted that the odometer read 32,900 mi. How many miles had the car been driven since January?

34. The circulation figures for three teenager magazines were: 967,249; 1,100,237; and 1,766,161. Round each number to the nearest ten thousand. Then estimate the total sum for the three magazines.

35. The length of a paramecium is about 0.238 mm and an amoeba is about 0.472 mm long. Round each number to the nearest tenth. Then estimate the difference in their lengths.

36. Order the following islands' areas from greatest to least. Find the total area of the five islands: Ireland, 32,597 sq mi; Greenland 839,999 sq mi; Honshu 88,925 sq mi; Luzon 40,420 sq mi; Sumatra 182,859 sq mi.

+10 BONUS

37. Imagine that you are on a committee that has volunteered to decorate a float for the Autumn Parade, and that the committee has a total of $300 to spend. Prepare a list showing the items you will need and the amount you can spend on each.

3

Multiplication and Division of Whole Numbers and Decimals

In this chapter you will:

- Multiply and divide by multiples of 10, 100, 1000
- Multiply and divide whole numbers and decimals
- Use estimation strategies
- Multiply and divide by powers of ten
- Use mental computation strategies
- Solve problems: identifying extra information
- Use technology: computers

Do you remember?

Powers of ten:

$$10 = (10 \times 1)$$
$$100 = 10 \times (10 \times 1)$$
$$1000 = 10 \times (10 \times 10)$$
$$10,000 = 10 \times (10 \times 10 \times 10)$$

The multiples of 10 are:
  10, 20, 30, 40, 50, ...

The multiples of 100 are:
  100, 200, 300, 400, 500, ...

RESEARCHING TOGETHER

Modular Matrix

The matrix below is an example of multiplication in modular arithmetic.

⊗	1	2	3	4	5	6	7	8	9	10	11	12
1	1	2	3	4	5	6	7	8	9	10	11	12
2	2	4	6	8	10	12	2	4	6	8	10	12
3	3	6	9	12	3	6	9	12	3	6	9	12

You can use a clock to multiply $3 \otimes 6$. Begin at 12 and move 6 places 3 times. You land at 6. In modular, remainder, arithmetic you can multiply and divide to compute $3 \otimes 6$ ($3 \times 6 = 18$; $18 \div 12 = 1$ R6). Copy and complete the matrix. Describe the patterns and symmetry in the matrix.

3-1 Multiplying with Multiples of 10, 100, or 1000

Is there a pattern for multiplying with 10, 100, or 1000? ———→

21 × 10	=	210	50 × 10	=	500
21 × 100	=	2100	50 × 100	=	5000
21 × 1000	=	21,000	50 × 1000	=	50,000

Shortcut for multiplying 10, 100, or 1000:

Multiply by 1, leaving out the zeros. Then attach as many zeros as were left out.

Use the shortcut for multiplying by *multiples* of 10, 100, and 1000.

80 × 30	=	2400
80 × 300	=	24,000
80 × 3000	=	240,000

Look at the examples again. Did you find another shortcut for multiplying? ———→

Shortcut for multiplying when both factors end in one or more zeros:

Multiply the digits, leaving out the zeros. Then attach as many zeros as were left out.

Multiplier × Multiplicand = Product

20 × 70 = ?

2 × 7 = 14   Two zeros were left out.
20 × 70 = 1400 ←——— Write the two zeros.

200 × 70 = ?

2 × 7 = 14   Three zeros were left out.
200 × 70 = 14,000 ←——— Write the three zeros.

Multiply each number in exercises 1–4 by 100.
Read aloud each multiplicand and product.

1. 56   **2.** 234   **3.** 1642   **4.** 22,357

Multiply.*

5.   10 × 20   **6.** 75 × 10   **7.**   10 × 3270   **8.** 90 × 50
      1000 × 20   75 × 1000   1000 × 3270   90 × 50,000

Write <, =, or >.

9. 3427 × 10 __?__ 347 × 100   **10.** 2250 × 10 __?__ 100 × 225

Which would you rather have:

11. $400 a day for 5 days or $40 a day for 10 days?

12. $20 an hour for 8 hours or $10 a day for 20 days?

13. 10 sets of 4 quarters or 4 sets of 10 half dollars?

14. Five sets of 100 pennies or one set of 100 dimes?

3-2 Estimating Whole-Number Products

To estimate products:

- Round each *factor* (number) to its greatest place-value position.

- Then use the rules for multiplying by multiples of 10, 100, and 1000.

Estimate: 27 × 82 = ?

$$\begin{array}{r} 82 \longrightarrow 80 \\ \times 27 \longrightarrow \times 30 \end{array} \quad 30 \times 80 = 2400$$

> 27 was rounded *up* more than 82 was rounded *down*. The exact product is *less than* 2400.

Estimate: 79 × 492 = ?

$$\begin{array}{r} 492 \longrightarrow 500 \\ \times\ 79 \longrightarrow \times\ 80 \end{array} \quad 80 \times 500 = 40{,}000$$

> Both factors (79 and 492) were rounded *up*, so the exact product is *less than* 40,000.

To estimate products when **both** factors have a 5 in the ones place:

- Get a first estimate by rounding one factor up and the other down.

$$\begin{array}{r} 45 \longrightarrow 50 \\ \times 25 \longrightarrow 20 \end{array} \quad 20 \times 50 = 1000$$

- Get a second estimate by reversing the rounding.

$$\begin{array}{r} 45 \longrightarrow 40 \\ \times 25 \longrightarrow 30 \end{array} \quad 30 \times 40 = 1200$$

- The product is between the two.

The product is *between* 1000 and 1200.

Estimate each product.

1. 73 × 14   **2.** 81 × 914   **3.** 73 × 126   **4.** 123 × 821   **5.** 768 × 3331

Between what two estimated products does each actual product lie?

6. 65 × 15   **7.** 150 × 350   **8.** 25 × 650   **9.** 250 × 650

Estimate. Write < or >.*

10. 36 × 89 _?_ 3600
11. 31 × 185 _?_ 6000
12. 320 × 411 _?_ 120,000
13. 37 × 2645 _?_ 120,000
14. 69 × 498 _?_ 35,000
15. 935 × 6471 _?_ 5,400,000

CALCULATOR ACTIVITY

Estimate the answer.
Then find it on a calculator and compare your answers.

16. Precision Tools manufactures a rotating drill that spins 65 times per second. How many revolutions is this per hour?

*See page 448 for more practice.

3-3 Multiplying by Two or Three Digits

To multiply by two or three digits:
- Estimate the product.
- Multiply by ones, then by tens, then by hundreds.
- Add the partial products.
- Check, using a calculator.

By two digits:

Factors
$$
\begin{array}{r}
8\ 2 \\
\times\ 2\ 9 \\
\end{array}
$$

Partial Products
$$
\begin{array}{r}
7\ 3\ 8 = \boxed{9 \times 82} \\
1\ 6\ 4\ 0 = \boxed{20 \times 82} \\
\hline
2\ 3\ 7\ 8 \quad \text{Actual Product}
\end{array}
$$

Estimate.

$30 \times 80 = 2400$ Estimated Product

Compare Actual Product to Estimate.

2378 is about 2400.

By three digits:

$$
\begin{array}{r}
7\ 4\ 3 \\
\times\ 3\ 7\ 4 \\
\hline
2\ 9\ 7\ 2 \\
5\ 2\ 0\ 1 \\
2\ 2\ 2\ 9 \\
\hline
2\ 7\ 7, 8\ 8\ 2 \quad \text{Actual Product}
\end{array}
$$

Start each partial product under the digit by which you are multiplying.

Estimate.

$400 \times 700 = 280,000$ Estimated Product

Compare Actual Product to Estimate.

277,882 is about 280,000.

When zero is one of the multiplying digits:

$$
\begin{array}{r}
4\ 8\ 6 \\
\times\ 2\ 0\ 9 \\
\hline
4\ 3\ 7\ 4 \\
9\ 7\ 2\ 0 \\
\hline
1\ 0\ 1, 5\ 7\ 4 \quad \text{Actual Product}
\end{array}
$$

Why are there only two partial products in this example?

Estimate.

$200 \times 500 = 100,000$ Estimated Product

Compare Actual Product to Estimate.

101,574 is about 100,000.

Estimate. Then find the product.

1. 641 × 8	**2.** 836 × 6	**3.** 1535 × 8	**4.** 6629 × 4	**5.** 509 × 18	**6.** 246 × 72
7. 582 × 19	**8.** 3166 × 24	**9.** 5389 × 57	**10.** 6153 × 82	**11.** 563 ×472	**12.** 942 ×313
13. 3657 × 977	**14.** 6398 × 805	**15.** 1483 × 709	**16.** 681 ×305	**17.** 4076 × 803	**18.** 9007 × 907

Estimate. Then multiply to find the exact product.

19. 8 × 5316 **20.** 3 × 6098 **21.** 5 × 8267 **22.** 7 × 3406

23. 75 × 823 **24.** 62 × 9109 **25.** 38 × 4526 **26.** 29 × 7829

27. 318 × 94,201 **28.** 646 × 81,765 **29.** 298 × 40,724 **30.** 945 × 85,097

Estimate to decide which products are correct.
Copy and correct any that seem to be incorrect.

31. 801 × 8096 = 6,484,896 **32.** 22 × 3208 = 700,576

33. 68 × 5309 = 36,652 **34.** 97 × 723 = 70,131

35. 637 × 205 = 130,585 **36.** 589 × 349 = 20,561

37. 985 × 398 = 126,180 **38.** 31 × 28,640 = 887,840

39. 759 × 5940 = 450,846 **40.** 612 × 9795 = 594,540

 **CALCULATOR ACTIVITY** **Multiply.** Check answers on a calculator.

41.
41,272
× 682

42.
60,721
× 826

43.
93,692
× 830

44.
80,900
× 509

45.
77,060
× 868

Estimate. Then find the product. (How did the zeros affect the multiplication?)

46.
607
× 45

47.
590
× 89

48.
789
×740

49.
234
×506

50.
8096
× 881

Copy and complete these estimates.
To get a closer estimate, use only the rounded factor shown.

51. 81 × 235 = ___?___

Estimates:
235
× 80

52. 592 × 416 = ___?___

416
×600

53. 145 × 309 = ___?___

300
×145

Solve.

54. The aquarium tickets cost $4.00, $5.00, and $6.00. If 232 four-dollar tickets, 175 five-dollar tickets, and 354 six-dollar tickets were sold one morning, how much money was collected through ticket sales?

55. In just three minutes, see how many sets of two whole-number factors you can find that will equal 4800. Check your answers on a calculator. Share your shortcuts.

3-4 Multiplying Decimals

To multiply decimals:

- Multiply as you would multiply whole numbers.
- Write the product.
- Then count the number of decimal places in both factors.
- Mark off that number of decimal places in the product.
- Check, using a calculator.

A decimal by a whole number:

19 × 0.843 = ?

```
    0.8 4 3  ←——— 3 decimal places
 ×      1 9.  ←——— no decimal places
    7 5 8 7
    8 4 3
  1 6.0 1 7  ←——— 3 decimal places
```

A decimal by a decimal:

0.275 × 75.6 = ?

```
        7 5.6  ←— 1 place
    ×0.2 7 5  ←— 3 places
      3 7 8 0
    5 2 9 2
  1 5 1 2
  2 0.7 9 0 0  ←— 4 places
```

1.78 × 0.0079 = ?

```
     0.0 0 7 9  ←——— 4 places
  ×      1.7 8  ←——— 2 places
       6 3 2
     5 5 3
     7 9
  0. 0 1 4 0 6 2  ←——— 6 decimal places
                     (Insert a zero.)
```

Find the product.

1. 3.4 × 7	**2.** 4.06 × 9	**3.** 1.532 × 8	**4.** 7.351 × 0.2

1. 3.4 × 7   **2.** 4.06 × 9   **3.** 1.532 × 8   **4.** 7.351 × 0.2   **5.** 0.034 × 0.4   **6.** 1.5 × 0.2

7. 66.3 × 0.82   **8.** 8.46 × 0.05   **9.** 563 × 0.015   **10.** 2.105 × 0.516   **11.** 5.003 × 747   **12.** 0.0079 × 1.87

Multiply.*

13. 0.42 × 0.08   **14.** 0.072 × 1.003   **15.** 0.981 × 0.006   **16.** 145 × 0.09

17. 1.98 × 0.462   **18.** 2.08 × 4.007   **19.** 20.63 × 54.07   **20.** 0.981 × 0.6

A NEW LOOK   SUPPOSE THAT...

21. If it costs $35.50 an hour to rent a sailboat for windsurfing, how much will 1.25 hours of windsurfing cost?

22. If the Surf Shoppe rents the sailboat to each of 3 people for 1.25 hours, how much money does it make? If the shop rents to 6 people, how much money does it make? Find the pattern.

3-5 Estimating Decimal Products

To estimate decimal products:

- Round the factors either to the nearest ten, whole number, or tenth.
- Then multiply to estimate the products.
- To get a closer estimate, round only one factor to a convenient place.
- Then multiply.

Nearest Tenth	Nearest One	Nearest Ten
$0.9\,\textcircled{0}\,4 \longrightarrow 0.9$	$5.\textcircled{0}\,0\,3 \longrightarrow 5$	$2\,\textcircled{3}.4\,7 \longrightarrow 20$
$\times 0.4\,\textcircled{2}\,5 \longrightarrow \times 0.4$	$\times\quad 7.\textcircled{1}\,5 \longrightarrow \times 7$	$\times 4\,\textcircled{6}.0\,9 \longrightarrow \times 50$
0.36	35	1000

Try rounding each example above to a different place.
Compare your answers.

Write < or >.

1. $0.3 \times 16 \underline{\ ?\ } 16$
2. $2.5 \times 25 \underline{\ ?\ } 25$
3. $1.5 \times 0.003 \underline{\ ?\ } 0.003$
4. $1.5 \times 2.5 \underline{\ ?\ } 2.5$
5. $0.7 \times 0.4 \underline{\ ?\ } 0.4$
6. $1.2 \times 0.0375 \underline{\ ?\ } 0.0375$

Find the product.
Write any needed zeros in the product. Put the decimal point in the correct place.

7. 2.3×3.25
8. 0.5×0.08
9. 5.1×23.003
10. 14.5×3.03
11. 0.034×2.4
12. 0.0012×0.012
13. 0.016×0.35
14. 0.002×0.13
15. 17×0.0003

Use your rounding skills to choose the best answer.*

16. 26.1×5.5 **a.** 14355   **b.** 14.355   **c.** 1.4355   **d.** 143.55
17. 4.7×5.71 **a.** 26.837   **b.** 2.6837   **c.** 26837   **d.** 268.37
18. 96×0.54 **a.** 5184   **b.** 0.5184   **c.** 5.184   **d.** 51.84
19. 19×0.017 **a.** 3.23   **b.** 0.323   **c.** 0.0323   **d.** 32.3
20. 7.5×1.6 **a.** 0.12   **b.** 1.2   **c.** 12   **d.** 120

Solve.

21. Show all possible arrangements of the decimal points in the factors 5678 and 243 that give 13.79754 as the product.

22. Tina spends $8.95 each time she phones home. Estimate the cost of 4 long-distance phone calls.

*See page 449 for more practice.

Dividing with Multiples of 10, 100, or 1000

Look at this pattern for dividing by 10, 100, or 1000.

$960 \div 10 = 96$	$9600 \div 100 = 96$	$96{,}000 \div 1000 = 96$
$9600 \div 10 = 960$	$96{,}000 \div 100 = 960$	$960{,}000 \div 1000 = 960$
$96{,}000 \div 10 = 9600$	$960{,}000 \div 100 = 9600$	$9{,}600{,}000 \div 1000 = 9600$

Apply this pattern when dividing by multiples of 10, 100, or 1000.

- Count the zeros in the divisor. $\qquad$ $4800 \div 40$ ⟵ One zero in divisor.
- Cross out that number of zeros in the divisor *and* the dividend. $\qquad$ $480\cancel{0} \div 4\cancel{0}$ ⟵ Cross out one zero in both divisor and dividend.
- Divide the remaining digits. $\qquad$ $480 \div 4 = 120$ ⟵ Divide what is left.

To Divide	Think	Quotient
$48{,}000 \div 400$	$48{,}0\cancel{0}\cancel{0} \div 4\cancel{0}\cancel{0}$	120
$48{,}000 \div 4000$	$48{,}\cancel{0}\cancel{0}\cancel{0} \div 4\cancel{0}\cancel{0}\cancel{0}$	12
$48{,}000 \div 40$	$48{,}00\cancel{0} \div 4\cancel{0}$	1200
$4800 \div 400$	$48\cancel{0}\cancel{0} \div 4\cancel{0}\cancel{0}$	12
$480{,}000 \div 400$	$480{,}0\cancel{0}\cancel{0} \div 4\cancel{0}\cancel{0}$	1200

Quotient

$$4\,\cancel{0}\,\cancel{0}\,)\overline{4\,8{,}0\,\cancel{0}\,\cancel{0}} \quad 1\,2\,0$$

Divisor $\qquad$ Dividend

Copy these charts. Complete them.*

	To Divide	Think	Quotient
1.	$480 \div 80$	$48\cancel{0} \div 8\cancel{0}$	6
2.	$7200 \div 90$		
3.	$36{,}000 \div 400$		
4.	$5400 \div 900$		
5.	$24{,}000 \div 600$		
6.	$90{,}000 \div 3000$		
7.	$68{,}500 \div 500$		

	To Divide	Think	Quotient
8.	$42{,}000 \div 700$		
9.	$63{,}000 \div 70$		
10.	$33{,}000 \div 11{,}000$		
11.	$93{,}000 \div 300$		
12.	$81{,}000 \div 900$		
13.	$150{,}000 \div 30$		
14.	$272{,}000 \div 800$		

There are three ways to show division:

$$640 \div 80 \qquad 80\overline{)640} \qquad \frac{640}{80}$$

*See page 449 for more practice.

3-7 Estimating Whole-Number Quotients

Communications satellites help relay signals over great distances. One of these satellites travels 12 457 kilometers every 24 hours. Estimate the number of kilometers the satellite travels in one hour.

Estimate: 12 457 ÷ 24 = __?__

When estimating quotients, remember these *rounding strategies:*

▶ **First way** – Round to the *greatest place-value position.*

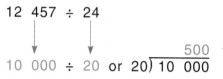

12 457 ÷ 24

10 000 ÷ 20 or 20) 10 000 = 500

▶ **Second way** – Round the dividend to the *greatest period.* Round the divisor to the *greatest place-value position.*

12 457 ÷ 24

12 000 ÷ 20 or 20) 12 000 = 600

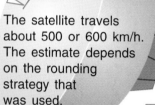

The satellite travels about 500 or 600 km/h. The estimate depends on the rounding strategy that was used.

Round the dividend and the divisor.

1. 79 ÷ 19 is about   **a.** 50   **b.** 5   **c.** 4   **d.** 40
2. 2987 ÷ 243 is about   **a.** 300   **b.** 150   **c.** 15   **d.** 30
3. 41,482 ÷ 367 is about   **a.** 10   **b.** 100   **c.** 140   **d.** 1000
4. 290,720 ÷ 512 is about   **a.** 6   **b.** 60   **c.** 600   **d.** 6000

Estimate each quotient.*

5. 6918 ÷ 73   6. 3672 ÷ 45   7. 7281 ÷ 66   8. 7918 ÷ 75
9. 87,550 ÷ 282   10. 16,475 ÷ 371   11. 91,472 ÷ 621   12. 71,841 ÷ 714

CALCULATOR ACTIVITY

Estimate each quotient. Then find it on a calculator and compare your answers.

13. 14)3192   14. 43)15,652   15. 258)88,236   16. 24)15,168

17. 36)32,220   18. 519)306,729   19. 98)301,350   20. 476)453,628

21. 623)285,957   22. 854)852,292   23. 108)231,012   24. 128)121,216

*See page 449 for more practice.   53

Find the quotient: 31,875 ÷ 7 = ___?___

To divide with 1-digit divisors:

- Estimate to find the first digit of the quotient.

  31 ÷ 7 is about 4.

  So, 31,000 ÷ 7 is about 4000.

  Try 4 as the first digit.
- Multiply: 4 × 7 = 28.
- Subtract: 31 − 28 = 3.
- Bring down 8. The *partial dividend* is 38.
- Repeat the steps for each partial dividend.

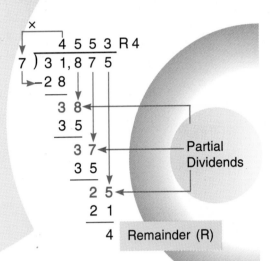

To divide, using short division:

- *Think* the steps.
- *Check* by multiplying and adding.

$$\begin{array}{r} 4\ 5\ 5\ 3\ \text{R}4 \\ 7\overline{)3\ 1{,}^3 8^3 7^2 5} \end{array}$$

Check: (7 × 4553) + 4 = 31,875 ⟵ Dividend

Divisor   Quotient   Remainder

You can estimate quotients using **compatible numbers**.

31,875 ÷ 7 ⟶ 28,000 ÷ 7 = 4000   The actual quotient
is greater than 4000.

Estimate each quotient by using compatible numbers.

1. 36,351 ÷ 4 > ___?___
2. 6262 ÷ 3 > ___?___
3. 43,216 ÷ 5 > ___?___
4. 48,291 ÷ 9 > ___?___
5. 199,697 ÷ 9 > ___?___
6. 726,692 ÷ 8 > ___?___
7. 28,528 ÷ 9 > ___?___
8. 25,890 ÷ 5 > ___?___
9. 45,621 ÷ 6 > ___?___

Use short division to find these quotients.*

10. 3)2256
11. 9)4626
12. 8)75,424
13. 7)65,909
14. 9)76,205

Choose the correct quotient.

15. 27,369 ÷ 9   **a.** 3041   **b.** 341   **c.** 3410   **d.** 30,041

16. 12,036 ÷ 4   **a.** 309   **b.** 3009   **c.** 3090   **d.** 30,090

3-9 Two-Digit Divisors

Find the quotient: $7446 \div 34 = \underline{\quad ? \quad}$

To divide with a 2-digit divisor:

- Estimate to find the first digit of the quotient.

$$7446 \div 34$$

$$6000 \div 30 = 200$$
Try 2 as the first digit.

- Multiply: $2 \times 34 = 68$.
- Subtract: $74 - 68 = 6$.
- Bring down 4.
- Estimate again with the *partial dividend* of 64.

$$64 \div 34$$

$$60 \div 30 = 2$$
Try 2 as the second digit.
But $2 \times 34 = 68$, and 68 is too large.
Try 1 as the second digit.

- Repeat the steps for each *partial dividend*.
- Check, using a calculator.

```
          2 1 9
    3 4 ) 7 4 4 6
         -6 8
          6 4
          3 4
          3 0 6
          3 0 6
```

Estimate:
$306 \div 34$

$300 \div 30$
Try 10.

But $10 \times 34 = 340$,
and $340 > 306$.

So try 9.

No digit in a quotient can be greater than 9.

Choose the best estimate.

1. $649 \div 11$ is about   **a.** 6   **b.** 60   **c.** 4   **d.** 40

2. $6045 \div 15$ is about   **a.** 35   **b.** 45   **c.** 400   **d.** 350

3. $29{,}602 \div 82$ is about   **a.** 308   **b.** 380   **c.** 3300   **d.** 480

4. $362{,}578 \div 61$ is about   **a.** 60   **b.** 600   **c.** 6000   **d.** 60,000

Estimate. Then find the quotient.*

5. $17\overline{)357}$ 6. $35\overline{)770}$ 7. $26\overline{)8086}$ 8. $35\overline{)3920}$

9. $28\overline{)5992}$ 10. $42\overline{)1512}$ 11. $19\overline{)1045}$ 12. $71\overline{)6035}$

13. $63\overline{)55{,}629}$ 14. $85\overline{)28{,}985}$ 15. $56\overline{)48{,}328}$ 16. $43\overline{)35{,}303}$

17. $34\overline{)32{,}912}$ 18. $46\overline{)22{,}034}$ 19. $58\overline{)42{,}108}$ 20. $94\overline{)30{,}550}$

*See page 449 for more practice.

3-10 | Three-Digit Divisors

Find the quotient: 24,518 ÷ 784 = ___?___

To divide by a three-digit number, follow the steps
for dividing by a one- or two-digit number:

- Estimate to find the first digit of the quotient.

  24,518 ÷ 784

  24,000 ÷ 800 = 30
  Try 3 as the first digit.

- Multiply: 3 × 784 = 2352.
- Subtract: 2451 − 2352 = 99.
- Bring down 8 to get a partial dividend of 998.
- Estimate again to find the next digit.

  998 ÷ 784

  900 ÷ 900 > 1
  Try 1 as the second digit.

- Multiply: 1 × 784 = 784
- Subtract: 998 − 784 = 214
- Check, using a calculator.

```
          ×
                  3 1 R214
      7 8 4 ) 2 4,5 1 8
           − 2 3 5 2
              9 9 8
              7 8 4
              2 1 4  Remainder (R)
```

Estimate. How many digits will the quotient have?

1. 941 ÷ 317
2. 1716 ÷ 143
3. 64,239 ÷ 482
4. 154,364 ÷ 518

What number should be the next try? In which place-value position? Complete.

5. $\dfrac{5}{168)8915}$
    840
    ‾‾51‾‾

6. $\dfrac{2}{367)9175}$
    734
    ‾‾183‾‾

7. $\dfrac{3}{726)23,958}$
    2178

8. $\dfrac{7}{406)292,726}$
    2842

Estimate. Then find the quotient.*

9. 169)3549
10. 211)3165
11. 432)9504
12. 182)4186

13. 211)8651
14. 152)5624
15. 178)14,596
16. 542)53,116

17. 672)64,512
18. 705)43,710
19. 444)19,536
20. 397)32,554

21. 263)16,589
22. 548)44,456
23. 972)28,192
24. 389)278,955

56   *See page 449 for more practice.

3-11 Zero in the Quotient

Write zero in the quotient whenever the divisor is *greater than* any of the partial dividends.

Find the quotient: 52,312 ÷ 52 = ?

- Estimate to find the first digit of the quotient.

  52,312 ÷ 52

  50,000 ÷ 50 = 1000
  Try 1 as the first digit.

- Bring down the next digit.
- Partial dividend is now 3.

> When the divisor is greater than the next partial dividend, put a zero in the quotient and bring down the next digit.

- Partial dividend is now 31.

> Put another zero in the quotient. Bring down the next digit.

- To get the last digit: Estimate. Try 6 as the last digit.
- Check, using a calculator.

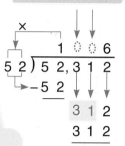

Neither "3" nor "31" divisible by 52, so zero in the quotient.

```
            1 0 0 6
    5 2 ) 5 2,3 1 2
        - 5 2
              3 1 2
              3 1 2
```

What is the next digit in the quotient? Why?

1.   18)1990   2.   11)23,133   3.   35)52,570   4.   102)10,914
     1?   2 ?   1 5?   1?
     18   22   35   10 2
     ───   ───   ───   ───
      1   1   17 5   71

Estimate; then divide. Use a calculator to check the answers.*

5.   3)9159   6.   6)12,450   7.   7)22,456   8.   82)24,928

9.   45)91,350   10. 956)19,120   11. 135)109,315   12. 198)99,396

Estimate and solve.

13. An astronaut spent a total of 19,520 minutes in space.
     During that time, his spacecraft completed 64 orbits.
     On the average, how many minutes long was each orbit?

14. In training for a dogsled race, a musher and her team traveled 1326 miles
     in 13 days. On the average, how many miles did they travel each day?

15. A store sold 82 car phones for $7421. What was the price of each phone?

*See page 449 for more practice.

3-12 Dividing a Decimal by a Whole Number

Find the quotient: 815.424 ÷ 24 = __?__

To divide a decimal by a whole number:

- Place the decimal point in the quotient just above the decimal point in the dividend.

- Divide as you would with whole numbers.

815.424 ÷ 24

800 ÷ 20 = 40

Try 4 as the first digit.

But 4 × 24 = 96

and 96 > 81, so try 3.

- Why is 3 also the second digit?
- Did you figure out why 9, 7, and 6 are the next digits in the quotient?
- Check, using a calculator.

```
              3 3.9 7 6
    2 4 ) 8 1 5.4 2 4
          - 7 2
              9 5
              7 2
              2 3 4
              2 1 6
                1 8 2
                1 6 8
                  1 4 4
                  1 4 4
```

Compatible numbers may be used in estimation.

0.718 ÷ 9

Round up or down to an easily divisible dividend.

0.72 ÷ 9 = 0.08 estimated quotient

A "compatible" number or dividend

Examine examples 1–4.

1. 725.424 ÷ 24 = __?__

```
           3 0.2 2 6
   2 4 ) 7 2 5.4 2 4
        - 7 2
            5 4
            4 8
            6 2
            4 8
            1 4 4
            1 4 4
```

Why was the zero placed in the quotient?

2. 0.10632 ÷ 12 = __?__

```
            0.0 0 8 8 6
   1 2 ) 0.1 0 6 3 2
        -   9 6
            1 0 3
              9 6
              7 2
              7 2
```

Why was one zero put before the decimal point and two zeros placed after it?

58

3. $7.95 \div 25 = $ ___?___

```
     0.3 1 8
2 5 )7.9 5 0
    -7 5
      4 5
      2 5
      2 0 0
      2 0 0
```

> Why do you keep writing zeros in a decimal dividend until there is no remainder?

4. $5 \div 8 = $ ___?___

```
     0.6 2 5
8 )5.0 0 0
  -4 8
    2 0
    1 6
      4 0
      4 0
```

> Why do you write a decimal point and zeros after the 5?

Write <, =, or > without dividing.

5. $8.1 \div 3$ ___?___ $8.10 \div 3$

6. $1.5 \div 5$ ___?___ $1.5 \div 6$

7. $0.6946 \div 23$ ___?___ $6.9 \div 23$

8. $0.726 \div 60$ ___?___ $72.6 \div 6$

Find the quotient. Place zeros where necessary.

9. $6)\overline{2.94}$ **10.** $4)\overline{1.4}$ **11.** $9)\overline{18.18}$ **12.** $7)\overline{0.0217}$

13. $77)\overline{45.584}$ **14.** $27)\overline{15.417}$ **15.** $15)\overline{0.2145}$ **16.** $32)\overline{3.008}$

17. $356)\overline{13.35}$ **18.** $122)\overline{30.7196}$ **19.** $132)\overline{670.626}$ **20.** $215)\overline{5.375}$

Solve. Round money answers to the nearest cent.

21. Jamie covered 502.68 km in a 4-hour race. How far did he travel each hour?

22. On her paper route Sharon collected $66.25 to cover deliveries to 53 families for one week. How much did each family pay?

23. In a fund-raising bike-a-thon Gary was pledged $4.05 per kilometer. How much should he collect for going 45 km?

One digit in each answer below is wrong. Correct it.

24. $5.6 \times 932.1 = 8219.76$ **25.** $72.24 \div 8 = 9.01$ **26.** $25 \times 1.89 = 49.25$

27. $69 \times 0.58 = 40.12$ **28.** $74.58 \div 6 = 12.33$ **29.** $31.92 \div 7 = 4.66$

Estimate each quotient and compare with a calculator's quotient.

30. $41.53 \div 6$ **31.** $0.347 \div 5$ **32.** $6.39 \div 8$ **33.** $0.192 \div 3$

34. $31.48 \div 15$ **35.** $2.51 \div 8$ **36.** $5.98 \div 12$ **37.** $56.1 \div 11$

SKILLS TO REMEMBER

Compute.

38. $8.6 - 2.5 + 2$ **39.** $3.5 + 1.6 - 4$ **40.** $4.8 \div 12 \times 0.2$

41. $1.5 \times 0.3 \div 9$ **42.** $300 \div 10 \times 0.2$ **43.** $2.8 \div 7 \times 20$

3-13 Multiplying and Dividing by Powers of Ten

Shortcut for multiplying by powers of ten (such as 10, 100, and 1000):
Move the decimal point as many places *to the right* as there are zeros in the multiplier.

$$4\ 4 . 8\ 7 \quad \times \quad 1\ 0 \quad = \quad 4\ 4\ 8 . 7$$

Check:
```
   44.87
 ×    10
 448.70
```

$$4\ 4 . 8\ 7 \quad \times \quad 1\ 0\ 0 \quad = \quad 4\ 4\ 8\ 7 .$$

```
    44.87
 ×    100
 4487.00
```

$$4\ 4 . 8\ 7\ 0 \quad \times \quad 1\ 0\ 0\ 0 \quad = \quad 4\ 4\ 8\ 7\ 0 .$$

```
     44.87
 ×    1000
 44870.00
```

Shortcut for dividing by powers of ten:
Move the decimal point as many places *to the left* as there are zeros in the divisor.

$$4\ 4 . 8\ 7 \quad \div \quad 1\ 0 \quad = \quad 4 . 4\ 8\ 7$$

Check:
```
      4.487
 10)44.870
```

$$4\ 4 . 8\ 7 \quad \div \quad 1\ 0\ 0 \quad = \quad 0 . 4\ 4\ 8\ 7$$

```
       0.4487
 100)44.8700
```

$$4\ 4 . 8\ 7 \quad \div \quad 1\ 0\ 0\ 0 \quad = \quad 0 . 0\ 4\ 4\ 8\ 7$$

```
        0.04487
 1000)44.87000
```

Shortcut for multiplying by decimal powers of ten (such as 0.1, 0.01, and 0.001)
is the same as the shortcut for dividing by powers of ten.

$$4\ 4 . 8\ 7 \quad \times \quad 0 . 1 \quad = \quad 4 . 4\ 8\ 7$$

Check:
```
  44.87
×  0.1
 4.487     (3 places)
```

$$4\ 4 . 8\ 7 \quad \times \quad 0 . 0\ 1 \quad = \quad 0 . 4\ 4\ 8\ 7$$

```
   44.87
 × 0.01
 0.4487    (4 places)
```

$$4\ 4 . 8\ 7 \quad \times \quad 0 . 0\ 0\ 1 \quad = \quad 0 . 0\ 4\ 4\ 8\ 7$$

```
    44.87
 ×0.001
 0.04487   (5 places)
```

Remember: $0.1 = \dfrac{1}{10} = 1 \div 10$ and $0.001 = \dfrac{1}{1000} = 1 \div 1000$.

Choose the correct answer.

1. Dividing by 100 is the same as:
   a. multiplying by 100.
   b. multiplying by 0.01.
   c. dividing by 0.01.

2. To multiply a decimal by 1000:
   a. move the decimal point five places to the right.
   b. move the decimal point five places to the left.
   c. neither of these

3. Moving the decimal point six places to the right is the same as:
   a. dividing by one million.
   b. multiplying by one million.
   c. neither of these

4. To divide a decimal by 1000:
   a. move the decimal point three places to the right.
   b. move the decimal point three places to the left.
   c. neither of these

Find the product or quotient.

5. 100×0.6671 6. 0.006×100 7. $20.8 \div 10$ 8. 0.56×1000

9. 223.3×0.001 10. $161.3 \div 100$ 11. $256 \div 100$ 12. 39.2×100

13. $0.651 \div 100$ 14. $4.5 \div 1000$ 15. 0.01×30.5 16. $1.9684 \div 10$

By what power of ten must you multiply to get the product?

17. $\underline{\ ?\ } \times 0.2 = 2$ 18. $\underline{\ ?\ } \times 1.5 = 15$ 19. $\underline{\ ?\ } \times 0.03 = 3$

20. $\underline{\ ?\ } \times 0.05 = 5$ 21. $\underline{\ ?\ } \times 0.009 = 9$ 22. $\underline{\ ?\ } \times 0.4 = 4$

23. $\underline{\ ?\ } \times 1.04 = 104$ 24. $\underline{\ ?\ } \times 1.15 = 11.5$ 25. $\underline{\ ?\ } \times 0.025 = 25$

Write the product or quotient, putting the decimal point in the correct position.

26. $84.32 \times 200 = 16864$

27. $0.035 \times 3000 = 105$

28. $0.251 \times 6000 = 1506$

29. $36.36 \div 20 = 1818$

30. $650{,}280 \div 2000 = 32514$

31. $454.8 \div 600 = 758$

Look at this pattern. Copy and complete the chart.

		16	2.5	0.72	1.36	0.209	1.061
32.	÷ 0.1	160	25	7.2	13.6	?	?
33.	÷ 0.01	1 600	250	?	?	20.9	?
34.	÷ 0.001	16,000	?	?	?	?	?
35.	÷ 0.0001	160,000	?	?	?	?	?

36. Discuss what happens to the quotient as the divisor gets smaller.

3-14 Dividing by a Decimal

Find the quotient: $64.452 \div 2.46 =$ ___?___

To divide by a decimal:

- Change the divisor to a whole number by moving the decimal point the necessary number of places *to the right*.
- Move the decimal point in the dividend the same number of places *to the right*.
- Place the decimal point in the quotient.
- Divide as usual.

Move the decimal point two places to the right.

Divide.

$$2.46\overline{\smash{)}64.45.2}$$

2 places   2 places

$$|64.452 \div 2.46|$$

$$\frac{64.452}{2.46} \times \frac{100}{100} = \frac{6445.2}{246.}$$

$$
\begin{array}{r}
26.2 \\
246{\overline{\smash{)}6445.2}} \\
492 \\
\hline
1525 \\
1476 \\
\hline
492 \\
492 \\
\hline
\end{array}
$$

- Check, using a calculator.

Examine each of these examples. Explain the zeros.

1. $0.0168 \div 5.6 =$ ___?___

$$
\begin{array}{r}
0.003 \\
5.6\overline{\smash{)}0.0168} \\
168 \\
\hline
\end{array}
$$

Why were zeros written in the quotient?

2. $3 \div 0.075 =$ ___?___

$$
\begin{array}{r}
40. \\
0.075\overline{\smash{)}3.000} \\
300 \\
\hline
0
\end{array}
$$

Why were 3 zeros written in the dividend?

Compute mentally.

3. $42 \div 0.6 = 420 \div$ ___?___

5. $42 \div 0.006 =$ ___?___ $\div 6$

7. $4.2 \div 0.06 =$ ___?___ $\div 6$

9. $0.42 \div 0.6 =$ ___?___ $\div 6$

4. $42 \div 0.06 =$ ___?___ $\div 6$

6. $4.2 \div 0.6 =$ ___?___ $\div 6$

8. $4.2 \div 0.006 =$ ___?___ $\div 6$

10. $0.42 \div 0.06 =$ ___?___ $\div 6$

Choose the best estimate.

11. $7741.9 \div 515.3$ is about

    a. 15   **b.** 10   **c.** 150   **d.** 200   **e.** 250

12. $65.615 \div 0.05$ is about

    a. 13   **b.** 130   **c.** 1300   **d.** 150   **e.** 1500

13. $8.742 \div 0.047$ is about

    a. 18   **b.** 180   **c.** 1800   **d.** 20   **e.** 2000

14. $27.734 \div 7$ is about

    a. 27.000   **b.** 3.000   **c.** 4.000   **d.** 2.700   **e.** none of these

Write the quotient, putting the decimal point in the correct position.

15. $72 \div 0.09 = 8000$ **16.** $0.3 \div 0.08 = 375$

17. $7.488 \div 1.3 = 576$ **18.** $0.2160 \div 0.72 = 30$

19. $0.6867 \div 6.3 = 109$ **20.** $0.1925 \div 7.7 = 25$

Estimate and divide. Then check by multiplying.

21. $2.8\overline{)9.8}$ **22.** $0.6\overline{)7.47}$ **23.** $0.46\overline{)0.805}$ **24.** $0.08\overline{)64}$

25. $0.5\overline{)219}$ **26.** $0.16\overline{)1}$ **27.** $0.96\overline{)3.984}$ **28.** $0.72\overline{)0.0216}$

Divide. Round the quotient to the nearest hundredth.

29. $3.26 \div 0.06$ **30.** $5.485 \div 1.13$ **31.** $0.8654 \div 0.24$

32. $5.7 \div 0.07$ **33.** $5 \div 0.03$ **34.** $7 \div 5.16$

35. $0.412 \div 0.13$ **36.** $34.12 \div 8.7$ **37.** $0.1252 \div 24.6$

Solve.

38. A car averages 6.3 kilometers per liter of gasoline. How much fuel will be needed for a 110.25-km trip?

39. Mr. Shaw paid $16.05 for 11.5 gallons of gasoline. How much did he pay per gallon?

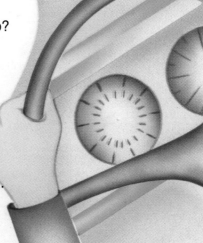

MAKE UP YOUR OWN...

40. Oil for an automobile's engine is $2.66 a quart. An oil change requires over 4 quarts of oil. Use this information to make up your own word problem.

It All Adds Up to Computers

What a world we live in! Push a button and calculate at lightning speed. Computers touch every part of our lives from the moment we are born. But are YOU ready for a computer in your future? You will be, if you know where computers came from and where they are going.

The idea of computers was born long ago. For example, the ancient Chinese used the abacus — a simple computing tool — to keep track of large numbers. Over the centuries men and women created more efficient ways of processing ever-growing quantities of information. Then, with the advance of the 20th century, the age of the modern computer arrived.

It was clumsy at first. Tubes and transistors made computers as large as rooms. Today, with integrated chips, we can build supercomputers and microcomputers that can perform fifty million calculations in a second. This is a long way from the abacus!

The computer is a powerful tool, and its use will be limited only by our imaginations and needs. Thanks to the hard work of computer pioneers, you will be free for even greater thoughts on your own voyage of discovery.

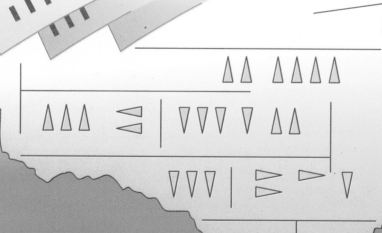

1. Choose a computer pioneer. Research his or her life and contribution to the world of computers. Do not forget to use a computer to prepare your report!

2. Create a computer-history timeline. Leave plenty of space on your timeline for recent computer discoveries.

3. Imagine that you are a computer wizard with a crystal ball to look into the future. Write what you see computers doing for our world in the 21st century.

4. Open a *Pioneers of Computers* file. See if you can organize your file about computer pioneers alphabetically, by nationality, and chronologically.

5. Plan a technology center for your school. Using a spreadsheet program, prepare a budget for all the materials (hardware and software) that you will need. You might prepare three different budgets based on such total amounts as $30,000; $60,000; and $100,000.

6. In the computer world we use acronyms for easy recall. For example: BASIC is an acronym for Beginners All-purpose Symbolic Instruction Code. Find the meaning of the following terms and where they came from. Which are acronyms?

COBOL	bit	ROM
Fortran	ENIAC	Logo
ASCII	Modem	RAM
CPU	Pascal	UNIVAC

7. Now try these.

ALGOL	PL/I	LISP
Ada	Forth	PROLOG
C	SNOBOL	SIMSCRIPT

STRATEGY

3-16 | Problem Solving: Identifying Extra Information

Problem: Ashley makes $.02 on each circular she delivers. Last week she made $4.28. This week she has made $4.62. How many circulars did she deliver last week?

1 IMAGINE You record the number of circulars delivered each week.

2 NAME

Facts: $ .02 — earned on each circular
$4.28 — last week's earnings
$4.62 — this week's earnings

Question: How many circulars did Ashley deliver last week?

3 THINK Look at your *Imagine* picture.
Write a number sentence to show what you will do to find the number of circulars, *c*, delivered.

$$ \$ .02 \quad \times \quad c \quad = \quad \$4.28 $$

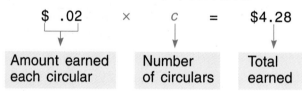

| Amount earned each circular | Number of circulars | Total earned |

Use the opposite of multiplication, division, to find the missing factor.

$$ c = \$4.28 \div \$ .02 $$

Do you need to know how much Ashley earned *this* week to find out how many circulars she delivered last week? — No! This is extra information, which you can ignore.

4 COMPUTE Solve: $c = \$4.28 \div \$ .02$
$c = 214$ circulars delivered

5 CHECK Estimate. Use the compatible numbers 2 and 400 to *divide* mentally.

$$ \frac{4.28}{0.02} \longrightarrow \frac{428}{2} \approx \frac{400}{2} = 200 \text{ (Estimated Quotient)} $$

The answer should be a little more than 200 because 428 was rounded down to 400 to estimate the quotient.

66

Solve these problems. Identify any extra information.

1. The *Town Crier* lists 450 want ads each day. The circulation of the *Town Crier* is 12,000. The paper charges $7.50 per day to run a want ad. What is the cost of a regular want ad if it is run for 6 days?

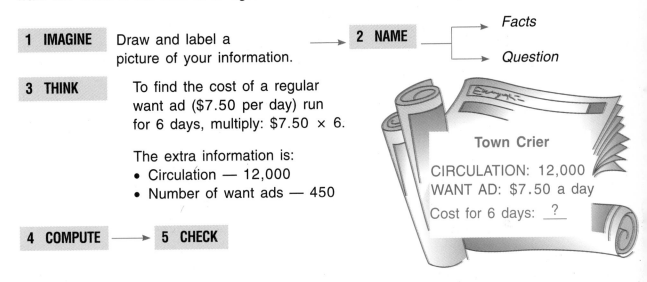

1 IMAGINE Draw and label a picture of your information.

2 NAME ⟶ Facts / Question

3 THINK To find the cost of a regular want ad ($7.50 per day) run for 6 days, multiply: $7.50 × 6.

The extra information is:
- Circulation — 12,000
- Number of want ads — 450

4 COMPUTE ⟶ **5 CHECK**

Town Crier
CIRCULATION: 12,000
WANT AD: $7.50 a day
Cost for 6 days: __?__

2. If 934 people in a town of 1450 people watch *Sideline News* for an hour each day, how many hours of *Sideline News* will they view in a year?

3. The *New York News* has a circulation of 1,636,600. The *San Antonio Express-News* has a circulation that is 0.125 times as much. These figures are based on the average circulation of each paper for 6 months. What is the circulation of the *San Antonio Express-News*?

4. The circulation of a monthly magazine in the United States is about 18,179,600 or 11.1 times its circulation in Canada. The circulation of the same magazine in Canada is 1.8 times larger than its nearest competitor. What is the circulation of the monthly magazine in Canada?

5. A certain industry spends five times more on newspaper advertising than on magazine advertising. Of $2,162,000 budgeted for magazine advertising, $903,000 is spent on news magazines. How much money is spent on newspaper advertising?

6. The communication satellite used by *Nightly News* travels 6875.4 miles per hour at an altitude of 22,300 miles. How far will it travel at that altitude in 48.5 hours?

7. It takes Maui 32 minutes to deliver the daily papers on his 3 paper routes and 55 minutes to deliver the Sunday papers. How much less time does it take him to deliver the daily papers?

Problem Solving: Applications

Solve.

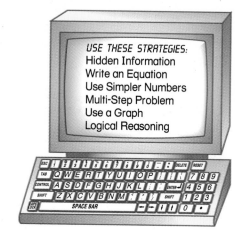

USE THESE STRATEGIES:
Hidden Information
Write an Equation
Use Simpler Numbers
Multi-Step Problem
Use a Graph
Logical Reasoning

1. On the average a person's heart pumps about 18 quarts of blood per minute. A trained athlete's heart pumps 30 quarts per minute. How many more quarts of blood can a trained athlete's heart pump per minute than the heart of an average person?

2. If the heart rate of a human is 80 beats per minute and the heart rate of a bird is 1.875 times faster, what is the heart rate of the bird?

3. Alyssa's resting heart rate is 80 beats per minute. After exercising, her heart rate is 180 beats per minute. How many times faster is her heart rate just after exercising than when she is at rest?

4. A red blood cell, which carries oxygen, can live 120 days. A platelet, which helps the blood to clot, can live only 0.075 times that long. How long can a platelet live?

5. Of the total volume of blood in the body, about 3.3 liters consists of a fluid called plasma. If the plasma represents 0.55 of the total, what is the total volume of blood in the body?

6. On the average, a microliter of blood contains 5 million red blood cells. This is 20 times the average number of platelets. On the average, how many platelets are there in a microliter of blood?

7. The human body requires 0.25 gram of iron to produce a pint of blood. How many grams of iron are needed to produce a gallon of blood? (Hint: 8 pints = 1 gallon.)

8. On Tuesday 72 volunteers visited the Ridgeway Center to donate blood. If it takes 20 minutes to donate a pint of blood and the Center can accommodate 4 donors at a time, how many hours did it take for all of the donors to give blood?

9. Gerry weighs 110 pounds. If 0.08 of his total body weight is the weight of his blood, how much does the blood in his body weigh?

10. A medical technician's job is to find the blood type of samples submitted to the lab. If it takes 0.25 hour to finish one sample, how many samples can be completed in 7.5 hours?

11. A red blood cell makes 43,750 trips between the lungs and the body tissues in 10 weeks. How many trips does it make in 1 day?

12. The human heart pumps 680,000 gallons of blood in a year. Over a 20-year period, how many gallons of blood will the human heart pump?

13. The diameter of the largest capillary in the human body is only 0.2 mm. The diameter of the aorta, the largest artery, is 1250 times larger. What is the diameter of the aorta?

Dr. Heartbeat graphed the results of his experiment on the effect of temperature on heart rate. Use his graph to answer these questions.

14. If Dr. Heartbeat later found that the heart rate at 102 °F was 1.3 times faster than the heart rate at 98 °F, what was the rate at 102 °F?

15. Based on his experiment, what did Dr. Heartbeat discover?

    a. Heart rate is constant at all body temperatures.

    b. Heart rate increases as temperature increases.

    c. Heart rate decreases as temperature increases.

16. Two girls, Wanda and Xanthe, and two boys, Yul and Zorba, each have a different blood type (A, B, AB, or O). Use the clues to complete the chart and match each person with his/her blood type.

    I. Xanthe and the boy who has type-B blood live near Zorba.

    II. Zorba does not have type-O blood.

    III. The girl who has type-AB blood does not live near Zorba.

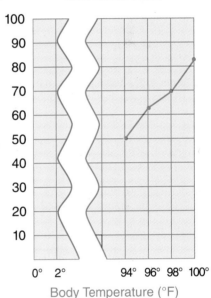

Body Temperature and Heart Rate

Beats per Minute

Body Temperature (°F)

Exercise 16

	A	B	AB	O
W				
X				
Y				
Z				

More Practice

Choose the best answer.

1. 1000 × 200   **a.** 2000   **b.** 20,000   **c.** 200,000   **d.** 2,000,000

2. 20 × 600   **a.** 120,000   **b.** 12,000   **c.** 200   **d.** 120

3. 36,000 ÷ 900   **a.** 40   **b.** 400   **c.** 4000   **d.** 40,000

4. 870 ÷ 1000   **a.** 8.7   **b.** 87   **c.** 0.087   **d.** 0.87

5. 132.5 ÷ 100   **a.** 13.25   **b.** 1.325   **c.** 0.1325   **d.** 0.01325

Find the product.

6. 807
   × 73

7. 86.99
   × 0.2

8. 0.438
   × 0.7

9. 1.087
   ×0.056

10. 22.08
    ×0.0752

Choose the best estimate. Explain your strategy.

11. 4830 ÷ 13   **a.** 200   **b.** 2000   **c.** 400   **d.** 4000   **e.** 480

12. 80.95 ÷ 8.9   **a.** 9   **b.** 10   **c.** 0.9   **d.** 0.10   **e.** 100

13. 300 ÷ 29.74   **a.** 100   **b.** 0.50   **c.** 1.0   **d.** 10   **e.** 20

Find the quotient. Round to the nearest hundredth.

14. 7)4.571   15. 32)0.72   16. 1.6)7.712   17. 0.21)4.25   18. 1.56)0.546

Solve.

19. The tide at Blancha Beach rises 7.5 m over a 12-hour period.
    How many meters does the tide rise each hour?

20. At the school's annual book fair the committee
    made an error in calculating the money earned. On the
    first count, the cashiers counted $273.50. Their actual
    earnings were $263.50. Examine the calculations.
    Then find and correct the mistakes.

    45 paperbacks at $4.50 each ⟶ $212.50
     7 posters at $5.00 each ⟶ 35.00
    65 bookmarks at $.40 each ⟶ 26.00

21. If 20 cans of juice cost $38.60, what is
    the cost of 15 cans at the same rate?

Math Probe

BE A MATHLETE! USE PATTERN POWER!

Study these five problems.

What is the relationship between the two factors?
How can the product be found mentally?

```
    1 2          1 3          1 6          1 9          1 5
  × 1 8        × 1 7        × 1 4        × 1 1        × 1 5
    2 1 6        2 2 1        2 2 4        2 0 9        2 2 5
```

Solve these, using the pattern. (The first one is done.)

1. 22   **2.** 23   **3.** 26   **4.** 29   **5.** 25
    × 28   × 27   × 24   × 21   × 25
    ‾‾‾
    616 | 30 × 20 = 600
         | AND
         | 8 × 2 = 16

Blast off by solving these:

6. 37 × 33   **11.** 2.5 × 2.5

7. 45 × 45   **12.** 8.4 × 8.6

8. 62 × 68   **13.** 3.8 × 3.2

9. 56 × 54   **14.** 9.7 × 9.3

10. 79 × 71   **15.** 0.11 × 0.19

16. Write some problems of your own that can be solved using this pattern.

> Be careful placing the decimal point in each product.

These problems use a different pattern.
Solve. Compute the sums within the parentheses first.

17. (3.03 + 4.04) ÷ (3 + 4)   **18.** (7.07 + 2.02) ÷ (7 + 2)

19. (6.06 + 50.5) ÷ (6 + 50)   **20.** (8.08 + 9.09) ÷ (8 + 9)

21. (10.1 + 5.05) ÷ (10 + 5)   **22.** (12.12 + 6.06) ÷ (12 + 6)

23. (15.15 + 20.2) ÷ (15 + 20)   **24.** (16.16 + 30.3) ÷ (16 + 30)

25. Write some problems of your own that can be solved using the pattern in exercises 17–24.

PATTERN POWER

Check Your Mastery

Find the product.

See pp. 46–49

1.	797	2.	1246	3.	13,807	4.	23,328
	× 100		× 95		× 619		× 804

Place the decimal point in each product.

See p. 50

5.	4.3	6.	847	7.	0.204	8.	2.504
	× 6		×0.09		× 0.52		× 3.05
	258		7623		10608		763720

Estimate. Use <, =, or >.

See p. 51

9. 9.2×150 __?__ 92×15 10. 0.8×3.25 __?__ 8×0.325

11. 250×73 __?__ 73×25 12. 225×25 __?__ 2.25×2.5

Estimate.

See pp. 52–53

13. $8094 \div 4$ __?__   14. $625 \div 25$ __?__

15. $6014 \div 20$ __?__   16. $92,756 \div 5$ __?__

Find the quotient.

See pp. 54–57

17. $4408 \div 8$ 18. $6840 \div 95$ 19. $26,102 \div 421$

20. $10,038 \div 14$ 21. $9592 \div 88$ 22. $76,050 \div 325$

Place the decimal point in each quotient.

See pp. 58–63

23. $9\overline{)6.3}$ = 7   24. $0.9\overline{)0.63}$ = 7   25. $0.009\overline{)6.3}$ = 7   26. $9\overline{)0.63}$ = 7   27. $0.09\overline{)63}$ = 7

Find the quotient to the nearest hundredth.

See pp. 62–63

28. $74 \div 3.9$ 29. $60.5 \div 0.07$ 30. $2.984 \div 18.2$ 31. $571.77 \div 21.7$

Choose the best estimate. Explain your strategy.

32. Bob earns $6.72 an hour. If he worked for 29 hours, how much did he earn?
   a. $90   **b.** $100   **c.** $180   **d.** $210

33. Together, fourteen employees earn $12,644 in a week. On the average, what does each employee earn per week?
   a. $600   **b.** $700   **c.** $800   **d.** $900

34. Karen earned $178 for 31 hours of work. Her hourly pay was:
   a. $3.50   **b.** $6.00   **c.** $7.50   **d.** $8.00

4 Expressions, Equations, and Inequalities

In this chapter you will:

- Write and evaluate mathematical expressions
- Examine equations and inequalities
- Identify properties for addition and multiplication
- Write and solve equations
- Identify order of operations
- Solve problems: using equations

Do you remember?

= stands for: is equal to, same as, is

≠ stands for: is not equal to, not the same as, is not

< stands for: is less than

> stands for: is greater than

RESEARCHING TOGETHER

From Whence Came I ???

Go through the "Imagination Math Machine" and come out as one of the symbols below. Who are you? When and where were you "born"? When first born long ago, what did you mean? Name how you are used today. Is it the same or different? Without naming yourself, share your life story. See if others can guess what symbol you are.

=   ÷   ×   −   <   >

Mathematical Expressions

In mathematical expressions letters, called **variables**, are used to represent numbers.

Mathematical Expression	Meaning
$d + 5$ ⟶	Add 5 to whatever value is assigned to the variable d.
$r - 8$ ⟶	Subtract 8 from the variable r.
$7s$ ⟶	Multiply the variable s by 7.
$\dfrac{z}{4}$ ⟶	Divide the variable z by 4.

Any letter can be used as a variable.

To evaluate a mathematical expression is to find the number the expression represents. The **value** of the entire mathematical expression depends on the numbers substituted for the variables.

**Evaluate d + 5 = ? **

When $d = 6$ ⟶ $d + 5 = 6 + 5 = 11$

When $d = 9$ ⟶ $d + 5 = 9 + 5 = 14$

Remember: Multiply or divide left to right, then Add or subtract left to right.

Evaluate $r - 8$ when $r = 10$. The answer is 2. Why?

Evaluate $t + 6$ when t is:
1. 7
2. 10
3. 21
4. 65
5. 0.06

Evaluate $18 + m$ when m is:
6. 17
7. 0
8. 6
9. 0.18
10. 7.3

Evaluate $n - 23$ when n is:
11. 47
12. 23
13. 150
14. 26.5
15. 172.9

Evaluate. (The first one is done.)
16. $a + a$ when $a = 5$ ⟶ $a + a = 5 + 5 = 10$

17. $b + b + b$ when $b = 7$
18. $x - x$ when $x = 3$

19. $5 + z + 3 + z$ when $z = 10$
20. $y + 5 - y$ when $y = 4$

Evaluate. (The first one is done.)
21. $r + s$ when $r = 7$ and $s = 8$ ⟶ $r + s = 7 + 8 = 15$

22. $b - q$ when $b = 18$ and $q = 9$
23. $4 + a - b$ when $a = 1$ and $b = 5$

24. $s - t + 3$ when $s = 27$ and $t = 11$
25. $y + x - y$ when $x = 3$ and $y = 4.1$

Evaluate. (The first two are done.)

For exercises 38–41, remember to *multiply* or *divide before* you add or subtract.

26. $3x$ when $x = 4 \longrightarrow 3x = 3 \times 4 = 12$

27. $\dfrac{d}{9}$ when $d = 27 \longrightarrow \dfrac{d}{9} = \dfrac{27}{9} = 27 \div 9 = 3$

28. $11s$ when $s = 10$

29. $5z$ when $z = 20$

30. $\dfrac{18}{m}$ when $m = 9$

31. $\dfrac{6}{r}$ when $r = 2$

32. $\dfrac{42}{y}$ when $y = 6$

33. $\dfrac{d}{10}$ when $d = 80$

34. $8x$ when $x = 8$

35. $1.50m$ when $m = 2$

36. $\dfrac{y}{x}$ when $y = 72$ and $x = 8$

37. $\dfrac{2r}{s}$ when $r = 10$ and $s = 4$

38. $3x + 4$ when $x = 5$

39. $4z - 8$ when $z = 2$

40. $\dfrac{m}{n} + 3$ when $m = 8$ and $n = 2$

41. $\dfrac{c}{d} - 5$ when $c = 20$ and $d = 2$

True or false. Explain.

42. $a + b$ always has the same value as $b + a$.

43. $a - b$ always has the same value as $b - a$.

44. xy always has the same value as yx.

45. $\dfrac{c}{d}$ always has the same value as $\dfrac{d}{c}$.

46. Which expression has a value of 9 when $n = 4$?

    a. $36n$ **b.** $\dfrac{36}{n}$ **c.** $\dfrac{n}{36}$ **d.** none of these

47. Which expression has a value of 4 when $n = 2$?

    a. $2n + 5$ **b.** $2 + n$ **c.** $\dfrac{10}{n}$ **d.** all of these

48. Which expression has a value of 5 when $n = 5$?

    a. $10n$ **b.** $10 + n$ **c.** $10 - n$ **d.** $\dfrac{10}{n}$

49. Which expression has a value of 20 when $r = 2$ and $s = 6$?

    a. $rs + 8$ **b.** $\dfrac{s}{r} + 17$ **c.** $r + s + 12$ **d.** all of these

MAKE UP YOUR OWN...

50. Write your own mathematical expressions that will have values equal to 10, 5, 25, and 1 when $a = 2$, $b = 5$, and $c = 0$.

Translating Expressions

English Expression	Mathematical Expression
Five more than a number	Let n be the number: $\quad n + 5$
Three times Rosa's weight	Let w be the weight: $\quad 3w$
A pizza pie divided into eighths	Let p be the pie: $\quad \dfrac{p}{8}$

English Sentence	Mathematical Sentence
Five more than a number *is equal to* 13.	$n + 5 = 13$
Three times Rosa's weight *is less than* 300 pounds.	$3w < 300$
A pizza pie divided into eighths *is not equal* to 2 pies.	$\dfrac{p}{8} \neq 2$

See the charts on page 77 to help you translate mathematical expressions.

All mathematical expressions may be classified as:

- **Numerical expressions,** which contain only numbers.
    two times the sum of 8 and 10 $\longrightarrow$ $2 \times (8 + 10)$

- **Algebraic expressions,** which contain numbers and variables.
    half the difference between 16 and a number $\longrightarrow$ $\dfrac{(16 - n)}{2}$

Write an expression for each. Explain. (The first three are done.)

1. 3 less than the variable p $\longrightarrow$ $p - 3$ 2. 8 times the variable m $\longrightarrow$ $8m$

3. a number increased by 15 $\longrightarrow$ $m + 15$ 4. a number added to 12

5. 6 subtracted from a number   6. a number divided by 10

7. 5 added to twice a number   8. 6 less than half a number

9. 6 plus a number doubled   10. 10 divided by twice a number

11. three more than twice a number   12. two less than a number doubled

Translate each, using grouping symbols.
(The first one is done.)

13. 4 multiplied by the sum of a number and 5
    $[4(m + 5)]$

14. the difference when a number doubled
    is subtracted from 3

15. the quotient when the sum of 5 and
    a number is divided by 10

16. the product when 15 minus 3
    is multiplied by a number

17. the quotient when 12 added
    to a number is divided by 10

18. a number times the sum of 5 and 8

Translate each into an English expression.

19. $a + 15$ 20. $a \times a$ 21. $a \div 5$

22. $a + a$ 23. $2a - 1$ 24. $a(4 - 2)$

25. $5(a + 1)$ 26. $\dfrac{a + 2}{12}$ 27. $\dfrac{20 - a}{10}$

28. Evaluate each expression in
    exercises 19 – 27 when $a = 10$.

Translate.

29. two years younger than Juanita
    Let d be Juanita's age now.

30. a width doubled
    Let m be the width.

31. one third of the students
    Let c be the students.

32. half her sister's height
    Let h be her sister's height.

33. plus a tax of $1.12
    Let c be the list price.

Mathematical Translations

Addition

$3 + a$ is the same as:

the sum of 3 and a
a added to 3
3 plus a
a more than 3
3 increased by a

Subtraction

$b - 7$ is the same as:

the difference between b and 7
7 subtracted from b
b minus 7
7 less than b
b decreased by 7

Multiplication

$5n$ is the same as:

the product of 5 and n
n multiplied by 5
5 times n

Division

$\dfrac{6}{x}$ is the same as:

the quotient of 6 and x
6 divided by x

Special Phrases

$2n$ = twice, or double,
   a number, n

$\dfrac{c}{2}$ = half a number, c

$3d$ = a number, d, tripled,
   or three times
   a number, d

4-3 | Equations and Inequalities

> **Equation:** a mathematical sentence that expresses an equality. (Symbol: "=")

$n + 5 = 13$ says "5 more than the number, n, *is equal to* 13."

> **Inequality:** a mathematical sentence that expresses an inequality. (Symbols: "<" or ">" or "≠")

$d - 6 < 16$ says "A number, d, decreased by 6 *is less than* 16."

$2x + 4 \neq 8$ says "Twice a number, x, increased by 4 *is not equal to* 8."

> **True and False Sentences** The value of the number that is substituted for a variable makes the mathematical sentence *true* or *false*.

Let $d = 8$

Equation: ⟶ $d + 7 = 15$ ⟶ $8 + 7 = 15$
Since $15 = 15$,
this is a *true* sentence.

Inequality: ⟶ $d - 4 > 5$ ⟶ $8 - 4 > 5$
But $4 \ngtr 5$, so this is
a *false* sentence.

$\ngtr$ ⟶ is not **greater than**
$\nless$ ⟶ is not **less than**

Translate each as a mathematical sentence. Identify each as an *equation* or *inequality*. (The first three are done.)

1. Six less than a certain number is equal to thirteen.

Let n be the number:
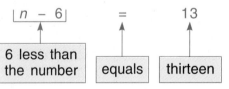
$n - 6$ — 6 less than the number
$=$ — equals
13 — thirteen

This is an *equation*.

2. Ten years from now Louis will be older than 21.

Let a be Louis' age now:
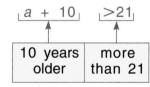
$a + 10$ — 10 years older
>21 — more than 21

This is an *inequality*.

3. Two dollars off the regular price of the calculator is $9.95.

$$c - \$2.00 = \$9.95 \quad \text{(Explain.)}$$

This is an *equation*.

4. Two less than twice a number is equal to ten.

5. One more than twice a number is greater than three.

6. Eight times a number is not equal to ten.

7. Six less than three times a number is equal to twenty.

8. Six is less than three times a number.

9. One fourth the cost of a pizza pie is less than $2.00.

10. One sixth of a number equals 7.

11. One sixth of a number is less than two.

12. Toni is 3 feet tall. She is half the height of her father.

13. One third of the price is less than $20.

14. Sue is five years younger than Jim. She is twelve.

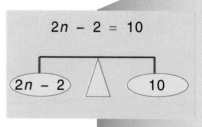
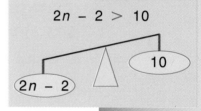
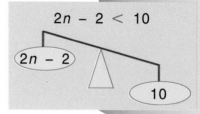

Write <, =, or > to make each sentence true.

15. 5×6 __?__ $5 + 6$

16. $5n \times 6$ __?__ $5n + 6$

17. $\dfrac{120}{10}$ __?__ $\dfrac{1200}{100}$

18. $\dfrac{120t}{10}$ __?__ $\dfrac{1200t}{100}$

19. $87 + 87$ __?__ 2×87

20. $87a + 87a$ __?__ $2 \times 87a$

21. $395 + 0$ __?__ 395×0

22. $395s + 0$ __?__ $395s \times 0$

23. Do you see a pattern between the answers to the questions on the left and those on your right? Can you explain?

Mathematical sentences can have more than one variable. Find some values of the variables in each sentence that will make it true.

24. $x + y = 10$

25. $x + y < 10$

26. $x + y > 10$

27. $2x + y = 10$

28. $2x + y < 10$

29. $2x + y > 10$

30. $\dfrac{x}{2} + y = 10$

31. $\dfrac{x}{2} + y < 10$

32. $\dfrac{x}{2} + y > 10$

 CRITICAL THINKING Can you explain why more than one value can be substituted for each variable in exercises 24–32? Is this true for the following?

33. $a \times a = 64$

34. $y - y = 0$

35. $\dfrac{t \times t}{t} = 4$

Properties of Addition and Multiplication

Let a, b, and c be any numbers:

Commutative Property...	of Addition	of Multiplication
Think: "order"	$a + b = b + a$ $6 + 2 = 2 + 6$	$a \times b = b \times a$ $6 \times 2 = 2 \times 6$

Changing the *order* of the addends or the factors does *not* change the sum or the product.

Associative Property...	of Addition	of Multiplication
Think: "grouping"	$(a + b) + c = a + (b + c)$ $(6 + 2) + 3 = 6 + (2 + 3)$ $8 + 3 = 6 + 5$	$(a \times b) \times c = a \times (b \times c)$ $(6 \times 2) \times 3 = 6 \times (2 \times 3)$ $12 \times 3 = 6 \times 6$

Changing the *grouping* of the addends or the factors does *not* change the sum or the product.

Identity Property...	of Addition	of Multiplication
Think: 0 for "+" 1 for "×"	$0 + a = a + 0 = a$ $0 + 6 = 6 + 0 = 6$ └Identity element	$1 \times a = a \times 1 = a$ $1 \times 6 = 6 \times 1 = 6$ └Identity element

Distributive Property of Multiplication over Addition

Think: "same factor across addends"

$6 \times (2 + 3) = (6 \times 2) + (6 \times 3)$

$6 \times 5 = 12 + 18$

$30 = 30$

Distribute the factor 6 across both addends, 2 and 3.

Zero Property of Multiplication

Think: $n \times 0 = 0$

$7 \times 0 = 0$

$0 \times 11 = 0$

What number makes each sentence true? Name the property used.*

1. $15 \times t = 4 \times 15$

2. $20 + s = 7 + 20$

3. $14 + n = 14$

4. $25y = 25$

5. $8 \times (2 + 12) = (p \times 2) + (p \times 12)$

6. $(7 + 8) + x = 7 + (8 + 3)$

7. $(15 \times 3) + (15 \times 7) = 15 \times (3 + y)$

8. $16 \times a = 0$

9. $5 \times (6 \times 8) = (5 \times 6) \times b$

10. Let **a** be any number. Complete and explain: $a \times 0 = \underline{\ ?\ }$

4-5 Inverse Operations

Addition and subtraction are inverse operations:

If $2 + 3 = 5$, then $5 - 3 = 2$
If $3 + 2 = 5$, then $5 - 2 = 3$
If $5 - 3 = 2$, then $2 + 3 = 5$
If $5 - 2 = 3$, then $3 + 2 = 5$

Addition and subtraction "undo" each other.

So, subtracting the number undoes the addition of the number.

Adding the number undoes the subtraction of the number.

Let a, b, and c be any numbers:
If $a + b = c$, then $c - b = a$
If $c - b = a$, then $a + b = c$

Multiplication and division are inverse operations:

If $2 \times 3 = 6$, then $6 \div 3 = 2$
If $3 \times 2 = 6$, then $6 \div 2 = 3$
If $6 \div 3 = 2$, then $2 \times 3 = 6$
If $6 \div 2 = 3$, then $3 \times 2 = 6$

Multiplication and division "undo" each other.

So, dividing by the number undoes the multiplication by the number.

Multiplying by the number undoes the division by the number.

Let a, b, and c be any numbers:
If $a \times b = c$, then $c \div b = a$
If $c \div b = a$, then $a \times b = c$

Replace each variable to form a true sentence.

1. $3 + y = 8$
so, $8 - y = 3$

2. $9 - x = 2$
so, $2 + x = 9$

3. $7 \times 8 = r$
so, $r \div 8 = 7$

4. $13 \times t = 26$
so, $26 \div t = 13$

5. $12 \div 3 = d$
so, $d \times 3 = 12$

6. $3r = 18$ → $3 \times r = r \times 3$ Why?
so, $\dfrac{18}{3} = r$

For each statement write a sentence that will "undo" the operation.*

7. $b - 4 = 8$

8. $c + 17 = 25$

9. $5d = 45$

10. $\dfrac{n}{10} = 4$

11. $t - 10 = 18$

12. $\dfrac{r}{2} = 7$

13. $f + 9 = 18$

14. $3q = 27$

15. $16k = 48$

CHALLENGE Can you "undo" these sentences? Explain.

16. $b - 4 > 8$

17. $5d < 45$

18. $t - 10 \neq 18$

*See page 450 for more practice.

4-6 Solving Addition and Subtraction Equations

In a basketball game a team's star scored a total of 32 points. 12 of these points were scored in the last half of the game. How many points did he score in the first half?

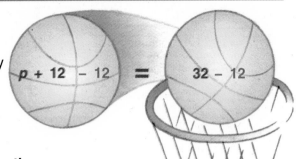

$$p + 12 - 12 = 32 - 12$$

Let p = the number of points scored in the first half ⟶ $p + 12 = 32$

To solve equations with addition and subtraction:

- See what operation is used with the variable.   (Adding 12)

$$p + 12 = 32$$

- What is the inverse of this operation?   (Subtracting 12) Do this on *both* sides of the equation.

$$p + 12 - 12 = 32 - 12$$

Subtract 12 from both sides.

- Solve.

$$p = 20$$

- Check.

$$p + 12 = 32$$

$$(20) + 12 = 32$$
$$32 = 32$$

Study these examples.

Solve:
$$t - 20 = 4$$
$$t - 20 + 20 = 4 + 20$$
$$t = 24$$

Check:
$$t - 20 = 4$$
$$(24) - 20 = 4$$
$$4 = 4$$

Solve:
$$m + 6 = 10$$
$$m + 6 - 6 = 10 - 6$$
$$m = 4$$

Check:
$$m + 6 = 10$$
$$4 + 6 = 10$$
$$10 = 10$$

Solve:
$$7 + y = 8$$

Since $(7 + y) = (y + 7)$, rewrite as:
$$y + 7 = 8$$
$$y + 7 - 7 = 8 - 7$$
$$y = 1$$

Check:
$$7 + y = 8$$ ← You can solve some equations mentally.
$$7 + (1) = 8$$
$$8 = 8$$

Solve and check.

1. $r + 8 = 10$
2. $t + 7 = 12$
3. $x - 2 = 21$
4. $x - 9 = 21$
5. $k + 13 = 20$
6. $m + 18 = 100$
7. $r - 100 = 13$
8. $b - 6 = 28$
9. $t + 8 = 15$
10. $s + 4 = 6$
11. $b - 7 = 10$
12. $c - 12 = 35$

Solve and check. (Remember: $7 + m = 12$ is the same as $m + 7 = 12$.)

13. $7 + m = 12$
14. $12 = 7 + m$
15. $19 = m - 25$
16. $9 + x = 21$
17. $k - 6 = 6$
18. $11 = x - 66$
19. $14 = 6 + p$
20. $15 = t + 8$
21. $20 + a = 31$
22. $7 + y = 11$
23. $35 = d - 12$
24. $9 = x - 9$

Which equations are equivalent because they have the same solution?

25. **a.** $x + 7 = 12$ **b.** $7 = x - 12$ **c.** $x - 5 = 0$ **d.** all of these
26. **a.** $x - 17 = 17$ **b.** $x - 7 = 17$ **c.** $7 + x = 17$ **d.** none of these
27. **a.** $8 + y = 20$ **b.** $4 = y - 8$ **c.** $y - 20 = 8$ **d.** none of these
28. **a.** $y + 12 = 26$ **b.** $y - 12 = 2$ **c.** $y = 14$ **d.** all of these
29. **a.** $28 = a - 5$ **b.** $a - 5 = 28$ **c.** $a = 23$ **d.** all of these
30. **a.** $z + z = 8$ **b.** $z = 4$ **c.** $z - 8 = z$ **d.** none of these

Write a mathematical sentence for each. Solve and check.

31. Charlene is twenty-five years older than Nina. Charlene is thirty-nine. How old is Nina?

32. Two groups of tourists flew to Japan. The first group took 4 hours less than the second to fly there. If the first group flew for 27 hours, how many hours did the second group fly?

33. George has three fewer subjects in school than Steve. If George has six subjects, how many does Steve have?

34. Fred scored the same number of points in the first two computer games he played. In the third game he scored 205 points. If Fred's total score was 725 points, what was his score in the first game?

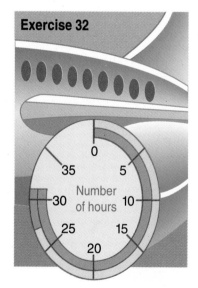

Exercise 32

CHALLENGE

Catch Some Z's. Solve. (Hint: Some equations have no solution; some have more than one solution.)

35. $z + 2 + z = 2$
36. $z + 10 = z$
37. $4 + z = z + 4$
38. $z - 1 = z$
39. $10 - z = z$
40. $z + 5 - z = 5$

83

Solving Multiplication and Division Equations

This graph shows the projected number of trees a reforestry group hopes to plant over a 3-year period. If the projected number of trees planted in the third year is three times the number planted in the first year, how many trees will be planted in the first year?

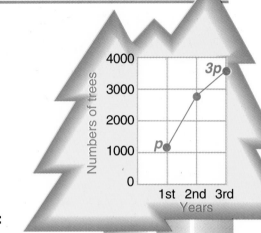

Let p = trees planted in the 1st year
so, $3p$ = projected number planted in the 3rd year
so, $3p = 3600$ (projected number of trees planted in the 3rd year)

To solve equations with multiplication and division:

- See what operation is used with the variable. (Multiplying by 3) $3p = 3600$

- What is the inverse of this operation? (Dividing by 3)
- Do this on *both* sides of the equation. $\dfrac{3p}{3} = \dfrac{3600}{3}$

- Solve. $p = 1200$
- Check.   So, $3p = 3600$

$$3 \times 1200 = 3600$$
$$3600 = 3600$$

Study these examples:

Solve: $\dfrac{n}{5} = 90$

$\dfrac{n}{5} \times 5 = 90 \times 5$ | Multiply *both* sides by 5.

$\dfrac{n}{\cancel{5}} \times \dfrac{\cancel{5}}{1} = 90 \times 5$

$n = 450$

Check: $\dfrac{n}{5} = 90 \rightarrow \dfrac{450}{5} = 90$

$90 = 90$

Solve: $96 = 8t$

$\dfrac{96}{8} = \dfrac{8t}{8}$ | Divide *both* sides by 8.

$12 = t$

Check: $96 = 8t$

$96 = 8(12)$

$96 = 96$

Solve and check.

1. $8x = 56$

2. $3n = 39$

3. $\dfrac{s}{3} = 9$

4. $\dfrac{y}{5} = 60$

5. $14t = 280$

6. $15t = 15$

7. $\dfrac{a}{7} = 3$

8. $\dfrac{x}{9} = 10$

9. $\dfrac{c}{.25} = 6$

10. $\dfrac{d}{20} = 20$

11. $\dfrac{m}{21} = 7$

12. $\dfrac{r}{12} = 12$

Choose the correct solution.

13. $15t = 1500$ **a.** $t = 10$ **b.** $t = 100$ **c.** $t = 0.1$ **d.** $t = 1$

14. $s \div 200 = 2$ **a.** $s = 400$ **b.** $s = 0.1$ **c.** $s = 100$ **d.** $s = 200$

15. $2r = 48$ **a.** $r = 24$ **b.** $r = 96$ **c.** $r = 42$ **d.** $r = 240$

16. $\dfrac{m}{21} = 7$ **a.** $m = 3$ **b.** $m = 7$ **c.** $m = 14$ **d.** $m = 147$

17. $5 = \dfrac{x}{3}$ **a.** $x = 25$ **b.** $x = 5$ **c.** $x = 15$ **d.** $x = 10$

18. $2d = 84$ **a.** $d = 168$ **b.** $d = 84$ **c.** $d = 42$ **d.** $d = 4$

19. $\dfrac{x}{6} = 12$ **a.** $x = 6$ **b.** $x = 2$ **c.** $x = 24$ **d.** $x = 72$

20. $3 = \dfrac{w}{13}$ **a.** $w = 3$ **b.** $w = 39$ **c.** $w = 10\dfrac{3}{10}$ **d.** $w = 10$

Which of these equations have the same solution?

21. **a.** $7x = 14$ **b.** $\dfrac{x}{14} = 7$ **c.** $x = 2$ **d.** all of these

22. **a.** $3s = 30$ **b.** $\dfrac{s}{30} = 3$ **c.** $3s = 10$ **d.** none of these

23. **a.** $4t = 36$ **b.** $t = \dfrac{36}{4}$ **c.** $4t = \dfrac{36}{4}$ **d.** all of these

24. **a.** $\dfrac{n}{2} = 8$ **b.** $n = 8 \times 2$ **c.** $n = 16$ **d.** all of these

25. **a.** $\dfrac{r}{7} = 63$ **b.** $7r = 63$ **c.** $r = 9$ **d.** none of these

Complete these.

THINK

26. $(3 + 1 - 1)p = 15$	$(3 + 1 - 1) = 3 + 0 = 3$	$3p = 15$	$p = \underline{\ ?\ }$
27. $[(2 + 4) \times 3]r = 36$	$[(2 + 4) \times 3] = [6 \times 3] = 18$	$18r = 36$	$r = \underline{\ ?\ }$
28. $t - [(3 + 2) \div 5] = 7$	$[(3 + 2) \div 5] = [5 \div 5] = 1$	$t - 1 = 7$	$t = \underline{\ ?\ }$
29. $m + [5 \times (2 + 3)] = 50$	$[5 \times (2 + 3)] = \underline{\ ?\ } = \underline{\ ?\ }$	$m + \underline{\ ?\ } = 50$	$m = \underline{\ ?\ }$

Solve.

30. One third of a number is 33. What is the number? (Hint: $\dfrac{n}{3} = 33$)

31. Six times a number is 84. What is the number?

SKILLS TO REMEMBER

Compute.

32. $\dfrac{0.9}{0.3}$

33. $\dfrac{0.126}{0.21}$

34. $\dfrac{5.8}{1.45}$

35. $\dfrac{297.92}{9.8}$

36. $0.3 \div 0.5$

37. $4.2 \div 2.8$

38. $53\overline{)66.78}$

39. $0.25\overline{)35.125}$

4-8   Order of Operations

$$36 \div [(4 + 8) - 3] = \underline{\quad?\quad}$$

Which answer is correct?   **a** = 14   **b** = 0   **c** = 4

To work with several operations in the same mathematical sentence, use these rules for the Order of Operations:

First, do everything inside the grouping symbols: ( ) or [ ] or fraction line. ⟶ $36 \div \underline{[(4 + 8)} - 3] = \underline{\quad?\quad}$
Do first.

Do ( ) before [ ]. ⟶ $36 \div [(4 + 8) - 3] = \underline{\quad?\quad}$
$36 \div [\quad 12 \quad - 3] = \underline{\quad?\quad}$
$36 \div \quad\quad 9 \quad\quad = \underline{\quad?\quad}$

Do "×" or "÷" working from *left* to *right*. $(36 \div \quad 9) \quad\quad = \underline{\quad?\quad}$
$36 \div \quad\quad 9 \quad\quad = 4$ Answer: **c**

Lastly, do "+" or "−" working from *left* to *right*.   No other "+" or "−" to be done.

Study this example.

$56 - 8 \times 2 + 9 \div 3 \times 5 \quad = \underline{\quad?\quad}$
$56 - (8 \times 2) + [(9 \div 3) \times 5] = \underline{\quad?\quad}$
$56 - \quad 16 \quad + [\quad 3 \quad \times 5] = \underline{\quad?\quad}$
$56 - \quad 16 \quad + [\quad 3 \quad \times 5] = \underline{\quad?\quad}$
$56 - \quad 16 \quad + \quad 15 \quad = \underline{\quad?\quad}$
$\quad\quad 40 \quad\quad + \quad 15 \quad = \underline{\quad?\quad}$
$\quad\quad 40 \quad\quad + \quad 15 \quad = 55$

Order of Operations

( ) before [ ]
"×" or "÷" left to right
"+" or "−" left to right

Remember:
$\dfrac{6 + 8}{9 - 2}$ The fraction line is a grouping symbol.

$\dfrac{14}{7}$ = 2 Simplify the numerator.

Evaluate the expressions.

1. $18 \div (4 + 5)$
2. $7 (4 - 2)$
3. $4 + (20 \div 4)$
4. $3 + (3 \times 3)$
5. $(3 \times 6) + (15 \div 3)$
6. $(3 \times 8) - (8 \div 2)$
7. $26 - 7 \times 2 + 3$
8. $3 \times 9 + 10 \div 2$
9. $51 \div (9 + 8) - 2$
10. $(2 \times 8) + (16 \div 2)$
11. $12 - (8 \div 4) - 5$
12. $(3 \times 4) + (16 - 2)$
13. $(40 - 8) \div 2 + (12 \div 2)$
14. $(3 \times 7) - 1 + (0 \div 5)$
15. $(20 + 2 \times 8 - 6) \div 6$
16. $6 + 2 - (4 \div 4 \times 2)$
17. $36 \div 3 - [4 + (8 \div 8)]$
18. $36 - [(6 \times 2) + 4]$
19. $[4 + (3 + 6)] \div 13 + 2$
20. $15 \div [9 + (3 \times 2)]$
21. $18 - [7 \times (6 - 4)]$

86

Which expression, a or b, has the same value as the one at the left?

22. $3 \times 2 + 5 \div 5 - 3$ **a.** $3 \times [2 + (5 \div 5) - 3]$ **b.** $(3 \times 2) + (5 \div 5) - 3$

23. $\dfrac{72 - 6 \times 6 + 12 \div 2}{7}$ **a.** $\dfrac{[72 - (6 \times 6) \div 12] \div 2}{7}$ **b.** $[72 - (6 \times 6) + (12 \div 2)] \div 7$

Complete each equation to make it true.

24. $[85 - (6 \div 6)] \div 12 = 9$ _____

25. $(17 - 2) \div 5 + 6 \times 2 = 3$ _____

26. $18 - 5 - 4 \div 4 \times 0 = 8$ _____

27. $(37 \div 37) \times (18 \div 9) + 17 = 1$ _____

28. $(100 \times 100) \div (100 \div 10) = 100$ _____

29. $100 + (0 \div 17) + 176 \div 176 = 2$ _____

By using ( ) and [ ] differently, create two answers for each.

30. $3 + 4 \times 5 \div 1 - 3 \times 2 + 2 \div 2 \times 1 + 0 - 6$

31. $10 \div 2 \times 5 + 1 - 0 \times 2 \div 2 + 4 - 1 \times 6$

32. $27 \div 3 \times 3 \times 2 - 1 + 4 - 0 \div 2 \times 1 \div 2 + 3$

Rewrite the equation,   using ( ) and [ ] to make each sentence true.

33. $12 - 4 \times 6 \div 2 = 0$

34. $17 - 5 + 3 \div 8 = 16$

35. $7 + 3 \times 3 \div 8 = 2$

36. $7 + 5 \div 12 - 6 = 2$

37. $2 + 7 \times 3 + 3 = 30$

38. $8 - 2 \times 1 + 3 = 3$

39. $6 + 24 - 4 \div 4 + 6 = 8$

40. $32 \div 3 + 5 - 2 \times 3 = 6$

41. $6 \div 3 + 2 \times 5 \div 5 + 7 \times 2 = 18$

42. $2 + 4 - 3 \div 3 + 3 \times 3 + 10 = 24$

CALCULATOR ACTIVITY

Some calculators "know" the correct order of operations.
Test your calculator by entering this problem:

`20` `+` `18` `÷` `3` . The correct answer is 26.
If your calculator reads 26, then you have a "smart," or scientific, calculator.
Scientific calculators have keys for grouping symbols. Use a calculator to solve:

43. $45 - 16 \div (3 + 1)$ 44. $60 + (18 - 6) \times 0.5$ 45. $(5.5 + 1.7) \div 9$

46. $54 + [18 \div (3 + 6)]$ 47. $[(11 - 2) \div (2 + 1)]$ 48. $[1.2 \times (11 - 1.5)]$

4-9 Formulas

Formulas: mathematical sentences that we memorize to solve some word problems quickly.

> How many feet of edging is needed to trim a banner 18 feet by 15 feet?

Edging goes *around* the banner, so the *perimeter* of the banner must be found.

The formula for perimeter is: $P = 2\ell + 2w$.

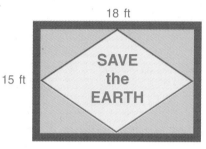

To solve problems using a formula:

- Write the formula.

$$P = 2 (\ell + w)$$

- Substitute the values for the variables into the formula.

$$P = 2 (18 + 15)$$

- Solve the equation.

$$P = 2 (33)$$
$$P = 66$$

66 feet of edging is needed.

Use the given value to solve each formula.

$b = 3$ ft $e = 2$ m $h = 10$ ft $\ell = 6$ ft $s = 5$ in. $w = 4$ ft

1. $A = bh$

2. $P = 6s$

3. $V = e \times e \times e$

4. $A = \ell w$

5. $A = s \times s$

6. $A = \dfrac{bh}{2}$

7. $P = 3s$

8. $V = \ell w h$

9. $P = 2\ell + 2w$

10. $P = 8s$

11. $S = 6 \times e \times e$

12. $P = 2(\ell + w)$

Missing Dimensions

Formulas can be used to solve problems involving missing dimensions.

To find a missing dimension:

- Write a formula. ⟶ $P = a + b + c$

- Substitute. ⟶ $24 = 6 + 8 + c$

- Solve the equation. ⟶
$$24 = 14 + c$$
$$24 - 14 = 14 - 14 + c$$
$$10 = c, \text{ or } c = 10$$

The missing side measures 10 cm.

$a = 6$ cm

$c = \underline{\ ?\ }$

$b = 8$ cm
$P = 24$ cm

Solve for the missing dimension.

13. $P = 4s$ when $P = 12$ *m*

14. $A = s \times s$ when $A = 4$ sq ft

15. $d = rt$ when $d = 600$ mi and $t = 8$ hr

16. $A = \ell w$ when $A = 72$ cm^2 and $w = 4$ cm

17. $A = bh$ when $A = 84$ m^2 and $h = 14$ m

18. $A = \dfrac{bh}{2}$ when $A = 40$ in.^{2} and $b = 10$ in.

Solve. Write the formula, substitute, and simplify.

19. Kurt bought shoes listed at \$59.95 for \$15 less. If the sales tax was \$3.90, what did Kurt pay for the shoes? ($C = LP - D + T$)

20. A pilot flies his jet for 2.5 hours at an average speed of 370 mph. How far does the plane travel?

21. The radius of the classroom clock is 7 inches. What is the circumference of the clock? ($C = 2\pi r$)

22. The speed of sound through steel is 16,400 feet per second. How far will the sound of a hammer striking a railroad track travel in half a minute?

23. Joan paid \$20.02 for a centerpiece. If this included the tax of \$1.32, what was the price of the centerpiece?

24. A rectangular garden covers 15 square meters. If it is 2.5 m wide, how long is the garden?

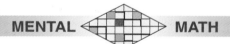

MENTAL MATH

Use inverse operations to solve for *n*. (The first one is done.)

$$6 - 6 = 0$$

25. $\overset{\downarrow\downarrow}{6} + n - 6 = 4, \quad n = 4$

26. $17 + n - 6 = 17$

27. $5 - 3 + n = 5$

28. $6 + 2 - n = 2$

29. $4 + n - 4 = 11$

30. $13 - n + 4 = 13$

Finding Together

Make up a problem using each of these formulas. Have a partner solve it.

31. Sale price: $SP = LP - D$

32. Diameter of a circle: $d = 2r$

33. Cost: $TC = LP + T$

34. Sale price: $SP = LP - D$

35. Circumference: $C = \pi d$ ($\pi \approx 3.14$)

36. Interest: $I = PRT$

STRATEGY
4-10 Problem Solving: Writing and Using Equations

Problem: Nadine and Leon work at the refreshment stand at the pool. The owner of the stand pays them $\frac{1}{6}$ of whatever they sell. Last Saturday, Nadine sold $42 worth of refreshments. If her earnings that day were twice those of Leon, how much did Leon earn on Saturday?

1 IMAGINE

Imagine what is happening. Draw and label a picture of the facts.

Nadine (n)	**Leon** (ℓ)
$n = \frac{1}{6} \times \$42$	$\ell = \frac{1}{2} n$
$n = 2\ell$	$\ell = \underline{\ ?\ }$

2 NAME

Facts: Nadine (n): $2\ell = \frac{1}{6} \times \$42$ sold

Leon (ℓ): $\frac{1}{2} n =$ earnings

Question: How much money did Leon earn?

3 THINK

Look at your *Imagine* picture and write equations representing Nadine's earnings (n) and Leon's earnings (ℓ).

$$n = \frac{1}{6} \times \$42.00$$

$$n = 2\ell$$

You are asked to find Leon's earnings, which are:

$\ell = \frac{n}{2}$ (One half of Nadine's earnings is the same as saying Nadine's are twice Leon's.)

You cannot solve this equation because it has two variables or unknowns, n and ℓ. But you can find the value of n because:

$$n = \frac{1}{6} \times 42 \longrightarrow n = \$7 \text{ (Nadine's Earnings)}$$

Now substitute the value of n in Leon's equation.

$$\ell = \frac{n}{2} \longrightarrow \ell = \frac{\$7}{2}$$

4 COMPUTE

Solve: $\ell = \frac{\$7}{2}$

$\ell = \$3.50$ (Leon's Earnings)

5 CHECK

Are Nadine's earnings twice Leon's? $\$7 \stackrel{?}{=} 2(\$3.50)$. Yes.

Write an equation for each problem. Then solve and check.

1. Milo swam 3 laps more than Carly at practice.
   If Milo swam 10 laps, how many laps did Carly swim?

| 1 IMAGINE | Imagine this situation.
Draw and label a picture of the facts. | → | 2 NAME | ⌐→ Facts
└→ Question |

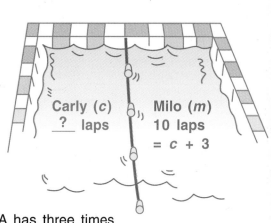

3 THINK

Let m be the number of laps
that Milo swims. ——→ $m = 10$

Let c be the number of laps
that Carly swims. ——→ $m = c + 3$

Substitute 10 for m in the above.
$10 = c + 3$

Carly (c)
___? laps

Milo (m)
10 laps
$= c + 3$

4 COMPUTE ——→ **5 CHECK**

2. Nickels were separated into two jars, A and B. Jar A has three times
   the number of nickels in jar B. There are 60 nickels in jar A.
   How many nickels are there in jar B?

3. Adam is 8 years younger than his brother John.
   If Adam is 37, how old is John?

4. Dana's sister is 0.5 older than she is. If Dana is 8, how old is her sister?

5. Wendy Lee received 180 votes. That was twice as many as her
   opponent received. How many votes did her opponent receive?

6. Sandy earned $24 less than Kristen last year. If Sandy earned
   $315, how much did Kristen earn?

7. Each of the 23 students in a class gave an equal amount of
   money to the town charity collection. If a total of $161 was
   collected from the class, how much did each student contribute?

8. The width of a rectangle is 5 cm less than the length. If the
   rectangle has a length of 9 cm, what is its area? ($A = \ell w$)

9. Two chicken farms entered their prize chickens in a weight
   contest. The entry from the *Happy Chicken Farm* weighed twelve
   ounces less than the entry from the *Plump Chicken Farm*. The
   Happy Chicken entry weighed 10 pounds. How much did the
   Plump Chicken entry weigh? (Hint: 16 ounces = 1 pound.)

4-11 | Problem Solving: Applications

Problem: Sixteen is four more than half a number.
Find the number.

To solve this two-step equation, use these steps:

- Write an equation.

$$16 = \frac{n}{2} + 4$$

- Subtract 4 from both sides of the equation.

$$16 - 4 = \frac{n}{2} + 4 - 4$$

$$12 = \frac{n}{2}$$

- Multiply by 2 on both sides of the equation.

$$12 \times 2 = \frac{n}{\cancel{2}} \times \cancel{2}$$

$$24 = n \text{ or } n = 24$$

Study this example:

Problem: Ten less than twice a number is 8.
Find the number.

- Write an equation.

$$2n - 10 = 8$$

- Add 10 to both sides.

$$2n - 10 + 10 = 8 + 10$$

- Divide by 2 on both sides.

$$\frac{2n}{2} = \frac{18}{2}$$

$$n = 9$$

Solve these two-step equations.

1. If Marty doubles a number and then subtracts six, the result is 42. What is Marty's number?

   Let n = the number $\longrightarrow$ $2n - 6 = 42$

2. Five less than six times a number is 73. What is the number?

   Let x = the number $\longrightarrow$ $6x - 5 = 73$

3. Felice is 13 years old. This is six years less than half her mother's age. How old is Felice's mother?

   Let m = mother's age $\longrightarrow$ $\frac{m}{2} - 6 = 13$

4. During the summer, Bianca read 11 novels. This was three more than twice the number of nonfiction books she read. How many nonfiction books did she read during the summer?

   Let b = nonfiction books $\longrightarrow$ $2b + 3 = 11$

Write an equation and solve.

5. Matt completed 7 book reports. This is 5 more than half the number of reports required. How many book reports are required?

6. Tina used 5 science magazine articles for her science project. This was 1 less than three times the number of science encyclopedias she used. How many science encyclopedias did Tina use?

7. The circulation at the public library on Friday was 56 books. This was 8 books less than four times the circulation on Thursday. What was the circulation on Thursday?

8. Four more than one-third of the books Celeste read are Newberry Award winners. If 8 of the books she read have won the Newberry Award, how many books did she read?

9. Twelve seventh graders entered the library poster contest. This was 3 less than 5 times the number of eighth graders who entered. How many eighth graders entered the contest?

10. Santos won the $50 first-place prize for his short story. This was $10 more than twice the amount of money for third place. How much money did the third-place winner receive?

11. There are 3250 books in the fiction section of the library. One-fifth of them are Caldecott winners. How many books in the fiction section have won the Caldecott Award?

12. Bonita read 15 of the books in the classroom paperback library. This is 42 less than the total number of paperbacks in the library. How many paperback books are in the class library?

13. The general encyclopedia has 21 volumes. This is 1 more than twice the number of volumes in the science encyclopedia. How many volumes are there in the science encyclopedia?

14. The library fine for an overdue lending book is $.10. The fine for an overdue reference book is 10 times more than the fine for a lending book. Ms. Wordy collected 5 fines for reference books and three times this number of fines for lending books. How much money in fines did she collect?

More Practice

Use the chart to find the value of each expression.

s	t	v	a	b	c	n
1	3	9	20	5	6	0.5

1. $4 + t$ **2.** $b + c$ **3.** $a - 14$ **4.** $20 + n$

5. $c - t$ **6.** $c + 16$ **7.** $t - s$ **8.** $s - n$

9. $a + v$ **10.** $a + b - c$ **11.** $v + t$ **12.** $t + 7 - n$

13. Write a mathematical expression for "seven more than two times a number."

14. Write an English expression for $\dfrac{s}{3}$.

15. Write a mathematical sentence for:
  a. Sue's brother is 5 years older than Sue. He is 18. How old is Sue?
  b. Sixteen is less than one more than a number.

16. Classify each sentence in exercise 15 as an equation or inequality.

Write the property that is illustrated by each.

17. $8 \times 3 = 3 \times 8$ **18.** $5 \times 16 = (5 \times 10) + (5 \times 6)$

19. $11 + (4 + 16) = (11 + 4) + 16$ **20.** $478 \times 216 \times 0 = 0$

21. $a + 5 = 5 + a$ **22.** $(b + 8) + c = b + (8 + c)$

23. $a + 0 = a$ **24.** $a(c + 3) = ac + 3a$

25. $a \times 3 = 3$ **26.** $d \times 0 = 0$

Solve and check.

27. $s + 10 = 16$ **28.** $w + 8 = 35$ **29.** $3 + a = 17$

30. $b - 13 = 8$ **31.** $x - 7 = 23$ **32.** $14 = d - 2$

33. $3x = 27$ **34.** $9t = 54$ **35.** $63 = 7r$

36. $d \div 4 = 9$ **37.** $a \div 11 = 10$ **38.** $\dfrac{x}{25} = 3$

Rewrite, using grouping symbols to show the correct order of operations. Then evaluate.

39. $3 \times 9 + 6 \div 3$ **40.** $6 + 6 \times 6 \div 6$ **41.** $18 \div 2 - 2 \times 2 + 1$

Use the order of operations in computing.

42. $47 - (20 + 8) \div 7$ **43.** $(7 + 23) \div (16 - 11)$ **44.** $[(16 + 8) - 10] \div 7$

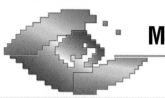

Math Probe

MATHEMATICAL SYSTEMS

This table shows how an operation in a number system works.
Certain properties can be discovered by studying the table.

Operation: Addition

+	0	1	2
0	0	1	2
1	1	2	3
2	2	3	4

Write the nine additions from the table.
How would you find the identity element?
How would you discover whether the
operation is commutative?

Here are three new operations on numbers. Find the identity element if
there is one. Tell whether the operation is commutative. Can you describe
a pattern for each operation?

Operation: Infinity

∞	1	2	3
1	3	1	2
2	1	2	3
3	2	3	1

1 ∞ 1 = __?__   1 ∞ 2 = __?__   1 ∞ 3 = __?__

2 ∞ 1 = __?__   2 ∞ 2 = __?__   2 ∞ 3 = __?__

3 ∞ 1 = __?__   3 ∞ 2 = __?__   3 ∞ 3 = __?__

Operation: Square-dot

⊡	0	1	2	3
0	0	1	2	3
1	1	2	3	0
2	2	3	0	1
3	3	0	1	2

Operation: Spy

🔍	1	2	3	4
1	3	4	5	6
2	4	5	6	7
3	5	6	7	8
4	6	7	8	9

Check Your Mastery

Evaluate each expression when $a = 9$ and $b = 16$.   See pp. 74–77

1. $36 - b$ 2. $\dfrac{a}{3}$ 3. $b - a$ 4. $(2a - b)5$

5. Write an English expression for: **a.** $\dfrac{s}{25}$ **b.** $3c - 4$

Write a mathematical expression or sentence for: See pp. 76–79

6. fourteen less than a number.

7. Ten doubled is less than thirty.

8. a number divided by nine.

9. Twice a number is greater than 10.

10. six times a number added to one.

11. Which value of n makes $19 - n \leq 4$ a true sentence?

    a. 3   **b.** 4   **c.** 15   **d.** 9

Write true or false.   See pp. 80–81

12. In the equation $r - 7 = 11$, the inverse of subtracting 7 is adding 7.

13. The commutative property lets you write $n \times 3$ as $3n$.

14. $10 + (7 + 4) = (10 + 7) + 4$ illustrates the distributive property.

Solve and check.   See pp. 82–85

15. $21 + a = 33$ 16. $10x = 120$ 17. $t - 10 = 70$

18. $\dfrac{r}{8} = 6$ 19. $14 = p + 2$ 20. $4x = 4$

21. $\dfrac{n}{4} = 36$ 22. $26 = n - 4$ 23. $4a = 44$

Use the order of operations in computing.   See pp. 86–87

24. $42 + (16 - 8)$ 25. $(12 - 3) \div (1 + 2)$

26. $[7 + (6 + 5 + 4)] \div 2$ 27. $2[60 - (35 - 6 \times 5) \times 8]$

Use the formula to solve.   See pp. 88–89

28. The area of a rectangle is 64.8 cm^2.
The length is 12 cm. Find the width. ($A = \ell w$)

Cumulative Review

Choose the correct answer.

1. How many times one hundred million is a billion?

   a. 1000 b. 10

   c. 10,000 d. 100

2. 160,000 is the estimated product for which of the following:

   a. 4710 × 43 b. 135 × 992

   c. 259 × 845 d. 2106 × 76

3. Add 5 to the sum of m and n when $m = 0.5$ and $n = 12$.

   a. 7 b. 17.5

   c. 0.3 d. 22

4. If $4 + 7 = 11$, then the inverse operation is:

   a. $11 - 7 = 4$ b. $7 + 4 = 11$

   c. $11 = 4 + 7$ d. $11 - 7 + 4 = 11$

5. Which formula will help you to find the length of the rim of a tire?

   a. $C = \pi d$
   b. $P = 2(\ell + w)$
   c. $A = \dfrac{b \times h}{2}$
   d. $C = \pi r^2$

6. How many times greater than 0.000005 is 0.5?

   a. 10,000
   b. 100,000
   c. 1,000,000
   d. 1,000,000,000

7. Find the difference between the estimated products of 4076×803 and 681×305.

   a. 3,000,000 b. 2,990,000

   c. 2,000,000 d. 2,600,000

8. Lou's weight of 132 lb is 3 times Jon's weight. What is Jon's weight?

   a. 34 lb b. 44 lb

   c. 135 lb d. 84 lb

9. Add x to $(y)z$ when $x = 0.3$, $y = 13$, and $z = 0$.

   a. 13.3 b. 0

   c. 0.3 d. 16.0

10. If $20r = 40$, then $5r + 13 = \underline{\quad?\quad}$

    a. 10 b. 40

    c. 20 d. 23

11. Find the product to the nearest ten thousandth of 0.035 and 0.0216.

    a. 0.0007 b. 0.0006

    c. 0.0076 d. 0.0008

12. $3(t + 5) = 3t + 15$ is an example of what property?

    a. commutative b. associative

    c. identity d. distributive

13. Sue is 6 years older than Traci. Together their ages equal 32. Which equation can be used to find their ages?

    a. $t + 6 = 32$
    b. $2t + 6 = 32$
    c. $t - 6 = 32$
    d. $6 + (t \div 2) = 32$

14. One solution for the inequality $2c + 3 > 7$ is:

    a. $c = 2$
    b. $c = 0$
    c. $c = 4$
    d. $c = 1$

Solve.

15. 40 × 300

16. 65 × 200

17. 100 × 0.62

18. 4.32 × 1000

19. $\begin{array}{r} 2736 \\ \times\ 820 \end{array}$

20. $\begin{array}{r} 5901 \\ \times\ 607 \end{array}$

21. $\begin{array}{r} 92.073 \\ \times\ \ \ 820 \end{array}$

22. $\begin{array}{r} 83.19 \\ \times\ 7.02 \end{array}$

23. 60,000 ÷ 30

24. 63.04 ÷ 1000

25. $4\overline{)1.6}$

26. $7\overline{)0.105}$

27. $72\overline{)24,840}$

28. $167\overline{)134,769}$

29. $3.4\overline{)0.85}$

30. $0.009\overline{)0.81}$

31. $74 = 9 + t$

32. $58 = 5 + h$

33. $y - 4 = 31$

34. $49 = n - 6$

35. $4m = 56$

36. $\frac{1}{3}x = 15$

37. $\frac{a}{5} = 60$

38. $\frac{2n}{3} = 18$

39. $3a + 5 = 20$

40. $\frac{m}{3} - 5 = 1$

41. $26 + 7 \times (5 - 3)$

42. $36 \div 3 - [(4 + 8) - 4]$

Write an equation and solve.

43. Bart earned $51.85 per day. How much did he earn in 25 days?

44. If Pam's intake of cholesterol last week was 1.918 grams, what was her average daily intake?

45. Tyron cuts an 8-meter piece of rope into four pieces. Three of the pieces are 2.6 m, 2.4 m, and 1.3 m long. How long is the fourth piece?

46. If it takes 0.09 second for a butterfly's wings to beat one stroke, how long does it take for a butterfly's wings to beat 100 strokes?

47. Su Ling is three years younger than her brother. If he is 17 years old, what is Su Ling's age?

48. Twelve is one half of a number. What is the number?

49. The perimeter of a square is 13.2 m. What is the length of each side?

50. Mr. Marra planted rose bushes along the length of the porch. The width of the porch is 2.25 m, and the perimeter is 16 m. Find the length of the porch.

51. A quarter is 0.18 cm thick. How thick would a $10 stack of quarters be?

52. The product of twice a number and 22 is half of 144. What is the number?

+10 BONUS

53. Write down how much time you spend on various activities on a normal weekday (for example: 8 hr, sleep; 2.5 hr, meals; etc.). Write formulas for discovering how much time you spend for each activity during the school week, for a month, and for a year. Write other formulas for figuring time spent for activities on the weekends and during vacation time. Solve each formula. Make a chart to show how you spend your time.

5 Integers

In this chapter you will:

- Use a number line to identify integers
- Compare and order integers
- Compute with integers
- Identify properties involving operations on integers
- Solve problems: working backwards
- Use technology: computation

The Wonder of Math

RESEARCHING TOGETHER

Sometimes math can seem like magic! The ancient Chinese thought this when they used these "magic squares."

+4	-1	+6
+5	+3	+1
0	+7	+2

You have used magic squares before. How can new magic squares be made by rotation or addition of a constant integer? Create magic squares using positive and negative integers. Challenge others to find your magic sums.

Do you remember?

Operations are related in this way:

$9 + 3 = 12 \longrightarrow 12 - 3 = 9$

$3 + 9 = 12 \longrightarrow 12 - 9 = 3$

$7 \times 2 = 14 \longrightarrow 14 \div 2 = 7$

$2 \times 7 = 14 \longrightarrow 14 \div 7 = 2$

The natural numbers include:
  1, 2, 3, . . .
The whole numbers include:
  0, 1, 2, 3, . . .

5-1 Using Integers

Integers: the set of whole numbers and their opposites.

The maximum elevation of the Rock of Gibraltar is 1396 feet *above* sea level, or $^+1396$ ft.

The maximum depth of the Straits of Gibraltar is 1200 feet *below* sea level, or $^-1200$ ft.

Integers can be shown on a number line.

Positive integers are to the *right* of 0. They are *greater than* 0.
Negative integers are to the *left* of 0. They are *less than* 0.

Every integer has an opposite. The opposite of zero is zero.
Zero is *neither* a positive integer *nor* a negative integer.

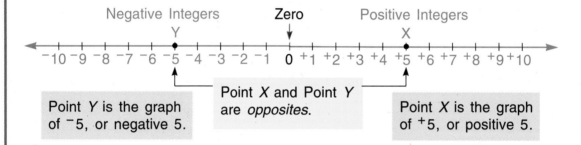

Point Y is the graph of $^-5$, or negative 5.

Point X and Point Y are *opposites*.

Point X is the graph of $^+5$, or positive 5.

The **absolute value** of a number is its distance from zero on the number line. The absolute value of $^-5$ equals the absolute value of $^+5$. $|^-5| = |^+5| = 5$

Name the letter that matches each integer on the number line.

1. $^-1$ 2. $^-6$ 3. $^-5$ 4. $^+2$ 5. $^-4$ 6. $^+3$

Name the integer that matches each letter on the number line.

7. *P*   8. *B*   9. *N*   10. *L*   11. *Q*   12. *A*

Write the opposite of each. Then write the absolute value of each.

13. $^+3$ 14. $^+7$ 15. $^-2$ 16. $^-8$ 17. $^+2$ 18. $^-32$

19. $^+100$ 20. $^-512$ 21. $^+486$ 22. $^-100$ 23. $^-1025$ 24. $^+4000$

Solve.

25. If you record a deposit of eighteen dollars as "⁺$18," how would you record a withdrawal of eighteen dollars?

Exercise 28

26. In a game, the card for ⁺7 says "Go Ahead 7 Steps." What would the card for ⁻7 say?

27. Begin at 0. What happens if you go 6 steps forward (⁺6) and then 6 steps backward (⁻6)?

28. On a thermometer, are the numbers below zero positive or negative?

29. On a vertical number line, are the numbers above zero positive or negative?

30. What is the opposite of ⁻17?

31. What is the opposite of *the opposite of* ⁻17?

32. What is the opposite of *the opposite of* ⁺10?

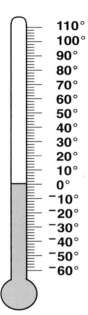

Express each as an integer. Then write its opposite.

33. 7 meters forward
34. 8°C below zero
35. 90 meters below sea level
36. $4 loss
37. 2 floors up
38. 15 kilometers west
39. lose 6 points
40. 12 seconds to blast-off
41. 10-point improvement
42. 32-pound weight loss
43. $4 profit
44. 6 weeks after planting
45. 10 days before the party
46. 3 miles north
47. 15 steps backward
48. 12°C decrease in temperature

CHALLENGE **Name the integer that matches each letter on this number line.**

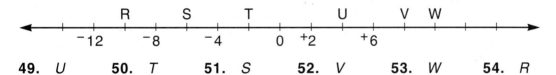

49. *U*   50. *T*   51. *S*   52. *V*   53. *W*   54. *R*

Name the integer that matches each letter on this number line.

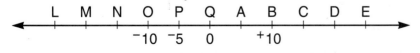

55. *A*   56. *C*   57. *N*   58. *E*   59. *L*   60. *D*

61. What integer is 2 units to the right of *E*?

62. What integer is 3 units to the left of *L*?

5-2 Comparing and Ordering Integers

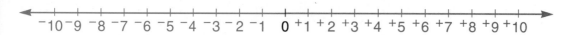

$$^-10\ ^-9\ ^-8\ ^-7\ ^-6\ ^-5\ ^-4\ ^-3\ ^-2\ ^-1\quad 0\quad ^+1\ ^+2\ ^+3\ ^+4\ ^+5\ ^+6\ ^+7\ ^+8\ ^+9\ ^+10$$

Use the number line to compare integers.

Any number on the number line is greater than a number to its left.

$^+7 > {}^+5$ $\qquad 0 > {}^-3$ $\qquad ^+1 > {}^-5$ $\qquad ^+4 < {}^+5$ $\qquad ^-7 < {}^-2$

Use the number line to order integers.

- Least to greatest → Begin with the integer farthest to the left.

- Greatest to least → Begin with the integer farthest to the right.

Order these integers: $^-2, {}^-8, {}^+6, {}^+4, {}^-10$

Least to greatest: $^-10, {}^-8, {}^-2, {}^+4, {}^+6$

Greatest to least: $^+6, {}^+4, {}^-2, {}^-8, {}^-10$

Compare integers. Write < or > .

1. $^+5\ \underline{\ ?\ }\ {}^+8$
2. $^+6\ \underline{\ ?\ }\ {}^+7$
3. $^+5\ \underline{\ ?\ }\ {}^-9$
4. $^-1\ \underline{\ ?\ }\ {}^+3$

5. $^+3\ \underline{\ ?\ }\ 0$
6. $0\ \underline{\ ?\ }\ {}^+1$
7. $^-4\ \underline{\ ?\ }\ {}^+2$
8. $^-8\ \underline{\ ?\ }\ {}^+1$

9. $^-6\ \underline{\ ?\ }\ {}^+7$
10. $^-5\ \underline{\ ?\ }\ {}^+3$
11. $^-4\ \underline{\ ?\ }\ {}^-2$
12. $^-5\ \underline{\ ?\ }\ {}^-8$

13. $^-5\ \underline{\ ?\ }\ {}^-9$
14. $0\ \underline{\ ?\ }\ {}^-6$
15. $^+1\ \underline{\ ?\ }\ {}^-2$
16. $^-8\ \underline{\ ?\ }\ {}^-7$

17. $^-28\ \underline{\ ?\ }\ {}^-27$
18. $^+100\ \underline{\ ?\ }\ {}^-1000$
19. $^-42\ \underline{\ ?\ }\ {}^-50$
20. $^+1\ \underline{\ ?\ }\ {}^-20$

21. $^+11\ \underline{\ ?\ }\ {}^-23$
22. $^-16\ \underline{\ ?\ }\ {}^+61$
23. $^+21\ \underline{\ ?\ }\ {}^-12$
24. $^-23\ \underline{\ ?\ }\ {}^-24$

CRITICAL THINKING True or false? Explain.

25. A positive integer is always greater than 0.

26. A negative integer is not less than 0.

27. A negative integer is always less than a positive integer.

28. A positive integer is never more than a negative integer.

Arrange in order from least to greatest.

29. $^{+}1$, $^{+}7$, $^{-}6$, $^{+}5$

30. $^{-}63$, $^{+}27$, $^{-}15$, $^{+}123$, $^{-}198$, $^{+}67$, 0, $^{+}11$, $^{-}11$

31. $^{-}2$, $^{-}8$, $^{+}7$, 0

32. $^{+}21$, $^{-}48$, $^{-}16$, $^{+}35$, $^{-}104$, $^{-}3$, $^{+}3$, $^{-}1$

33. 0, $^{-}6$, $^{+}5$, $^{+}3$, $^{-}3$

34. $^{-}321$, $^{-}867$, $^{-}415$, $^{+}163$, $^{+}367$, 0, $^{+}100$, $^{+}321$

Arrange in order from greatest to least.

35. $^{-}11$, $^{+}6$, 0, $^{-}5$

36. $^{+}8$, $^{+}18$, $^{-}28$, $^{-}10$, $^{+}20$, 0, $^{+}9$

37. $^{-}16$, $^{-}19$, $^{-}9$, $^{-}6$

38. $^{+}12$, 0, $^{-}4$, $^{+}3$, $^{-}16$, $^{+}15$, $^{-}12$

39. $^{+}32$, $^{-}18$, $^{+}40$, $^{-}20$

40. $^{-}15$, $^{-}35$, $^{-}10$, $^{+}20$, $^{-}25$, $^{+}40$, $^{-}20$

Solve.

41. The TV meteorologist recorded the daily low temperatures for the week: $^{-}10°C$, $^{+}5°C$, $^{+}8°C$, $^{-}4°C$, $^{-}11°C$, $^{+}4°C$, $^{-}1°C$. List these from highest to lowest.

42. A marine biologist recorded the depth of 5 species of underwater plants. List the data he collected from shallowest to deepest.

    Species: A = $^{-}85$ ft, B = $^{-}60$ ft, R = $^{-}78$ ft, N = $^{-}90$ ft, C = $^{-}54$ ft

43. From a wind-chill chart Doreen found that an air temperature of 20°F feels like 3°F at a wind speed of 10 mph, but the same air temperature feels like $^{-}10°F$ at 20 mph. Find the difference in the wind-chill temperatures.

44. After measuring the air temperature of 15°F, Sid found that due to the wind speed the air felt 18° colder. What was the wind-chill temperature?

45. On Wednesday the temperature was $^{-}8°C$. It rose 10°C on Thursday. What was the temperature on Thursday?

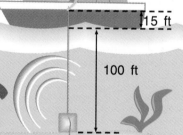

15 ft

100 ft

46. An oceanographer lowers a sonar device 100 feet below sea level. If her lowered hand was 15 feet above sea level, how far was her hand from where the sonar device finally landed?

```
┌──────────┐
│ Finding  │
│┌────────┐│
││Together││   How many integers are between the two given integers?
│└────────┘│
└──────────┘
```

For example: between 0 and $^{+}10$ there are nine integers;
between $^{+}10$ and $^{+}20$ there are nine integers; and
between $^{-}10$ and 0 there are nine integers.

Do you see a pattern? Try to use this pattern to find
how many integers are between the following:

47. $^{-}20$ and $^{+}20$ 48. $^{-}100$ and 0   49. $^{+}200$ and $^{+}300$

50. What is the pattern for exercises 47–49?

5-3 | Adding Integers

To add integers with *like signs*, find their sum and use the sign of the addends.

Add: $^+2 + {}^+3$

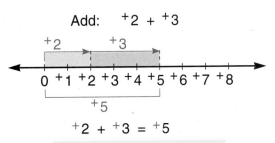

$$^+2 + {}^+3 = {}^+5$$

The sum of two positive integers is positive.

Add: $^-2 + {}^-4$

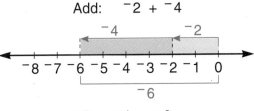

$$^-2 + {}^-4 = {}^-6$$

The sum of two negative integers is negative.

To add integers with *unlike signs*, find their difference and use the sign of the addend farther from 0.

Add: $^+2 + {}^-3$

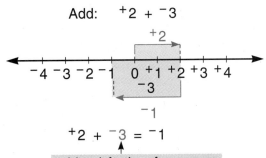

$$^+2 + {}^-3 = {}^-1$$

addend farther from zero

Add: $^-4 + {}^+1$

$$^-4 + {}^+1 = {}^-3$$

addend farther from zero

An integer plus its *opposite* is zero. This is the **inverse property** of addition.

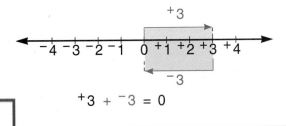

$$^+3 + {}^-3 = 0$$

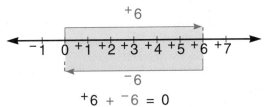

$$^+6 + {}^-6 = 0$$

Find the sum.

1. $^-6 + {}^-3$ 2. $^-3 + {}^-9$ 3. $^-16 + {}^-4$ 4. $^-8 + {}^-8$

5. $^+4 + {}^+8$ 6. $^+7 + {}^+4$ 7. $^+10 + {}^+3$ 8. $^+11 + {}^+5$

9. $^+4 + {}^+3$ 10. $^+7 + {}^+2$ 11. $^-2 + {}^-1$ 12. $^-5 + {}^-1$

13. $^+6 + {}^+5$ 14. $^-5 + {}^-3$ 15. $^-2 + {}^-9$ 16. $^+8 + {}^+3$

17. $^-4 + ^-2$	18. $^+7 + ^+7$	19. $^-5 + ^-6$	20. $^+3 + ^+7$
21. $^-2 + ^-5$	22. $^+4 + ^+4$	23. $^-3 + ^-3$	24. $^+7 + ^+6$
25. $^-6 + ^-6$	26. $^-3 + ^-6$	27. $^+9 + ^+8$	28. $^+1 + ^+6$
29. $^+5 + ^+5$	30. $^+6 + ^+8$	31. $^-9 + ^-9$	32. $^-7 + ^-8$
33. $^+42 + ^+21$	34. $^-23 + ^-35$	35. $^-16 + ^-9$	36. $^+50 + ^+25$

Add.

37. $^+5 + ^-6$	38. $^-2 + ^+2$	39. $^+4 + ^-2$	40. $^-1 + 0$
41. $^-6 + ^+7$	42. $^+6 + ^-8$	43. $^-9 + ^+5$	44. $^-5 + ^+2$
45. $^+3 + ^-8$	46. $^-7 + ^+7$	47. $^-2 + ^+9$	48. $^+6 + ^-1$
49. $^+8 + ^-8$	50. $^+7 + ^+9$	51. $^-4 + ^+10$	52. $^-12 + ^+3$
53. $^-9 + ^+8$	54. $^+4 + ^-10$	55. $^-15 + ^+3$	56. $^-9 + ^+9$
57. $0 + ^-2$	58. $^-5 + ^-4$	59. $^-8 + ^+5$	60. $^+2 + ^+15$
61. $^+3 + ^-6$	62. $^-6 + ^+2$	63. $0 + ^+3$	64. $^-6 + ^+1$
65. $^-3 + ^-7$	66. $^+2 + ^+2$	67. $^-7 + ^-7$	68. $^+4 + ^-8$
69. $^+10 + ^-12$	70. $^+15 + ^-20$	71. $^-41 + ^-9$	72. $^+32 + ^-2$

Solve.

73. My score in one game is $^-9$. If I make another score of $^-20$, what is my total score?

If B.C. years are represented by negative numbers and A.D. years
by positive numbers, how many years are there between:

74. 40 B.C. and 4 B.C.

75. 44 A.D. and 67 A.D.

76. 40 B.C. and 67 A.D.

77. 10 B.C. and 10 A.D.

78. From a deck printed with integers Tess and Todd
chose the following:
   Tess: $^+10, ^+5, ^-3, ^-10, ^-2, ^-8, ^+1$
   Todd: $^-10, ^-8, ^+7, 0, ^-3, ^+9, ^+20$
Who was farther from the starting point
at the end of the game?

SKILLS TO REMEMBER   Solve for *n*.

79. $1.2 + n = 1.2$

80. $3.7 + 4.1 = n + 3.7$

81. $(9.3 + n) + 6 = 9.3 + (4 + 6)$

82. $n + \frac{2}{3} = \frac{2}{3}$

83. $1\frac{1}{4} + n = 5\frac{1}{2} + 1\frac{1}{4}$

84. $\frac{1}{5} + (n + \frac{1}{3}) = (\frac{1}{5} + \frac{4}{5}) + \frac{1}{3}$

85. $^-10 + n = ^-10$

86. $^-1 + ^+7 = n + ^-1$

87. $(^+2 + ^-4) + n = ^+2 + (^-4 + ^+4)$

105

5-4 Subtracting Integers

To subtract an integer:

- Add its opposite.
- Rewrite as an addition.
- Then use the rules for adding integers.

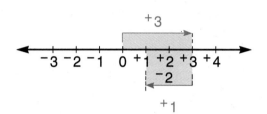

Subtract: $^+3 - {}^+2$

$$^+3 - {}^+2 = \underline{\ ?\ } \longrightarrow {}^+3 + {}^-2 = {}^+1$$

$$\underset{\text{opposite}}{\underbrace{}}$$

Look at these examples:

$$^-4 - {}^-1 = \underline{\ ?\ }$$
$$^-4 + {}^+1 = \underline{\ ^-3\ }$$

$$^+3 - {}^-5 = \underline{\ ?\ }$$
$$^+3 + {}^+5 = \underline{\ ^+8\ }$$

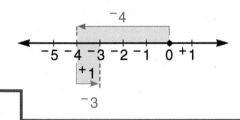

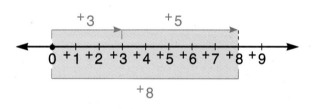

Subtract.

1. $^-9 - {}^+5$
2. $^+6 - {}^+5$
3. $^+13 - {}^-3$
4. $^-6 - {}^-6$

5. $^+3 - {}^-6$
6. $^-3 - {}^-2$
7. $^+4 - {}^+6$
8. $^-2 - {}^+8$

9. $^+2 - {}^-9$
10. $^+4 - {}^+2$
11. $^-6 - {}^-3$
12. $^-7 - {}^-7$

13. $^-7 - {}^+8$
14. $^-5 - {}^-4$
15. $^+4 - {}^-6$
16. $^-10 - {}^-3$

17. $^+5 - {}^+1$
18. $^-4 - {}^+1$
19. $^-3 - {}^-5$
20. $^+3 - {}^-7$

21. $^-9 - {}^-3$
22. $^+11 - {}^-6$
23. $^+14 - 0$
24. $0 - {}^-2$

25. $^-5 - {}^-5$
26. $^+6 - {}^-6$
27. $^-11 - {}^-22$
28. $0 - {}^+5$

29. $^-6 - 0$
30. $^+3 - {}^-2$
31. $^-4 - {}^+10$
32. $^-2 - {}^+11$

33. $^+2 - {}^+9$
34. $^-5 - {}^-1$
35. $^+3 - {}^-11$
36. $^-4 - {}^+12$

37. $0 - {}^-9$
38. $^+1 - {}^+2$
39. $^+10 - {}^+6$
40. $^-11 - {}^+7$

41. $^-12 - {}^+3$
42. $^+11 - {}^-6$
43. $^+15 - {}^-5$
44. $^-20 - {}^+5$

45. $^+3 - {}^+5$
46. $^-1 - {}^-9$
47. $^-1 - {}^+9$
48. $^-5 - {}^-5$

Find the difference.

49. $^+4 - {^+3}$ 50. $^+3 - {^+8}$ 51. $^+5 - {^+5}$ 52. $^+10 - {^+2}$

53. $^-5 - {^-6}$ 54. $^-2 - {^-8}$ 55. $^-9 - {^-5}$ 56. $^-7 - {^-4}$

57. $^-12 - {^+7}$ 58. $^-4 - {^+2}$ 59. $^-15 - {^+8}$ 60. $^-8 - {^+7}$

61. $^+9 - {^-4}$ 62. $^+5 - {^-5}$ 63. $^+6 - {^-2}$ 64. $^+7 - {^-9}$

65. $^+7 - {^+7}$ 66. $^-6 - 0$ 67. $^+12 - {^-12}$ 68. $^-25 - {^-2}$

69. $^+26 - {^+27}$ 70. $^-21 - {^-8}$ 71. $^+16 - {^-32}$ 72. $^-47 - {^-39}$

73. $^+100 - 0$ 74. $0 - {^+4}$ 75. $^-36 - 0$ 76. $0 - {^-25}$

Add or subtract.

77. $^+6 + {^+5}$ 78. $^+8 - {^-1}$ 79. $^-8 - {^-7}$ 80. $^+6 + {^-9}$

81. $^+9 + {^+6}$ 82. $^-6 - {^+2}$ 83. $^-7 + {^+8}$ 84. $^-8 + {^+8}$

85. $^+15 - {^+9}$ 86. $^-18 + {^-9}$ 87. $^-14 - {^-6}$ 88. $^+2 - {^-7}$

89. $^+5 - {^+8}$ 90. $^-16 + {^+8}$ 91. $^+15 - {^-9}$ 92. $^+20 + {^-10}$

93. $^-14 - {^+7}$ 94. $^-13 - {^-6}$ 95. $^-12 + {^-3}$ 96. $^+14 + {^-8}$

97. $^+1 - {^-9}$ 98. $^+3 + {^-5}$ 99. $^-14 - {^-2}$ 100. $^+5 - {^+10}$

101. $^-11 + {^+9}$ 102. $^+10 - {^-6}$ 103. $^+15 - {^+5}$ 104. $^-20 - {^-5}$

Solve.

105. Theo's original test score was 82, in part because 8 points had been deducted for a missing diagram. The teacher decided not to count the diagram in the final grade. What was Theo's final score?

106. The referee called for a penalty of 5 yards in addition to the loss of 3 yards on a play. If another referee gave an offsetting penalty of 10 yards against the other team, what integer would represent the net yardage?

Finding Together

Insert positive and negative signs and addition and subtraction signs to make each sentence true.

Example: $1 = {^+1} + {^-2} + {^-3} + {^+4} - {^+5} + {^+6}$

107. $7 =$ 1 2 3 4 5 6 108. $3 =$ 1 2 3 4 5 6

109. $9 =$ 1 2 3 4 5 6 110. $5 =$ 1 2 3 4 5 6

111. $^-1 =$ 1 2 3 4 5 6 112. $^-11 =$ 1 2 3 4 5 6

5-5 Multiplying Integers

The product of two integers with *like signs* is positive.

$$^+3 \times {}^+8 = {}^+24 \qquad\qquad {}^-5 \times {}^-9 = {}^+45$$

$$^+7 \times {}^+2 = {}^+14 \qquad\qquad {}^-6 \times {}^-1 = {}^+6$$

The product of two integers with *unlike signs* is negative.

$$^+3 \times {}^-8 = {}^-24 \qquad\qquad {}^-5 \times {}^+9 = {}^-45$$

$$^+7 \times {}^-2 = {}^-14 \qquad\qquad {}^-6 \times {}^+1 = {}^-6$$

Multiplication Rules
Like signs: (+, + or −, −)
+ × + = +
− × − = +
Unlike signs: (+, − or −, +)
+ × − = −
− × + = −

Remember — Multiplying by zero:

$$^-4 \times 0 = 0$$

So, $n \times 0 = 0$

Multiply.

1. $^+3 \times {}^+9$ 2. $^+8 \times {}^+2$ 3. $^+4 \times {}^+5$ 4. $^+5 \times {}^+7$

5. $^-5 \times {}^-6$ 6. $^-6 \times {}^-7$ 7. $^-4 \times {}^-6$ 8. $^-8 \times {}^-8$

9. $^+9 \times {}^-2$ 10. $^-4 \times {}^+7$ 11. $^-8 \times {}^+9$ 12. $^+9 \times {}^-9$

13. $^+9 \times {}^-7$ 14. $^+2 \times {}^+9$ 15. $^-5 \times {}^-5$ 16. $^-6 \times {}^+5$

17. $^+10 \times {}^-3$ 18. $^-1 \times {}^+12$ 19. $^+13 \times {}^-3$ 20. $^+6 \times {}^+6$

21. $^-10 \times {}^+2$ 22. $0 \times {}^-6$ 23. $^+4 \times 0$ 24. $^-3 \times {}^+5$

25. $^-2 \times {}^+2$ 26. $^+3 \times {}^-7$ 27. $^-1 \times {}^-8$ 28. $^+4 \times {}^-4$

29. $^-5 \times {}^+8$ 30. $^+10 \times {}^+2$ 31. $^+4 \times {}^+9$ 32. $^+2 \times {}^+3$

33. $^-6 \times {}^-7$ 34. $^-2 \times {}^-9$ 35. $^+8 \times {}^-3$ 36. $^-7 \times {}^-7$

37. $^+4 \times {}^-5$ 38. $^-2 \times {}^+7$ 39. $^-6 \times {}^-9$ 40. $^+7 \times {}^-8$

41. $^-3 \times {}^+3$ 42. $0 \times {}^-7$ 43. $^+4 \times {}^-6$ 44. $^+9 \times {}^-8$

45. $^+10 \times {}^-2$ 46. $^-10 \times {}^-6$ 47. $^+5 \times {}^-11$ 48. $^-8 \times {}^+8$

49. $^+4 \times {}^-2$ 50. $^-1 \times {}^-1$ 51. $^+9 \times {}^-3$ 52. $^-5 \times {}^-7$

53. $^+7 \times {}^-9$ 54. $^-7 \times {}^-10$ 55. $^+3 \times {}^-11$ 56. $^-12 \times {}^-3$

57. $^+6 \times {}^-9$ 58. $^-9 \times {}^-10$ 59. $^+1 \times {}^-3$ 60. $^-5 \times 0$

Find the product.

61. $^-2 \times {}^-8$ 62. $^+4 \times {}^-6$ 63. $^-7 \times {}^+4$ 64. $^-3 \times {}^+7$

65. $^-6 \times {}^-9$ 66. $^+5 \times {}^-8$ 67. $^-2 \times {}^+1$ 68. $^-5 \times {}^-5$

69. $^+4 \times {}^-4$ 70. $^-6 \times 0$ 71. $^+11 \times {}^-2$ 72. $^-8 \times {}^-6$

73. $^+3 \times {}^-10$ 74. $^+4 \times {}^-8$ 75. $^-7 \times {}^-9$ 76. $^-5 \times {}^+6$

77. $^+8 \times {}^+1$ 78. $^+3 \times {}^-5$ 79. $0 \times {}^+9$ 80. $^-2 \times {}^-2$

81. $^-1 \times {}^+1$ 82. $^-6 \times {}^+12$ 83. $^-22 \times {}^+4$ 84. $^-11 \times {}^+9$

Use any integers as factors to find each product.

85. $\underline{\ ?\ } \times \underline{\ ?\ } = {}^-6$ 86. $\underline{\ ?\ } \times \underline{\ ?\ } = {}^+14$

87. $\underline{\ ?\ } \times \underline{\ ?\ } \times \underline{\ ?\ } = {}^+6$ 88. $\underline{\ ?\ } \times \underline{\ ?\ } \times \underline{\ ?\ } \times \underline{\ ?\ } = {}^-10$

89. $\underline{\ ?\ } \times \underline{\ ?\ } \times \underline{\ ?\ } = {}^-12$ 90. $\underline{\ ?\ } \times \underline{\ ?\ } \times \underline{\ ?\ } = {}^+4$

91. $\underline{\ ?\ } \times \underline{\ ?\ } \times \underline{\ ?\ } = {}^+16$ 92. $\underline{\ ?\ } \times \underline{\ ?\ } \times \underline{\ ?\ } = {}^-20$

93. Find six factors for $^-36$.   94. Find six factors for $^+60$.

Compute; then compare. Write <, =, or >.

95. $(^-3 + {}^-4) + {}^+7 \ \underline{\ ?\ }\ 0$ 96. $^-3 \times {}^+6 \ \underline{\ ?\ }\ {}^+6 \times {}^-3 \times {}^-1$

97. $(^-3 - {}^+5) \ \underline{\ ?\ }\ (^-5 + {}^+3)$ 98. $0 \times {}^+1 \ \underline{\ ?\ }\ {}^-6$

99. $^-1 \times {}^-7 \ \underline{\ ?\ }\ {}^-1 \times {}^+7$ 100. $^-2 \times {}^+14 \ \underline{\ ?\ }\ {}^-6 \times {}^-7$

101. $^-1 \times {}^-3 \ \underline{\ ?\ }\ {}^+1 \times {}^+3$ 102. $^-3 \times (^-5 \times {}^+10) \ \underline{\ ?\ }\ (^-3 \times {}^-5) \times {}^-10$

103. $^-4 \times (^+2 + {}^+3) \ \underline{\ ?\ }\ (^-4 \times {}^+2) + (^-4 \times {}^+3)$

The Distributive Property

The **distributive property** of multiplication over addition can be shown using integers.

$$a \ (b + c) \ = \ (a \times b) \ + \ (a \times c)$$

$$^-3 \ (^-4 + {}^+5) \ = \ (^-3 \times {}^-4) \ + \ (^-3 \times {}^+5)$$

MAKE UP YOUR OWN...

Write an equation using integers to illustrate each property:

104. the commutative property of addition; of multiplication.

105. the associative property of addition; of multiplication.

106. the identity property of addition; of multiplication.

5-6 Dividing Integers

The quotient of integers with *like signs* is positive.

$$^+18 \div {}^+3 = {}^+6 \qquad ^-20 \div {}^-4 = {}^+5$$

$$^+15 \div {}^+5 = {}^+3 \qquad ^-12 \div {}^-3 = {}^+4$$

The quotient of integers with *unlike signs* is negative.

$$^-10 \div {}^+2 = {}^-5 \qquad ^-8 \div {}^+4 = {}^-2$$

$$^+16 \div {}^-4 = {}^-4 \qquad ^+24 \div {}^-3 = {}^-8$$

Think of dividing integers as *finding a missing factor*.

$$^-10 \div {}^+2 = \underline{\ ?\ } \longrightarrow {}^+2 \times \underline{\ ?\ } = {}^-10$$

$$^+2 \times {}^-5 = {}^-10$$

$$\text{So, } ^-10 \div {}^+2 = {}^-5$$

$$^+16 \div {}^-4 = \underline{\ ?\ } \longrightarrow {}^-4 \times \underline{\ ?\ } = {}^+16$$

$$^-4 \times {}^-4 = {}^+16$$

$$\text{So, } ^+16 \div {}^-4 = {}^+4$$

Division Rules
Same as Multiplication Rules
Like signs: (+, + or −, −)
+ ÷ + = +
− ÷ − = +
Unlike signs: (+, − or −, +)
+ ÷ − = −
− ÷ + = −

Remember: NEVER divide by zero or "nothing."
$$2 \div 0 = \text{IMPOSSIBLE}$$

Divide.

1. $^+9 \div {}^+3$
2. $^-45 \div {}^-9$
3. $^-18 \div {}^-2$
4. $^+28 \div {}^+7$

5. $^-55 \div {}^+5$
6. $^+30 \div {}^-5$
7. $^+63 \div {}^-9$
8. $^-21 \div {}^+7$

9. $^+36 \div {}^+6$
10. $^+36 \div {}^-6$
11. $^-42 \div {}^-7$
12. $^-42 \div {}^+7$

13. $^-72 \div {}^+8$
14. $^+18 \div {}^-1$
15. $^+49 \div {}^-7$
16. $^-18 \div {}^-9$

17. $^+10 \div {}^-10$
18. $^-32 \div {}^-32$
19. $0 \div {}^+5$
20. $0 \div {}^-12$

21. $^-48 \div {}^-6$
22. $^+24 \div {}^-3$
23. $^-14 \div {}^+7$
24. $^-54 \div {}^-6$

25. $^+36 \div {}^+6$
26. $^-50 \div {}^+10$
27. $^-8 \div {}^+2$
28. $^+12 \div {}^-2$

Multiply or divide.

29. $^+20 \div {}^-5$
30. $^+27 \div {}^+3$
31. $^-11 \times {}^-6$
32. $^-18 \div {}^+2$

33. $^-12 \times {}^+4$
34. $^-12 \div {}^+3$
35. $^+21 \div {}^-3$
36. $^+21 \times {}^-3$

37. $^-50 \div {}^+2$
38. $^-50 \div {}^-25$
39. $^+6 \times {}^+16$
40. $^+8 \times {}^-12$

Solve.

41. Multiply each number by ⁻1:   ⁺8, ⁻3, ⁻11, ⁺2, ⁻1.

42. Describe the products in exercise 41.

43. Divide each number by ⁻1:   ⁺5, ⁺12, ⁻8, ⁻3, ⁻1.

44. Describe the quotients in exercise 43.

True or false? Explain your answer.

45. The product of five negative numbers is always negative.

46. The product of ten negative numbers is always positive.

Name the next three numbers in each sequence.

47. ⁺2, ⁻4, ⁺8, ⁻16, __?__ , __?__ , __?__

48. ⁻128, ⁺64, ⁻32, ⁺16, __?__ , __?__ , __?__

49. ⁺1, ⁻5, ⁺25, ⁻125, __?__ , __?__ , __?__

50. ⁺20,000, ⁻2000, ⁺200, ⁻20, __?__ , __?__ , __?__

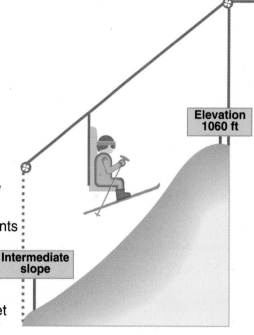

Elevation 1060 ft

Intermediate slope

Solve.

51. An oil rig is anchored to the sea floor 96 feet below sea level. A drill can cut through 24 feet of rock each hour. After a 4-hour shift, what integer represents the depth of a newly drilled well from sea level?

52. A chair lift at Ski Mount climbs at the rate of 30 feet per minute. If the elevation at the top of the mountain is 1060 feet, and it takes 7 minutes to get from the intermediate slope to the top, what integer represents the elevation of the intermediate slope?

Exercise 52

CALCULATOR **ACTIVITY**

Find the missing factors. Use the key to change a number to its opposite. Look at this example:

⁺13 × __?__ = ⁻78   78 ÷ 13 = ⁻6

53. ⁻21 × __?__ = ⁻630   **54.** ⁻18 × __?__ = ⁺57.6   **55.** ⁻27 × __?__ = ⁺135

56. ⁻23 × __?__ = ⁻9.2   **57.** ⁻1.5 × __?__ = ⁻10.5   **58.** ⁺16.5 × __?__ = ⁻36.3

TECHNOLOGY

More Than One Way to Compute

It is the finals of the *Be A Mathlete* competition. The four finalists are COMPUTER COURTNEY, CALCULATOR CHRIS, PAPER-AND-PENCIL PEDRO, and MENTAL MATH MEGAN.

Find the sum of the first 50 counting numbers:
$1 + 2 + 3 + 4 + \ldots + 47 + 48 + 49 + 50.$

Courtney writes a computer program in BASIC.

```
10 LET SUM = 0
20 FOR N = 1 TO 50
30 SUM = SUM + N
40 NEXT N
50 PRINT SUM
```

Chris begins entering each addend in a calculator.

As Pedro computes, he notices some patterns that would simplify his computation.

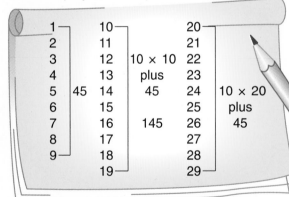

Megan pairs the numbers mentally. She realizes that there will be 25 pairs, each with a sum of 51. She computes mentally.

$1 + 2 + 3 + \ldots + 48 + 49 + 50$

51
51
51

$50 \times 25 = 1250$ plus
$1 \times 25 = 25$ equals 1275

- Which method of computing is the best way to find the answer? Why?
- Which one would you have picked? Why?

Think About It ...

1. Could Courtney have used the computer in a different way to solve the problem?

2. What number properties did Megan use in multiplying first by 50 and then by 1?

3. What difficulty might Chris experience with his method?

4. What is the sum of Pedro's next column? What pattern does he notice?

Solve each problem. What method of computation will you use? Why?

5. Find the sum: 2.3 + 0.04 + 0.007 + 6

6. Find the product: 21,000 × 46,000

7. Find the difference: $6\frac{7}{8} - 3$ 8. Find the difference: $6 - 3\frac{7}{8}$

9. Find the product: 48 × 0.5   10. Find the product: $\frac{12}{25} \times \frac{35}{36}$

11. Find three numbers whose product is 3553.

12. Which fraction has the greatest value? $\frac{1}{75}$, $\frac{19}{21}$, $\frac{301}{601}$, $\frac{96}{150}$

A **perfect number** is a number that is equal to the sum of all of its factors except itself.

     EXAMPLE: 6 is a perfect number because 6 = 1 + 2 + 3.

13. Is 24 a perfect number?

14. Is 28 a perfect number?

15. Is 100 a perfect number?

16. Is 256 a perfect number?

17. Find another perfect number.

18. Compute 6.5% novelty tax on each of the items sold for School Spirit Week.

19. The principal, Mr. Nguyen, bought one of each item. What will this cost him, including tax?

Spirit Week Sale!!

Pennants — small	$2.49
Pennants — large	$3.59
Rain poncho — child size	$1.98
Rain poncho — adult size	$3.49
Tote bag	$6.00
Duffel bag	$8.49
Paperweight	$2.39
Dozen pencils	$ .79
Pen	$1.25
Mug	$4.50
Glass	$2.25

6.5% NOVELTY TAX ON EACH ITEM

All proceeds will benefit Student Assistance Project.

Problem: Right now Angela has 11 fish in her aquarium. Since she first bought the tank and stocked it with fish, 5 baby fish have been born and 4 new ones have been purchased. She has given 3 to a friend and 2 have died. How many fish did Angela have in the tank originally?

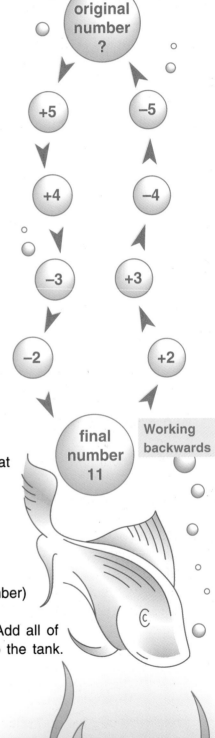

original number ?

+5   −5

+4   −4

−3   +3

−2   +2

final number 11

Working backwards

1 IMAGINE Imagine how the number of fish in Angela's aquarium is changing. Draw and label a picture.

2 NAME *Facts:* 5 — the number born
4 — the number bought
3 — the number given away
2 — the number that died
11 — the number in the tank

Question: How many fish were put in the tank originally?

3 THINK To find the original number of fish, work backwards.

Step 1 Start with the final number, 11. Put back (add) all of the fish that have been taken away.

Step 2 Take away (subtract) all of the fish that have been added.

4 COMPUTE Step 1 11 + 3 + 2 = 16

Step 2 16 − 5 − 4 = 7 (Original Number)

5 CHECK Step 1 Start with the original number. Add all of the fish that had been added to the tank.

7 + 5 + 4 = 16

Step 2 Subtract all of the fish that have been taken away.

16 − 3 − 2 = 11

Solve by working backwards.

1. Penny Guin, a scientist at the Antarctica station, records the hourly temperature on a weather chart. Somehow she has lost the reading for the noon temperature. The temperature at 6 P.M. was ⁻36 °C. Her record showed that the temperature had increased 4 °C each hour from noon to 3 P.M., but had decreased 6 °C each hour after 3 P.M. What was the noon temperature?

| 1 IMAGINE | Draw and label a picture. | → | 2 NAME | → Facts |
| | | | | → Question |

| 3 THINK | Start at ⁻36 °C and work backwards by adding 6 °C for each hour after 3 P.M. and then subtracting 4 °C for each hour from noon to 3 P.M. |

| 4 COMPUTE | → | 5 CHECK |

Time	Temp.
12 P.M.	
1 P.M.	
2 P.M.	
3 P.M.	
4 P.M.	
5 P.M.	
6 P.M.	⁻36 °C

2. Ms. Betsy Broker is studying a stock that she bought at $5 a share earlier in the day. In her research she finds that in the past 2 weeks the price of a share has gone up 25¢ twice, up 50¢ five times, down $1.25 twice, and down 75¢ once. What was the price of the stock 2 weeks ago?

3. After buying supplies at the Fish Forum to set up an aquarium, Scott had $1.42 left in change. He had bought 8 lb of gravel at $.69 a pound, a filter for $35, a heater for $13, and cotton for $1.19. The tax was $4.38. How much money had he brought to the Fish Forum?

4. Mr. and Mrs. Keough dined out last night. When the bill came, Mr. Keough gave the waiter $45 and told him to keep the change as his tip. Mr. Keough had had the crab special for $17.95 and Mrs. Keough had had the flounder for $13.95. Both had had dessert, which cost $3.50 each. How much money had Mr. Keough left to the waiter as a tip?

5. A veterinarian recommended that Doug put his dog on a diet. The first week the dog lost 2 lb. The second week the dog gained 1.25 lb. The third week the dog lost 2.5 lb. The fourth week the dog lost 1.75 lb. If the dog's weight after four weeks was 58 lb, how much did the dog weigh before the diet began?

5-9   Problem Solving: Applications

Solve. Let m = the missing integer in exercises 1–8.

USE THESE STRATEGIES:
Write an Equation
Use a Model/Drawing
Multi-Step Problem

1. What integer is $^{+}2$ more than $^{-}3$ doubled?

   Hint: $m = (2 \times {}^{-}3) + {}^{+}2$

2. What integer is $^{+}3$ less than the product of $^{-}4$ and $^{+}6$?

3. What integer added to $^{-}16$ equals $^{-}4$ divided by $^{+}2$?

   Hint: $^{-}16 + m = \underline{{}^{-}4 \div {}^{+}2}$
   $\qquad\quad\; {}^{-}16 + m = \quad\;\; {}^{-}2$

   Subtract $^{-}16$ from both sides to find the missing addend.

4. What integer added to $^{+}15$ equals $^{+}10$ multiplied by $^{-}3$?

5. What integer divided by $^{-}3$ equals the sum of $^{+}12$ and $^{-}20$?

   Hint: $\dfrac{m}{^{-}3} = {}^{+}12 + {}^{-}20$

6. What integer divided by $^{+}5$ equals the difference when $^{+}2$ is subtracted from $^{-}4$?

7. What integer multiplied by $^{+}5$ equals the difference when $^{+}18$ is subtracted from $^{-}2$?

   Hint: $^{+}5m = {}^{-}2 - {}^{+}18$

8. What integer multiplied by $^{-}6$ equals the sum of $^{+}27$ and $^{-}3$?

9. What integer divided by the sum of $^{-}4$ and $^{+}6$ equals $^{-}15$?

   Hint: $\dfrac{m}{^{-}4 + {}^{+}6} = {}^{-}15 \longrightarrow \dfrac{m}{^{+}2} = {}^{-}15$

10. What integer divided by the sum of $^{-}5$ and $^{-}1$ equals $^{+}3$?

11. What integer multiplied by the difference when $^{+}1$ is subtracted from $^{-}6$ equals $^{+}28$?

    Hint: $m(^{-}6 - {}^{+}1) = {}^{+}28$
    $\qquad\qquad\;\; {}^{-}7\, m = {}^{+}28$

12. What integer multiplied by the difference when $^{+}4$ is subtracted from $^{-}3$ equals $^{-}35$?

13. In one 24-hour period the temperature in Browning, Montana, fell from 44 °F to ⁻56 °F. Write an integer to express this change.

14. In Granville, North Dakota, the temperature once rose from ⁻33 °F to ⁺50 °F in 12 hours. What integer can be used to express the number of degrees the temperature changed?

15. Records for a town in Montana show that one day in 1892 the temperature rose 40 ° in 15 minutes. If the thermometer read ⁻5 °F when the temperature began to rise, what did it read after 15 minutes?

16. In 1911 the residents of Rapid City, South Dakota, one day saw the temperature drop 63 ° in 15 minutes. If the thermometer read 55 °F when the fall began, what did it read 15 minutes later?

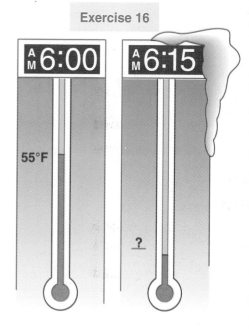
Exercise 16

17. One day in 1924 the temperature in Fairfield, Montana, dropped 84 ° in 12 hours. It was 63 °F when the plunge began. What was the temperature 12 hours later?

18. The highest temperature ever recorded in Nevada was 122 °F on June 23, 1954. The lowest temperature recorded in that state was ⁻50 °F on January 8, 1937. What is the difference between the two temperatures?

19. The nighttime temperature on the moon is ⁻261 °F. The daytime temperature on the moon is 504 ° higher. What is the daytime temperature on the moon?

20. In a laboratory test technicians lowered the temperature of an experimental compound 2 °C per minute until it froze. If its starting temperature was 20 °C and it took 18 minutes to freeze, what was the final temperature of the compound?

21. Fluorine and carbon compounds are used to make nonstick substances to coat pots and pans. The melting point of fluorine is ⁻223 °C. The boiling point of fluorine is 36 ° higher. What is the boiling point of fluorine?

◄ **MAKE UP YOUR OWN...** ►

22. Make up an original equation for which you must use two inverse operations to solve. Check your solution.

More Practice

Write the opposite. Then order the opposites from greatest to least.

1. $^-7$ 2. $^+5$ 3. $^-6$ 4. $^+352$

5. 0 6. $^-35$ 7. $^+16$ 8. $^-3001$

Compare. Write <, =, or >.

9. $^+8 \underline{\ ?\ } ^-8$ 10. $0 \underline{\ ?\ } ^-4$ 11. $^-5 \underline{\ ?\ } ^-11$ 12. $^-3 \underline{\ ?\ } ^+1$

13. $^-7 \underline{\ ?\ } ^-9$ 14. $^-2 \underline{\ ?\ } ^-1$ 15. $^+1 \underline{\ ?\ } 0$ 16. $^+2 \underline{\ ?\ } 2$

Compute.

17. $^+5 + ^+8$ 18. $^-12 + ^+4$ 19. $^+3 + ^-12$ 20. $^-12 + ^-12$

21. $^-9 + ^-3$ 22. $^-20 + ^+20$ 23. $^-3 - ^+4$ 24. $^-8 - ^-11$

25. $^-13 + ^+12$ 26. $^+12 - ^+4$ 27. $^+4 - ^+12$ 28. $^-5 - ^+5$

29. $^-4 \times ^+7$ 30. $^-12 \times ^+4$ 31. $^-1 \times ^-18$ 32. $^+6 \times 0$

33. $^+9 \times ^-9$ 34. $^+3 \times ^-10$ 35. $^-45 \div ^-9$ 36. $^+48 \div ^-8$

37. $^-39 \div ^+3$ 38. $^-18 \div ^-2$ 39. $^-16 \div ^-1$ 40. $^-25 \div ^+5$

41. $^-20 + ^+5$ 42. $^-2 - ^-7$ 43. $^+12 \times ^-5$ 44. $^-16 \div ^+4$

45. $^-21 - ^-21$ 46. $^+10 + ^+7$ 47. $^-32 \div ^-4$ 48. $0 \times ^-3$

49. $^-60 - ^+50$ 50. $^-15 \div ^-3$ 51. $^-5 \times ^+4$ 52. $^-15 \times ^-4$

53. $^-63 \div ^+7$ 54. $^-72 \div ^+6$ 55. $^-50 \times ^-10$ 56. $^-83 + ^-16$

57. $^+7 + ^-8 - ^+20$ 58. $^-3 \times ^-6 - ^+10$ 59. $^-48 \div ^+4 + ^+7$

60. $(^+3 - ^-7) \div (^-3 + ^-2)$ 61. $[(^-20 + ^+5) \div ^-3] + ^-2$ 62. $^+8 - [^+2 + (^-2 \times ^-9)]$

Copy and complete. Name the property shown.

63. $^-3 + ^+4 = ^+4 + \underline{\ ?\ }$ 64. $^+6 \times \underline{\ ?\ } = ^+6$ 65. $\underline{\ ?\ } + ^-7 = ^-7$

66. $^-7 \times ^+3 = \underline{\ ?\ } \times ^-7$ 67. $^-6 \times 0 = \underline{\ ?\ }$ 68. $^-6 + \underline{\ ?\ } = ^-6$

69. $^+6 \times (^-3 + ^+2) = ^-18 + \underline{\ ?\ }$ 70. $^-4 \times (^+1 \times ^-3) = (\underline{\ ?\ } \times ^-3) \times ^-4$

71. Catherine received a bill for $76, then a commission of $112.
    Write an equation to show the state of her finances after
    she pays the bill.

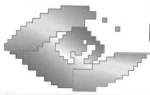

Math Probe

ABSOLUTE VALUE

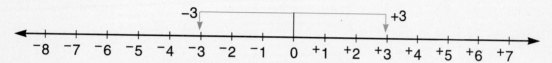

The *absolute value* of a number is its distance from zero on the number line.

The absolute value of $^-3$ is: $|-3| = 3$ ⟵⎤ 3 units is the distance
The absolute value of $^+3$ is: $|^+3| = 3$ ⟵⎦ from zero.

When computing the absolute value, solve the problem within the absolute value symbol first.

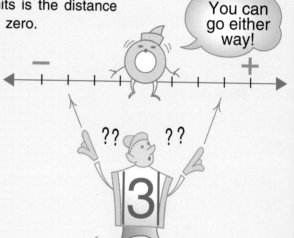

You can go either way!

Compare. $|^-5| - |^+2|$ __?__ $|^-5 - ^+2|$

Think: $5 - 2$ __?__ $|^-5 + ^-2|$

3 __?__ $|^-7|$

3 __<__ 7

Find the absolute value.

1. $|^+4|$ 2. $|^-6|$ 3. $|^-8|$ 4. $|^+9|$ 5. $|0|$

Solve.

6. $|^-2| + |^-1|$ 7. $|^-2 + ^-1|$ 8. $|^+8| - |^+5|$ 9. $|^+8 - ^+5|$

10. $|^-5| + |^+2|$ 11. $|^-5 + ^+2|$ 12. $|^-8| - |^+5|$ 13. $|^-8 - ^+5|$

14. Compare exercises 10–11 and 12–13. What do you notice about these problems?

15. $|^-5| \times |^+2|$ 16. $|^-5 \times ^+2|$ 17. $|^-15| \div |^-3|$ 18. $|^-15 \div ^-3|$

Compare. Write <, =, or >.

19. $|^-8|$ __?__ $|^+8|$ 20. $|^-10|$ __?__ $|^+2|$ 21. $|^+6|$ __?__ $|^-8|$

22. $|^+2 + ^-1|$ __?__ $|^+2| + |^-1|$ 23. $|^+3 + ^-5|$ __?__ $|^+3| + |^-5|$

24. Use $^+4$ and $^+5$ and addition and subtraction in absolute value to equal 1 and 9.

25. How many degrees separate a temperature of $^-10$ from $^+10$? (Explain why absolute value helps you answer this question.)

Check Your Mastery

Write an integer to represent each of the following.   See pp. 100–101

1. a withdrawal of $46
2. the opposite of the opposite of $^-5$
3. a penalty of 15 yards
4. the opposite of $^-19$
5. 11 seconds to lift-off
6. 80 m above sea level

Compare integers. Write < or >.   See pp. 102–103

7. $0 \underline{\ ?\ } {}^-3$
8. $0 \underline{\ ?\ } {}^+7$
9. $^-11 \underline{\ ?\ } {}^-13$

Compute.   See pp. 104–111

10. $^+5 + {}^-10$
11. $^-6 + {}^-11$
12. $^+9 + {}^-6$
13. $^-8 + {}^+8$
14. $^+6 - {}^-4$
15. $^-6 - {}^-5$
16. $0 - {}^-9$
17. $^+5 - 0$
18. $^-8 \times {}^-4$
19. $^+9 \times {}^-6$
20. $^+10 \times {}^-2$
21. $^-16 \times {}^-2$
22. $0 \div {}^-4$
23. $^-64 \div {}^-1$
24. $^-48 \div {}^+6$
25. $^+35 \div {}^+7$

Copy and complete. Name the number property used.

26. $^+2 \times ({}^-4 + {}^-3) = ({}^+2 \times \underline{\ ?\ }) + ({}^+2 \times \underline{\ ?\ })$
27. $\underline{\ ?\ } \times {}^-44 = 0$
28. $^-6 + \underline{\ ?\ } = 0$

Solve.

29. Order the melting point of these gases from least to greatest: hydrogen, $^-259°C$; chlorine, $^-103°C$; helium, $^-272°C$; oxygen, $^-218°C$.

30. Some homeowners test for the presence of radon. How many degrees separate the melting point of radon ($^-110°C$) from its boiling point ($^-62°C$)?

31. 56 people were in the movie theater lobby. 5 people went outside for air, and 3 ushers went into the office. How many people were left in the lobby?

32. An underwater camera filmed a rock formation at 350 meters below sea level. 75 meters deeper than the formation, the camera filmed a sunken ship. At what depth below sea level was the ship?

33. In the early afternoon 24 people were ice skating on the pond. 6 people left at 4:00 P.M. 10 skaters arrived about 7:00 P.M. How many people were at the pond after 7:00 P.M.?

34. Tracey's savings account had $226 on April 11. On July 1 she had $242. On August 1 her balance was $231. Write integers to express the changes from April to July and from July to August.

6 Number Theory

In this chapter you will:

Work with sets and set operations

Solve inequalities

Work with exponents

Use divisibility rules

Identify prime and composite numbers

Use greatest common factor; least common multiple

Work with scientific notation

Solve problems: tables and drawings

Do you remember?

6 is *divisible* by 3 because

6 ÷ 3 = 2 R 0

6 is *not divisible* by 4 because

6 ÷ 4 = 1 R 2

$0.01 = \dfrac{1}{100} = \dfrac{1}{10} \times \dfrac{1}{10}$

Primetime!

You are creative television game producers. You come up with a game you call *Primetime*. Make up your own prime-number chart with questions and answers about prime numbers. For example: I am the only even prime number. Who am I? Challenge your group to play the game. Give out "prime" prizes!

6-1 Facts About Groups

Set: a collection, or group, of objects.
The objects in a set are called *members*.

These three objects are a "set of masks."
A group of numbers, such as the whole numbers from
0 through 9, can be identified as a *set* in different ways.
Two of these ways are:

Robot mask

Japanese *Kabuki* mask

Animal mask

with words the "set of whole numbers up to and including 9"

with numbers "$A = \{0, 1, 2, 3, 4, 5, 6, 7, 8, 9\}$"

Use a *capital letter* to name a set.

List the members of the set within *braces.*

Empty set: a set with *no* members.
Because there are *no* even numbers between 6 and 8:
"Set of even numbers between 6 and 8" = the empty set
$$= \{ \ \}$$
$$= \phi$$

Write { } or ϕ to express the empty set.

Infinite set: a set that has *no end* to the number of its members.
The set of whole numbers has a never-ending number of members. This is written as:
$$W = \{0, 1, 2, 3, 4, 5, \ldots\}$$

Three dots mean that the numbers continue without ending.

One set, set A, is a **subset** of another set, set B, if all the members of set A are also members of set B.

The set of odd numbers 1 through 7 is a subset of the set of odd numbers. This is written as:
$$\{1, 3, 5, 7\} \subset \{1, 3, 5, 7, 9, 11, \ldots\}$$

The symbol $\subset$ means "is a subset of."

The empty set is a subset of *every* set.

Describe or name each set.

1. test tube, beaker, flask

2. rose, petunia, marigold

3. car, truck, wagon, bus

4. eagle, vulture, hawk, falcon

5. breakfast, lunch, supper

6. heart, arteries, veins, capillaries

7. penny, nickel, dime, quarter

8. Monday, Tuesday, Wednesday, Thursday, Friday

Use braces { } to list the members of each set.

9. *E* is the set of even numbers 2 through 12.

10. *O* is the set of odd numbers less than 11.

11. *W* is the set of whole numbers greater than 25 and less than or equal to 32.

12. *M* is the set of multiples of ten between 100 and 160.

13. *F* is the set of multiples of five between 35 and 55.

14. *Q* is the set of numbers that solve the equation *q* + 2 = 7.

Which are empty sets?

15. the set of whole numbers less than one

16. the set of odd numbers that are multiples of 2

17. the set of multiples of 2 between 10 and 20

18. the set of 1000-pound chickens on a farm

19. the set of students in your class

20. the set of 90-year-old students in your class

Write "true" or "false."

21. The set notation for multiples of five is { 0, 5, 10, 15, ... }.

22. The set of even numbers is a subset of the set of whole numbers.

23. The solution set of 2 + *x* = 2 is the empty set.

24. { 2, 4, 6 } is a subset of the set of whole numbers between 1 and 7.

25. { 0 } ⊂ { 9, 3, 2, 0, 5 }   26. φ ⊂ { 1, 3, 5, ... }

27. { 0 } represents the empty set.   28. { 0, 2, 4, ... , 10 } is an infinite set.

Equal or Equivalent Sets

Equal sets have exactly the *same* members, no matter the order of their listing.

{ A, B, C } is equal to { C, A, B }.

Equivalent sets have the same *number* of members.

{ A, B } is equivalent to { 1, 2 }.

Write an equivalent set for each set.

29. { R, O, Y, G, B, I, V }   30. { 1, 2, 5, 10 }

31. the set of items in a dozen   32. the set of negative integers

6-2 | Working with Number Groups

These pennants are entries in an art contest.

Set *D* is the set of pennants having designs:
   D = { 1, 2, 3, 4, 5 }

Set *C* is the set of pennants having colored backgrounds:
   C = { 4, 5, 6 }

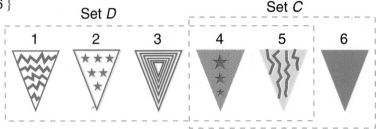

The **union,** (∪), of sets *D* and *C* is the set of pennants having *either* designs *or* colored backgrounds.

   D ∪ *C* = { 1, 2, 3, 4, 5, 6 }

The members of set *D* ∪ *C* belong to set *D* *or* to set *C* *or* to both.

The **intersection,** (∩), of sets *D* and *C* is the set of pennants having *both* designs **and** colored backgrounds.

   D ∩ *C* = { 4, 5 }

The members of set *D* ∩ *C* belong to set *D* *and* to set *C*.

List the members of the set that is the *union* of:

1. the set of pennants above having colored backgrounds or the set of pennants having lines.

2. the set of pennants above having lines or the set of pennants having stars.

3. *A* = { 3, 0, 4, 9 }   or   *B* = { 2 }

4. *X* = { 2, 4, 6, 8 }   or   *Y* = { 4, 8, 10, 12 }

5. *M* = { a, b, c }   or   *N* = { d, e, g }

List the members of the set that is the *intersection* of:

6. the set of pennants having white backgrounds and the set of pennants having designs.

7. the set of pennants having white backgrounds and the set of pennants having stars.

8. *S* = { 7, 1 }   and   *T* = { 3, 4, 17, 1 }

9. *W* = { 8, 10, 12 }   and   *R* = { 8, 12, 14, 16, 18 }

10. *V* = { a, e, i, o, u }   and   *C* = { b, c, d, f, g }

Venn Diagrams

The union and intersection of sets can also be shown with Venn diagrams.

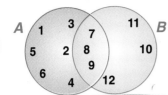

$A = \{1, 2, 3, 4, 5, 6, 7, 8, 9\}$
$B = \{7, 8, 9, 10, 11, 12\}$

$A \cup B = \{1, 2, 3, 4, 5, 6, 7, 8, 9, 10, 11, 12\}$
$A \cap B = \{7, 8, 9\}$

Use { } to list the members of each set.
Then use $\cup$ and $\cap$ to list the members of the union and the intersection of the sets.

11.

R   3   6   7   4   S
8   2   9
5

12.

X   2   4   Y
3
8   1

13.

L   0   1   2   M

14.

D   7
6
5   4   3 E
  2
1

(Set *E* is a subset of set *D*.)

15.

T   1   5   2   4   W
3   6   8
9

Disjoint sets have *no* members in common. (Set *T* and set *W* are disjoint sets.)

Write each answer in set notation.

16. $\{3, 6, 9, 12, \ldots\} \cap \{1, 2, 3, 4, 5, 6\}$

17. $\{2, 4, 6, 8, \ldots\} \cup \{1, 3, 5, 7, \ldots\}$

18. $\{10, 20, 30, \ldots\} \cap \{5, 10, 15, 20, \ldots\}$

19. $\{1, 3, 5, 7, \ldots\} \cap \{2, 4, 6, 8, \ldots\}$

20. $\phi \cap \{0, 1\}$

Write $\cup$ or $\cap$ in place of ? .

21. $\{\text{Al, Ted, Alice}\} \underline{\ \ ?\ \ } \{\text{Ed, Al, Bob}\} = \{\text{Al}\}$

22. $\{1, 11, 121\} \underline{\ \ ?\ \ } \{2, 22, 242\} = \{\ \}$

23. $\{10, 20, 30\} \underline{\ \ ?\ \ } \{5, 10, 15, 25\} = \{5, 10, 15, 20, 25, 30\}$

24. $\{0.1, 0.2, 0.3, 0.4\} \underline{\ \ ?\ \ } \{0, 1, 2, 3, 4, 5\} = \phi$

25. $\{25, 50, 75, 100\} \underline{\ \ ?\ \ } \{75, 100, 125\} = \{25, 50, 75, 100, 125\}$

6-3 Solving Inequalities

To solve inequalities, replace the variable with values that make the sentence true.

$b > 17$ ⟶ | What numbers are greater than 17?
18, 19, 20, 21, ...

$c < 12$ ⟶ | What numbers are less than 12?
11, 10, 9, 8, ...

$a \neq 100$ ⟶ | What numbers are not equal to 100?
All numbers except 100.

$m - m > 0$ ⟶ | What number minus itself is greater than 0?
There is no solution.

Replacement set (R): the set of numbers from which you replace the variable.

Some examples:

whole numbers:	{0, 1, 2, 3, 4, ...}
odd numbers:	{1, 3, 5, 7, ...}
even numbers:	{0, 2, 4, 6, ...}

Solution set (S): the subset of the replacement set that makes the inequality true.

When *no* number from the replacement set makes the inequality true, the solution set is the *empty* set.

---

1. Copy and complete this chart to solve inequalities.

Inequality	Replacement Set (R)	Solution Set (S)
$a < 0$	0, 1, 2, 3, ...	{ }
$x > 8$	0, 1, 2, 3, ...	9, 10, 11, 12, ...
$t < 25$	20, 21, 22, 23, ...	20, 21, 22, 23, 24
$a \neq 100$	2, 4, 6, 8, ...	2, 4, ..., 98, 102, ...
$r + 1 < 7$	0, 1, 2, 3, ...	0, 1, 2, 3, 4, 5
$s + 2 < 8$	0, 1, 2, 3, ...	?
$3t < 30$	0, 1, 2, 3, ...	?

2. Explain why the solution to an inequality can be either one number, several numbers, a set of numbers with no end (infinite set), or the empty set.

Find the solution set (S) for each inequality. The replacement set (R) is given for each.

3. $t + 1 > 5$ $R = \{1, 2, 3, 4, \ldots\}$

4. $x - 4 < 10$ $R = \{13, 14, 15, 16\}$

5. $x + 2 > 8$ $R = \{5, 6, 7, 8, 9\}$

6. $t + 1 < 25$ $R = \{22, 24, 26\}$

7. $a - 5 \geq 0$ $R = \{5, 10, 15, \ldots\}$

8. $b - 2 < 100$ $R = \{100, 99, 98, \ldots\}$

9. $3 + r \leq 17$ $R = \{3, 6, 9, 12, 15\}$

10. $28 - n > 20$ $R = \{1, 2, 3, \ldots, 9, 10\}$

11. $19 - b > 10$ $R = \{9, 10, 11\}$

12. $2x + 5 \neq 5$ $R = \{0, 1, 2, 3, \ldots\}$

13. $3n - 4 < 14$ $R = \{3, 4, 5, 6, 7, 8, 9\}$

14. $\frac{s}{2} + 3 \geq 20$ $R = \{2, 4, 6, 8, \ldots\}$

$\leq$ means "is less than or equal to."

$\geq$ means "is greater than or equal to."

Describe these replacement sets in words.

15. 10, 20, 30, 40, ...
16. 1, 3, 5, ...
17. 8, 10, 12

18. 99, 98, 97, ... , 2, 1
19. 3, 6, 9, 12, ...
20. 5, 10, 15, 20, 25

21. 4, 8, 12, 16, ...
22. 6, 12, 18, 24
23. 11, 13, 15, 17, 19

Write an inequality for each. Let n be the variable.

24. An auto-parts store has more than ten thousand items in stock.

25. Lisa did not get 98 on the test.
26. All children under 12 fly free.

What integers are members of each solution set? (The first one is done.)

27. $^-2 < x < {^+3}$ Draw a number line. Color the space *between* $^-2$ and $^+3$. The integers in this colored screen are the solution set.

$S = \{^-1, 0, {^+1}, {^+2}\}$ Why are $^-2$ and $^+3$ *not* part of the solution set?

28. $^+2 < x \leq {^+5}$ (Hint: $x \leq {^+5}$ includes $^+5$.)
29. $^-3 \leq x \leq 0$

30. $^-1 \leq x < {^+6}$
31. $^-1 < x \leq {^+1}$
32. $^-5 < x < {^-2}$

33. $^-4 < x$ and $x \leq {^+1}$
34. $0 \leq x$ and $x < {^+7}$
35. $^-6 \leq x$ and $x \leq {^+2}$

6-4 Exponents: Powers of Ten

Exponent: a number that tells how many times a number, called the **base**, is used as a factor. Exponents are used to express numbers that are products of the same factor.

exponent → 10^3 ← base

10 used 3 times

$$10^3 = 10 \times 10 \times 10 = 1\,000$$

3 zeros

10^3 is read: "ten to the *third* power" or "ten cubed."

10^2 is read: "ten to the *second* power" or "ten squared."

Any number to the first power is that number: $10^1 = 10$.
Any number, except zero, to the zero power is one: $10^0 = 1$.

Copy and complete the chart.

	Exponent Form	Ten as a Factor	Standard Numeral
1.	10^6	$10 \times 10 \times 10 \times 10 \times 10 \times 10$	?
2.	10^5	?	100,000
3.	10^4	?	?
4.	10^3	$10 \times 10 \times 10$	?
5.	10^2	?	?
6.	10^1	10	?

Write each power of ten as a standard numeral.

7. 10^0 8. 10^9 9. 10^1 10. 10^{11} 11. 10^8

Write each as a power of ten (in exponent form).

12. 10,000,000,000
13. $10 \times 10 \times 10 \times 10 \times 10$
14. ten squared
15. $10 \times 10 \times 10 \times 10$
16. 1
17. 1,000,000,000,000
18. 1,000,000
19. $10 \times 10 \times 10 \times 10 \times 10 \times 10 \times 10$
20. ten cubed
21. 1 thousand
22. 1 billion
23. 1 million
24. 1 ten
25. 1 hundred thousand
26. 1 hundred
27. 1 ten million
28. 1 ten thousand
29. one

Complete.

30. If $3 \times 10^y = 3,000,000$, then $y = 6$. Why?

31. If $5 \times 10^y = 50,000$, then $y = \underline{\ ?\ }$.   **32.** If $8 \times 10^y = 8000$, then $y = \underline{\ ?\ }$.

33. If $6 \times 10^y = 60$, then $y = \underline{\ ?\ }$.   **34.** If $7 \times 10^y = 700$, then $y = \underline{\ ?\ }$.

Study the pattern found in exercises 1–6 on page 128.

35. What happens to the standard numeral as the exponent decreases by 1?

Copy the chart. Use the pattern found in exercise 35 to complete it.

	Exponent Form	Fractional Form	As a Decimal Numeral
	10^{-1}	$\dfrac{1}{10}$	0.1
36.	10^{-2}	$\dfrac{1}{10 \times 10} = \dfrac{1}{?}$	0.01
37.	10^{-3}	$\dfrac{1}{10 \times 10 \times 10} = \dfrac{1}{1000}$	?
38.	10^{-4}	$\dfrac{1}{?} = \dfrac{1}{10,000}$	?
39.	10^{-5}	$\dfrac{1}{?} = ?$	?

40. Write one millionth in exponential (exponent) form.

CALCULATOR ACTIVITY

Other numbers can be used as the base.

$6^7 = 6 \times 6 \times 6 \times 6 \times 6 \times 6 \times 6 = 279,936$

6^7 is read "six to the seventh power."

A scientific calculator has a key y^x for raising a base to a power.

Use a calculator to find a standard numeral for each.

41. 2^5 **42.** 3^6 **43.** 10^2 **44.** 4^3 **45.** 8^1

46. 7^5 **47.** 6^4 **48.** 3^9 **49.** 5^4 **50.** 9^7

51. Which power of 2 is the last one that fits on your calculator?

SKILLS TO REMEMBER   **Write the numeral in expanded form.**

52. 0.201   **53.** 0.065   **54.** 0.0007   **55.** 0.00018

56. 1.006   **57.** 40.03   **58.** 8.00109   **59.** 6.03007

6-5 Expanded Form: Using Powers of Ten

The value of each place in a decimal numeral is a power of ten.

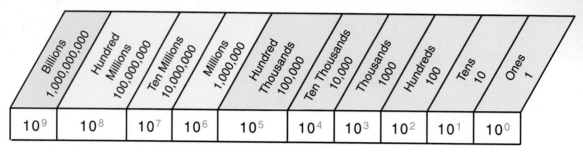

Exponents may be used to write numerals in expanded form.

$36{,}509 = (3 \times 10{,}000) + (6 \times 1000) + (5 \times 100) + (0 \times 10) + (9 \times 1)$
$\phantom{36{,}509} = (3 \times 10^4) + (6 \times 10^3) + (5 \times 10^2) + (0 \times 10^1) + (9 \times 10^0)$

Zero is *not* needed as a placeholder in an expanded numeral.

$So \; 36{,}509 = (3 \times 10^4) + (6 \times 10^3) + (5 \times 10^2) + (9 \times 10^0)$

Copy and fill in the exponents.

1. $57{,}054 \quad = (5 \times 10^?) + (7 \times 10^?) + (5 \times 10^?) + (4 \times 10^?)$
2. $830{,}602 \quad = (8 \times 10^?) + (3 \times 10^?) + (6 \times 10^?) + (2 \times 10^?)$
3. $2{,}102{,}020 \quad = (2 \times 10^?) + (1 \times 10^?) + (2 \times 10^?) + (2 \times 10^?)$
4. $70{,}096 \quad = (7 \times 10^?) + (9 \times 10^?) + (6 \times 10^?)$
5. $80{,}402{,}000 = (8 \times 10^?) + (4 \times 10^?) + (2 \times 10^?)$

Write in expanded form using exponents.

6. 7395
7. 6870
8. 56,035
9. 30,741
10. 48,029
11. 395,007
12. 104,320
13. 8,010,176
14. 91,000,001
15. 164,000
16. 275,100
17. 10,101,101

Write the standard numeral.

18. $(7 \times 10^7) + (1 \times 10^6) + (3 \times 10^5) + (5 \times 10^4) + (2 \times 10^3)$
$ + (6 \times 10^2) + (4 \times 10^1) + (8 \times 10^0)$
19. $(4 \times 10^4) + (6 \times 10^2) + (3 \times 10^1)$
20. $(9 \times 10^6) + (9 \times 10^0)$
21. $(2 \times 10^5) + (8 \times 10^1)$
22. $(8 \times 10^6) + (2 \times 10^4)$
23. $(6 \times 10^7) + (1 \times 10^6) + (4 \times 10^2)$
24. $(3 \times 10^5) + (8 \times 10^3) + (6 \times 10^1)$
25. $(5 \times 10^6) + (5 \times 10^3) + (5 \times 10^0)$
26. $(1 \times 10^5) + (2 \times 10^4) + (3 \times 10^3)$
27. $(9 \times 10^7) + (6 \times 10^4) + (3 \times 10^2)$
28. $(4 \times 10^8) + (2 \times 10^4)$

29. Which of these numbers is less than 10^4?
  a. 10,000   **b.** 10,004   **c.** 1004   **d.** none of these

30. Which of these numbers is greater than 10^5?
  a. 10,005   **b.** 1100   **c.** 100,010   **d.** none of these

31. Which of these numbers is between 10^2 and 10^3?
  a. 100   **b.** 1000   **c.** 1010   **d.** none of these

Write <, =, or > to complete.

32. 6000 __?__ 6×10^4

33. 4,000,000 __?__ 4×10^9

34. 50,000 __?__ 5×10^5

35. 90,000,000 __?__ 9×10^7

Decimals and Expanded Notation

This place-value chart shows how numbers *less than one* may be written in expanded form using exponents.

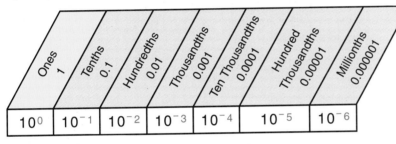

$$0.02003 = (2 \times 0.01) + (3 \times 0.00001)$$
$$= (2 \times 10^{-2}) + (3 \times 10^{-5})$$

Fill in the missing exponents. (The first one is done.)

36. $4.135 = (4 \times 10^0) + (1 \times 10^{-1}) + (3 \times 10^{-2}) + (5 \times 10^{-3})$

37. $40.1006 = (4 \times 10^?) + (1 \times 10^?) + (6 \times 10^?)$

38. $0.051002 = (5 \times 10^?) + (1 \times 10^?) + (2 \times 10^?)$

39. $0.30204 = (3 \times 10^?) + (2 \times 10^?) + (4 \times 10^?)$

Write <, =, or > to compare.

40. 4.004 __?__ $(4 \times 10^0) + (4 \times 10^{-3})$ **41.** 0.055 __?__ (5×10^{-2})

42. 0.0204 __?__ $(2 \times 10^{-3}) + (4 \times 10^{-5})$ **43.** 1.06 __?__ $(1 \times 10^0) + (6 \times 10^{-1})$

44. (8×10^{-2}) __?__ (8×10^{-3}) **45.** (7×10^{-3}) __?__ (6×10^{-2})

46. $(2 \times 10^{-2}) + (5 \times 10^{-3})$ __?__ $(5 \times 10^{-2}) + (2 \times 10^{-3})$

6-6 Divisibility

Here are some rules to check for divisibility:

Example	Number Is Divisible By	Rule	Check
Is 6<u>8</u> divisible by 2? (Is 2 a factor of 68?)	2	If the ones digit of a number is divisible by 2, then the number is divisible by 2.	$8 \div 2 = 4$ 68 is divisible by 2.
Is 1<u>20</u> divisible by 4? (Is 4 a factor of 120?)	4	If the tens and ones digits of a number together are divisible by 4, then the number is divisible by 4.	$20 \div 4 = 5$ 120 is divisible by 4.
Is 360 divisible by 3? (Is 3 a factor of 360?)	3	If the sum of the digits of a number is divisible by 3, then the number is divisible by 3.	$3 + 6 + 0 = 9$ $9 \div 3 = 3$ 360 is divisible by 3.
Is 855 divisible by 9? (Is 9 a factor of 855?)	9	If the sum of the digits of a number is divisible by 9, then the number is divisible by 9.	$8 + 5 + 5 = 18$ $18 \div 9 = 2$ 855 is divisible by 9.
Is 34<u>5</u> divisible by 5? (Is 5 a factor of 345?)	5	If the ones digit of a number is 0 or 5, then the number is divisible by 5.	Ones digit is 5. ↓ 34<u>5</u> is divisible by 5.
Is 14<u>0</u> divisible by 10? (Is 10 a factor of 140?)	10	If the ones digit of a number is 0, then the number is divisible by 10.	Ones digit is 0. ↓ 14<u>0</u> is divisible by 10.

Which numbers are divisible by 2? Which are divisible by 4?

1. 916
2. 656
3. 675
4. 1827
5. 192
6. 154
7. 5317
8. 7038
9. 423
10. 1632
11. 6784
12. 4450
13. 2669
14. 780
15. 68
16. 2045
17. 4166
18. 3472
19. 1560
20. 295

21. Are all the numbers above that are divisible by 2 also divisible by 4? Explain.
22. Are all the numbers above that are divisible by 4 also divisible by 2? Explain.

Which numbers are divisible by 3? Which are divisible by 9?

23. 78	**24.** 234	**25.** 3627	**26.** 501	**27.** 4063
28. 384	**29.** 2435	**30.** 896	**31.** 9081	**32.** 345
33. 3571	**34.** 7440	**35.** 27,072	**36.** 7103	**37.** 8631
38. 17,184	**39.** 36,882	**40.** 26,451	**41.** 21,662	**42.** 1276

Look at your answers for exercises 23–42.

43. Are all the numbers that are divisible by 3 also divisible by 9? Which are? Which are not?

44. Are all the numbers that are divisible by 9 also divisible by 3? Explain your answer.

Which numbers are divisible by 5? Which are divisible by 10?

45. 146	**46.** 287	**47.** 295	**48.** 6480	**49.** 3761
50. 7135	**51.** 114	**52.** 6080	**53.** 3672	**54.** 110
55. 8450	**56.** 6955	**57.** 139	**58.** 7604	**59.** 290
60. 41,390	**61.** 16,275	**62.** 52,770	**63.** 24,098	**64.** 3005

Look at your answers for exercises 45–64.

65. Are all the numbers that are divisible by 5 also divisible by 10? Which are? Which are not?

66. Are all the numbers that are divisible by 10 also divisible by 5? Explain your answer.

Copy and complete. Use a check (✓) to show divisibility.

	Number	Divisible By			
		2	3	4	5
67.	18				
68.	48				
69.	30				
70.	102				

	Number	Divisible By			
		2	3	4	5
71.	24				
72.	42				
73.	138				
74.	174				

CRITICAL THINKING

75. Test each of the numbers in exercises 67–74 for divisibility by 6. The numbers that are divisible by 6 are divisible by which other two numbers?

> **Factors of a product:** the numbers that are multiplied to find that product.
>
> The factors of 9 are 1, 3, and 9.
>
> **Prime number:** a whole number greater than 1 that has *only* two factors, itself and 1.
>
> The factors of 3 are 1 and 3, so 3 is a prime number.
>
> **Composite number:** a whole number, other than 0, that has *more than* two factors.
>
> The factors of 4 are 1, 2, and 4, so 4 is a composite number.

> The numbers 0 and 1 are *neither* prime *nor* composite.

1. Copy and complete the chart for the whole numbers 1 through 20.

If the same factor is used twice, it counts as only one factor.

Number	Factors	Number of Factors
1	1	1
2	1, 2	2
3	1, 3	2
4	1, 2, 4	3
5	1, 5	2

2. Which numbers between 1 and 20 are prime?
3. Which numbers between 1 and 20 are composite?
4. Name the only even prime number.
5. Which number between 0 and 20 is neither prime nor composite? Explain.
6. Write the numbers from 1 to 100 on a chart.

1	2	3	4	5	6	7	8	9	10
11	12	13	14	15	16	17	18	19	20

a. Cross out 1.
b. Circle 2 and cross out all multiples of 2.
c. Circle 3 and cross out all multiples of 3.
d. Circle 5 and cross out all multiples of 5.
e. Circle 7 and cross out all multiples of 7.

All numbers *not* crossed out are prime numbers.

Tell whether each number is prime or composite.*

7. 41
8. 17
9. 64
10. 54
11. 31
12. 99
13. 82
14. 47
15. 50
16. 15
17. 27
18. 48
19. 85
20. 7
21. 29
22. 125
23. 111
24. 300
25. 211
26. 121

6-8 Prime Factorization

Factor tree: used to find prime numbers. Every composite number may be written as the product of *two or more* prime numbers.

Prime factorization: a way of showing a number as the product of prime numbers.

Find the prime factors of 12 and 40.

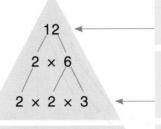

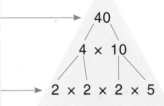

Number to be factored	
Continue factoring until only prime numbers remain.	
Write repeated primes in exponential form.	

The *prime factorization* of 12 is 2 × 2 × 3 or 2^2 × 3.

The *prime factorization* of 40 is 2 × 2 × 2 × 5 or 2^3 × 5.

Use a factor tree to find the *prime factorization* of each number.*

1. 16	**2.** 24	**3.** 48	**4.** 30	**5.** 18
6. 34	**7.** 81	**8.** 42	**9.** 36	**10.** 63
11. 25	**12.** 20	**13.** 88	**14.** 50	**15.** 10
16. 72	**17.** 55	**18.** 75	**19.** 56	**20.** 45
21. 100	**22.** 90	**23.** 180	**24.** 240	**25.** 150

Draw two different factor trees for each number.
Then, using exponents, write the prime factorization for each.

26. 70	**27.** 200	**28.** 32	**29.** 54	**30.** 80
31. 78	**32.** 64	**33.** 96	**34.** 60	**35.** 84

Rewrite, using exponents. Then find the product.

36. 2 × 2 × 2 × 3	**37.** 2 × 2 × 5	**38.** 2 × 2 × 2 × 2
39. 2 × 2 × 5 × 5	**40.** 3 × 3 × 3	**41.** 2 × 2 × 3 × 7
42. 3 × 3 × 3 × 5	**43.** 2 × 2 × 3 × 5 × 5	**44.** 2 × 3 × 3 × 7
45. 5 × 5 × 5	**46.** 2 × 7 × 7	**47.** 2 × 3 × 5 × 5

48. The prime factorization for 72 is:

    a. 3^3 × 2^3 **b.** 3^2 × 2^2 **c.** 2^2 × 3^3 **d.** 3^2 × 2^3

49. Why do you think 25, 100, 49, 81, and 36 are called *squared* numbers?

6-9 | Greatest Common Factor

Factors of a product: the numbers that are multiplied to find that product.

To find the factors of 8, show 8 as a product:	To find the factors of 20, show 20 as a product:
1 × 8 = 8 2 × 4 = 8	1 × 20 = 20 2 × 10 = 20 4 × 5 = 20

The set of factors of 8 is: {1, 2, 4, 8}.

The set of factors of 20 is: {1, 2, 4, 5, 10, 20}.

Common factors: those factors that are the same for a given pair of numbers.

The set of common factors of 8 and 20 is: {1, 2, 4}.

Greatest common factor, or **GCF**, of 8 and 20 is 4.

Write the set of factors for each number.

1. 6   2. 18   3. 27   4. 12   5. 36   6. 42   7. 10   8. 9

9. 13   10. 63   11. 30   12. 50   13. 28   14. 7   15. 14   16. 21

Write the set of common factors for each pair of numbers.

17. 24 and 32   18. 13 and 52   19. 18 and 27   20. 36 and 63

21. 11 and 22   22. 25 and 50   23. 14 and 28   24. 3 and 9

25. Find the GCF for each pair of numbers in exercises 17–24.

Find the GCF for each set of numbers.

26. 8 and 20   27. 17 and 19   28. 50 and 75   29. 35 and 28

30. 56, 72, and 40 31. 32, 16, and 48 32. 45, 36, and 63 33. 56, 21, and 7

34. How many factors does a prime number have?

35. Can two numbers have no common factors? Explain your answer.

36. The set of common factors of two given numbers is the same as:
    a. the union of the two sets of factors **b.** the intersection of the two sets of factors.
    Explain your answer.

Find the GCF of each pair of numbers in the columns below. (Look for a pattern.)

	Column I		Column II		Column III
37.	2 and 3	**41.**	2 and 4	**45.**	2 and 7
38.	3 and 4	**42.**	3 and 9	**46.**	3 and 11
39.	5 and 6	**43.**	5 and 10	**47.**	5 and 13
40.	7 and 8	**44.**	7 and 28	**48.**	7 and 11

49. In which column(s) does the GCF for each pair of numbers equal 1?

50. In which column(s) does the GCF for each pair of numbers equal one of the two numbers?

Which column above best matches each description of the relationship between the pairs of numbers?

51. Both numbers in each pair are primes.

52. The numbers in each pair are consecutive numbers.

53. The numbers in each pair consist of a prime number and one of its multiples.

54. When the GCF of two numbers is equal to 1, the numbers are called **relatively prime**. In which of the three columns above are the pairs of numbers *relatively prime*?

The GCF and Simplest Form

The **GCF** is used to express fractions in *simplest form*.

Fractions are in simplest form when the GCF of the numbers in both the numerator and the denominator is equal to 1.

$\dfrac{3}{8}$ ◄——— The GCF of 3 and 8 is 1, so $\dfrac{3}{8}$ is in simplest form.

$\dfrac{15}{18}$ ◄——— The GCF of 15 and 18 is 3, so $\dfrac{15}{18}$ is *not* in simplest form.

To express fractions in simplest form, divide *both* the numerator *and* the denominator by the GCF.

$$\dfrac{15 \div 3}{18 \div 3} = \dfrac{5}{6} \quad ◄——— \text{simplest form}$$

Find the GCF of each fraction. Then tell which fractions are in simplest form.

55. $\dfrac{8}{9}$ **56.** $\dfrac{2}{11}$ **57.** $\dfrac{7}{21}$ **58.** $\dfrac{15}{20}$ **59.** $\dfrac{5}{17}$

60. $\dfrac{12}{24}$ **61.** $\dfrac{4}{5}$ **62.** $\dfrac{6}{18}$ **63.** $\dfrac{16}{20}$ **64.** $\dfrac{11}{12}$

6-10 Least Common Multiple

> **Multiples** of a number: the products of that number and 0, 1, 2, 3, ...
>
> Multiples of 3 = {0, 3, 6, 9, 12, 15, 18, 21, 24, ...}
>
> Multiples of 4 = {0, 4, 8, 12, 16, 20, 24, 28, ...}

> **Common multiples:** nonzero multiples that are the *same* for a pair of numbers.
>
> The set of common multiples of 3 and 4 is: {12, 24, ...}.

> **Least common multiple**, or **LCM**, of 3 and 4 is 12.
>
> This is true because 12 is the least number that is a common multiple of both 3 and 4.

Write the first six nonzero multiples of each number.

1. 4	**2.** 10	**3.** 8	**4.** 6	**5.** 19
6. 2	**7.** 5	**8.** 7	**9.** 15	**10.** 12
11. 3	**12.** 14	**13.** 20	**14.** 13	**15.** 16

16. List the multiples of 4 greater than 50 and less than 75.

17. List the multiples of 10 greater than 200 and less than 400.

Find the LCM for each set of numbers.

18. 3 and 12	**19.** 2 and 5	**20.** 6 and 9	**21.** 4 and 7
22. 5 and 8	**23.** 9 and 18	**24.** 8 and 12	**25.** 6 and 10
26. 6 and 2	**27.** 10 and 15	**28.** 3 and 9	**29.** 9 and 5
30. 3 and 10	**31.** 2 and 3	**32.** 6 and 4	**33.** 4 and 8
34. 3 and 7	**35.** 5 and 4	**36.** 3 and 5	**37.** 2 and 7
38. 5 and 12	**39.** 7 and 8	**40.** 9 and 12	**41.** 8 and 6
42. 3 and 8	**43.** 7 and 5	**44.** 10 and 8	**45.** 4 and 10
46. 2, 3, and 8	**47.** 6, 9, and 12	**48.** 2, 4, and 6	**49.** 6, 8, and 9
50. 4, 6, and 8	**51.** 4, 5, and 10	**52.** 3, 4, and 5	**53.** 2, 3, 4, and 8

Find the LCM of each pair of numbers in the columns below. (Look for a pattern.)

Column I	Column II	Column III
54. 2 and 3	58. 2 and 4	62. 2 and 7
55. 3 and 4	59. 3 and 9	63. 3 and 11
56. 5 and 6	60. 5 and 10	64. 5 and 13
57. 7 and 8	61. 7 and 28	65. 7 and 11

66. In which column(s) is the LCM for each pair of numbers found by multiplying the numbers?

67. In which column(s) is the LCM for each pair of numbers one of the two numbers?

68. Describe the relationship between the two numbers in each pair for the three columns.

69. The pairs of numbers in columns I and III are relatively prime (GCF = 1). The LCM of a pair of numbers that are relatively prime is:
    a. the larger of the two numbers   **b.** the product of the two numbers
    c. the smaller of the two numbers. Explain.

Unit Fractions

Unit fractions are fractions like $\frac{1}{3}$, $\frac{1}{4}$, $\frac{1}{5}$, and $\frac{1}{10}$.

To compute the sum of unit fractions having denominators that are relatively prime, use the method shown below.

$$\frac{1}{2} + \frac{1}{3} = \frac{?}{\_\_}$$
$$(3 \times 1) + (2 \times 1)$$
$$\frac{1}{2} \times \frac{1}{3} = \frac{5}{6} \leftarrow \boxed{\text{LCM} \; (2 \times 3)}$$

$$\frac{1}{3} + \frac{1}{5} = \frac{?}{\_\_}$$
$$(5 + 3)$$
$$\frac{1}{3} \times \frac{1}{5} = \frac{8}{15} \leftarrow \boxed{\text{LCM} \; (3 \times 5)}$$

To compute the difference between unit fractions having denominators that are relatively prime, use the method shown below.

$$\frac{1}{3} - \frac{1}{5} = \frac{?}{\_\_}$$
$$(5 \times 1) - (3 \times 1)$$
$$\frac{1}{3} \times \frac{1}{5} = \frac{2}{15} \leftarrow \boxed{\text{LCM} \; (3 \times 5)}$$

$$\frac{1}{2} - \frac{1}{7} = \frac{?}{\_\_}$$
$$(7 - 2)$$
$$\frac{1}{2} \times \frac{1}{7} = \frac{5}{14} \leftarrow \boxed{\text{LCM} \; (2 \times 7)}$$

MENTAL MATH   **Find the sums and differences of these unit fractions mentally. Time yourself.**

70. $\frac{1}{2} + \frac{1}{7}$ 71. $\frac{1}{3} + \frac{1}{8}$ 72. $\frac{1}{4} + \frac{1}{5}$ 73. $\frac{1}{9} + \frac{1}{10}$

74. $\frac{1}{3} - \frac{1}{7}$ 75. $\frac{1}{4} - \frac{1}{5}$ 76. $\frac{1}{2} - \frac{1}{9}$ 77. $\frac{1}{4} - \frac{1}{7}$

6-11 Scientific Notation

Scientific notation: a way of writing *very* large numbers in a shortened form.

Scientific notation shows a number as a product of a number greater than or equal to 1 but less than 10, and a power of ten.

Standard Numeral		Scientific Notation
42,000,000	=	4.2×10^7

Number greater than or equal to 1 but less than 10

Power of ten

To write a number in scientific notation:

- Move the decimal point to the *left* to show a number greater than or equal to 1 but less than 10.

- Multiply by the power of ten that corresponds to the *number of places* the decimal point was moved.

$$3\,.\,0\,7\,,\,0\,0\,0\,,\,0\,0\,0 = 3.07 \times 10^8$$ ← Number of places

moved 8 places

Copy and complete with the correct power of ten.

1. $5{,}146{,}000 = 5.146 \times 10^?$
2. $12{,}000{,}000 = 1.2 \times 10^?$
3. $210{,}000 = 2.1 \times 10^?$
4. $5264 = 5.264 \times 10^?$
5. $63{,}000 = 6.3 \times 10^?$
6. $8{,}000{,}000{,}000 = 8 \times 10^?$
7. $761{,}000 = 7.61 \times 10^?$
8. $3124.9 = 3.1249 \times 10^?$
9. $493{,}000{,}000 = 4.93 \times 10^?$
10. $1{,}020{,}000 = 1.02 \times 10^?$

Copy and complete with the correct number.

11. $94{,}000{,}000 = \underline{\,?\,} \times 10^7$
12. $6{,}800{,}000 = \underline{\,?\,} \times 10^6$
13. $1400 = \underline{\,?\,} \times 10^3$
14. $710{,}000{,}000 = \underline{\,?\,} \times 10^8$
15. $4{,}360{,}000 = \underline{\,?\,} \times 10^6$
16. $55{,}500 = \underline{\,?\,} \times 10^4$
17. $830{,}200{,}000 = \underline{\,?\,} \times 10^8$
18. $101{,}000 = \underline{\,?\,} \times 10^5$
19. $2{,}012{,}000{,}000 = \underline{\,?\,} \times 10^9$
20. $36{,}170{,}000 = \underline{\,?\,} \times 10^7$

Write in scientific notation.

21. 531,000   **22.** 82,000,000   **23.** 6,100,000   **24.** 92,100

25. 914,000   **26.** 2,201,600,000   **27.** 40,040,000   **28.** 370,000,000

29. 52,900,000   **30.** 11,220.9   **31.** 2,105,000,000   **32.** 1,110,000

33. 3000.1   **34.** 800,000,000   **35.** 1,010,100,000   **36.** 680,000

Standard Numeral

To write the standard numeral for a number written in scientific notation, move the decimal point to the *right* the number of places shown by the power of ten.

$$5.1 \times 10^8 = 5.1\ 0\ 0\ 0\ 0\ 0\ 0\ 0. = 510,000,000$$

moved 8 places

Write the standard numeral.

37. 3.7×10^4 **38.** 8.31×10^5 **39.** 2.93×10^3 **40.** 8.614×10^2

41. 7×10^7 **42.** 3.9×10^8 **43.** 5.41×10^5 **44.** 9.32×10^9

45. 1.01×10^6 **46.** 4.312×10^2 **47.** 6.6×10^7 **48.** 7×10^{10}

49. 1×10^{10} **50.** 3.471×10^3 **51.** 2.01×10^7 **52.** 7.34×10^6

Which is greater?

53. 5.4×10^3 or 4.5×10^4 **54.** 7.6×10^5 or 7.06×10^5

55. 9×10^8 or 8×10^9 **56.** 6.9×10^3 or 8.09×10^2

Express each number in scientific notation.

57. Voyager Two traveled 4,400,000,000 miles before reaching Neptune.

58. Scientists believe they have observed the birth of a new galaxy 65,000,000 light-years from Earth.

59. The average distance from the sun to Earth is 92,900,000 miles.

60. Another way to write $4^2 \times 3^2 \times 10^5$ is:
    a. 6.4×10^6 **b.** 1.44×10^7 **c.** 1.44×10^5 **d.** 6.4×10^7

CALCULATOR ACTIVITY

To change to scientific notation, use the ENG key on a scientific calculator. Enter `0.0005` `ENG`. The display `5.⁻04` means 5.0×10^{-4}.

Express each in scientific notation. Check, using a calculator.

61. 0.000012   **62.** 0.0025   **63.** 0.00027

64. 0.0000075   **65.** 0.0000081   **66.** 0.0000009

Animal	Speed (mph)
Spider	1.17
Giant tortoise	0.1
Three-toed sloth	0.
Garden snail	

Problem: José drew a table for his science report. But his dog ate part of it. Only this much was left. Help José reconstruct his table by finding the missing information.

"…The three-toed sloth travels at 0.15 mile per hour, which is five times faster than the garden snail. The giant tortoise travels one mile per hour slower than the spider."

Snail (n) = ___?___
Sloth (s) = 0.15 mph or 5n
Spider (p) = ___?___
Tortoise (t) = p – 1

1 IMAGINE Make and label a table for the report.

2 NAME

Facts:

| Three-toed sloth's speed | — 0.15 mph or 5 times snail's pace |
| Tortoise's speed | — 1 mph less than spider's speed |

Question: How fast do the snail and tortoise travel?

3 THINK Use the chart and labels above to create a number sentence:

Spider's speed (p) – 1 mph = Tortoise's speed (t)
$$p - 1 = t$$

5 × Snail's speed (n) = Sloth's speed (s)
$$5n = s \longrightarrow 5n = 0.15$$

4 COMPUTE Solve the two equations to find the tortoise's (t) and the snail's speed (n) to complete the chart.

Substitute
$p = 1.17$ $\longrightarrow$

$p - 1 = t$
$1.17 - 1 = 0.17$ Tortoise's speed

$$5n = 0.15$$

Use inverse operation.
Divide by 5.

$$\frac{5n}{5} = \frac{0.15}{5} \longrightarrow n = 0.03 \text{ Snail's speed}$$

Complete the chart with all the correct speeds.

5 CHECK Compare the results to the facts.

Is 0.17 mph, the tortoise's speed, one mph less than 1.17 mph, the spider's speed? Yes.

Is 0.15 mph, the sloth's speed, five times 0.03 mph, the snail's speed? Yes.

Solve by using tables and drawings.

1. Twelve athletes took part in a track meet. Three competed only in the cross-country race. Two of the five 800-meter runners also entered the mile relay race. How many athletes competed only in the mile relay race?

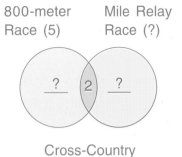

800-meter Race (5)   Mile Relay Race (?)

?   2   ?

Cross-Country Race

3

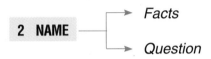

1 IMAGINE   Draw and label a picture.

2 NAME   → *Facts*
         → *Question*

3 THINK   Use your picture to see how the 12 athletes competed in 3 different categories. First find the number who competed only in the 800-meter race. Then find the number who competed only in the mile relay race.

4 COMPUTE ⟶ **5 CHECK**

2. Three pieces of string measure 3 cm, 8 cm, and 12 cm. How can all three strings be used to measure a distance of: 1 cm? of 17 cm? of 7 cm?

3. Use the following information to complete the table and exercises a and b.
The maximum distance to Jupiter is 24 times the minimum distance from Earth to Venus. The minimum distance to Saturn is 7.1×10^8 miles farther than the minimum distance from Earth to Mars. The minimum distance to Pluto is 4.0×10^6 miles less than the minimum distance between Earth and Neptune.

a. Write the maximum and minimum distance for Saturn in scientific notation.

b. What planet is an average distance of 93 million miles from Earth?

Distance to Earth In Millions of Miles		
Planet	Maximum	Minimum
Mercury	136	50
Venus	161	25
Mars	248	35
Jupiter	?	368
Saturn	1031	?
Uranus	1953	1606
Neptune	2915	2667
Pluto	4644	?

6-13 Problem Solving: Applications

Solve.

1. What is the *greatest* three-digit number divisible by 5 and 3? (Hint: 105, 120, ..., __?__ greatest)

2. What is the *least* four-digit number divisible by 4 and 3? (Hint: __?__ least, ..., 9984, 9996, ...)

3. What is the *greatest* five-digit number divisible by 6 and 4?

4. What is the *least* six-digit number divisible by 9 and 5?

5. What is the least number that when divided by either 2, 3, 4, or 8 has a remainder of 1?

6. What is the least number that when divided by either 2, 3, 4, or 5 has a remainder of 1?

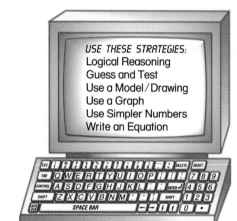

USE THESE STRATEGIES:
Logical Reasoning
Guess and Test
Use a Model / Drawing
Use a Graph
Use Simpler Numbers
Write an Equation

7. Beginning at the starting line of a racetrack, blue markers are placed every 20 yards and green markers are placed every 15 yards. If the racetrack is 200 yards long, how many times will the two markers be placed together?

8. At a nursery Polly places a special fertilizer on tree seedlings every 6 weeks. Ollie adds new soil to these same seedlings every 4 weeks. If Polly and Ollie begin the same week, how many times during the year (52 weeks) will the seedlings receive both special fertilizer and new soil during the same week?

9. The amount of water that a tree of average size returns to Earth's atmosphere during a year is about 30,000 gallons. Write this number in scientific notation.

10. Some pollutants have become concentrated in living things at levels 10^7 times their original concentration in the environment. Write 10^7 as a standard numeral.

11. It is estimated that lead amounting to three million tons can be found in fifty-seven million American homes. Write each number in expanded form.

12. It has been estimated that in the United States lead poses a hazard in about 3.8×10^6 homes in which there are small children. Write 3.8×10^6 as a standard numeral.

This chart shows the projected population of five large cities for the year 2000. Use the chart to answer exercises 13–14.

13. Round each number on the chart to the greatest period.

14. Write in scientific notation each rounded number in exercise 13.

15. Each year industry uses 160,000 gallons of water for each person in the United States. If there are 250,000,000 people in the United States, how many gallons of water does industry in the United States use in one year? (Express the answer in scientific notation.)

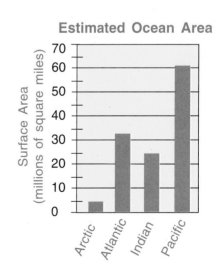

City	Projected Population
Tokyo, Japan	29,971,000
Mexico City, Mexico	27,872,000
Moscow, Russia	11,121,000
São Paulo, Brazil	12,911,000
Lagos, Nigeria	12,528,000

16. To produce a ton of steel, 6×10^4 gallons of water are used. To produce a ton of corn, 2.4×10^5 gallons of water are used. How many more gallons of water are used to produce a ton of corn than a ton of steel? Explain why more water is used to produce the corn.

17. The average family uses 370 gallons of water a day. How many gallons of water will the average family use in the month of June?

18. Scientists estimate that the volume of flow over Niagara Falls is 500,000 tons a minute. How many tons of water flow over Niagara Falls in an hour? (Express your answer in scientific notation.)

The graph shows the approximate surface area of the world's oceans. Use the graph for exercises 19–21.

19. List the oceans in order of increasing surface area.

20. By about how many square miles does the surface area of the Pacific exceed that of the Atlantic?

21. What is the total surface area covered by all the oceans? (Express the total as a standard numeral.)

Estimated Ocean Area

Surface Area
(millions of square miles)

70
60
50
40
30
20
10
0

Arctic Atlantic Indian Pacific

More Practice

1. $S = \{20, 22, 24, 26, \ldots\}$ describes __?__

   a. the set of even numbers between 20 and 26.

   b. the set of even numbers 20 or greater.

2. List the members that are the union and intersection of sets A and B.

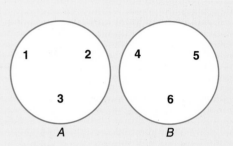

Find the solution set for each inequality. $R = \{0, 2, 4, 6, 8\}$

3. $n < 10$ 4. $n + 3 \leq 5$ 5. $n - 10 > 20$

Write in expanded form, using exponents.

6. 104,321   7. 6060   8. 65,503   9. 8,906,577   10. 9943

Write a standard numeral for each.

11. 9^2 12. 10^0 13. 4^2 14. 5^3 15. 10^8 16. 10^{-2}

17. $(3 \times 10^5) + (5 \times 10^0)$ 18. $(6 \times 10^8) + (8 \times 10^3) + (2 \times 10^1)$

19. $(2 \times 10^0) + (4 \times 10^{-1})$ 20. $(7 \times 10^{-1}) + (1 \times 10^{-2})$

Check these numbers for divisibility by 2, 3, 4, 5, 9, and 10.

21. 127   22. 92   23. 40   24. 81   25. 386

Write the prime factorization for each composite number. Identify any prime numbers.

26. 80   27. 48   28. 9   29. 11   30. 40

31. 13   32. 21   33. 12   34. 32   35. 94

Write the set of factors for each number.

36. 40   37. 56   38. 18   39. 72   40. 39

Find the GCF for each set of numbers.

41. 14 and 12   42. 15 and 20   43. 32 and 16   44. 25 and 10

Find the LCM for each set of numbers.

45. 2 and 9   46. 7 and 6   47. 10 and 4   48. 12 and 10

Write in scientific notation.

49. 800,000,000   50. 10.03   51. 7,021,000   52. 72.04

SQUARE ROOT BY PRIME FACTORIZATION

What number multiplied by itself equals 196?

When the product of a number times itself is given, calculate the square root to find the number.

$\sqrt{196}$ means "the square root of 196" or
"What number multiplied
by itself is equal to 196?"

196
2 x 98
2 x 2 x 49
2 x 2 x 7 x 7

196

Study the examples below.
What pattern is used to find the square root?

$$\sqrt{4} = \sqrt{2 \times 2} = 2 \qquad\qquad \sqrt{9} = \sqrt{3 \times 3} = 3$$

$$\sqrt{16} = \sqrt{2 \times 2 \times 2 \times 2} = 4 \qquad \sqrt{25} = \sqrt{5 \times 5} = 5$$

Use prime factorization to find the number that when multiplied by itself is equal to the given number.

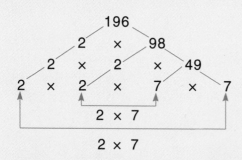

So, $\sqrt{196} = \sqrt{2 \times 2 \times 7 \times 7}$

$\sqrt{196} = 2 \times 7$

$\sqrt{196} = 14$

Calculators also may be used to find square roots.
Check by using a calculator.   196 √   14

Try these. (Hint: The prime numbers are {2, 3, 5, 7, 11, 13,...}.)

1. $\sqrt{100}$ 2. $\sqrt{324}$ 3. $\sqrt{144}$ 4. $\sqrt{441}$

5. $\sqrt{784}$ 6. $\sqrt{3025}$ 7. $\sqrt{484}$ 8. $\sqrt{1764}$

9. $\sqrt{225}$ 10. $\sqrt{1225}$ 11. $\sqrt{4356}$ 12. $\sqrt{1936}$

Make up your own perfect squares. Be sure to multiply a number by itself.
For example: $2 \times 2 \times 17 \times 17 = 1156$

Check Your Mastery

1. List the members of the set of odd numbers greater than 100, using set notation. See pp. 122-127

2. Describe this replacement set in words: {11, 13, 15, 17, 19, . . .}

3. The intersection of Set M = {0, 12, 17, 18} and Set N = {0, 13, 15} is __?__
   a. {0, 12, 13, 15, 17, 18}   b. ø   c. {0}   d. {12, 13, 15, 17, 18}

4. Given the replacement set 0, 2, 4, ..., 10, find the solution set for the expression: $m - 2 \geq 5$.

Write each as a power of ten (in exponent form). See pp. 128-129

5. 1 billion   6. 1 million   7. 1 ten thousand   8. 1 tenth

Write in expanded form, using exponents. See pp. 130-131

9. 56,010   10. 7000   11. 404,104   12. 90.25

Which numbers are divisible by both 2 and 4? See pp. 132-133

13. 4702   14. 9660   15. 7524   16. 8818

Which numbers are divisible by both 3 and 9?

17. 1602   18. 8061   19. 2622   20. 7317

Use exponents to write the prime factorization for each number. See p. 135

21. 36   22. 40   23. 55   24. 17

Find the GCF for each set of numbers. See pp. 136-137

25. 45 and 27   26. 24 and 48   27. 28 and 35   28. 56 and 40

Find the LCM for each set of numbers. See pp. 138-139

29. 7 and 4   30. 3 and 5   31. 9 and 7   32. 6 and 8

Write in scientific notation. See pp. 140-141

33. 97,000,000   34. 700,000   35. 542,000,000   36. 50,000

Write the standard numeral.

37. 2.101×10^5 38. 1.4×10^9 39. 8×10^6 40. 5.1345×10^3

Cumulative Review

Choose the correct answer.

1. The set of prime numbers between 20 and 30 are:
   a. {21, 23, 29}
   b. {23, 29}
   c. {21, 23, 27, 29}
   d. {21, 29}

2. What integer is 5 units to the right of $^-1$?
   a. 0
   b. 4
   c. 5
   d. $^-6$

3. Write 10,000,000 as a power of ten.
   a. 10^4
   b. 10^8
   c. 10^9
   d. 10^7

4. Find the greatest common factor for 45 and 36.
   a. 3
   b. 6
   c. 15
   d. 9

5. $^-2(^+4 + {}^-3) = \underline{\ ?\ }$
   a. $^-8 + {}^-6$
   b. $^-8 + {}^+3$
   c. $^-2$
   d. $^+2$

6. Write 2.631×10^5 in standard form.
   a. 263,100
   b. 263,100,000
   c. 2,631,000
   d. 26,310,000

7. The prime factorization of 240 is:
   a. $2^2 \times 5 \times 6$
   b. $2^3 \times 5 \times 3$
   c. $2^4 \times 3 \times 5$
   d. $2^3 \times 5 \times 6$

8. Which number is divisible by 2, 3, 4, and 5?
   a. 3672
   b. 6480
   c. 1635
   d. 1265

9. What is the sum of the prime numbers between 10 and 20?
   a. 75
   b. 64
   c. 60
   d. 43

10. Which number is less than 10^{-8}?
    a. 0.000001
    b. 0.0000001
    c. 0.00000001
    d. 0.000000001

11. If the replacement set is 0, 1, 2, 3, 4, 5, what is the solution set for $\dfrac{m}{2} + 4 \le 6$?
    a. {0, 1}
    b. {0, 1, 2}
    c. {2, 3, 5}
    d. {0, 1, 2, 3, 4}

12. What is the greatest prime number that is a factor of 140?
    a. 14
    b. 5
    c. 7
    d. 20

13. $\dfrac{^-48}{^-12} \times {}^+6 + {}^-5 = \underline{\ ?\ }$
    a. 24
    b. 19
    c. 21
    d. 37

14. Name the next numbers in the sequence 5, 7, 2, 4, $^-1$, $\underline{\ ?\ }$, $\underline{\ ?\ }$, $\underline{\ ?\ }$.
    a. $^-6, {}^+2, {}^-4$
    b. $2, {}^-4, {}^-1$
    c. $0, {}^-2, 8$
    d. $1, {}^-4, {}^-2$

Arrange in order from least to greatest.
15. $^-3$, 1, $^-2$, 0, 2, $^-4$
16. 8, $^-3$, 5, $^-8$, 0
17. 1000, 0.1, 101.01, 10, 0.01
18. 0.01, 0, 0.011, 1, 0.101, 11

Write each answer in set notation.
19. { 2, 4, 6 } ∪ { 1, 2, 3 }
20. { 2, 4, 6 } ∪ { 1, 2 }
21. { 5, 10, 15 } ∩ { 10, 15, 20 }
22. { 0, 1, 2 } ∩ φ

Compute.
23. $^-8$ + $^-7$
24. $^-6$ + $^+9$
25. $^-7$ − $^-3$
26. $^-5$ − $^+6$
27. $^+4$ × $^-5$
28. $^-9$ × $^-9$
29. $^-6$ × 0 × $^+2$
30. $^-12$ × $^+10$ × $^-2$
31. $^-50$ ÷ $^-5$
32. $^+42$ ÷ $^-3$
33. $^+16$ ÷ $^-16$
34. 0 ÷ $^-25$
35. $^+6$ + ($^-3$ × $^-2$)
36. $^-8$ − ($^-16$ ÷ $^-4$)
37. [($^-2$ + $^-7$) ÷ $^+3$]
38. [$^-40$ ÷ ($^-3$ − $^+2$)]
39. 10^4 = ___?___
40. 10^8 = ___?___
41. 10 × 10 × 10 = $10^?$
42. 10,000,000 = $10^?$

Solve for _n_. Identify the property used.
43. $^+5$ + $^-11$ = n + $^+5$
44. n + $^-3$ = $^-3$
45. ^-7n = $^-7$
46. $^+17$ + n = 0

Write the prime factorization for each number, using exponents.
47. 70
48. 200
49. 144
50. 96

Find the greatest common factor.
51. 18 and 24
52. 56, 40, and 72

Find the least common multiple.
53. 15 and 6
54. 2, 3, and 5

Compare. Write <, =, or >.
55. 8,000,000 ___?___ 8 × 10^7
56. 4.5 × 10^5 ___?___ 4500
57. 3^2 × 2^2 × 5^2 ___?___ 900
58. 17.8 × 10^{-2} ___?___ 0.178

Solve.
59. A seagull is flying 50 ft above the ocean's surface. A scuba diver is exploring 15 ft below the surface. How far apart are the seagull and the diver?

60. The temperature at noon was 13°F. It dropped 17 degrees. What is the temperature now?

61. A hummingbird's wings beat 70 times a second. Use scientific notation to express the number of beats per minute.

62. On one expedition a spacecraft traveled 9,872,350,000 km. Write this distance in expanded form, using exponents.

63. Greta is organizing a luncheon for 48 guests. She plans to have 6 tables seating 8 guests each. Name two other seating arrangements she might use.

64. The Wayne Company has an order from a hardware store chain for 9360 light bulbs. Can the light bulbs be divided evenly among the chain's 10 stores? Will each store be able to receive an even number of bulbs if they are packaged 4 to a box?

+10 BONUS

65. Imagine that you are a time traveler. Take a trip to a specific year in the past. Record specific information in scientific notation (the year, distance traveled, etc.). Then take a trip to a specific year in the future. Use scientific notation to record specific information (population of the colony, distance to nearest planet, etc.).

7 Fractions: Addition and Subtraction

In this chapter you will:

- Find equivalent fractions
- Express fractions in simplest form
- Compare and order fractions
- Use estimation strategies
- Add and subtract fractions and mixed numbers
- Solve problems: logical reasoning
- Use technology: spreadsheets

Do you remember?

The **numerator** of a fraction names the number of parts being considered.

$$\frac{3}{5}$$

The **denominator** of a fraction names the total number of congruent or equal parts.

When the numerator and denominator are identical, the fraction is equal to 1. Some fractions equal to 1 are:

$$\frac{3}{3}, \frac{9}{9}, \frac{18}{18}, \frac{36}{36}$$

RESEARCHING TOGETHER

Lost over the Nile!

The Mystery of the Lost Papyrus! Find out all you can about the mysterious Rhind Papyrus. How did Egyptians work with fractions? Show one of their fraction problems and how it was solved. Did the Papyrus mathematician ever make an error? What was it? Act it out on "Lost over the Nile!"

151

Fractions, Mixed Numbers, Equivalent Fractions

Fraction: describes part of a region, an object, or a set.

It has two parts: $\dfrac{\text{Numerator}}{\text{Denominator}}$ ◄ The fraction bar means division.

7 of the 12 quilt parts are completed.

$\dfrac{7}{12}$ of the quilt is completed.

What part of the quilt is *not* completed?

Mixed number: has a whole-number part and a fraction part.

$1\dfrac{3}{4}$ skeins of yarn were used.

What part of 2 skeins was *not* used? $\dfrac{1}{4}$ not used

Equivalent fractions: name the *same* part of a region, an object, or a set.

$$\dfrac{3}{4} = \dfrac{6}{8} \qquad \dfrac{10}{15} = \dfrac{2}{3}$$

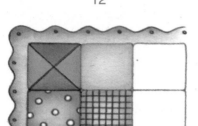

$\dfrac{7}{12}$ completed

$\dfrac{5}{12}$ *NOT* completed

Write the equivalent fraction.

1.

$\dfrac{1}{3} = \dfrac{?}{6}$

2.

$\dfrac{1}{2} = \dfrac{?}{4}$

3.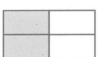

$\dfrac{3}{6} = \dfrac{?}{2}$

4.

$\dfrac{1}{2} = \dfrac{?}{10}$

5.

$\dfrac{4}{12} = \dfrac{?}{6}$

6.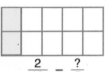

$\dfrac{2}{10} = \dfrac{?}{5}$

7.

$\dfrac{12}{8} = \dfrac{?}{4} = \dfrac{?}{2}$

8.

$\dfrac{4}{8} = \dfrac{?}{4} = \dfrac{?}{2}$

Write the mixed number that represents the shaded area.

9.

10.

11.

12.

13. Is every mixed number greater than 1? Why or why not?

Finding Equivalent Fractions

Multiply the fraction by names for 1:

$$1 = \frac{2}{2} = \frac{3}{3} = \frac{4}{4} = \frac{5}{5} \cdots$$

This is the same as saying:

Multiply *both* the numerator and the denominator by the *same* number.

Some equivalent fractions for $\frac{5}{6}$:

$$\frac{5}{6} = \frac{10}{12} = \frac{15}{18} = \frac{20}{24} = \frac{25}{30}$$

Equivalent Fractions

$$\frac{5}{6} \times \frac{2}{2} = \frac{10}{12}$$
$$\frac{5}{6} \times \frac{3}{3} = \frac{15}{18}$$
$$\frac{5}{6} \times \frac{4}{4} = \frac{20}{24}$$
$$\frac{5}{6} \times \frac{5}{5} = \frac{25}{30}$$

Number Line: shows equivalent fractions.

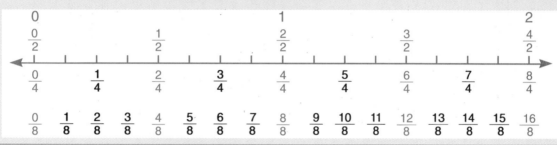

Use the number line to name an equivalent fraction for each.

14. $\frac{3}{4} = \frac{?}{8}$ **15.** $\frac{5}{4} = \frac{?}{8}$ **16.** $\frac{4}{8} = \frac{?}{4}$ **17.** $\frac{12}{8} = \frac{?}{2}$

18. $\frac{0}{4} = \frac{?}{8}$ **19.** $\frac{4}{2} = \frac{?}{4}$ **20.** $\frac{2}{8} = \frac{?}{4}$ **21.** $\frac{0}{2} = \frac{?}{4}$

22. $\frac{0}{2}$ means $0 \div 2 = 0$. Give other fraction names for zero.

23. $\frac{4}{4}$ means $4 \div 4 = 1$. Give other fraction names for one.

Write an equivalent fraction.

24. $\frac{1}{4} = \frac{?}{16}$ **25.** $\frac{1}{4} = \frac{?}{20}$ **26.** $\frac{1}{8} = \frac{?}{16}$ **27.** $\frac{1}{5} = \frac{?}{20}$

28. $\frac{3}{8} = \frac{?}{16}$ **29.** $\frac{2}{3} = \frac{?}{9}$ **30.** $\frac{6}{7} = \frac{?}{14}$ **31.** $\frac{5}{3} = \frac{?}{15}$

32. $\frac{3}{4} = \frac{?}{24}$ **33.** $\frac{3}{5} = \frac{?}{45}$ **34.** $\frac{5}{6} = \frac{?}{18}$ **35.** $\frac{7}{4} = \frac{?}{28}$

Find three more equivalent fractions in each set.

36. $\{\frac{2}{7}, \frac{4}{14}, \frac{?}{\_}, \frac{?}{\_}, \frac{?}{\_}\}$ **37.** $\{\frac{5}{8}, \frac{10}{16}, \frac{?}{\_}, \frac{?}{\_}, \frac{?}{\_}\}$

38. $\{\frac{10}{3}, \frac{20}{6}, \frac{?}{\_}, \frac{?}{\_}, \frac{?}{\_}\}$ **39.** $\{\frac{5}{4}, \frac{?}{\_}, \frac{?}{\_}, \frac{?}{\_}, \frac{25}{20}\}$

7-2 Fractions, Simplest Form

Simplest form, or lowest terms: means that a fraction's numerator *and* denominator have NO common factor other than 1.

To change a fraction to its simplest form, find its simplest, or lowest, equivalent fraction. Divide the fraction by names for one by dividing the numerator *and* denominator by any common factor.

Write $\frac{18}{30}$ in simplest form.

2 is a common factor of both 18 and 30.

There are 30 videos, and 18 are rented. $\frac{18}{30}$ of the videos are rented.

$\dfrac{18 \div 2}{30 \div 2} = \dfrac{9}{15}$ ← $\frac{9}{15}$ and $\frac{6}{10}$ are both simpler forms of $\frac{18}{30}$.

$\dfrac{18 \div 3}{30 \div 3} = \dfrac{6}{10}$ ← But both fractions still have other common factors.

$\dfrac{18 \div 6}{30 \div 6} = \dfrac{3}{5}$ ← $\frac{3}{5}$ is the *simplest form* of $\frac{18}{30}$.

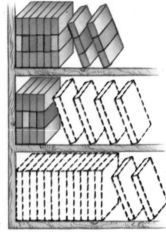

▶ A Shortcut to Find Simplest Form

Divide both the numerator and the denominator by the GCF *(greatest common factor)* of the fraction.

$\dfrac{18}{36} = \dfrac{?}{?}$ →→ Factors of 18 = { 1, 2, 3, 6, 9, **18** }
→→ Factors of 36 = { 1, 2, 3, 4, 6, 9, 12, **18** , 36 }

The common factors of 18 and 36 are: 1, 2, 3, 6, 9, and 18.

The greatest of these is 18, the GCF. $\dfrac{18 \div 18}{36 \div 18} = \dfrac{1}{2}$

Express in simplest form or lowest terms.*

1. $\frac{8}{16}$ 2. $\frac{9}{18}$ 3. $\frac{9}{24}$ 4. $\frac{6}{10}$ 5. $\frac{14}{18}$ 6. $\frac{19}{38}$ 7. $\frac{48}{60}$ 8. $\frac{24}{72}$

9. $\frac{30}{45}$ 10. $\frac{8}{12}$ 11. $\frac{40}{45}$ 12. $\frac{30}{35}$ 13. $\frac{81}{99}$ 14. $\frac{48}{60}$ 15. $\frac{27}{42}$ 16. $\frac{35}{50}$

17. $\frac{3}{1}$ means 3 ÷ 1 = 3. What is the denominator for any whole number?

18. Can zero be the denominator for any number? Explain.

19. Do the numerator and denominator of every fraction have a GCF? Explain.

20. Which fractions are in simplest form?

   a. $\frac{9}{10}$ **b.** $\frac{12}{15}$ **c.** $\frac{2}{8}$ **d.** $\frac{25}{27}$ **e.** none of these

Changing Improper Fractions to Mixed Numbers

Improper fraction, or mixed number: used to express a value equal to or greater than 1.

An improper fraction has a numerator that is greater than or equal to its denominator.

To change an improper fraction to a mixed number:

- Divide.
- Write any remainder as a fraction.

Improper Fractions

$\frac{11}{8}$ because $11 > 8$

$\frac{22}{6}$ because $22 > 6$

$\frac{14}{7}$ because $14 > 7$

$\frac{8}{8}$ because $8 = 8$

$\frac{11}{8}$ means $11 \div 8$

$8\overline{)11}$ = $1\frac{3}{8}$ ← remainder ← divisor, 1 R3

$\frac{11}{8} = 1\frac{3}{8}$

$\frac{22}{6}$ means $22 \div 6$

$6\overline{)22}$ = $3\frac{4}{6}$ = $3\frac{2}{3}$, 3 R4

Always express in lowest terms. $\frac{22}{6} = 3\frac{2}{3}$

$\frac{14}{7}$ means $14 \div 7$

$7\overline{)14}$, 2

$\frac{14}{7} = 2$

Which are improper fractions?

1. $\frac{7}{10}$ 2. $\frac{4}{7}$ 3. $\frac{13}{8}$ 4. $\frac{5}{4}$ 5. $\frac{9}{12}$

6. $\frac{18}{3}$ 7. $\frac{3}{3}$ 8. $\frac{17}{17}$ 9. $\frac{475}{575}$ 10. $\frac{21}{20}$

Change to a whole or mixed number.*

11. $\frac{14}{3}$ 12. $\frac{19}{9}$ 13. $\frac{15}{4}$ 14. $\frac{13}{5}$ 15. $\frac{30}{6}$

16. $\frac{28}{7}$ 17. $\frac{20}{19}$ 18. $\frac{17}{17}$ 19. $\frac{81}{9}$ 20. $\frac{13}{13}$

21. $\frac{66}{9}$ 22. $\frac{10}{3}$ 23. $\frac{10}{5}$ 24. $\frac{32}{4}$ 25. $\frac{27}{6}$

26. $\frac{58}{8}$ 27. $\frac{50}{9}$ 28. $\frac{32}{8}$ 29. $\frac{1}{1}$ 30. $\frac{54}{9}$

31. Which fraction above is a name for 2? Name another improper fraction that names 2.

32. Write three improper fractions that name the whole number 3.

33. Write an improper fraction with 7 as a denominator that is equivalent to a mixed number between 1 and 2.

34. Is $0 \div 1$ an improper fraction? Explain.

7-4 Comparing and Ordering Fractions

To compare fractions with *like denominators*:

Compare the numerators.

Compare $\frac{4}{5}$ and $\frac{3}{5}$ $\longrightarrow$ $4 > 3$ so, $\frac{4}{5} > \frac{3}{5}$

To compare fractions with *unlike denominators*:   Compare $\frac{4}{5}$ and $\frac{5}{8}$.

- Find the LCD.   LCD of $\frac{4}{5}$ and $\frac{5}{8}$:

  LCD is the least common multiple (LCM) of the denominators.

  Multiples of 5:   5, 10, 15, 20, 25, 30, 35, 40, ...

  Multiples of 8:   8, 16, 24, 32, 40, ...

  LCD of $\frac{4}{5}$ and $\frac{5}{8}$ is 40.

- Use the LCD to rewrite the fractions so the denominators are the same.

  Rewrite $\frac{4}{5}$ and $\frac{5}{8}$ with denominators of 40.

  $\frac{4}{5} \times \frac{8}{8} = \frac{32}{40}$ $\frac{5}{8} \times \frac{5}{5} = \frac{25}{40}$

- Compare the numerators. $32 > 25 \longrightarrow \frac{32}{40} > \frac{25}{40} \longrightarrow \frac{4}{5} > \frac{5}{8}$

Find the LCD for each pair of fractions.

1. $\frac{1}{3}$, $\frac{2}{5}$ 2. $\frac{3}{5}$, $\frac{2}{7}$ 3. $\frac{4}{9}$, $\frac{7}{16}$ 4. $\frac{5}{12}$, $\frac{7}{30}$ 5. $\frac{1}{6}$, $\frac{1}{10}$

Rewrite each pair so the denominators are the same.

6. $\frac{3}{8}$, $\frac{3}{16}$ 7. $\frac{3}{2}$, $\frac{7}{6}$ 8. $\frac{9}{3}$, $\frac{10}{9}$ 9. $\frac{3}{4}$, $\frac{3}{5}$ 10. $\frac{2}{3}$, $\frac{3}{4}$

Compare. Write $<$ or $>$.

11. $\frac{3}{8}$? $\frac{5}{8}$ 12. $\frac{19}{16}$? $\frac{10}{16}$ 13. $\frac{5}{30}$? $\frac{15}{30}$ 14. $\frac{12}{12}$? $\frac{21}{12}$ 15. $\frac{3}{5}$? $\frac{1}{5}$

16. $\frac{3}{4}$? $\frac{1}{4}$ 17. $\frac{7}{9}$? $\frac{2}{3}$ 18. $\frac{3}{2}$? $\frac{13}{6}$ 19. $\frac{7}{10}$? $\frac{4}{5}$ 20. $\frac{5}{12}$? $\frac{5}{8}$

21. $\frac{4}{5}$? $\frac{5}{7}$ 22. $\frac{2}{3}$? $\frac{3}{4}$ 23. $\frac{4}{5}$? $\frac{3}{4}$ 24. $\frac{7}{8}$? $\frac{8}{9}$ 25. $\frac{11}{11}$? $\frac{11}{12}$

26. $\frac{5}{9}$? $\frac{7}{12}$ 27. $\frac{8}{12}$? $\frac{5}{8}$ 28. $\frac{3}{10}$? $\frac{8}{25}$ 29. $\frac{20}{24}$? $\frac{20}{36}$ 30. $\frac{13}{14}$? $\frac{14}{28}$

Ordering Fractions

To order fractions:

- Rewrite the fractions so the denominators are the same.

- Compare the numerators.

Order: $\dfrac{2}{3}, \dfrac{5}{6}, \dfrac{1}{2}$

$\downarrow \quad \downarrow \quad \downarrow$

LCD is 6. $\dfrac{4}{6}, \dfrac{5}{6}, \dfrac{3}{6}$

From *least to greatest*:

$3 < 4 < 5 \longrightarrow \dfrac{3}{6} < \dfrac{4}{6} < \dfrac{5}{6}$

or $\dfrac{1}{2} < \dfrac{2}{3} < \dfrac{5}{6}$

From *greatest to least*:

$5 > 4 > 3 \longrightarrow \dfrac{5}{6} > \dfrac{4}{6} > \dfrac{3}{6}$

or $\dfrac{5}{6} > \dfrac{2}{3} > \dfrac{1}{2}$

Write in order from least to greatest.

31. $\dfrac{4}{5}, \dfrac{5}{8}, \dfrac{3}{4}$

32. $\dfrac{3}{5}, \dfrac{5}{3}, \dfrac{1}{2}, \dfrac{11}{10}$

33. $\dfrac{3}{4}, \dfrac{1}{2}, \dfrac{5}{6}, \dfrac{2}{3}$

34. $\dfrac{1}{3}, \dfrac{5}{9}, \dfrac{1}{6}, \dfrac{1}{2}$

35. $\dfrac{1}{2}, \dfrac{3}{10}, \dfrac{1}{4}, \dfrac{2}{5}$

36. $\dfrac{2}{3}, \dfrac{4}{5}, \dfrac{1}{2}, \dfrac{3}{4}$

Solve.

37. Mrs. Link bought stock at $8\frac{1}{2}$. Later she sold it at $8\frac{5}{8}$. Did she gain or lose on the sale?

38. A wheel on a tricycle turns $17\frac{2}{5}$ times over a certain distance. A wheel on another tricycle turns $17\frac{4}{9}$ times over the same distance. Which tricycle wheel is larger?

39. Ross painted three signs. On one sign he used letters $1\frac{3}{4}$ in. tall. On the second sign he used letters $\frac{17}{8}$ in. tall, and on the third the letters were $\frac{31}{16}$ in. tall. List the heights of the letters in order from shortest to tallest.

Exercise 39

$1\frac{3}{4}''$ letters

$\frac{17''}{8}$ letters

$\frac{31''}{16}$ letters

CRITICAL THINKING   **True or false? Explain your answers.**

40. When the denominators are the same, the fraction with the greater numerator is greater.

41. You can always find a common denominator by multiplying the two denominators.

42. You always get the LCD when you multiply the two denominators.

7-5 Fraction Sense

To determine whether a fraction is close to 0, $\frac{1}{2}$, or 1, examine these boxes.

Close to 0

A fraction *close to 0* has a very small number for a numerator when compared to its denominator.

$\frac{1}{10}$ $\frac{3}{20}$

1 is very small compared to 10.

3 is very small compared to 20.

Close to 1

A fraction *close to 1* has a numerator that is a little less than its denominator.

$\frac{9}{10}$ $\frac{17}{20}$

9 is a little less than 10.

17 is a little less than 20.

Close to $\frac{1}{2}$

A fraction *a little less than* $\frac{1}{2}$ has a numerator that is a little less than its denominator divided by 2.

$\frac{4}{10}$ $\frac{8}{20}$

4 is a little less than 10 divided by 2.

8 is a little less than 20 divided by 2.

Close to $\frac{1}{2}$

A fraction *a little more than* $\frac{1}{2}$ has a numerator that is a little more than its denominator divided by 2.

$\frac{6}{10}$ $\frac{12}{20}$

6 is a little more than 10 divided by 2.

12 is a little more than 20 divided by 2.

Which fractions are close to 0? Explain.

1. $\frac{3}{28}$ 2. $\frac{2}{19}$ 3. $\frac{9}{17}$ 4. $\frac{2}{3}$ 5. $\frac{5}{61}$ 6. $\frac{3}{40}$

Which fractions are close to 1? Explain.

7. $\frac{9}{11}$ 8. $\frac{5}{8}$ 9. $\frac{3}{40}$ 10. $\frac{12}{13}$ 11. $\frac{13}{15}$ 12. $\frac{11}{90}$

Which fractions are a little less than $\frac{1}{2}$? Explain.

13. $\frac{9}{14}$ 14. $\frac{10}{21}$ 15. $\frac{15}{32}$ 16. $\frac{5}{6}$ 17. $\frac{12}{26}$ 18. $\frac{23}{50}$

Which fractions are a little more than $\frac{1}{2}$? Explain.

19. $\frac{14}{25}$ 20. $\frac{7}{18}$ 21. $\frac{7}{12}$ 22. $\frac{23}{24}$ 23. $\frac{10}{95}$ 24. $\frac{43}{80}$

Replace the "?" with a number so that the resulting fraction is close to 1.

25. $\dfrac{?}{8}$ **26.** $\dfrac{12}{?}$ **27.** $\dfrac{?}{27}$ **28.** $\dfrac{90}{?}$ **29.** $\dfrac{15}{?}$ **30.** $\dfrac{?}{33}$

Replace the "?" with a number so that the resulting fraction is a little less than $\dfrac{1}{2}$.

31. $\dfrac{?}{24}$ **32.** $\dfrac{?}{7}$ **33.** $\dfrac{19}{?}$ **34.** $\dfrac{?}{60}$ **35.** $\dfrac{35}{?}$ **36.** $\dfrac{?}{150}$

Replace the "?" with a number so that the resulting fraction is a little more than $\dfrac{1}{2}$.

37. $\dfrac{?}{28}$ **38.** $\dfrac{21}{?}$ **39.** $\dfrac{?}{19}$ **40.** $\dfrac{30}{?}$ **41.** $\dfrac{44}{?}$ **42.** $\dfrac{?}{75}$

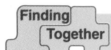 **Finding Together**   **How sharp is your fraction sense?**

Use a box of paper clips. Decide on a strategy you can use to determine:

43. How many paper clips you should link together to use a little less than half the box.

44. How many paper clips you should link together to use a little more than half the box.

45. How many paper clips you should link together to use almost the whole box. Think of another way you can solve this problem.

Write in order from least to greatest. Do not compute.

46. $\dfrac{12}{13}, \dfrac{5}{42}, \dfrac{10}{19}$ **47.** $\dfrac{8}{35}, \dfrac{19}{22}, \dfrac{21}{40}$ **48.** $\dfrac{27}{28}, \dfrac{11}{70}, \dfrac{31}{65}$

49. $\dfrac{3}{26}, \dfrac{11}{20}, \dfrac{14}{29}$ **50.** $\dfrac{18}{23}, \dfrac{4}{9}, \dfrac{13}{24}$ **51.** $\dfrac{24}{49}, \dfrac{8}{71}, \dfrac{30}{59}$

 **CALCULATOR ACTIVITY**   **Tell whether the fractions below are close to 0, $\dfrac{1}{2}$, or 1. Use a calculator to check your estimate. (The first one is done.)**

52. $\dfrac{9}{17}$ (close to $\dfrac{1}{2}$?) $9 \div 1 \; 7 = 0.5294118$ $0.5294118 \approx 0.5$

So, $\dfrac{9}{17}$ is close to $\dfrac{1}{2}$.

53. $\dfrac{2}{13}$ **54.** $\dfrac{12}{23}$ **55.** $\dfrac{27}{28}$ **56.** $\dfrac{4}{99}$ **57.** $\dfrac{22}{45}$ **58.** $\dfrac{87}{170}$

59. $\dfrac{73}{75}$ **60.** $\dfrac{3}{40}$ **61.** $\dfrac{18}{33}$ **62.** $\dfrac{19}{21}$ **63.** $\dfrac{7}{80}$ **64.** $\dfrac{24}{55}$

▶ **To add fractions with *like* denominators:**

- Add the *numerators.*
- Write the sum over the common denominator.
- Express in lowest terms.

Add: $\dfrac{2}{9} + \dfrac{4}{9} = \underline{\ ?\ }$

$$\dfrac{2}{9} + \dfrac{4}{9} = \dfrac{2+4}{9} = \dfrac{6}{9} = \dfrac{2}{3} \text{ (lowest terms)}$$

$$\dfrac{6}{9}$$

$$\dfrac{2}{9} \qquad \dfrac{4}{9}$$

$$\dfrac{2}{9} + \dfrac{4}{9} = \dfrac{6}{9}$$

▶ **To add fractions with *unlike* denominators:**

- Find the LCD.
- Rewrite the fractions with a common denominator.
- Add and express in lowest terms.

Add: $\dfrac{7}{10} + \dfrac{5}{6} = \underline{\ ?\ }$

Find the LCD of $\dfrac{7}{10}$ and $\dfrac{5}{6}$:

Multiples of 10: 10, 20, **30**, 40, . . .

Multiples of 6: 6, 12, 18, 24, **30**, 36, . . .

LCD is **30.**

Rewrite the fractions using the LCD of 30 and add.

$$\dfrac{7}{10} = \dfrac{21}{30}$$
$$+\dfrac{5}{6} = +\dfrac{25}{30}$$
$$\overline{\dfrac{46}{30}}$$

$$1\dfrac{16}{30}$$
$$30\overline{)46}$$

$$\dfrac{46}{30} = 1\dfrac{16}{30} = 1\dfrac{8}{15} \text{ (lowest terms)}$$

Always express the answer in simplest form or lowest terms.

Find the sum. Express the answer in lowest terms or simplest form.

1. $\dfrac{1}{12} + \dfrac{7}{12}$ 2. $\dfrac{7}{16} + \dfrac{5}{16}$ 3. $\dfrac{3}{10} + \dfrac{11}{10}$ 4. $\dfrac{5}{6} + \dfrac{5}{6}$

5. $\dfrac{8}{15} + \dfrac{7}{15}$ 6. $\dfrac{3}{7} + \dfrac{9}{7}$ 7. $\dfrac{13}{10} + \dfrac{7}{10}$ 8. $\dfrac{9}{14} + \dfrac{17}{14}$

Find the LCD for each set of fractions.

9. $\dfrac{2}{3}, \dfrac{1}{12}$ 10. $\dfrac{1}{4}, \dfrac{5}{8}$ 11. $\dfrac{3}{7}, \dfrac{2}{3}$ 12. $\dfrac{1}{3}, \dfrac{5}{8}$ 13. $\dfrac{2}{5}, \dfrac{3}{10}$

14. $\dfrac{3}{5}, \dfrac{5}{6}$ 15. $\dfrac{1}{2}, \dfrac{5}{8}, \dfrac{2}{3}$ 16. $\dfrac{1}{4}, \dfrac{2}{5}, \dfrac{4}{15}$ 17. $\dfrac{2}{5}, \dfrac{3}{8}, \dfrac{7}{10}$ 18. $\dfrac{1}{6}, \dfrac{5}{9}, \dfrac{7}{18}$

Find the sum. Write your answer in lowest terms or simplest form.

19. $\frac{5}{9} + \frac{1}{3}$

20. $\frac{2}{3} + \frac{11}{15}$

21. $\frac{5}{8} + \frac{7}{24}$

22. $\frac{6}{12} + \frac{4}{6}$

23. $\frac{1}{3} + \frac{3}{16}$

24. $\frac{7}{8} + \frac{2}{5}$

25. $\frac{2}{3} + \frac{3}{4}$

26. $\frac{8}{9} + \frac{1}{8}$

27. $\frac{1}{2} + \frac{4}{9}$

28. $\frac{4}{5} + \frac{1}{4}$

29. $\frac{5}{8} + \frac{1}{3}$

30. $\frac{5}{6} + \frac{4}{5}$

31. $\frac{7}{8} + \frac{3}{10}$

32. $\frac{9}{10} + \frac{4}{15}$

33. $\frac{4}{9} + \frac{5}{12}$

34. $\begin{array}{r} \frac{1}{15} \\ \frac{7}{10} \\ +\frac{2}{15} \\ \hline \end{array}$

35. $\begin{array}{r} \frac{2}{7} \\ \frac{5}{14} \\ +\frac{3}{7} \\ \hline \end{array}$

36. $\begin{array}{r} \frac{1}{3} \\ \frac{3}{5} \\ +\frac{4}{5} \\ \hline \end{array}$

37. $\begin{array}{r} \frac{1}{6} \\ \frac{4}{9} \\ +\frac{1}{3} \\ \hline \end{array}$

38. $\begin{array}{r} \frac{5}{12} \\ \frac{1}{18} \\ +\frac{5}{6} \\ \hline \end{array}$

Solve.

39. If one book is $\frac{5}{8}$ inch thick, a second is $\frac{3}{4}$ inch thick, and a third is $\frac{5}{16}$ inch thick, what will be the width of the three books?

Exercise 39

40. A high jump bar is set at 61 inches. The bar was raised $\frac{1}{2}$ inch two times. How high is the bar set?

41. A shuttle astronaut worked 1 hour repairing a satellite on the first day. On each of the next two days he worked $\frac{5}{6}$ hour repairing it. How long did the astronaut work on the satellite?

42. In preparing a fruit tray a chef used $\frac{3}{4}$ lb of strawberries, 1 lb of grapes, and $\frac{7}{8}$ lb of bananas. How many pounds of fruit did the chef use?

43. A frog leaped $\frac{2}{3}$ yd, $\frac{3}{4}$ yd, and $\frac{5}{6}$ yd. A different frog leaped a total of $2\frac{1}{2}$ yd. Which frog leaped farther?

$\frac{5''}{8} \quad \frac{3''}{4} \quad \frac{5''}{16}$

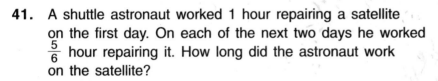

MENTAL ◆ MATH

Write <, =, or > to make a true sentence.

44. $\frac{1}{8} + \frac{5}{8}$ __?__ 1

45. $\frac{1}{6} + \frac{5}{6}$ __?__ 1

46. $\frac{1}{10} + \frac{2}{5} + \frac{3}{10}$ __?__ $\frac{4}{10} + \frac{2}{5}$

47. $\frac{1}{10} + \frac{2}{5} + \frac{3}{10}$ __?__ $\frac{7}{10}$

48. $\frac{7}{24} + \frac{5}{8}$ __?__ $\frac{7}{24}$

49. $\frac{3}{4} + \frac{1}{2} + \frac{1}{4}$ __?__ $1\frac{3}{4}$

50. $\frac{3}{10} + \frac{9}{20}$ __?__ $\frac{9}{20}$

51. $\frac{3}{8} + \frac{1}{8}$ __?__ $\frac{1}{2}$

52. $\frac{2}{9} + \frac{1}{3} + \frac{5}{9}$ __?__ $\frac{7}{9}$

161

Subtracting Fractions

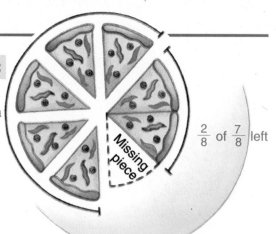

$\frac{5}{8}$ of pizza taken

$\frac{2}{8}$ of $\frac{7}{8}$ left

Missing piece

$\frac{7}{8}$ of a pizza on a plate

To subtract fractions with *like denominators*:

- Subtract the *numerators*.
- Write the difference over the common denominator.
- Express in lowest terms.

$\frac{7}{8}$ of a pizza is on a plate. $\frac{5}{8}$ is taken. How many eighths are left?

Subtract: $\dfrac{7}{8} - \dfrac{5}{8} = \underline{\ ?\ }$

$$\frac{7}{8} - \frac{5}{8} = \frac{7-5}{8} = \frac{2}{8} = \frac{1}{4} \text{ (lowest terms)}$$

To subtract fractions with *unlike denominators*:

- Find the LCD.
- Rewrite the fractions with a common denominator.
- Subtract and express in lowest terms.

Subtract: $\dfrac{17}{18} - \dfrac{4}{9} = \underline{\ ?\ }$

Find the LCD of $\dfrac{17}{18}$ and $\dfrac{4}{9}$:

Multiples of 18: 18, 36, …

Multiples of 9: 9, 18, 27, 36, …

LCD is 18.

Rewrite the fractions, using the LCD of 18 and subtract.

$$\frac{17}{18} = \frac{17}{18}$$
$$-\frac{4}{9} = -\frac{8}{18}$$
$$\overline{\phantom{-\frac{4}{9}=}\ \frac{9}{18}} = \frac{1}{2} \text{ (lowest terms)}$$

Find the difference. Express in lowest terms.

1. $\dfrac{5}{6} - \dfrac{1}{6}$
2. $\dfrac{11}{12} - \dfrac{4}{12}$
3. $\dfrac{9}{15} - \dfrac{7}{15}$
4. $\dfrac{9}{10} - \dfrac{6}{10}$
5. $\dfrac{5}{7} - \dfrac{2}{7}$

Find equivalent fractions mentally. (This will help you estimate sums and differences.)

6. $\dfrac{1}{2} = \dfrac{?}{4}$
7. $\dfrac{1}{3} = \dfrac{?}{9}$
8. $\dfrac{1}{5} = \dfrac{?}{10}$
9. $\dfrac{1}{4} = \dfrac{?}{12}$
10. $\dfrac{1}{6} = \dfrac{?}{18}$

11. $\dfrac{1}{2} = \dfrac{?}{12}$
12. $\dfrac{1}{3} = \dfrac{?}{12}$
13. $\dfrac{1}{2} = \dfrac{?}{10}$
14. $\dfrac{1}{2} = \dfrac{?}{8}$
15. $\dfrac{1}{7} = \dfrac{?}{21}$

Find the difference and express in lowest terms.

16. $\frac{4}{5} - \frac{1}{4}$ **17.** $\frac{1}{3} - \frac{1}{11}$ **18.** $\frac{2}{3} - \frac{1}{7}$ **19.** $\frac{5}{7} - \frac{2}{3}$

20. $\frac{7}{8} - \frac{3}{10}$ **21.** $\frac{7}{9} - \frac{1}{6}$ **22.** $\frac{2}{9} - \frac{2}{15}$ **23.** $\frac{8}{9} - \frac{1}{3}$

24. $\begin{array}{r} \frac{5}{14} \\ -\frac{1}{8} \\ \hline \end{array}$ **25.** $\begin{array}{r} \frac{3}{8} \\ -\frac{4}{24} \\ \hline \end{array}$ **26.** $\begin{array}{r} \frac{1}{2} \\ -\frac{5}{12} \\ \hline \end{array}$ **27.** $\begin{array}{r} \frac{9}{10} \\ -\frac{4}{15} \\ \hline \end{array}$

28. $\begin{array}{r} \frac{3}{4} \\ -\frac{2}{5} \\ \hline \end{array}$ **29.** $\begin{array}{r} \frac{4}{5} \\ -\frac{5}{7} \\ \hline \end{array}$ **30.** $\begin{array}{r} \frac{7}{8} \\ -\frac{2}{3} \\ \hline \end{array}$ **31.** $\begin{array}{r} \frac{1}{2} \\ -\frac{5}{11} \\ \hline \end{array}$

Solve.

32. If a pattern requiring $\frac{7}{8}$ of a yard of material is cut from a 1-yard piece, how much material is left?

33. Karen has $\frac{5}{8}$ of a cup of raisins. If she puts $\frac{1}{4}$ of a cup in her cake, what part of the cup of raisins will she have left?

34. Jeoff bought $\frac{3}{4}$ lb of nuts. He ate $\frac{1}{2}$ lb. What part of a pound is left?

35. A house had 18 windows. On Friday David washed 6 windows and on Saturday he washed 4 windows. What fractional part of all the windows did he wash each day? What fractional part was still *unwashed* at the end of the day on Saturday?

Exercise 33

$\frac{5}{8}$ cup of raisins

$\frac{1}{4}$ cup of raisins used in cake

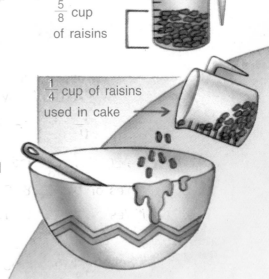

MENTAL ◆ MATH

Write <, =, or > to make a true sentence.

36. $\frac{7}{8} - \frac{1}{2} \underline{\ ?\ } \frac{1}{2}$ **37.** $\frac{1}{2} - \frac{1}{4} \underline{\ ?\ } \frac{1}{2}$ **38.** $\frac{7}{12} - \frac{1}{2} \underline{\ ?\ } \frac{1}{2}$ **39.** $\frac{6}{8} - \frac{1}{4} \underline{\ ?\ } \frac{1}{2}$

40. $\frac{11}{12} - \frac{1}{3} \underline{\ ?\ } \frac{1}{2}$ **41.** $\frac{9}{10} - \frac{1}{5} \underline{\ ?\ } \frac{1}{2}$ **42.** $\frac{9}{12} - \frac{1}{3}, \underline{\ ?\ } \frac{1}{2}$ **43.** $\frac{5}{6} - \frac{1}{3} \underline{\ ?\ } \frac{1}{2}$

Subtract. (Hint: Remember all the fraction names for one.)

44. $1 - \frac{2}{5}$ **45.** $1 - \frac{7}{9}$ **46.** $1 - \frac{4}{7}$ **47.** $1 - \frac{17}{20}$

48. $1 - \frac{3}{4}$ **49.** $1 - \frac{7}{12}$ **50.** $1 - \frac{14}{16}$ **51.** $1 - \frac{25}{50}$

Write $2\frac{4}{5}$ as an improper fraction.

$$2\frac{4}{5} = 2 \text{ and } \frac{4}{5} = 2 + \frac{4}{5}$$

Think: How many fifths are in 2?

If $\frac{5}{5} = 1$, then $\frac{10}{5} = 2$.

So, $2\frac{4}{5} = \boxed{2} + \frac{4}{5}$

$$2\frac{4}{5} = \frac{\boxed{10}}{5} + \frac{4}{5} = \frac{14}{5}$$

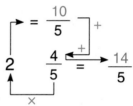

To express a mixed number as an improper fraction:
- Multiply the denominator of the fraction and the whole number.
- Add the numerator of the fraction to this product.

Or, $2\frac{4}{5} = \frac{(5 \times 2) + 4}{5} = \frac{14}{5}$

> 5 fifths times 2 is 10 fifths; plus 4 fifths equals 14 fifths.

Express $3\frac{5}{8}$ as an improper fraction.

$$3\frac{5}{8} = \frac{29}{8} \quad \text{or} \quad 3\frac{5}{8} = \frac{(8 \times 3) + 5}{8} = \frac{29}{8}$$

Copy and complete. Write the improper fraction.

1. $2\frac{3}{7} = \frac{(7 \times 2) + 3}{7}$
   $2\frac{3}{7} = \frac{?}{7}$

2. $4\frac{2}{5} = \frac{(5 \times 4) + 2}{5}$
   $4\frac{2}{5} = \frac{?}{5}$

3. $5\frac{2}{3} = \frac{(3 \times 5) + 2}{3}$
   $5\frac{2}{3} = \frac{?}{3}$

Change the mixed number to an improper fraction.*

4. $1\frac{2}{5}$ 5. $1\frac{1}{2}$ 6. $1\frac{5}{8}$ 7. $1\frac{7}{12}$ 8. $1\frac{3}{10}$ 9. $1\frac{2}{3}$

10. $2\frac{2}{9}$ 11. $5\frac{1}{2}$ 12. $6\frac{5}{8}$ 13. $10\frac{7}{12}$ 14. $17\frac{3}{10}$ 15. $25\frac{1}{2}$

Compare. Write <, =, or >.

16. $3\frac{1}{2}$ ___?___ $\frac{14}{4}$ 17. $\frac{25}{3}$ ___?___ $8\frac{1}{3}$ 18. $4\frac{2}{5}$ ___?___ $\frac{28}{10}$ 19. $5\frac{1}{4}$ ___?___ $\frac{21}{8}$

20. $\frac{13}{6}$ ___?___ $3\frac{2}{3}$ 21. $\frac{15}{7}$ ___?___ $2\frac{1}{7}$ 22. $3\frac{3}{4}$ ___?___ $7\frac{1}{2}$ 23. $\frac{18}{8}$ ___?___ $2\frac{2}{16}$

24. $5\frac{1}{5}$ ___?___ $\frac{52}{10}$ 25. $9\frac{1}{6}$ ___?___ $\frac{26}{3}$ 26. $10\frac{1}{2}$ ___?___ $\frac{42}{4}$ 27. $\frac{33}{3}$ ___?___ $11\frac{1}{3}$

Subtracting Fractions from Whole Numbers

To subtract a fraction from a whole number:

- Rewrite the whole number as a mixed number.
- The fraction part of the mixed number should have the same denominator as the fraction.
- Then subtract.

Subtract: $7 - \dfrac{3}{5} = $ ___?___

Think: $7 = 6 + 1$

$7 = 6 + \dfrac{5}{5}$

Regroup 7 as $6\dfrac{5}{5}$

$$
\begin{array}{r}
7 = 6\dfrac{5}{5} \\
- \dfrac{3}{5} = -\dfrac{3}{5} \\
\hline
6\dfrac{2}{5}
\end{array}
$$

Copy and complete the regrouping.

1. $6 = 5\dfrac{?}{7}$ 2. $11 = 10\dfrac{?}{8}$ 3. $5 = 4\dfrac{?}{16}$ 4. $2 = 1\dfrac{?}{12}$ 5. $25 = \dfrac{?}{}\,\dfrac{?}{2}$

Subtract.*

6. $\begin{array}{r} 6 \\ -\dfrac{1}{3} \\ \hline \end{array}$ 7. $\begin{array}{r} 4 \\ -\dfrac{1}{8} \\ \hline \end{array}$ 8. $\begin{array}{r} 10 \\ -\dfrac{1}{5} \\ \hline \end{array}$ 9. $\begin{array}{r} 8 \\ -\dfrac{5}{6} \\ \hline \end{array}$ 10. $\begin{array}{r} 14 \\ -\dfrac{2}{3} \\ \hline \end{array}$

11. $\begin{array}{r} 7 \\ -\dfrac{4}{5} \\ \hline \end{array}$ 12. $\begin{array}{r} 13 \\ -\dfrac{2}{9} \\ \hline \end{array}$ 13. $\begin{array}{r} 5 \\ -\dfrac{1}{4} \\ \hline \end{array}$ 14. $\begin{array}{r} 11 \\ -\dfrac{3}{7} \\ \hline \end{array}$ 15. $\begin{array}{r} 16 \\ -\dfrac{6}{11} \\ \hline \end{array}$

16. $24 - \dfrac{4}{9}$ 17. $12 - \dfrac{1}{4}$ 18. $9 - \dfrac{1}{6}$ 19. $14 - \dfrac{2}{3}$ 20. $18 - \dfrac{7}{8}$

21. $3 - \dfrac{1}{7}$ 22. $17 - \dfrac{3}{23}$ 23. $33 - \dfrac{11}{19}$ 24. $29 - \dfrac{5}{14}$ 25. $46 - \dfrac{8}{9}$

Solve.

26. Mario designed a stained-glass window with three sections. Each section was made from 12 pieces of glass. He still has 6 pieces of glass to fill in. How much of his window is finished?

27. Ms. Oslow set her alarm to get 8 hours of sleep. If it took her 1 hour to fall asleep and she woke up 30 minutes before the alarm rang, how long did she sleep?

 ◁▷◁ **MAKE UP YOUR OWN...** ▷◁▷

28. Write a word problem in which you subtract a fraction from a whole number.

7-10 Mixed Numbers: Rounding and Estimating

To round a mixed number when its fraction part is *greater than or equal to* $\frac{1}{2}$, round the whole-number part UP to the next whole number and drop the fraction.

Round $2\frac{11}{18}$.

$\frac{11}{18} > \frac{1}{2}$ So, round 2 UP to 3.

$2\frac{11}{18}$ is rounded to 3.

To round a mixed number when its fraction part is *less than* $\frac{1}{2}$, leave the the whole-number part alone and drop the fraction.

Round $4\frac{7}{17}$.

$\frac{7}{17} < \frac{1}{2}$ So, leave 4 unchanged.

$4\frac{7}{17}$ is rounded to 4.

To estimate sums and differences with mixed numbers:

- Round the mixed numbers to the nearest whole number.
- Then add or subtract.

Estimate: $5\frac{2}{7} + 1\frac{5}{6} \approx$?

  $\quad\quad\quad 5 + 2 = 7$ Estimate

Estimate: $6\frac{3}{4} - 3\frac{2}{5} \approx$?

  $\quad\quad\quad 7 - 3 = 4$ Estimate

Round each mixed number to the nearest whole number.

1. $4\frac{1}{3}$ 2. $5\frac{7}{9}$ 3. $8\frac{3}{7}$ 4. $7\frac{1}{2}$ 5. $3\frac{3}{5}$ 6. $9\frac{4}{11}$

Estimate.*

7. $\begin{array}{r} 6\frac{3}{4} \\ +9\frac{2}{3} \\ \hline \end{array}$ 8. $\begin{array}{r} 3\frac{5}{9} \\ +4\frac{5}{8} \\ \hline \end{array}$ 9. $\begin{array}{r} 7\frac{5}{12} \\ -4\frac{7}{10} \\ \hline \end{array}$ 10. $\begin{array}{r} 6\frac{1}{5} \\ -2\frac{5}{6} \\ \hline \end{array}$ 11. $\begin{array}{r} 10\frac{8}{9} \\ +16\frac{2}{3} \\ \hline \end{array}$ 12. $\begin{array}{r} 33\frac{1}{6} \\ +17\frac{6}{15} \\ \hline \end{array}$

13. $\begin{array}{r} 8\frac{8}{9} \\ +2\frac{1}{8} \\ \hline \end{array}$ 14. $\begin{array}{r} 9\frac{1}{2} \\ +4\frac{3}{10} \\ \hline \end{array}$ 15. $\begin{array}{r} 6\frac{9}{10} \\ -3\frac{1}{11} \\ \hline \end{array}$ 16. $\begin{array}{r} 11\frac{5}{6} \\ -7\frac{1}{9} \\ \hline \end{array}$ 17. $\begin{array}{r} 8\frac{1}{12} \\ -5\frac{1}{15} \\ \hline \end{array}$ 18. $\begin{array}{r} 9\frac{1}{7} \\ -1\frac{2}{12} \\ \hline \end{array}$

SKILLS TO REMEMBER

Compute using the order of operations.

19. $3.05 + (2.6 - 1.9)$ 20. $9.1 - (3.4 + 1.5)$ 21. $(5.02 - 1.6) + 2.3$

22. $(4.25 - 0.7) - 1.2$ 23. $(0.9 + 6.3) - 1.7$ 24. $6 - (3.7 - 2.9)$

7-11 Adding Mixed Numbers

To add mixed numbers when the denominators are the *same*:

- Add the fractions.
- Then add the whole numbers.

Add: $4\frac{3}{8} + 2\frac{1}{8} = $ __?__

$4\frac{3}{8} \longrightarrow 4 \quad \frac{3}{8}$

$+2\frac{1}{8} \longrightarrow +2 \quad +\frac{1}{8}$

$\overline{6 \quad \frac{4}{8}} = 6 + \frac{1}{2} = 6\frac{1}{2}$

> Always express in lowest terms.

When the denominators are *different*:

- Find the LCD of the fractions.
- Express the fractions as equivalent fractions with the LCD as the denominator.
- Add the fractions.
- Then add the whole numbers.

Add: $5\frac{1}{5} + 3\frac{2}{3} = $ __?__   **Estimate:** $5 + 4 = 9$

$5\frac{1}{5} \longleftarrow \boxed{\text{Find LCD (15)}} = 5\frac{3}{15} \longrightarrow 5\frac{3}{15}$

$+3\frac{2}{3} \longleftarrow \phantom{\boxed{\text{Find LCD (15)}}} = +3\frac{10}{15} \longrightarrow +3\frac{10}{15}$

$\overline{\phantom{+3\frac{10}{15}}8\frac{13}{15}} = 8\frac{13}{15}$

Add: $4\frac{3}{6} + 2\frac{7}{9} = $ __?__   **Estimate:** $5 + 3 = 8$

$4\frac{3}{6} \longleftarrow \boxed{\text{Find LCD (18)}} = 4\frac{9}{18} \longrightarrow 4\frac{9}{18}$

$+2\frac{7}{9} \longleftarrow \phantom{\boxed{\text{Find LCD (18)}}} = +2\frac{14}{18} \longrightarrow +2\frac{14}{18}$

$\overline{\phantom{+2\frac{14}{18}}6\frac{23}{18}} = 6 + \left(1\frac{5}{18}\right) = 7\frac{5}{18}$

> Change the improper fraction to a mixed number or a whole number. Then add.

Estimate. Then find the sum.* Write the answer in simplest form.

1. $3\frac{2}{7}$
   $+10\frac{3}{7}$

2. $19\frac{2}{5}$
   $+ 6\frac{2}{5}$

3. $1\frac{5}{18}$
   $+7\frac{7}{18}$

4. $7\frac{1}{4}$
   $+8\frac{2}{4}$

5. $12\frac{3}{8}$
   $+ 5\frac{7}{8}$

6. $49\frac{5}{12}$
   $+33\frac{8}{12}$

7. $52\frac{2}{15}$
   $+23\frac{4}{5}$

8. $72\frac{7}{36}$
   $+29\frac{5}{18}$

9. $88\frac{4}{11}$
   $+44\frac{3}{44}$

10. $100\frac{9}{10}$
    $+ 20\frac{13}{20}$

11. $18\frac{1}{2}$
    $+ 7\frac{3}{8}$

12. $26\frac{1}{6}$
    $+13\frac{2}{3}$

13. $42\frac{1}{3}$
    $+28\frac{4}{7}$

14. $29\frac{3}{4}$
    $+17\frac{3}{16}$

15. $34\frac{5}{6}$
    $+10\frac{2}{9}$

7-12 Subtracting Mixed Numbers

To subtract mixed numbers when the denominators are the *same*:

- Subtract the fractions.
- Then subtract the whole numbers.
- Express the difference in lowest terms.

Subtract: $9\frac{7}{16} - 5\frac{3}{16} = $ **?**

Estimate: $9 - 5 = 4$

$$9\frac{7}{16} \longrightarrow 9 \qquad \frac{7}{16}$$
$$-5\frac{3}{16} \longrightarrow -5 \qquad -\frac{3}{16}$$
$$\overline{} \qquad \overline{}$$
$$4 \qquad \frac{4}{16} = 4\frac{4}{16} = 4\frac{1}{4}$$

Subtract the fractions first.

When the denominators are *different*:

- Find the LCD of the fractions.
- Change the fractions to equivalent fractions with LCD as denominator.
- Subtract the fractions.
- Then subtract the whole numbers.

Subtract: $2\frac{7}{10} - 1\frac{1}{2} = $ **?**

Estimate: $3 - 2 = 1$

$$2\frac{7}{10} \qquad \text{Find} \qquad 2\frac{7}{10} = 2 \qquad \frac{7}{10}$$
$$-1\frac{1}{2} \qquad \text{LCD} \qquad -1\frac{5}{10} = -1 \qquad -\frac{5}{10}$$
$$\qquad (10) \qquad \overline{} \qquad \overline{}$$
$$1 \qquad \frac{2}{10} = 1\frac{1}{5}$$

lowest terms

Subtract: $7\frac{1}{4} - 3\frac{2}{3} = $ **?**

$$7\frac{1}{4} \qquad \text{Find} \quad 7\frac{3}{12} = 7 \qquad \frac{3}{12}$$
$$-3\frac{2}{3} \qquad \text{LCD} \quad -3\frac{8}{12} = -3 \qquad -\frac{8}{12}$$
$$\qquad (12)$$

Need to regroup 7.

$$7 = 6 + 1 = 6 + \frac{12}{12}$$
$$\frac{3}{12} + \frac{12}{12} = \frac{15}{12}$$

$$6 \qquad \frac{15}{12}$$
$$-3 \qquad -\frac{8}{12}$$
$$\overline{} \qquad \overline{}$$
$$3 \qquad \frac{7}{12} = 3\frac{7}{12}$$

Estimate. Then find the difference. Write the answer in simplest form.

1. $3\frac{5}{6}$
$-1\frac{1}{6}$

2. $7\frac{4}{7}$
$-2\frac{1}{7}$

3. $10\frac{3}{12}$
$-\ 3\frac{1}{10}$

4. $17\frac{7}{9}$
$-\ 5\frac{4}{9}$

5. $34\frac{3}{4}$
$-18\frac{1}{8}$

6. $33\frac{4}{5}$
$-14\frac{6}{15}$

Complete the regrouping.

7. $3\frac{3}{5} = 2\frac{?}{5}$

8. $7\frac{5}{8} = 6\frac{?}{8}$

9. $4\frac{4}{9} = \underline{\ ?\ }\frac{13}{9}$

10. $12\frac{1}{4} = \underline{\ ?\ }\frac{5}{4}$

11. $4\frac{4}{5} = 4\frac{?}{10} = 3\frac{?}{10}$

12. $3\frac{5}{6} = 3\frac{?}{12} = 2\frac{?}{12}$

13. $7\frac{3}{8} = 7\frac{?}{24} = 6\frac{?}{24}$

Subtract.

14. $8\frac{5}{10}$
$-2\frac{7}{10}$

15. $15\frac{3}{8}$
$-\ 9\frac{7}{8}$

16. $23\frac{5}{12}$
$-22\frac{7}{12}$

17. $18\frac{1}{2}$
$-\ 9\frac{3}{4}$

18. $26\frac{3}{5}$
$-10\frac{4}{15}$

19. $28\frac{1}{7}$
$-17\frac{3}{4}$

Use familiar rounding strategies to estimate differences.

20. $12\frac{3}{8} - 7\frac{5}{6}$ is about ___?___   **a.** 4   **b.** 5   **c.** 6   **d.** 7

21. $6\frac{7}{10} - 4\frac{3}{5}$ is about ___?___   **a.** 2   **b.** 3   **c.** 4   **d.** $3\frac{1}{2}$

22. $26\frac{5}{6} - 10\frac{3}{12}$ is about ___?___   **a.** $15\frac{1}{2}$ **b.** 16   **c.** $16\frac{1}{2}$ **d.** 17

Estimate. Then subtract and compare.

23. $4\frac{1}{2}$ 24. $6\frac{2}{5}$ 25. $10\frac{5}{6}$ 26. $5\frac{2}{3}$ 27. $9\frac{2}{5}$ 28. $8\frac{5}{6}$ 29. $12\frac{4}{7}$
   $-1\frac{1}{3}$ $-2\frac{1}{4}$ $-2\frac{5}{9}$ $-4\frac{3}{4}$ $-7\frac{2}{3}$ $-7\frac{4}{5}$ $-7\frac{5}{6}$

30. $8\frac{3}{4}$ 31. $4\frac{1}{5}$ 32. $7\frac{3}{8}$ 33. $6\frac{1}{2}$ 34. $10\frac{2}{9}$ 35. $7\frac{2}{5}$ 36. $9\frac{1}{6}$
   $-4\frac{5}{6}$ $-3\frac{1}{2}$ $-4\frac{2}{5}$ $-1\frac{3}{7}$ $-7\frac{2}{5}$ $-5\frac{3}{7}$ $-8\frac{3}{8}$

37. $10\frac{5}{12} - 8\frac{2}{9}$ 38. $6\frac{4}{5} - 3\frac{2}{3}$ 39. $7\frac{1}{10} - 3\frac{5}{6}$ 40. $8\frac{1}{10} - 2\frac{3}{8}$

Compare. Write <, =, or >. (Hint: Use order of operations.)

41. $3\frac{1}{4} + (6\frac{1}{2} - 3\frac{3}{4})$ ___?___ $3\frac{1}{4} + 6\frac{1}{2} - 3\frac{3}{4}$ 42. $7\frac{4}{5} + 8\frac{1}{3} - 7\frac{4}{5}$ ___?___ $8\frac{2}{3}$

43. $9\frac{1}{2} - 6\frac{1}{4} + 2\frac{3}{8}$ ___?___ $9\frac{1}{2} - (6\frac{1}{4} + 2\frac{3}{8})$ 44. $1\frac{1}{5} + 2\frac{1}{8} + 1\frac{4}{5}$ ___?___ 5

45. $(5\frac{5}{9} - 2\frac{7}{18}) - 1\frac{1}{3}$ ___?___ $5\frac{5}{9} - (2\frac{7}{18} - 1\frac{1}{3})$ 46. $(2\frac{1}{10} + 3\frac{2}{5}) - \frac{2}{5}$ ___?___ $5\frac{1}{10}$

Solve.

47. Subtract 3 from the sum of $8\frac{5}{6}$ and $4\frac{5}{9}$.

48. Add $1\frac{5}{8}$ to the difference between 6 and $2\frac{7}{12}$.

49. The Urban Bank's money market rate is $\frac{3}{5}$ higher than Suburban Bank's and $\frac{1}{8}$ lower than Farmer's Bank. If Urban Bank's rate is $9\frac{1}{4}$, what are the rates of the other two banks?

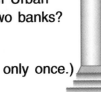

CHALLENGE   Solve. (A number from each replacement set may be used only once.)

50. $\dfrac{1}{\square} + \dfrac{\square}{\square} \approx \dfrac{1}{2}$ when $\square$ = {2, 3, 4, 5}

51. $\dfrac{9}{\square} - \dfrac{\square}{\square} \approx \dfrac{1}{2}$ when $\square$ = {5, 6, 7, 8}

52. $\dfrac{\square}{\square} + \dfrac{\square}{\square} \approx 1$ when $\square$ = {1, 3, 5, 7}

53. $\dfrac{\square}{\square} - \dfrac{\square}{\square} \approx 0$ when $\square$ = {3, 4, 6, 7}

TECHNOLOGY

Spreadsheet Application

The Bayou Reunion

Five former Bayou residents traveled to the Bayou Reunion.
Swamp Fox used a spreadsheet program to find their arrival time.

A	B	C	D
Friend	**Distance to Travel**	**Rate of Speed**	**Time/Hours**
Al Ligator	50 miles	0.075 mph	666.7 h
Frida Frog	175 miles	1.2 mph	145.8 h
Creanne Crawfish	96 miles	3.6 mph	26.7 h
Ted Turtle	115 miles	0.23 mph	500.0 h
Lonnie Lizard	520 miles	12 mph	43.3 h

Swamp Fox then used the data in the spreadsheet to estimate the arrival
time of each guest. Use the spreadsheet information to answer the following.

1. If all the animals left their homes at the same time on the same day,
   what was the order of their arrival?

2. Was the first guest to arrive the fastest-moving animal?

3. Was the last guest to arrive the slowest-moving animal?

4. Did the first arrival have the greatest distance to travel?

5. Use a calculator or spreadsheet to determine how many days it took
   each guest to arrive at the reunion.

Al Ligator invited the other four traveling friends to stop at his home for a visit.
Use a calculator or spreadsheet program to determine how many days it will take
each friend to travel to Al Ligator's house. Complete the chart below.

	A	B	C	D
	Friend	**Distance**	**Rate of Speed**	**Time/Hours**
6.	Frida Frog	120 miles	1.2 mph	?
7.	Creanne Crawfish	100 miles	3.6 mph	?
8.	Ted Turtle	92 miles	0.23 mph	?
9.	Lonnie Lizard	450 miles	12 mph	?

The Gulf States

Five states border the Gulf of Mexico. The following information was entered into a spreadsheet program.

A	B	C	D	E	F
State	**Total Area**	**Land Area**	**%**	**Water Area**	**%**
Florida	58,664 sq mi	54,153 sq mi	92.3	4511 sq mi	7.7
Alabama	51,704 sq mi	50,708 sq mi	98.1	996 sq mi	1.9
Mississippi	47,689 sq mi	47,233 sq mi	99.0	456 sq mi	1.0
Louisiana	47,752 sq mi	44,521 sq mi	93.2	3231 sq mi	6.8
Texas	266,807 sq mi	262,017 sq mi	98.2	4790 sq mi	1.8

Use the spreadsheet to answer exercises 10–12.

10. Which state has the smallest land area?
Does it have the lowest percentage of land?

11. Which state has the greatest water area?
Does it have the highest percentage of water?

12. Which two states are likely to have many water sports businesses?

13. Find the sum of the percent of land area and water area for each state.

14. Explain why the sums in exercise 13 equal 100%.

Read the following chart. Then use a calculator or spreadsheet program to compute the number of square miles of forested land for each state.
Remember: 640 acres = 1 sq mi

	State	**Forest Area/Acres**	**Forest Area/Sq Mi**
15.	Florida	17,039,700	?
16.	Alabama	21,361,100	?
17.	Mississippi	16,715,600	?
18.	Louisiana	14,558,100	?
19.	Texas	23,279,300	?

STRATEGY
7-14 Problem Solving: Logical Reasoning

Problem: A math magazine offers a reward for solving this cryptic message. Help Toni win the prize.

Win a Computer!
2 grumps equal 4 snooks and 3 trells equal 12 snooks. How many trells make 16 grumps?

1 IMAGINE Instead of inches and feet, or pints and quarts, or kilometers and meters, you have grumps, snooks, and trells. Draw and label them.

2 NAME

Facts: 2 grumps = 4 snooks
3 trells = 12 snooks

Question: __?__ trells = 16 grumps

TRELL

3 THINK

To solve the problem, set up this table and use logical reasoning:
Let G = grumps, T = trells, and S = snooks.

$$2G = 4S$$
Since $2G = 4S$ (Divide by 2.)
then $1G = 2S$

$$3T = 12S$$
Since $3T = 12S$ (Divide by 3.)
then $1T = 4S$ (Divide by 2.)
and $\frac{1}{2}T = 2S$

GRUMP

SNOOK

4 COMPUTE Combine both equations: $\frac{1}{2}T = 2S = 1G$

So, $\frac{1}{2}T = 1G$ (Multiply both sides by 16.)

$8T = 16G$

So, 8 trells make 16 grumps.

5 CHECK Does your answer make sense?
Use the original equations and substitute to check.

$$\frac{1}{2}T = 1G$$

If 8 trells = 16 grumps,
then 1 trell = 2 grumps, and
$\frac{1}{2}$ trell = 1 grump
So, $\frac{1}{2}T$ = 1G

Solve using logical reasoning.

1. Mr. Dibs put up four shelves to hold his books. He had four types of books: biographies, history books, mysteries, and reference books. Find the order from bottom to top in which he placed his books, given that each type of book occupies one shelf and: the reference books are on the bottom shelf; the mysteries are above the biographies; the history books are two shelves above the reference books.

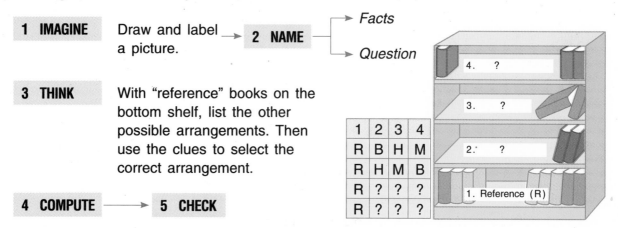

| 1 IMAGINE | Draw and label a picture. | → 2 NAME | → Facts
→ Question |

| 3 THINK | With "reference" books on the bottom shelf, list the other possible arrangements. Then use the clues to select the correct arrangement. |

1	2	3	4
R	B	H	M
R	H	M	B
R	?	?	?
R	?	?	?

| 4 COMPUTE | → 5 CHECK |

2. The girls' softball teams from East High and West High were mingling on the field. Six of the players wore jerseys numbered 9, 11, 18, 19, 22, and 23 respectively. Name the number of the pitcher, catcher, and shortstop for each team, given that: the number worn by East's catcher is twice the number worn by the pitcher from her school; the number on the jersey of the pitcher from West High is ten more than the rival school's pitcher; the number worn by the shortstop from West is half the number worn by the shortstop from East.

3. Find the weights of Maryellen, her brother, and her pet from the following clues: Maryellen, her brother, and her pet altogether weigh 155 lb; the weight of her brother and her pet together, subtracted from Maryellen's weight, is 37 lb; Maryellen and her pet together weigh 112 lb.

Exercise 4

4. Solve this puzzle: Four munks equal six rells and three rells equal eight borks. How many munks equal 24 borks?

5. Anna is trying to determine the lengths of three differently colored blocks: one red, one white, and one blue. She has only a string one foot long to work with. By comparing the string and the blocks, Anna finds that: the blue block and the red block together are as long as the string; the blue block is longer than the red block by half the length of the string; the red and blue blocks together are three times longer than the white block. How long is each block?

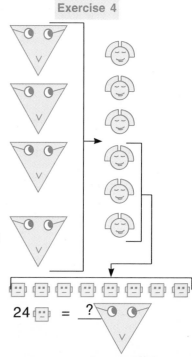

$$24 \; \boxed{\cdot\cdot} \; = \; \underline{?}$$

173

Match each sentence with the correct problem. Then solve.

1. From $4\frac{3}{5}$ take $2\frac{7}{8}$.

2. Take 7 from the sum of $4\frac{3}{5}$ and $2\frac{7}{8}$.

3. Find the difference between 7 and $4\frac{3}{5}$.

4. Find the sum of 7 and $4\frac{3}{5}$.

5. One addend is $2\frac{7}{8}$. The other addend is $4\frac{3}{5}$. Find the sum.

a. $7 + 4\frac{3}{5}$

b. $4\frac{3}{5} - 2\frac{7}{8}$

c. $2\frac{7}{8} + 4\frac{3}{5}$

d. $7 - 4\frac{3}{5}$

e. $2\frac{7}{8} - 4\frac{3}{5}$

f. $\left(4\frac{3}{5} + 2\frac{7}{8}\right) - 7$

Solve.

6. Eight out of the 32 triangles within a design are painted black. What part of the design is *not* painted black?

USE THESE STRATEGIES:
Write an Equation
Use Simpler Numbers
Use a Model/Drawing
Multi-Step Problem
Logical Reasoning

7. To make some purple paint, Elena mixed $1\frac{2}{3}$ pints of red paint with $1\frac{3}{5}$ pints of blue paint.

   a. Of which color did she use the least?

   b. How much purple paint did she make?

8. To make orange paint, Les needs to mix equal parts of red and yellow. If Les has *two* jars of red each holding $\frac{7}{8}$ pint of paint, how much yellow paint must he use?

9. An artist has a set of calligraphy pens for lettering. Their widths are: $\frac{5}{8}$ in., $\frac{1}{2}$ in., $\frac{7}{16}$ in., $\frac{3}{4}$ in., $\frac{9}{32}$ in., and $\frac{3}{16}$ in.

   a. Write the widths in order from least to greatest.

   b. How much wider is the greatest than the least?

10. It took a leather worker $\frac{5}{6}$ hour to cut and shape a belt. The worker then spent $1\frac{3}{5}$ hours decorating the belt.

    a. How much longer did it take to decorate the belt than to make it?

    b. How long did it take the worker to complete the belt?

11. The craft class bought 20 kg of clay. Of this, $8\frac{4}{5}$ kg was used to make mugs, and $10\frac{3}{10}$ kg was used to make flower pots. How much of the clay is left?

The students in an art class were surveyed to determine which mediums they preferred to use for drawing. The survey results are listed on this frequency table. Use the table to complete exercises 12–16.

12. What part of the art class prefers pastels?

13. What part of the art class prefers pen and ink?

14. What part of the art class prefers acrylics?

15. What part of the art class does *not* prefer acrylics?

16. What part of the class prefers either pen and ink or pastels?

Frequency Table

Medium	Number of Students
charcoal	𝍷𝍷𝍷𝍷 ||||
pen and ink	𝍷𝍷𝍷𝍷
acrylics	𝍷𝍷𝍷𝍷	
pastels	𝍷𝍷𝍷𝍷 𝍷𝍷𝍷𝍷

17. Conchita bought $5\frac{1}{3}$ yards of fabric to make a skirt and vest. The vest required $1\frac{7}{8}$ yards. Estimate how much fabric the skirt requires.

18. A metal worker ordered 20 pounds of gold. He used $5\frac{5}{8}$ pounds to make necklaces, $2\frac{2}{3}$ pounds to make rings, and $8\frac{3}{16}$ pounds to make belts. About how many pounds does he have left to make bracelets?

19. This triangular piece of glass is used to make a sun catcher. How much lead is needed around the outer edge to hold the sun catcher together?

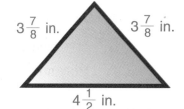

$3\frac{7}{8}$ in. $3\frac{7}{8}$ in.

$4\frac{1}{2}$ in.

20. A jewelry maker makes a pendant in the shape of a kite. The two shorter sides each measure $\frac{7}{16}$ in. and the two longer sides each measure $\frac{3}{4}$ in. How much gold coil does the jeweler need to edge the finished kite?

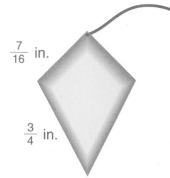

$\frac{7}{16}$ in.

$\frac{3}{4}$ in.

21. Mr. Hawk is going to use a $2\frac{3}{4}$-in. mat as a border around his pen-and-ink sketch. If the sketch is $15\frac{1}{2}$ in. long on each side, how long is the matted sketch?

22. The directions for making a macramé belt state that $13\frac{1}{3}$ yards of string are needed. The string is sold in 4-yard and 10-yard spools. What combination of spools should be purchased in order to ensure the least waste?

   a. four 4-yard spools

   b. one of each size

   c. two 4-yard spools and one 10-yard spool

More Practice

Complete.

1. $\frac{8}{12} = \frac{?}{3}$ **2.** $\frac{4}{5} = \frac{?}{40}$ **3.** $\frac{?}{45} = \frac{48}{9}$ **4.** $\frac{5}{?} = \frac{25}{15}$ **5.** $\frac{16}{48} = \frac{?}{3}$ **6.** $\frac{21}{49} = \frac{3}{?}$

Compare. Write < or >.

7. $\frac{2}{5}$ __?__ $\frac{3}{10}$ **8.** $\frac{5}{6}$ __?__ $\frac{11}{12}$ **9.** $\frac{9}{10}$ __?__ $\frac{8}{9}$ **10.** $\frac{6}{13}$ __?__ $\frac{9}{25}$

Order from least to greatest.

11. $\frac{1}{3}$, $\frac{1}{2}$, $\frac{1}{6}$ **12.** $\frac{7}{10}$, $\frac{1}{5}$, $\frac{3}{4}$ **13.** $\frac{6}{7}$, $\frac{5}{6}$, $\frac{1}{3}$, $\frac{9}{14}$

Compute mentally. Write <, =, or >.

14. $\frac{1}{2}$ __?__ $\frac{7}{15}$ **15.** $\frac{5}{32}$ __?__ $\frac{1}{2}$ **16.** $\frac{3}{4} + \frac{1}{2} + \frac{5}{8}$ __?__ 2 **17.** $\frac{8}{14} - \frac{1}{2}$ __?__ $\frac{1}{2}$

Complete.

18. $1\frac{5}{7} = \frac{?}{7}$ **19.** $6\frac{7}{9} = \frac{?}{9}$ **20.** $12\frac{4}{5} = \frac{?}{5}$ **21.** $\frac{18}{10} = \frac{?}{\underline{\quad}}$

Find the sum or difference.

22. $18\frac{2}{9} + 6\frac{5}{12}$ **23.** $32\frac{4}{10} + 18\frac{1}{3}$ **24.** $\frac{3}{12} - \frac{1}{6}$ **25.** $\frac{5}{9} - \frac{1}{3}$ **26.** $3\frac{1}{5} - 1\frac{4}{7}$

Subtract.

27. $6 - \frac{1}{4}$ **28.** $2 - \frac{4}{5}$ **29.** $8 - \frac{3}{8}$ **30.** $7 - \frac{9}{10}$

Estimate sums and differences. Then add or subtract.

31. $\begin{array}{r} 2\frac{1}{5} \\ +1\frac{1}{5} \\ \hline \end{array}$ **32.** $\begin{array}{r} 2\frac{1}{3} \\ +2\frac{4}{7} \\ \hline \end{array}$ **33.** $\begin{array}{r} 3\frac{4}{8} \\ -1\frac{1}{8} \\ \hline \end{array}$ **34.** $\begin{array}{r} 12\frac{1}{10} \\ -2\frac{3}{4} \\ \hline \end{array}$ **35.** $\begin{array}{r} 16\frac{1}{6} \\ -15\frac{2}{3} \\ \hline \end{array}$

Solve.

36. $7 - 1\frac{1}{5} - 1\frac{4}{5}$ **37.** $\left(2\frac{1}{10} + 3\frac{1}{6}\right) - \frac{1}{6}$ **38.** $5\frac{3}{8} - 3 + 5\frac{5}{8}$

39. Sid skated for $2\frac{1}{2}$ hours on Friday and $3\frac{3}{4}$ hours on Saturday. If his hockey coach requires 8 hours of skating practice weekly, how much longer must Sid practice?

40. Faith made $3\frac{2}{3}$ dozen cookies. Her brother ate $1\frac{1}{2}$ dozen. If she needs 5 dozen cookies for her party, how many more dozens of cookies must she bake?

Math Probe

MÖBIUS STRIP

August F. Möbius, a 19th-century German mathematician, was a pioneer in a special branch of geometry called **topology**.

He investigated and wrote a paper about the one-sided surface formed by taking a long rectangular band and joining the ends after giving it a twist. Today this is known as the **Möbius strip**.

Conduct your own investigation.
Make a chart to record your observations.

Construction

Materials: 8 in. by 11 in. paper, scissors, ruler, pencil, tape or glue

Directions:

Step 1: Cut the paper into strips measuring 1 in. by 11 in.

Step 2: Glue or tape the ends of each strip together in these ways:

    a. without twisting the first strip.

    b. after giving the second strip one twist.

    c. after giving the third strip two twists.

    d. after giving the fourth strip three twists.

Step 3: Investigate the number of surfaces and edges of each strip. Record your observations.

Step 4: Cut each strip down the center. Record your observations.

Step 5: Make another set of strips as described in Step 2.

Step 6: Cut each strip in thirds. Record your observations.

1. Summarize your investigation and explain your findings to the class.

2. Look up *topology* in an encyclopedia. Write about how this branch of geometry is useful today.

3. Have any practical uses been found for the Möbius strip?

Check Your Mastery

See pp. 152–154

1. Which fraction is equivalent to the shaded area?

   a. $\frac{7}{16}$ b. $\frac{2}{3}$ c. $\frac{4}{7}$ d. $\frac{1}{2}$

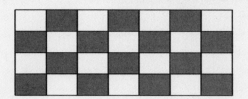

2. The simplest form of $\frac{18}{72}$ is __?__

   a. $\frac{2}{8}$ b. $\frac{3}{12}$ c. $\frac{1}{4}$ d. $\frac{4}{1}$

3. Tracy practiced guitar for $\frac{3}{4}$ of an hour. How many twelfths of an hour is that?

Which is the greater fraction in each pair?

See pp. 155–157

4. $\frac{4}{7}$ or $\frac{2}{3}$ 5. $\frac{6}{7}$ or $\frac{4}{5}$ 6. $\frac{3}{8}$ or $\frac{2}{7}$ 7. $\frac{15}{16}$ or $\frac{7}{8}$ 8. $\frac{2}{9}$ or $\frac{1}{8}$

Order these fractions from least to greatest.

9. $\frac{4}{5}$, $\frac{9}{10}$, $\frac{7}{8}$

10. $\frac{1}{3}$, $\frac{3}{8}$, $\frac{1}{4}$, $\frac{1}{12}$

11. $\frac{7}{18}$, $\frac{1}{2}$, $\frac{5}{6}$, $\frac{2}{9}$

Change to mixed numbers.   **Change to improper fractions.**

12. $\frac{9}{2}$ 13. $\frac{86}{7}$ 14. $\frac{98}{9}$ 15. $2\frac{3}{5}$ 16. $3\frac{3}{16}$ 17. $12\frac{1}{8}$

Find the sum or difference. Express in simplest form.

See pp. 160–165

18. $\frac{3}{14}$ $+\frac{7}{14}$ 19. $\frac{4}{3}$ $+\frac{2}{5}$ 20. $\frac{9}{16}$ $+\frac{5}{6}$ 21. $\frac{8}{9}$ $-\frac{2}{9}$ 22. $\frac{15}{10}$ $-\frac{6}{5}$ 23. $\frac{3}{8}$ $-\frac{1}{3}$ 24. 10 $-\frac{6}{7}$

Estimate. Then find the sum or difference.

See pp. 166–169

25. $2\frac{5}{8}$ $+1\frac{3}{4}$ 26. $1\frac{2}{3}$ $+4\frac{1}{5}$ 27. $6\frac{9}{10}$ $-2\frac{3}{4}$ 28. $10\frac{1}{6}$ $-1\frac{2}{3}$ 29. $7\frac{5}{6}$ $-6\frac{6}{7}$ 30. $20\frac{1}{5}$ $-19\frac{7}{9}$

Solve.

31. Julio bicycled $5\frac{2}{3}$ miles on Saturday and $2\frac{5}{6}$ miles on Sunday. How many more miles does he need to bicycle to cover a distance of 10 miles?

32. Rosa took $1\frac{2}{3}$ pounds of bird seed from a 5-pound bag. Then the squirrels ate $2\frac{1}{4}$ pounds of seed from the bag. How many pounds of seed are left?

More Fractions and Decimals

Do you remember?

A whole number may be written as a fraction with a denominator of 1. For example, 4 may be written as $\frac{4}{1}$.

A mixed number has a whole-number part and a fraction part. For example, $6\frac{3}{8}$.

In this chapter you will:

- Multiply and divide fractions and mixed numbers
- Use the cancellation shortcut
- Work with complex fractions
- Find part of a number; find a number, given a part
- Change fractions, mixed numbers, and decimals
- Identify terminating and repeating decimals
- Solve problems: interpreting remainders

Notable Fractions

RESEARCHING TOGETHER

The value of the notes in music today are based on fractional parts of a whole, whereas the fractions in a time signature are not associated with a part/ whole relationship. Find examples from printed music of different time signatures and copy one measure of music from each. Research when and why this notation system began.

8-1 Multiplying Fractions

For an arts-and-crafts project, Van and Rita made a kite.

$\frac{1}{2}$ of it had a blue background.

$\frac{2}{5}$ of the blue background had stars.

What part of the kite had stars?

$\frac{2}{5}$ of $\frac{1}{2}$ of the kite had stars.

To multiply fractions:

- Multiply the numerators.
- Multiply the denominators.
- Write the product in simplest form.

$\frac{2}{5}$ of $\frac{1}{2}$ means $\frac{2}{5} \times \frac{1}{2} = \frac{2 \times 1}{5 \times 2} = \frac{2}{10} = \frac{1}{5}$

So, $\frac{1}{5}$ of the kite had stars.

$\frac{2}{5}$ of $\frac{1}{2}$ had stars.

Remember: "of" means multiply.

Always write the product in simplest form by dividing by the GCF.

Try this:

$\frac{4}{3}$ of $\frac{5}{6}$ means $\frac{4}{3} \times \frac{5}{6} = \frac{4 \times 5}{3 \times 6} = \frac{20}{18} = 1\frac{2}{18} = 1\frac{1}{9}$

Find the product.*

1. $\frac{3}{5} \times \frac{3}{5}$ 2. $\frac{3}{4} \times \frac{5}{8}$ 3. $\frac{2}{3} \times \frac{1}{8}$ 4. $\frac{3}{6} \times \frac{1}{7}$ 5. $\frac{8}{7} \times \frac{1}{2}$ 6. $\frac{2}{5} \times \frac{1}{2}$

7. $\frac{1}{2} \times \frac{21}{5}$ 8. $\frac{10}{3} \times \frac{6}{5}$ 9. $\frac{11}{4} \times \frac{4}{3}$ 10. $\frac{5}{4} \times \frac{2}{3}$ 11. $\frac{2}{5} \times \frac{7}{8}$ 12. $\frac{7}{4} \times \frac{4}{11}$

13. Which examples above have an improper fraction as one factor?

14. When an improper fraction is a factor, can the product be either a fraction, a whole number, or a mixed number? Why or why not?

Solve.

15. Ms. Ruiz worked $\frac{2}{3}$ of an hour arranging silk flowers.
She worked $\frac{1}{2}$ of the time on a centerpiece.
What part of an hour did she work on the centerpiece?

16. Donny bought $\frac{3}{4}$ yd of felt. He used $\frac{5}{6}$ of it to make a banner.
How much of the felt did he use for the banner?

8-2 Multiplying Fractions and Whole Numbers

To multiply a fraction and a whole number:

- Rewrite the whole number as an improper fraction with 1 in the denominator.
- Then multiply.

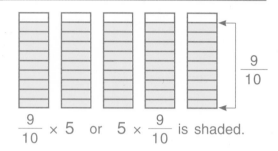

$\frac{9}{10} \times 5$ or $5 \times \frac{9}{10}$ is shaded.

$$\frac{9}{10} \times 5 = \frac{9}{10} \times \frac{5}{1} = \frac{9 \times 5}{10 \times 1} = \frac{45}{10} = 4\frac{5}{10} = 4\frac{1}{2}$$

OR

$$5 \times \frac{9}{10} = \frac{5}{1} \times \frac{9}{10} = \frac{5 \times 9}{1 \times 10} = \frac{45}{10} = 4\frac{5}{10} = 4\frac{1}{2}$$

Find the product.*

1. $\frac{1}{6} \times 24$ 2. $\frac{3}{5} \times 25$ 3. $16 \times \frac{3}{4}$ 4. $16 \times \frac{3}{8}$ 5. $\frac{5}{7} \times 6$ 6. $\frac{3}{8} \times 7$

7. $\frac{2}{3} \times 12$ 8. $\frac{2}{5} \times 15$ 9. $10 \times \frac{4}{5}$ 10. $16 \times \frac{1}{3}$ 11. $25 \times \frac{2}{8}$ 12. $37 \times \frac{2}{5}$

Write <, =, or > to make true statements.

13. $\frac{3}{4} \times 2 \underline{\ ?\ } \frac{4}{3} \times 2$ 14. $\frac{2}{5} \times 45 \underline{\ ?\ } \frac{4}{5} \times 45$ 15. $\frac{2}{5} \times 20 \underline{\ ?\ } \frac{2}{5} \times 10$

16. $\frac{2}{5} \times 5 \underline{\ ?\ } \frac{2}{5} \times 1$ 17. $14 \times \frac{7}{10} \underline{\ ?\ } \frac{7}{10} \times 14$ 18. $8 \times \frac{7}{16} \underline{\ ?\ } 9 \times \frac{2}{3}$

Multiply. (Remember: "of" means multiply.)

19. $\frac{3}{4}$ of 2 20. $\frac{7}{6}$ of 3 21. $\frac{2}{3}$ of 4 22. $\frac{6}{8}$ of 1 23. $\frac{1}{3}$ of 3

24. $\frac{2}{3}$ of 5 25. $4 \times \frac{3}{8} \times \frac{2}{3}$ 26. $\frac{1}{2} \times \frac{1}{4} \times 8$ 27. $7 \times \frac{2}{5} \times 8$ 28. $16 \times \frac{3}{7} \times \frac{15}{16}$

Solve.

29. Marsha had 80 stamps in her collection. $\frac{3}{8}$ of them were foreign stamps. How many stamps were foreign?

30. $\frac{3}{5}$ of the 150 stamps in Mr. Sim's collection are worth over $200. How many of his stamps are worth over $200?

 MENTAL MATH

31. $\frac{2}{3} \times \frac{1}{2} \times 1$ 32. $\frac{3}{8} \times \frac{3}{6} \times 2$ 33. $3 \times \frac{5}{6} \times \frac{2}{9}$ 34. $\frac{3}{5} \times \frac{5}{8} \times \frac{8}{3}$

35. $2 \times \frac{2}{3} \times 6$ 36. $7 \times \frac{1}{8} \times \frac{1}{14}$ 37. $\frac{3}{7} \times 2 \times \frac{4}{3}$ 38. $\frac{2}{9} \times 5 \times \frac{9}{20}$

*See page 454 for more practice. 181

8-3 Multiplying Mixed Numbers

To multiply mixed numbers:

- Rewrite the numbers as improper fractions.
- Then multiply. (The product can be
  a mixed number, a fraction, or a whole number.)

Mixed-Number Product

$$1\frac{2}{5} \times 3\frac{1}{3} = \underline{?}$$

$$\frac{7}{5} \times \frac{10}{3} = \frac{7 \times 10}{5 \times 3} = \frac{70}{15} = 4\frac{10}{15} = 4\frac{2}{3}$$

Always simplify products.

Fraction Product

$$\frac{1}{6} \times 2\frac{1}{3} = \underline{?}$$

$$\frac{1}{6} \times \frac{7}{3} = \frac{1 \times 7}{6 \times 3} = \frac{7}{18}$$

Whole-Number Product

$$3\frac{1}{3} \times 9 = \underline{?}$$

$$\frac{10}{3} \times \frac{9}{1} = \frac{10 \times 9}{3 \times 1} = \frac{90}{3} = 30$$

Find the product.

1. $16 \times 3\frac{1}{2}$
2. $1\frac{1}{3} \times 9$
3. $2\frac{2}{5} \times 10$
4. $6 \times 3\frac{5}{6}$

5. $\frac{3}{4} \times 4\frac{3}{4}$
6. $\frac{3}{4} \times 12\frac{1}{2}$
7. $\frac{1}{3} \times 8\frac{1}{2}$
8. $1\frac{3}{4} \times \frac{3}{10}$

9. $18\frac{1}{2} \times \frac{2}{7}$
10. $16\frac{1}{3} \times \frac{1}{2}$
11. $1\frac{1}{2} \times 1\frac{1}{4}$
12. $2\frac{7}{8} \times 1\frac{3}{4}$

13. $11\frac{2}{3} \times \frac{1}{5} \times 10$
14. $\frac{2}{5} \times 1\frac{1}{2} \times 20$
15. $16 \times \frac{3}{4} \times 1\frac{2}{3}$
16. $3\frac{1}{6} \times 2\frac{2}{3}$

Estimate products by rounding both factors to the nearest whole number.

17. $10\frac{2}{3} \times 1\frac{1}{8}$
18. $2\frac{3}{4} \times 1\frac{3}{5}$
19. $5 \times 3\frac{3}{10}$
20. $2\frac{1}{2} \times 1\frac{3}{8}$

21. $6 \times 8\frac{2}{3}$
22. $4\frac{1}{5} \times 1\frac{1}{3}$
23. $\frac{7}{8} \times 5\frac{1}{8}$
24. $2\frac{7}{12} \times 2\frac{1}{7}$

25. $2\frac{4}{5} \times 3\frac{1}{3}$
26. $5\frac{2}{5} \times 2\frac{3}{8}$
27. $5\frac{5}{6} \times 1\frac{4}{5}$
28. $9\frac{1}{7} \times 6\frac{1}{9}$

Name the mathematical property illustrated by each statement.

29. $3\frac{1}{4} \times 0$ is the same as 0.

30. $\frac{7}{3} \times \frac{3}{8}$ is the same as $\frac{3}{8} \times \frac{7}{3}$.

31. $\frac{3}{3} \times 7$ is the same as 7.

32. $\left(\frac{1}{3} \times \frac{1}{2}\right) \times \frac{1}{4}$ is the same as $\frac{1}{3} \times \left(\frac{1}{2} \times \frac{1}{4}\right)$.

Using the Distributive Property

The distributive property may be used as a shortcut when multiplying mixed numbers. Study this example: $2\frac{1}{2} \times 14 = \underline{\ ?\ }$

Think: $2\frac{1}{2} \times 14 = (2 + \frac{1}{2})14 = (2 \times 14) + (\frac{1}{2} \times 14) = 28 + 7 = 35$

Multiply. (Hint: Use the distributive property.)

33. $6\frac{2}{3} \times 9$

34. $8\frac{4}{5} \times 10$

35. $2\frac{4}{9} \times 18$

36. $7\frac{1}{8} \times 4$

37. $5\frac{3}{4} \times 16$

38. $2 \times 3\frac{1}{2}$

39. $15 \times 10\frac{1}{3}$

40. $10\frac{1}{9} \times 27$

41. $16 \times 4\frac{3}{8}$

42. $10\frac{3}{5} \times 20$

43. $14 \times 5\frac{5}{7}$

44. $10\frac{5}{6} \times 24$

Write < or > to make true statements.

45. $2\frac{2}{5} \times 10 \ \underline{\ ?\ }\ 20$

46. $3\frac{1}{2} \times 20 \ \underline{\ ?\ }\ 60$

47. $6\frac{1}{2} \times 3 \ \underline{\ ?\ }\ 4\frac{2}{3} \times 5$

48. $\frac{3}{10} \times 2\frac{1}{2} \ \underline{\ ?\ }\ \frac{2}{10} \times 2\frac{1}{2}$

49. $\frac{1}{2} \times 4\frac{1}{6} \ \underline{\ ?\ }\ \frac{1}{3} \times 4\frac{1}{6}$

50. $\frac{1}{3} \times 2\frac{1}{6} \ \underline{\ ?\ }\ \frac{2}{3} \times \frac{3}{4}$

51. $1\frac{1}{2} \times 1\frac{1}{2} \ \underline{\ ?\ }\ \frac{3}{2}$

52. $2\frac{5}{6} \times 1\frac{2}{3} \ \underline{\ ?\ }\ \frac{5}{3}$

53. $7\frac{2}{7} \times 6\frac{1}{9} \ \underline{\ ?\ }\ 6\frac{2}{9} \times 7\frac{1}{7}$

Solve.

54. Jamie built a dinosaur model $7\frac{1}{2}$ in. tall. His sister's model is $1\frac{1}{5}$ times taller. How tall is his sister's model?

55. Mr. Maio spent $1\frac{1}{4}$ hours each day building a cabinet. He finished the cabinet in 5 days. How many hours did it take him to build the cabinet?

56. James walks $6\frac{1}{3}$ blocks on his paper route in one day. How many blocks does he walk in $4\frac{1}{2}$ days?

 SUPPOSE THAT...

Joshua designed a paper mosaic shaped like this butterfly.

57. Suppose he made his design 4 times larger. What would each new dimension be?

58. Suppose he made his design $2\frac{1}{2}$ times larger. What would each new dimension be?

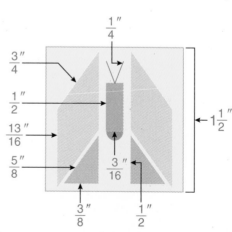

183

8-4 Multiplying Using Cancellation

To multiply fractions or mixed numbers quickly, use *cancellation:*

Divide both the numerator and denominator by any common factors before you multiply.

$\frac{2}{3} \times \frac{3}{5} = $ _?_

3 is a factor of both numerator and denominator.

Cancel once

$\frac{2}{\overset{1}{\cancel{3}}} \times \frac{\overset{1}{\cancel{3}}}{5}$

Multiply

$= \frac{2 \times 1}{1 \times 5} = \frac{2}{5}$

$\frac{5}{8} \times 2\frac{2}{15} = $ _?_

$\frac{5}{8} \times \frac{32}{15}$

5 is a factor of 5 and 15.
8 is a factor of 8 and 32.

Cancel twice

$\frac{\overset{1}{\cancel{5}}}{\underset{1}{\cancel{8}}} \times \frac{\overset{4}{\cancel{32}}}{\underset{3}{\cancel{15}}}$

Multiply

$= \frac{1 \times 4}{1 \times 3} = \frac{4}{3} = 1\frac{1}{3}$

If you cancel until there are no more common factors in both the numerator and the denominator, the product will be in lowest terms.

Name the common factor used to cancel.

1. $\frac{5}{6} \times 24 = \frac{5}{\underset{1}{\cancel{6}}} \times \frac{\overset{4}{\cancel{24}}}{1}$

2. $\frac{2}{5} \times 25 = \frac{2}{\underset{1}{\cancel{5}}} \times \frac{\overset{5}{\cancel{25}}}{1}$

3. $15 \times \frac{2}{3} = \frac{\overset{5}{\cancel{15}}}{1} \times \frac{2}{\underset{1}{\cancel{3}}}$

4. $16 \times \frac{3}{4} = \frac{\overset{4}{\cancel{16}}}{1} \times \frac{3}{\underset{1}{\cancel{4}}}$

Multiply. Use cancellation, whenever possible.

5. $\frac{1}{8} \times 32$ 6. $\frac{1}{3} \times 24$ 7. $\frac{3}{4} \times \frac{4}{3}$ 8. $\frac{5}{7} \times \frac{7}{5}$ 9. $\frac{1}{8} \times 2\frac{7}{9}$

10. $4\frac{1}{6} \times \frac{1}{2}$ 11. $8 \times 2\frac{1}{4}$ 12. $2\frac{2}{3} \times \frac{3}{5}$ 13. $\frac{1}{6} \times 3\frac{3}{4}$ 14. $9\frac{1}{7} \times \frac{1}{28}$

Multiply. Cancel more than once.

15. $\frac{4}{15} \times \frac{3}{10}$ 16. $\frac{12}{15} \times \frac{10}{24}$ 17. $\frac{3}{7} \times 3\frac{1}{9}$ 18. $2\frac{4}{10} \times \frac{5}{8}$ 19. $6\frac{1}{2} \times 4\frac{4}{8}$

20. $3\frac{1}{7} \times 1\frac{3}{11}$ 21. $4\frac{1}{2} \times 9\frac{1}{3}$ 22. $1\frac{1}{4} \times 4\frac{4}{5}$ 23. $3\frac{1}{3} \times 3\frac{3}{5}$ 24. $2\frac{2}{9} \times 9\frac{3}{5}$

25. $16 \times \frac{5}{6} \times \frac{3}{4}$ 26. $2\frac{1}{2} \times 1\frac{2}{10} \times \frac{1}{3}$ 27. $3\frac{1}{8} \times 32 \times \frac{1}{5}$ 28. $4\frac{1}{6} \times \frac{9}{10} \times 12$

29. $\frac{1}{6} \times 12 \times \frac{2}{3}$ 30. $8 \times \frac{1}{8} \times 3\frac{1}{5}$ 31. $\frac{3}{5} \times \frac{2}{3} \times 3\frac{3}{4}$ 32. $\frac{4}{7} \times \frac{5}{7} \times 1\frac{3}{4}$

Multiply.

33. $4\frac{1}{4} \times 3\frac{2}{3}$

34. $4\frac{2}{7} \times 6\frac{2}{9}$

35. $6\frac{1}{8} \times 9\frac{1}{2}$

36. $8\frac{1}{3} \times 7\frac{1}{4}$

37. $\frac{20}{23} \times 4\frac{1}{4}$

38. $9\frac{1}{11} \times 6\frac{2}{5}$

39. $7\frac{1}{2} \times \frac{6}{25}$

40. $12\frac{2}{7} \times 5\frac{3}{5}$

41. $9\frac{1}{16} \times 1\frac{3}{5}$

42. $16\frac{1}{3} \times 2\frac{5}{14}$

43. $8\frac{1}{10} \times 5\frac{5}{21}$

44. $9\frac{7}{12} \times 1\frac{13}{15}$

45. $7\frac{1}{9} \times \frac{18}{41}$

46. $13\frac{5}{7} \times 5\frac{1}{18}$

47. $1\frac{1}{5} \times 3\frac{3}{14}$

48. $6\frac{2}{9} \times 1\frac{11}{16}$

49. $2\frac{5}{8} \times 3 \times 6\frac{6}{7}$

50. $8\frac{1}{6} \times 4\frac{1}{2} \times 1\frac{1}{35}$

51. $2\frac{1}{7} \times 8\frac{5}{9} \times 4$

52. $7\frac{3}{11} \times 8\frac{1}{15} \times \frac{1}{4}$

True or false? Explain your answer.

53. $2\frac{1}{2} \times 2\frac{2}{3} = \frac{5}{1} \times \frac{4}{3}$

54. $2\frac{2}{9} \times 3\frac{3}{10} = \frac{2}{3} \times \frac{11}{5}$

55. $5\frac{1}{3} \times 3\frac{3}{4} \times 2\frac{1}{12} \times 4\frac{1}{5} > 5 \times 3 \times 2 \times 4$

56. $6\frac{1}{2} \times 1\frac{1}{3} = 6\frac{1}{3} \times 1\frac{1}{2}$

57. $\frac{3}{11} \times 44 \times \frac{3}{4} > 44$

58. $\frac{1}{8} \times 66 \times \frac{4}{7} < \frac{1}{2} \times 66 \times \frac{1}{8}$

Using Compatible Numbers

Compatible numbers: two numbers that you can compute with easily. They are used in estimation.

$\frac{3}{20} \times 201 = \underline{\ ?\ }$

20 and 201 are *not* compatible.
But 20 and 200 *are,* because
you can compute with them mentally.

$\frac{3}{20} \times 201 \approx \frac{3}{20} \times \overset{10}{\cancel{200}} = 30$
$\phantom{\frac{3}{20} \times 201 \approx \frac{3}{20}} {\scriptstyle 1}$

So, $\frac{3}{20} \times 201 \approx 30$

$1\frac{2}{5} \times 24 = \underline{\ ?\ }$

5 and 24 are *not* compatible.
But 5 and 25 *are,* because
you can compute with them mentally.

$1\frac{2}{5} \times 24 = \frac{7}{5} \times 24 \approx \frac{7}{5} \times 25$

$\frac{7}{\cancel{5}} \times \frac{\overset{5}{\cancel{25}}}{1} = 35 \quad$ So, $1\frac{2}{5} \times 24 \approx 35$
${\scriptstyle 1}$

Restate each multiplication by using compatible numbers. Then estimate.

59. $\frac{5}{6} \times 41$

60. $\frac{7}{15} \times 46$

61. $\frac{9}{11} \times 45$

62. $\frac{7}{8} \times \frac{2}{3}$

63. $\frac{4}{5} \times \frac{10}{17}$

64. $\frac{3}{7} \times \frac{20}{27}$

65. $\frac{5}{16} \times \frac{2}{9}$

66. $\frac{10}{17} \times \frac{3}{31}$

67. $1\frac{1}{4} \times \frac{13}{20}$

68. $2\frac{1}{3} \times \frac{6}{13}$

69. $2\frac{1}{2} \times \frac{7}{24}$

70. $7\frac{1}{6} \times \frac{7}{11}$

Solve.

71. A man-made pond holds $40\frac{1}{2}$ gal of water. During a drought, $\frac{1}{3}$ of the water evaporated. How many gallons of water are left in the pond?

72. After stocking a man-made pond with 24 fish, it was projected that the fish population would increase $1\frac{1}{2}$ times in a year. What would the fish population be in a year?

73. On a store shelf there are 92 boxes of fish food. If $\frac{1}{4}$ of them are the store's own brand, how many boxes of other brands are there?

8-5 Reciprocals

Reciprocals: two numbers whose product is 1.

$\frac{2}{7}$ and $\frac{7}{2}$ are reciprocals because:

$$\frac{2}{7} \times \frac{7}{2} = 1$$

5 and $\frac{1}{5}$ are reciprocals because:

$$\frac{5}{1} \times \frac{1}{5} = 1$$

$1\frac{1}{3}$ and $\frac{3}{4}$ are reciprocals because:

$$\frac{4}{3} \times \frac{3}{4} = 1$$

To find the reciprocal of a number:
Change the number to a fraction; then invert it, that is, turn it upside down.

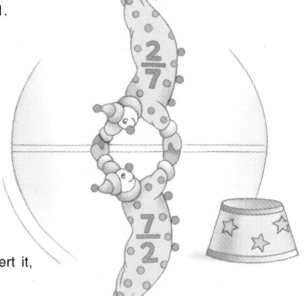

Write the reciprocal.

1. $\frac{1}{2}$ 2. $\frac{1}{9}$ 3. $\frac{6}{11}$ 4. $\frac{8}{3}$ 5. $\frac{2}{8}$ 6. $\frac{5}{7}$ 7. $\frac{10}{3}$

8. $\frac{6}{7}$ 9. $\frac{5}{9}$ 10. $\frac{4}{16}$ 11. 16   12. 9   13. $\frac{12}{5}$ 14. $\frac{23}{10}$

15. $\frac{15}{4}$ 16. $\frac{6}{12}$ 17. $\frac{19}{10}$ 18. $\frac{9}{6}$ 19. 64   20. $\frac{3}{5}$ 21. $\frac{8}{19}$

Complete.*

22. $\frac{5}{6} \times \underline{\ ?\ } = 1$ 23. $\frac{1}{4} \times \underline{\ ?\ } = 1$ 24. $\frac{6}{5} \times \underline{\ ?\ } = 1$ 25. $\frac{7}{7} \times \underline{\ ?\ } = 1$

26. $\underline{\ ?\ } \times 9 = 1$ 27. $\underline{\ ?\ } \times \frac{3}{10} = 1$ 28. $\underline{\ ?\ } \times \frac{16}{1} = 1$ 29. $\underline{\ ?\ } \times \frac{6}{7} = 1$

30. $\frac{13}{8} \times \underline{\ ?\ } = 1$ 31. $\underline{\ ?\ } \times 12 = 1$ 32. $\frac{5}{12} \times \underline{\ ?\ } = 1$ 33. $\frac{1}{16} \times \underline{\ ?\ } = 1$

34. $\frac{1}{3} \times \underline{\ ?\ } = 1$ 35. $3\frac{1}{2} \times \underline{\ ?\ } = 1$ 36. $\underline{\ ?\ } \times 1\frac{2}{5} = 1$ 37. $8\frac{1}{2} \times \underline{\ ?\ } = 1$

38. $\underline{\ ?\ } \times 4\frac{1}{4} = 1$ 39. $6\frac{1}{8} \times \underline{\ ?\ } = 1$ 40. $9\frac{1}{3} \times \underline{\ ?\ } = 1$ 41. $\underline{\ ?\ } \times 10\frac{1}{3} = 1$

42. $2\frac{4}{5} \times \underline{\ ?\ } = 1$ 43. $\underline{\ ?\ } \times 8\frac{1}{2} = 1$ 44. $\underline{\ ?\ } \times 6\frac{1}{2} = 1$ 45. $12\frac{1}{3} \times \underline{\ ?\ } = 1$

Which numbers are reciprocals of each other?

46. $3\frac{1}{5}$ and $\frac{5}{16}$ 47. $2\frac{1}{3}$ and $\frac{7}{3}$ 48. $\frac{5}{2}$ and $\frac{2}{5}$ 49. $\frac{1}{3}$ and $\frac{3}{4}$

CRITICAL THINKING

50. What number has no reciprocal? Why?

51. What number is its own reciprocal? Why?

Jan has 4 feet of wrapping paper. She needs pieces of paper $\frac{1}{3}$ foot long. How many pieces of paper can be cut?

How many $\frac{1}{3}$'s are in 4 is the same as asking: $4 \div \frac{1}{3} = \underline{\ ?\ }$

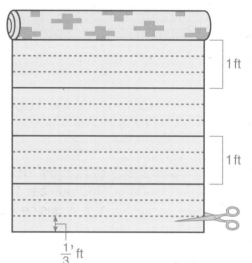

▶ **To divide a *whole number by a fraction*:**

Multiply by the reciprocal of the fraction.

$$4 \div \frac{1}{3}$$

$$\downarrow$$

$$4 \times \frac{3}{1} = \frac{4}{1} \times \frac{3}{1} = \frac{12}{1} = 12$$

Try these:

$$5 \div \frac{3}{4} = 5 \times \frac{4}{3} = \frac{5}{1} \times \frac{4}{3} = \frac{20}{3} = 6\frac{2}{3}$$

$$16 \div \frac{8}{5} = 16 \times \frac{5}{8} = \frac{\overset{2}{\cancel{16}}}{1} \times \frac{5}{\underset{1}{\cancel{8}}} = \frac{10}{1} = 10$$

Complete these division examples.

1. $3 \div \frac{1}{6} = \frac{3}{1} \times \underline{\ ?\ } = \underline{\ ?\ }$

2. $3 \div \frac{2}{3} = \frac{3}{1} \times \underline{\ ?\ } = \underline{\ ?\ }$

3. $10 \div \frac{1}{10} = \frac{10}{1} \times \underline{\ ?\ } = \underline{\ ?\ }$

4. $9 \div \frac{3}{7} = \frac{9}{1} \times \underline{\ ?\ } = \underline{\ ?\ }$

Find the quotient.*

5. $3 \div \frac{9}{10}$

6. $15 \div \frac{3}{4}$

7. $18 \div \frac{3}{4}$

8. $8 \div \frac{12}{19}$

9. $48 \div \frac{7}{10}$

Solve.

10. If $\frac{8}{9}$ of the number of students in a school is 480, how many students are in the school?

11. If $\frac{1}{3}$ of José's weight is equal to $\frac{1}{2}$ of 70 pounds, how much does José weigh?

 MENTAL ◆ MATH

Write <, =, or >.

12. $5 \div \frac{1}{2} \ \underline{\ ?\ } \ 5 \times \frac{1}{2}$

13. $5 \div \frac{1}{2} \ \underline{\ ?\ } \ 5 \div \frac{1}{5}$

14. $5 \div \frac{1}{2} \ \underline{\ ?\ } \ 4 \div \frac{1}{2}$

15. $5 \div \frac{1}{2} \ \underline{\ ?\ } \ 5 \times \frac{2}{1}$

16. $5 \div \frac{1}{2} \ \underline{\ ?\ } \ 6 \div \frac{1}{2}$

17. $5 \div \frac{1}{2} \ \underline{\ ?\ } \ 4 \div \frac{2}{1}$

Dividing a Fraction by a Fraction or a Whole Number

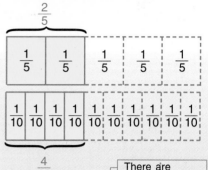

▶ **To divide a *fraction by a fraction*:**

- Multiply by the *reciprocal* of the divisor.

$$\frac{2}{5} \div \frac{1}{10} = \frac{2}{5} \times \frac{\overset{2}{\cancel{10}}}{1} = \frac{4}{1} = 4$$

reciprocals

$$\frac{14}{15} \div \frac{7}{10} = \frac{\overset{2}{\cancel{14}}}{\underset{3}{\cancel{15}}} \times \frac{\overset{2}{\cancel{10}}}{\underset{1}{\cancel{7}}} = \frac{2 \times 2}{3 \times 1} = \frac{4}{3} = 1\frac{1}{3}$$

$$\frac{2}{5} \div \frac{1}{10} = 4$$

There are four $\frac{1}{10}$'s in $\frac{2}{5}$.

▶ **To divide a *fraction by a whole number*:**

- Rewrite the whole number as an *improper* fraction with a denominator of 1.
- Then divide the fractions.

$$\frac{2}{7} \div 4 = \frac{2}{7} \div \frac{4}{1} = \frac{\overset{1}{\cancel{2}}}{7} \times \frac{1}{\underset{2}{\cancel{4}}} = \frac{1}{14}$$

Complete these division examples.

1. $\frac{7}{9} \div \frac{2}{3} = \frac{7}{9} \times \underline{\ ?\ } = \underline{\ ?\ }$

2. $\frac{6}{7} \div \frac{1}{8} = \frac{6}{7} \times \underline{\ ?\ } = \underline{\ ?\ }$

3. $\frac{5}{8} \div 15 = \frac{5}{8} \times \underline{\ ?\ } = \underline{\ ?\ }$

4. $\frac{1}{2} \div 4 = \frac{1}{2} \times \underline{\ ?\ } = \underline{\ ?\ }$

Write each quotient in lowest terms.*

5. $\frac{3}{4} \div \frac{1}{5}$ 6. $\frac{2}{5} \div \frac{1}{3}$ 7. $\frac{4}{5} \div \frac{1}{6}$ 8. $\frac{9}{10} \div \frac{5}{7}$ 9. $\frac{4}{11} \div \frac{1}{3}$

10. $\frac{5}{6} \div \frac{11}{12}$ 11. $\frac{2}{9} \div 3$ 12. $\frac{3}{8} \div 4$ 13. $\frac{7}{12} \div 3$ 14. $\frac{3}{7} \div \frac{2}{11}$

Write <, =, or >.

15. $\frac{4}{5} \div \frac{4}{5} \ \underline{\ ?\ } \ 1$ 16. $\frac{14}{15} \div \frac{7}{15} \ \underline{\ ?\ } \ 1$ 17. $\frac{3}{2} \div 3 \ \underline{\ ?\ } \ \frac{1}{2}$ 18. $\frac{4}{7} \div \frac{3}{7} \ \underline{\ ?\ } \ 1$

19. $\frac{7}{8} \div \frac{7}{10} \ \underline{\ ?\ } \ 1\frac{1}{2}$ 20. $\frac{2}{5} \div \frac{2}{7} \ \underline{\ ?\ } \ 1$ 21. $\frac{6}{5} \div \frac{2}{5} \ \underline{\ ?\ } \ 2\frac{1}{2}$ 22. $\frac{16}{19} \div \frac{2}{7} \ \underline{\ ?\ } \ \frac{1}{3}$

Solve.

23. The distance of a relay race was $\frac{4}{5}$ mile long. What distance did each of the 4 athletes on the relay team run?

24. If a watch loses $\frac{3}{20}$ of a minute in one day, in how many days will it have lost $\frac{3}{5}$ of a minute?

8-8 | Simplifying Complex Fractions

Complex fraction: a fraction that has a fraction in either the *numerator* or the *denominator* or *both*.

Examples: $\dfrac{\frac{3}{4}}{12}$, $\dfrac{9}{\frac{3}{5}}$, $\dfrac{\frac{8}{9}}{\frac{5}{6}}$

To simplify a complex fraction:
Divide the numerator by the denominator.

Fraction in Numerator	Fraction in Denominator	Fraction in Both Places
$\dfrac{\frac{3}{4}}{12} = \dfrac{3}{4} \div \dfrac{12}{1}$	$\dfrac{9}{\frac{3}{5}} = \dfrac{9}{1} \div \dfrac{3}{5}$	$\dfrac{\frac{8}{9}}{\frac{5}{6}} = \dfrac{8}{9} \div \dfrac{5}{6}$
$= \dfrac{3}{4} \times \dfrac{1}{12}$	$= \dfrac{9}{1} \times \dfrac{5}{3}$	$= \dfrac{8}{9} \times \dfrac{6}{5}$
$= \dfrac{\overset{1}{3}}{4} \times \dfrac{1}{\underset{4}{12}} = \dfrac{1}{16}$	$= \dfrac{\overset{3}{9}}{1} \times \dfrac{5}{\underset{1}{3}} = \dfrac{15}{1} = 15$	$= \dfrac{8}{\underset{3}{9}} \times \dfrac{\overset{2}{6}}{5} = \dfrac{16}{15} = 1\dfrac{1}{15}$

Simplify.*

1. $\dfrac{10}{\frac{2}{7}}$　　2. $\dfrac{9}{\frac{3}{8}}$　　3. $\dfrac{6}{\frac{2}{3}}$　　4. $\dfrac{4}{\frac{8}{9}}$　　5. $\dfrac{\frac{3}{4}}{\frac{1}{4}}$　　6. $\dfrac{\frac{3}{5}}{\frac{1}{10}}$

7. $\dfrac{\frac{6}{5}}{12}$　　8. $\dfrac{\frac{2}{5}}{20}$　　9. $\dfrac{\frac{2}{3}}{10}$　　10. $\dfrac{\frac{5}{6}}{12}$　　11. $\dfrac{9}{\frac{3}{4}}$　　12. $\dfrac{\frac{5}{7}}{\frac{1}{2}}$

13. $\dfrac{\frac{1}{7}}{\frac{1}{2}}$　　14. $\dfrac{\frac{7}{9}}{\frac{2}{3}}$　　15. $\dfrac{\frac{3}{7}}{\frac{9}{21}}$　　16. $\dfrac{\frac{1}{9}}{\frac{13}{9}}$　　17. $\dfrac{\frac{3}{2}}{\frac{21}{10}}$　　18. $\dfrac{8}{\frac{4}{3}}$

19. $\dfrac{\frac{3}{8}}{5}$　　20. $\dfrac{8}{\frac{1}{3}}$　　21. $\dfrac{\frac{2}{3}}{\frac{7}{9}}$　　22. $\dfrac{4}{\frac{9}{16}}$　　23. $\dfrac{19}{\frac{7}{8}}$　　24. $\dfrac{\frac{2}{3}}{\frac{5}{4}}$

Solve.

25. A bicycle path of $\frac{4}{5}$ of a mile is marked in tenths of a mile. How many markers are there?

26. If $\frac{2}{3}$ of the distance between two parks is 48 miles, how far apart are they?

27. If a bicycle rack $\frac{3}{4}$ full holds 18 bicycles, how many bicycles will the rack hold?

28. In a park $\frac{3}{8}$ of the trees or 27 trees are evergreens. How many trees are in the park?

Dividing Mixed Numbers

The Biking Club is planning a $6\frac{1}{4}$-hour trip. If they stop every $1\frac{1}{4}$ hours, how many times will they stop? $\longrightarrow$ $6\frac{1}{4} \div 1\frac{1}{4} = \underline{\ ?\ }$

To divide mixed numbers:

- Rewrite the numbers as improper fractions.
- Then divide. (The quotient can be a *whole number*, a *fraction*, or a *mixed number*.)

Whole-Number Quotient

$$6\frac{1}{4} \div 1\frac{1}{4} = \frac{25}{4} \div \frac{5}{4} = \underline{\ ?\ }$$
$$= \frac{25}{4} \times \frac{4}{5} = 5 \text{ stops}$$

Fraction Quotient

$$2\frac{2}{9} \div 3\frac{1}{3} = \frac{20}{9} \div \frac{10}{3} = \underline{\ ?\ }$$
$$= \frac{20}{9} \times \frac{3}{10} = \frac{2}{3}$$

Mixed-Number Quotient

$$3\frac{1}{3} \div 1\frac{1}{4} = \frac{10}{3} \div \frac{5}{4} = \underline{\ ?\ }$$
$$= \frac{10}{3} \times \frac{4}{5} = \frac{8}{3} = 2\frac{2}{3}$$

Find the quotient. Write each in lowest terms.

1. $2\frac{1}{3} \div 1\frac{1}{3}$
2. $8\frac{1}{4} \div 1\frac{1}{8}$
3. $6\frac{1}{3} \div 3\frac{4}{5}$
4. $3\frac{2}{5} \div 3\frac{2}{5}$

5. $4\frac{1}{2} \div 2$
6. $93 \div 7\frac{3}{4}$
7. $3\frac{5}{6} \div 3$
8. $1\frac{5}{7} \div 6\frac{3}{4}$

9. $9\frac{1}{2} \div 6\frac{1}{3}$
10. $10\frac{1}{3} \div 2\frac{1}{5}$
11. $6\frac{5}{8} \div 2\frac{1}{4}$
12. $9\frac{1}{21} \div 6\frac{2}{3}$

13. $\dfrac{19\frac{3}{8}}{6\frac{1}{5}}$
14. $\dfrac{11\frac{1}{2}}{8\frac{5}{8}}$
15. $\dfrac{1\frac{5}{8}}{1\frac{1}{2}}$
16. $\dfrac{20\frac{5}{6}}{11\frac{1}{9}}$

17. $\dfrac{7}{2\frac{5}{8}}$
18. $\dfrac{9\frac{1}{2}}{3\frac{1}{3}}$
19. $\dfrac{10\frac{2}{7}}{3\frac{5}{9}}$
20. $\dfrac{41\frac{2}{3}}{7\frac{1}{7}}$

Write <, =, or >.

21. $2\frac{1}{2} \div 2\frac{1}{2} \underline{\ ?\ } 1$
22. $3\frac{1}{3} \div 2\frac{1}{3} \underline{\ ?\ } 1$
23. $6\frac{1}{8} \div 2 \underline{\ ?\ } 4$

24. $1\frac{1}{6} \div 4\frac{1}{6} \underline{\ ?\ } 1$
25. $3\frac{1}{3} \div 6\frac{1}{3} \underline{\ ?\ } \frac{1}{2}$
26. $3\frac{1}{2} \div \frac{1}{6} \underline{\ ?\ } 2$

27. $5\frac{1}{5} \div 5\frac{1}{10} \underline{\ ?\ } 1$
28. $4\frac{2}{5} \div 8\frac{4}{5} \underline{\ ?\ } \frac{1}{2}$
29. $7\frac{1}{3} \div 6\frac{1}{6} \underline{\ ?\ } \frac{1}{2}$

Estimate quotients by rounding to the nearest whole number.

30. $8\frac{1}{2} \div 2\frac{1}{10}$

31. $7\frac{4}{5} \div 4\frac{1}{10}$

32. $6\frac{1}{8} \div \frac{15}{16}$

33. $12\frac{1}{3} \div 6\frac{1}{8}$

34. $7\frac{1}{2} \div 4\frac{1}{4}$

35. $3\frac{3}{4} \div 1\frac{7}{8}$

36. $15\frac{1}{10} \div 2\frac{4}{5}$

37. $27\frac{1}{2} \div 7\frac{1}{9}$

38. $16\frac{1}{8} \div 10\frac{1}{2}$

39. $\frac{19}{20} \div 1\frac{9}{10}$

40. $3\frac{1}{2} \div 3\frac{7}{10}$

41. $17\frac{2}{3} \div 9\frac{2}{5}$

Solve. Look for a shortcut.

42. $1\frac{1}{4} \times 2\frac{2}{9} \div 4\frac{1}{6}$

43. $3\frac{1}{5} \times \frac{7}{12} \div 2\frac{1}{3}$

44. $\frac{4}{5} \times 3\frac{1}{4} \div \frac{9}{10}$

45. $1\frac{3}{7} \times 2\frac{1}{4} \div \frac{9}{7}$

46. $\frac{9}{10} \times \frac{8}{9} \div \frac{7}{30}$

47. $7\frac{1}{3} \times 4\frac{1}{8} \div \frac{11}{12}$

48. $6\frac{7}{8} \times \frac{1}{2} \div \frac{5}{6}$

49. $9\frac{1}{3} \times 6 \div \frac{7}{3}$

50. $6\frac{1}{2} \times 2 \div \frac{3}{4} \times \frac{1}{2} \times \frac{2}{4}$

51. $5 \times \frac{3}{4} \div \frac{1}{8} \times \frac{5}{8} \div 2\frac{1}{2} \times 2$

Solve.

52. In a frog-leaping contest one frog covered a distance of $203\frac{1}{4}$ in. in three leaps. Find the average length of one leap.

53. Tim's frog leaped $6\frac{1}{8}$ in. This was $1\frac{3}{4}$ times farther than its last leap. How long was the frog's last leap?

54. The smallest frog in the contest is $\frac{2}{3}$ the size of the largest frog. The smallest frog is $6\frac{1}{4}$ in. long. How long is the largest frog?

55. Human hair grows at the rate of $\frac{1}{3}$ in. per month. Tara wants her hair to grow $4\frac{2}{3}$ in. longer. How many months will this take?

$203\frac{1}{4}$ in.

CHALLENGE

Compute. Use what you know about the order of operations.

56. $2\frac{1}{2} \times 3\frac{1}{3} + 4\frac{1}{6} - 1\frac{1}{2} \div \frac{3}{4}$

57. $\left(\frac{1}{2} + 3\frac{1}{4}\right) \div 3 + 2\frac{2}{5}$

58. $\left(3\frac{3}{4} - \frac{5}{8}\right) \div \frac{5}{6} + 1\frac{1}{5}$

59. $10 \div \left(8\frac{1}{3} - 6\frac{1}{4}\right) - \frac{3}{10}$

60. $\dfrac{5\frac{1}{4} - 2\frac{1}{2}}{7\frac{4}{5} - 6\frac{1}{3}}$

61. $\dfrac{4\frac{2}{3} + 1\frac{1}{6}}{1\frac{1}{4} + 1\frac{7}{8}}$

8-10 Finding Parts of a Number and the Number

To find a fractional part of a number:

Multiply the number by the fraction.

Two sevenths of the 35 minerals in the exhibit are gems. How many are gems?

$$\frac{2}{7} \text{ of } 35 \longrightarrow \frac{2}{7} \times 35 \longrightarrow \frac{2}{\cancel{7}_1} \times \frac{\cancel{35}^5}{1} = 10$$

10 minerals are gems.

Remember: "of" means multiply.

To find a number when a fractional part of it is given:

Divide the part by the fraction.

Ten minerals in the exhibit are gems.
This is $\frac{2}{7}$ of the minerals in the exhibit.
How many minerals are in the exhibit?

$$\frac{2}{7} \text{ of } \underline{\ ?\ } = 10 \longrightarrow \frac{2}{7} \times \underline{\ ?\ } = 10$$

Remember: Divide to find a missing factor or number.

$$\frac{2}{7} \times \underline{\ ?\ } = 10 \longrightarrow 10 \div \frac{2}{7} = \frac{\cancel{10}^5}{1} \times \frac{7}{\cancel{2}_1} = 35$$

35 minerals are in the exhibit.

Find the fractional part of the number.

1. $\frac{1}{2}$ of 18 = $\underline{\ ?\ }$ 2. $\frac{4}{5}$ of 35 = $\underline{\ ?\ }$ 3. $\frac{1}{4}$ of 80 = $\underline{\ ?\ }$ 4. $\frac{3}{5}$ of 25 = $\underline{\ ?\ }$

5. $\frac{2}{3}$ of 90 = $\underline{\ ?\ }$ 6. $\frac{3}{7}$ of 35 = $\underline{\ ?\ }$ 7. $\frac{1}{4}$ of 64 = $\underline{\ ?\ }$ 8. $\frac{1}{5}$ of 555 = $\underline{\ ?\ }$

9. $\frac{5}{6}$ of 48 = $\underline{\ ?\ }$ 10. $\frac{1}{3}$ of 45 = $\underline{\ ?\ }$ 11. $\frac{3}{8}$ of 72 = $\underline{\ ?\ }$ 12. $\frac{3}{8}$ of 64 = $\underline{\ ?\ }$

13. $\frac{2}{9}$ of 63 = $\underline{\ ?\ }$ 14. $\frac{3}{16}$ of 80 = $\underline{\ ?\ }$ 15. $\frac{5}{7}$ of 210 = $\underline{\ ?\ }$ 16. $\frac{7}{9}$ of 189 = $\underline{\ ?\ }$

Find each number, given the fractional part.

17. $\frac{1}{8}$ of ___?___ = 32

18. $\frac{2}{3}$ of ___?___ = 24

19. $\frac{1}{4}$ of ___?___ = 100

20. $\frac{2}{5}$ of ___?___ = 82

21. $\frac{5}{6}$ of ___?___ = 30

22. $\frac{2}{9}$ of ___?___ = 50

23. $\frac{3}{4}$ of ___?___ = 69

24. $\frac{4}{5}$ of ___?___ = 44

25. $\frac{1}{2}$ of ___?___ = 75

26. $\frac{3}{2}$ of ___?___ = 642

27. $\frac{5}{8}$ of ___?___ = 75

28. $\frac{4}{3}$ of ___?___ = 16

29. $\frac{3}{7}$ of ___?___ = 51

30. $\frac{7}{10}$ of ___?___ = 210

31. $\frac{8}{7}$ of ___?___ = 96

32. $\frac{3}{8}$ of ___?___ = 39

33. $\frac{3}{8}$ of ___?___ = 48

34. $\frac{5}{4}$ of ___?___ = 65

35. $\frac{3}{4}$ of ___?___ = 120

36. $\frac{7}{12}$ of ___?___ = 84

37. $\frac{2}{9}$ of ___?___ = 38

38. $\frac{3}{16}$ of ___?___ = 48

39. $\frac{2}{5}$ of ___?___ = 98

40. $\frac{4}{9}$ of ___?___ = 96

Solve.

41. In a collection of 144 minerals $\frac{3}{8}$ are metallic. How many are metallic?

Exercise 45

42. Ten members of the class chose fossils for their science report. This was $\frac{2}{9}$ of the total number in class. How many students are there in all?

43. The school bus has seats for 45 students. The bus is $\frac{2}{3}$ full. How many seats are empty?

44. 120 tickets to the museum show have been sold. This is $\frac{2}{5}$ of the number of tickets available. How many tickets are left to be sold?

45. This graph shows how Evelyn spent the $60 she earned by babysitting last month. How much money did Evelyn have left?

MAKE UP YOUR OWN...

46. Write a word problem in which you have to find the fractional part of a number.

47. Write a word problem in which you have to find a number when a fractional part of it is given.

8-11 Changing Fractions and Mixed Numbers to Decimals

Change fractions to decimals in *one* of these ways:

1. When the denominator is already a power of ten, use *inspection* to write the equivalent decimal.

$$\frac{7}{10} = 0.7 \qquad \frac{23}{100} = 0.23 \qquad \frac{69}{1000} = 0.069$$

2. Change the denominator to a power of ten.

$$\frac{3}{5} = \frac{3 \times 2}{5 \times 2} = \frac{6}{10} = 0.6 \qquad \frac{7}{25} = \frac{7 \times 4}{25 \times 4} = \frac{28}{100} = 0.28$$

3. Divide.

$$\frac{5}{8} \longrightarrow 8\overline{)5.000}^{\,0.625} \longrightarrow \frac{5}{8} = 0.625$$

Change mixed numbers to decimals by: $\qquad 3\frac{3}{4} = \underline{\ ?\ }$

- Changing the fraction part to a decimal. $\longrightarrow \dfrac{3}{4} = \dfrac{3 \times 25}{4 \times 25} = \dfrac{75}{100} = 0.75$

$$\frac{3}{4} = 0.75$$

- Attaching the decimal to the whole number. $\longrightarrow 3\frac{3}{4} = 3 + 0.75 = 3.75$

Study this: $2\frac{3}{8} = \underline{\ ?\ }$

Think: $\qquad \dfrac{3}{8} = 8\overline{)3.000}^{\,0.375} \longrightarrow$ So, $2\dfrac{3}{8} = 2.375$

Change improper fractions to decimals by changing to a mixed number and proceeding as above.

$$\frac{7}{5} = \underline{\ ?\ } \longrightarrow \frac{7}{5} = 1\frac{2}{5} \longrightarrow \text{Think: } \frac{2}{5} = 5\overline{)2.0}^{\,0.4} \longrightarrow 1\frac{2}{5} = 1.4$$

Change these fractions to decimals.

1. $\dfrac{3}{10}$ 2. $\dfrac{9}{100}$ 3. $\dfrac{3}{25}$ 4. $\dfrac{3}{50}$ 5. $\dfrac{9}{16}$

6. $\dfrac{7}{8}$ 7. $\dfrac{63}{100}$ 8. $\dfrac{8}{20}$ 9. $\dfrac{3}{4}$ 10. $\dfrac{3}{5}$

11. $\dfrac{3}{16}$ 12. $\dfrac{5}{25}$ 13. $\dfrac{1}{4}$ 14. $\dfrac{6}{16}$ 15. $\dfrac{40}{64}$

194

Change these improper fractions to decimals.

16. $\frac{20}{8}$ 17. $\frac{31}{5}$ 18. $\frac{17}{8}$ 19. $\frac{250}{200}$ 20. $\frac{35}{4}$

21. $\frac{17}{2}$ 22. $\frac{25}{4}$ 23. $\frac{44}{5}$ 24. $\frac{125}{10}$ 25. $\frac{225}{100}$

Change these to decimals. Compare them. Write <, =, or >.

26. $\frac{1}{2}$? $\frac{4}{8}$ 27. $\frac{3}{8}$? $\frac{5}{10}$ 28. $\frac{7}{8}$? $\frac{9}{10}$

29. $\frac{19}{8}$? $\frac{15}{6}$ 30. $\frac{9}{15}$? $\frac{5}{8}$ 31. $\frac{1}{16}$? $\frac{3}{20}$

32. $2\frac{1}{5}$? $2\frac{1}{4}$ 33. $8\frac{3}{4}$? $8\frac{6}{8}$ 34. $7\frac{2}{5}$? $7\frac{2}{16}$

35. $3\frac{3}{50}$? $3\frac{2}{100}$ 36. $4\frac{4}{200}$? $4\frac{2}{100}$ 37. $5\frac{4}{5}$? $5\frac{13}{16}$

38. $\frac{73}{8}$? $\frac{73}{4}$ 39. $\frac{97}{2}$? $\frac{94}{4}$ 40. $\frac{101}{20}$? $\frac{205}{40}$

Copy and complete this chart. Memorize it.
Leave space to add other fractions as you learn them.

	Common Fractions	Equivalent Decimals
41.	$\frac{1}{2}$	0.50
42.	$\frac{1}{4}$	
43.	$\frac{3}{4}$	
44.	$\frac{1}{5}$	
45.	$\frac{2}{5}$	
46.	$\frac{3}{5}$	

	Common Fractions	Equivalent Decimals
47.	$\frac{4}{5}$	0.80
48.	$\frac{1}{8}$	
49.	$\frac{3}{8}$	
50.	$\frac{5}{8}$	
51.	$\frac{7}{8}$	
52.	$\frac{1}{10}$	

CALCULATOR ACTIVITY

Change these mixed numbers to decimals.
Use your calculator as shown in exercise 53.

53. $1\frac{3}{5}$ ⟶ [5 × 1 + 3] ÷ 5 = 1.6   54. $2\frac{3}{4}$

55. $7\frac{1}{8}$ 56. $12\frac{4}{5}$ 57. $30\frac{3}{10}$ 58. $8\frac{2}{10}$ 59. $15\frac{5}{8}$ 60. $5\frac{3}{16}$

Terminating Decimals: decimals with NO remainders.

$$0.8 \qquad\qquad 0.02 \qquad\qquad 0.125$$

These are terminating decimals. When their fractional equivalents are divided to change each to a decimal, the remainder is always zero.

$$\frac{4}{5} \longrightarrow 5\overline{)4.0}^{\,0.8} \qquad \frac{1}{50} \longrightarrow 50\overline{)1.00}^{\,0.02} \qquad \frac{1}{8} \longrightarrow 8\overline{)1.000}^{\,0.125}$$

Remainders are zero.

Repeating Decimals: decimals whose last digit or digits repeat.

$$0.33333\ldots = 0.\overline{3} \qquad\qquad 0.363636\ldots = 0.\overline{36}$$

These are repeating decimals because the digits in the quotient repeat.

$$3\overline{)1.\,{}^10\,{}^10\,{}^10\,{}^10}^{\,0.\ 3\ 3\ 3\ 3\ldots} = 0.\overline{3}$$

Write a bar over the digit or digits that repeat.

$$11\overline{)4.\ 0\ 0\ 0\ 0}^{\,0.\ 3\ 6\ 3\ 6\ldots} = 0.\overline{36}$$

```
  0. 3 6 3 6...    = 0.36
11)4. 0 0 0 0
  -3 3
     7 0
    -6 6
       4 0
      -3 3
         7 0
        -6 6
           4
```

Remainders 7 and 4 repeat to yield "36" repeating in the quotient.

Write the next 4 digits for each decimal.

1. $0.13\overline{45}$
2. $0.\overline{1345}$
3. $0.134\overline{5}$
4. $0.1\overline{345}$
5. 0.1345

Rewrite each decimal using a bar over the repeating digits.

6. $0.155555\ldots$
7. $0.151515\ldots$
8. $0.151151\ldots$
9. $2.678787\ldots$
10. $2.678678\ldots$
11. $2.678888\ldots$

Write each fraction as a decimal. Which are repeating decimals?

12. $\frac{1}{2}$

13. $\frac{5}{16}$

14. $\frac{1}{8}$

15. $\frac{5}{11}$

16. $\frac{9}{22}$

17. $\frac{13}{20}$

18. $\frac{2}{15}$

19. $\frac{7}{8}$

20. $\frac{1}{9}$

21. $\frac{7}{18}$

22. $\frac{2}{3}$

23. $\frac{7}{40}$

24. $\frac{4}{5}$

25. $\frac{2}{7}$

26. $\frac{9}{20}$

27. $\frac{5}{12}$

28. $\frac{8}{15}$

29. $\frac{1}{32}$

Write <, =, or >.
Change each fraction to a decimal to compare.

30. $\frac{3}{5}$? $\frac{2}{3}$

31. $\frac{7}{9}$? $\frac{3}{4}$

32. $\frac{2}{3}$? $0.\overline{6}$

33. $\frac{4}{5}$? 0.7

34. $\frac{15}{100}$? $\frac{2}{13}$

35. $\frac{7}{6}$? $1.1\overline{6}$

36. 0.45 ? $\frac{4}{9}$

37. 0.704 ? $\frac{71}{100}$

38. $1.\overline{4}$? $1.\overline{41}$

39. $6.\overline{21}$? $6.2\overline{12}$

40. $7.\overline{03}$? $7.0\overline{3}$

41. $9.\overline{152}$? $9.1\overline{52}$

 **Finding Together**

Copy and complete each chart.
Use a calculator to help you.

42.

Fraction	$\frac{1}{6}$	$\frac{2}{6}$	$\frac{3}{6}$	$\frac{4}{6}$	$\frac{5}{6}$
Decimal	?	$0.\overline{3}$	?	?	?

43.

Fraction	$\frac{1}{9}$	$\frac{2}{9}$	$\frac{3}{9}$	$\frac{4}{9}$	$\frac{5}{9}$	$\frac{6}{9}$	$\frac{7}{9}$	$\frac{8}{9}$
Decimal	$0.\overline{1}$	?	?	?	?	?	?	?

44. Make two more charts showing equivalent fractions and
decimals for sevenths and for elevenths.
(Remember: Use a bar to show repeating decimals.)

Write the equivalent fraction for each decimal. Use your charts.

45. $0.\overline{3}$

46. $0.8\overline{3}$

47. $0.\overline{27}$

48. 0.3125

49. $0.\overline{7}$

50. 0.45

51. $0.1\overline{6}$

52. 0.26

53. $0.\overline{09}$

54. $0.\overline{1}$

55. 0.81

56. $0.\overline{5}$

57. 0.875

58. 0.125

59. $0.\overline{6}$

8-13 Changing Decimals to Fractions and Mixed Numbers

To change decimals to fractions and mixed numbers:

▶ **When the decimal is less than 1.**

- Read the decimal.
- Write the decimal as the numerator of the fraction.
- Write the power of ten named by the place value of the decimal as the denominator of the fraction.
- Express the fraction in lowest terms.

$$0.4 \longrightarrow 4 \text{ tenths} \quad = \frac{4}{10} = \frac{4 \div 2}{10 \div 2} = \frac{2}{5}$$

$$0.35 \longrightarrow 35 \text{ hundredths} = \frac{35}{100} = \frac{35 \div 5}{100 \div 5} = \frac{7}{20}$$

$$0.125 \longrightarrow 125 \text{ thousandths} = \frac{125}{1000} = \frac{125 \div 125}{1000 \div 125} = \frac{1}{8}$$

▶ **When the decimal is greater than 1.**

- Change the decimal part to a fraction as outlined above.
- Write the whole-number part and the fraction as a mixed number.

$$3.25 = \underline{\ ?\ }$$

Think: $3.25 = 3 + 0.25 \longrightarrow 0.25 = \frac{25}{100} = \frac{25 \div 25}{100 \div 25} = \frac{1}{4}$

$$3.25 = 3\frac{1}{4}$$

Write each decimal as a fraction or mixed number.
Express the fraction in lowest terms.

1. 0.6	**2.** 0.4	**3.** 0.05	**4.** 0.65	**5.** 0.750
6. 0.125	**7.** 0.045	**8.** 0.075	**9.** 0.500	**10.** 0.2500
11. 0.089	**12.** 0.625	**13.** 0.44	**14.** 0.875	**15.** 8.1875
16. 2.08	**17.** 3.2	**18.** 1.12	**19.** 5.55	**20.** 4.63
21. 8.015	**22.** 6.275	**23.** 4.0625	**24.** 7.125	**25.** 3.875
26. 4.625	**27.** 7.0875	**28.** 9.0375	**29.** 1.0065	**30.** 9.125

Copy and complete each table.

31.

Decimal	Fraction
0.25	?
0.50	$\frac{2}{4} = \frac{1}{2}$
0.75	?
1.00	?
1.25	$\frac{5}{4}$
1.50	?
1.75	?

32.

Decimal	Fraction
0.20	$\frac{1}{5}$
0.40	?
0.60	?
0.80	?
1.00	$\frac{5}{5} = \frac{1}{1}$
1.20	?
1.40	?
1.60	$\frac{8}{5}$
1.80	?

33.

Decimal	Fraction
0.125	?
0.250	?
0.375	?
0.500	$\frac{4}{8} = \frac{1}{2}$
0.625	?
0.750	?
0.875	?
1.000	$\frac{8}{8} = \frac{1}{1}$
1.125	?
1.250	?

Compare. Write <, =, or >.

34. $0.75 \underline{\ ?\ } \frac{6}{8}$ 35. $0.2 \underline{\ ?\ } \frac{3}{16}$ 36. $\frac{3}{5} \underline{\ ?\ } 0.5$

37. $\frac{3}{8} \underline{\ ?\ } 0.25$ 38. $0.05 \underline{\ ?\ } \frac{1}{10}$ 39. $0.125 \underline{\ ?\ } \frac{1}{16}$

40. $\frac{3}{2} \underline{\ ?\ } 1.3$ 41. $2.50 \underline{\ ?\ } \frac{9}{4}$ 42. $\frac{14}{5} \underline{\ ?\ } 2.4$

43. $1.07 \underline{\ ?\ } 1\frac{1}{8}$ 44. $1\frac{1}{5} \underline{\ ?\ } 1.40$ 45. $2.75 \underline{\ ?\ } 2\frac{1}{8}$

Which number in each set does *not* belong? Explain.

46. $\left\{0.70, \frac{35}{50}, 0.7, \frac{49}{70}, 0.9\right\}$ 47. $\left\{\frac{18}{16}, 1.125, \frac{45}{40}, 1.250, 1\frac{1}{8}\right\}$

48. $\left\{0.3125, \frac{1}{3}, \frac{10}{32}, \frac{25}{80}, \frac{15}{48}\right\}$ 49. $\left\{1.75, \frac{7}{4}, 1.250, 1\frac{6}{8}, 1\frac{3}{4}\right\}$

SKILLS TO REMEMBER

Name the quotient and remainder.

50. $805 \div 14$ 51. $1370 \div 16$ 52. $7834 \div 124$ 53. $5646 \div 132$

54. $3040 \div 28$ 55. $9246 \div 36$ 56. $22{,}500 \div 180$ 57. $65{,}490 \div 214$

STRATEGY
8-14 Problem Solving: Interpreting the Remainder

Problem: Mr. Valdez is taking his science classes on a trip to NASA. There are 132 people on the trip, including the chaperones. Each bus can hold 48 passengers. How many buses should he hire?

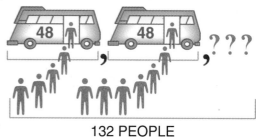

132 PEOPLE

1 IMAGINE Picture buses that hold 48 passengers each. Imagine the 132 people boarding the buses.

2 NAME *Facts:*   1 bus — 48 people
Total — 132 people

Question: ? buses for 132 people

3 THINK Since one bus holds 48 passengers, you should divide the 132 people by the 48 passengers.

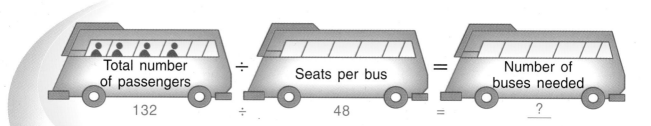

Total number of passengers	÷	Seats per bus	=	Number of buses needed
132	÷	48	=	?

4 COMPUTE $132 \div 48 = 2\frac{36}{48}$

But there is a remainder of 36 people. You must hire another bus to hold them.

So, Mr. Valdez must hire 3 buses.

5 CHECK Multiply to check your division:

$2\frac{36}{48} \times 48 = 2\frac{3}{4} \times 48 = 132$

Three buses are hired to have enough room for the 132 passengers. There will be extra seats.

STOP

Solve by interpreting the remainder.

1. NASA purchased 65 yards of material to make suits for the six new astronauts. If each suit requires $5\frac{3}{4}$ yards of material, can NASA make two suits for each astronaut? How much more material is needed?

| 1 IMAGINE | Draw and label a picture. | → | 2 NAME | → Facts → Question |

| 3 THINK | Multiply to find how much material is needed for 12 suits if $5\frac{3}{4}$ yd makes 1 suit: |

(6 astronauts × 2 suits each) × $5\frac{3}{4}$ yd per suit.

| 4 COMPUTE | → | 5 CHECK |

65 yards of material

2. Gerry's class ate lunch at NASA's cafeteria, which has tables each seating 13 persons. If there were 87 in her class, how many tables were used to seat the class for lunch?

3. The shuttle orbiter can carry a crew of seven and an additional 29 500 kilograms of equipment. The shuttle already has 27 625 kg of equipment on board and has to carry some satellites to be launched, each weighing 608 kg. How many satellites can the shuttle carry?

4. A total of 25,000 ceramic tiles are bonded to the body of the shuttle for use as heat shields. If 2 280 tiles need to be replaced after each flight and the tiles come in boxes of 250, how many boxes of tiles are needed?

5. The lunar rover must travel 27 km on a mission. One battery for the rover provides it with energy to travel 11.8 km without recharging. How many batteries should be loaded on the rover for this mission?

6. Zach's class decided to have a car wash to defray the expenses of the class trip. There are 32 students in the class and they washed 196 cars, charging $3.50 for each vehicle. If the cost of the bus for the trip is $800, will they have enough money for the trip if each student adds $4.00 to the proceeds of the car wash?

7. For extravehicular walks NASA provided a total of 24.5 hours of oxygen in three tanks. Two astronauts each used 7.25 hours of oxygen to repair a space satellite. Can they spend ten hours more working outside the spacecraft with the remaining oxygen supply?

8. NASA can accommodate 18 guests on each of their touring vehicles. If Hilda's class has 124 students, how many vehicles will they need to tour the facilities?

Match each question with the equation. Then solve.

1. What number is $\frac{3}{7}$ of 21?

2. Three sevenths of what number is 21?

3. What part of 3 is 7?

4. What part of 7 is 3?

a. $n \times 3 = 7$

b. $n = \frac{3}{7} \times 21$

c. $\frac{3}{7} \times n = 21$

d. $n \times 7 = 3$

e. $\frac{3}{7} \times n = 7$

True or false?

5. When a product is unknown, multiply.

6. When a factor is unknown, divide.

7. The reciprocal of every whole number greater than 1 is a unit fraction.

8. The sum of the reciprocals of 2 and 3 is $\frac{1}{5}$.

9. The sum of the reciprocals of 3 and 5 is $\frac{8}{15}$.

10. The reciprocal of a unit fraction is a whole number.

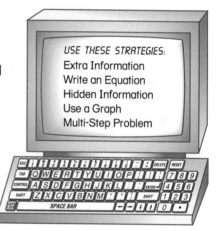

USE THESE STRATEGIES:
Extra Information
Write an Equation
Hidden Information
Use a Graph
Multi-Step Problem

Solve. You may combine strategies.

11. At the Ridley Park spring fair $\frac{2}{5}$ of the 350 young people who attended won prizes. How many young people won prizes?

12. The second-place finisher in the canoe race lost to the winner by only $\frac{1}{12}$ of the winner's time. If the winning time was $3\frac{1}{5}$ minutes, by how many seconds did the runner-up lose? (Hint: 60 sec = 1 min.)

13. Of the 42 young people who entered the bicycle race 35 finished. If the sum of all the finishing times was $8\frac{2}{5}$ hours, what was the average finishing time?

14. Sid's record-breaking jump was $1\frac{1}{8}$ times longer than the previous best in the broad-jump competition. If Sid's distance was $5\frac{2}{5}$ m, what was the old record?

15. During the kite-flying contest, Doreen's kite flew $6\frac{2}{3}$ times higher than Clara's and $3\frac{1}{2}$ times higher than Teri's. If Teri's kite reached a height of 84 feet, how high did Doreen's kite fly? How high did Clara's fly?

This graph shows the top six finishers in the fishing contest and the weight of the fish caught by each. Use the graph to complete exercises 16–19.

Fishing Contest

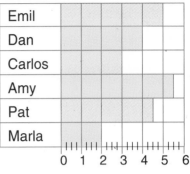

Weight in Kg

16. How many times larger than Dan's fish was Pat's fish?

17. What part of the total weight of the six fish did Dan's catch represent?

18. What was the average weight of the fish that were caught?

19. If the first prize was $.50 per kilogram, how much money did Amy win?

20. Sonja and Tara rented a paddle boat for $1\frac{1}{3}$ hours. If the rental charge was $5.25 per half hour and the girls split the cost, how much did each girl pay?

21. Jay and Steve competed in a pitching contest. Jay pitched three balls at a total speed of 207 mph. Steve's average speed for three pitches was $70\frac{1}{2}$ mph. By how many miles per hour did his average beat that of Jay?

Four teams participated in a potato sack race. The graph shows the times posted by each team. Use the graph to complete exercises 22–25.

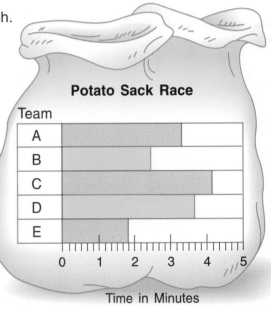

Potato Sack Race

Time in Minutes

22. Express each time as a decimal. Which is/are repeating decimals? Which is/are terminating decimals?

23. Team A's time is $\frac{4}{5}$ the time of Team __?__.

24. Team D's time is twice that of Team __?__.

25. Team C's time is $1\frac{2}{3}$ times faster than Team __?__.

There were 240 entries in the cooking contest. The circle graph shows the entry categories. Use the graph to complete exercises 26–29.

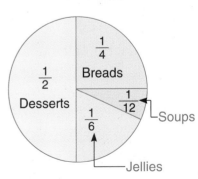

26. How many entries were there in the soup category?

27. How many entries were there in the jelly category?

28. If one twentieth of the dessert entries won prizes, how many entries won prizes in the dessert category?

29. Of the bread entries $\frac{3}{4}$ were sold for $3.00 a loaf and the rest were sold for $\frac{1}{3}$ of that price. How much money was raised by the sale of the bread entries?

More Practice

Find the product.

1. $\frac{3}{8} \times \frac{1}{8}$
2. $\frac{6}{9} \times \frac{3}{5}$
3. $\frac{2}{5} \times 25$
4. $\frac{7}{12} \times 7$
5. $5 \times \frac{3}{10} \times \frac{2}{3}$

6. $\frac{4}{5} \times \frac{20}{25} \times \frac{7}{8}$
7. $1\frac{2}{5} \times 7\frac{1}{2}$
8. $\frac{3}{4} \times 3\frac{1}{5}$
9. $7\frac{1}{12} \times 2\frac{1}{5}$

Compute mentally. Write <, =, or >.

10. $\frac{1}{2} \times \frac{4}{3}$? $\frac{3}{4} \times \frac{1}{2}$
11. $\frac{7}{9} \times \frac{3}{5}$? $\frac{8}{9} \times \frac{3}{5}$
12. $\frac{6}{7} \times \frac{3}{8}$? $\frac{1}{9} \times \frac{2}{3}$

Write the reciprocal.

13. $\frac{1}{8}$
14. $\frac{2}{9}$
15. $\frac{13}{2}$
16. 4
17. $\frac{5}{8}$
18. 6
19. $2\frac{7}{9}$
20. $\frac{16}{5}$

Find the quotient.

21. $8 \div \frac{1}{2}$
22. $\frac{7}{3} \div \frac{3}{8}$
23. $\frac{8}{9} \div 4$
24. $2\frac{1}{3} \div \frac{2}{8}$
25. $3\frac{1}{4} \div 2\frac{1}{2}$
26. $7\frac{1}{6} \div \frac{1}{6}$

Simplify these complex fractions.

27. $\dfrac{\frac{2}{5}}{10}$
28. $\dfrac{6}{1\frac{2}{3}}$
29. $\dfrac{10}{\frac{3}{4}}$
30. $\dfrac{8\frac{1}{2}}{4\frac{1}{8}}$
31. $\dfrac{9\frac{1}{9}}{5\frac{1}{3}}$
32. $\dfrac{7\frac{1}{2}}{3\frac{1}{4}}$
33. $\dfrac{10\frac{1}{3}}{12\frac{2}{3}}$

Write each fraction or mixed number as a decimal.

34. $\frac{5}{16}$
35. $\frac{26}{25}$
36. $\frac{3}{10}$
37. $\frac{7}{20}$
38. $\frac{11}{100}$
39. $1\frac{15}{20}$
40. $8\frac{7}{8}$

Write <, =, or >.
Change each to a decimal to compare.

41. $\frac{3}{4}$? $\frac{9}{10}$
42. $\frac{8}{20}$? $\frac{47}{100}$
43. $2\frac{3}{5}$? $2\frac{1}{2}$

Write each decimal as a fraction or mixed number in lowest terms.

44. 0.6
45. 0.275
46. 0.14
47. 0.0625
48. 0.35
49. 1.25

50. 9.9
51. 3.625
52. 1.80
53. 2.4375
54. 4.36
55. 5.22

Write each fraction as a decimal.
Use a bar to show repeating decimals.

56. $\frac{2}{3}$
57. $\frac{9}{11}$
58. $\frac{1}{15}$
59. $\frac{2}{7}$
60. $\frac{1}{6}$
61. $\frac{7}{40}$
62. $\frac{1}{8}$

Solve.

63. The English Channel is 33 kilometers wide. If a swimmer can swim only $\frac{1}{3}$ of the distance, how far will the athlete swim?

Math Probe

CONTINUED FRACTIONS

Special fractions like the one below are called *continued fractions*. Those having 1's in the numerators are called *simple continued fractions*.

$$\cfrac{1}{2 + \cfrac{1}{2 + \frac{1}{3}}}$$

Christian Huygens, a Dutch astronomer, developed this modern symbolism for these fractions.

To simplify a continued fraction:

- Begin with the last denominator and express the sum as an improper fraction.

$$\cfrac{1}{2 + \cfrac{1}{\boxed{2 + \frac{1}{3}}}}$$

$$2\frac{1}{3} = \frac{7}{3}$$

- Simplify the complex fraction. (Remember that dividing by a number is the same as multiplying by its reciprocal.)

$$\cfrac{1}{2 + \boxed{\cfrac{1}{\frac{7}{3}}}}$$

THINK:

$$\cfrac{1}{\frac{7}{3}} = 1 \times \frac{3}{7} = \frac{3}{7}$$

- Repeat this process until the fraction is simplified.

$$\cfrac{1}{2 + \frac{3}{7}} = \cfrac{1}{2\frac{3}{7}} = \cfrac{1}{\frac{17}{7}} = \frac{7}{17}$$

Simplify these continued fractions.

1. $\cfrac{1}{1 + \cfrac{1}{1 + \frac{1}{5}}}$

2. $\cfrac{1}{1 + \cfrac{1}{1 + \frac{1}{4}}}$

3. $\cfrac{1}{2 + \cfrac{1}{1 + \frac{1}{2}}}$

4. $\cfrac{1}{3 + \cfrac{1}{1 + \cfrac{1}{1 + \frac{1}{2}}}}$

5. $\cfrac{1}{1 + \cfrac{1}{2 + \cfrac{1}{1 + \frac{1}{2}}}}$

6. $\cfrac{1}{2 + \cfrac{1}{2 + \cfrac{1}{1 + \frac{1}{2}}}}$

---

7. **Make up your own** simple continued fraction and simplify it.

Check Your Mastery

Compute.

See pp. 180–191

1. $\frac{5}{8} \times \frac{1}{3}$
2. $6 \times \frac{8}{9}$
3. $\frac{1}{10} \times \frac{5}{12}$
4. $\frac{7}{8} \times \frac{9}{17}$
5. $6 \times \frac{1}{6}$

6. $\frac{2}{7} \times 28$
7. $\frac{5}{18} \times 24$
8. $2\frac{1}{2} \times 26$
9. $6\frac{7}{8} \times 1\frac{1}{7}$
10. $9 \times 4\frac{1}{6}$

11. $12 \div \frac{3}{7}$
12. $\frac{2}{9} \div \frac{2}{9}$
13. $\frac{6}{11} \div \frac{1}{22}$
14. $\frac{5}{8} \div \frac{5}{7}$
15. $\frac{9}{11} \div \frac{1}{3}$

16. $\dfrac{1}{2\frac{1}{2}}$
17. $\dfrac{1\frac{3}{5}}{8}$
18. $\dfrac{6\frac{2}{3}}{4}$
19. $\dfrac{2\frac{1}{2}}{1\frac{2}{3}}$
20. $\dfrac{2\frac{3}{4}}{3\frac{1}{7}}$

Solve.

See pp. 192–193

21. $\frac{6}{7}$ of $21 = $ ___?___
22. $\frac{8}{3}$ of ___?___ $= 40$
23. $\frac{7}{12}$ of ___?___ $= 63$

24. $\frac{5}{8}$ of $32 = $ ___?___
25. $\frac{3}{2}$ of $48 = $ ___?___
26. $\frac{15}{16}$ of ___?___ $= 18$

**Write each fraction or mixed number as a decimal.
Use a bar to show repeating decimals.**

See pp. 194–197

27. $\frac{9}{8}$
28. $\frac{9}{50}$
29. $\frac{73}{48}$
30. $\frac{5}{11}$
31. $\frac{23}{8}$

32. $\frac{4}{26}$
33. $\frac{20}{50}$
34. $\frac{7}{6}$
35. $\frac{9}{16}$
36. $\frac{83}{80}$

Write each decimal as a fraction or mixed number in lowest terms.

See pp. 198–199

37. 1.02
38. 0.4375
39. 3.625
40. 1.055
41. 0.60

42. 0.875
43. 2.44
44. 5.3125
45. 0.35
46. 2.375

Solve.

47. Ricky is saving money to buy 2 concert tickets, which each cost $30. If he has $\frac{2}{3}$ of the money needed, how much has he saved?

48. On Monday Mrs. Suarez's commute was $\frac{7}{10}$ hr. On Thursday it was $\frac{5}{8}$ hr. On which day was the commute faster?

49. Tina is reading a 748-page book. How many pages has she read if she has finished $\frac{3}{4}$ of the book?

50. A book store received 620 books. If this was $\frac{1}{2}$ the number ordered, how many would be shipped for a double order?

Cumulative Review

Choose the correct answer.

1. Which fractions are in order from greatest to least?

   a. $\frac{11}{9}, \frac{2}{3}, \frac{5}{6}, \frac{7}{8}$ b. $\frac{11}{9}, \frac{7}{8}, \frac{5}{6}, \frac{2}{3}$

   c. $\frac{2}{3}, \frac{5}{6}, \frac{7}{8}, \frac{11}{9}$ d. $\frac{5}{6}, \frac{3}{4}, \frac{1}{3}, \frac{1}{2}$

2. One half is greater than?

   a. $\frac{4}{7}$ b. $\frac{17}{30}$

   c. $\frac{15}{19}$ d. $\frac{6}{17}$

3. Which pair of numbers are reciprocals?

   a. $\frac{1}{10}$ and 0.1 b. $1\frac{2}{3}$ and 1.6

   c. $\frac{2}{3}$ and 1.5 d. $\frac{7}{8}$ and 8.75

4. Which fraction is a repeating decimal?

   a. $\frac{15}{12}$ b. $\frac{9}{11}$

   c. $\frac{13}{26}$ d. $\frac{3}{4}$

5. Which fraction is a terminating decimal?

   a. $\frac{2}{11}$ b. $\frac{1}{13}$

   c. $\frac{12}{30}$ d. $\frac{7}{9}$

6. Which fraction is closest to 4?

   a. $3\frac{1}{3}$ b. $3\frac{2}{3}$

   c. $\frac{39}{30}$ d. $4\frac{6}{8}$

7. Which is the best way to estimate $\frac{3}{11}$ of 29?

   a. $\frac{3}{12} \times 28$ b. $\frac{3}{12} \times 32$

   c. $\frac{3}{10} \times 30$ d. $\frac{3}{10} \times 35$

8. $6\frac{3}{8} - 2\frac{5}{6}$ is about:

   a. 4   b. 6

   c. 3   d. 5

9. $\frac{3}{8}$ of 32 is the same as:

   a. $2 \div \frac{1}{6}$

   b. $6 \times \frac{1}{2}$

   c. $6 \div 2$

   d. $\frac{1}{6} \div 2$

10. $\frac{3}{5} \times 10\frac{1}{3}$ is the same as:

    a. $(\frac{3}{5} \times 10) + (\frac{3}{5} \times \frac{1}{3})$

    b. $(\frac{3}{5} + 10) \times (\frac{3}{5} + \frac{1}{3})$

    c. $(\frac{3}{5} \times 10) \times (\frac{3}{5} \times \frac{1}{3})$

    d. $(\frac{3}{5} \times 10) + (\frac{1}{3} \times 10)$

11. Take $\frac{3}{4}$ from the product of $1\frac{1}{3}$ and $\frac{9}{10}$.

    a. $\frac{19}{20}$ b. $\frac{13}{20}$

    c. $\frac{9}{20}$ d. $1\frac{1}{5}$

12. Divide the sum of $2\frac{1}{4}$ and $\frac{7}{8}$ by $\frac{2}{3}$.

    a. $1\frac{5}{8}$ b. $3\frac{1}{8}$

    c. $2\frac{1}{12}$ d. $4\frac{11}{16}$

13. Which is equal to 1?

    a. $2.5 \div 2$ b. $3\frac{1}{3} \div \frac{10}{3}$

    c. $\frac{5}{8}$ of 16   d. $\frac{5}{8} \div \frac{8}{16}$

14. What number completes the sentence: $1\frac{3}{4} \div n = 0.25$

    a. 4   b. 7

    c. $\frac{1}{7}$ d. $\frac{1}{4}$

Compute.

15. $\dfrac{2}{7} + \dfrac{3}{7} + 1\dfrac{1}{7}$ 16. $\dfrac{1}{12} + \dfrac{5}{6} + 3$ 17. $\dfrac{3}{4} - \dfrac{1}{8}$ 18. $\dfrac{3}{5} - \dfrac{1}{3}$

19. $8 - \dfrac{3}{5}$ 20. $7 - 1\dfrac{2}{3}$ 21. $2 - (\dfrac{1}{4} + \dfrac{1}{2})$ 22. $3\dfrac{1}{3} - (\dfrac{1}{2} + \dfrac{1}{6})$

23. $3\dfrac{4}{10}$ 24. $9\dfrac{5}{6}$ 25. $5\dfrac{1}{3}$ 26. $10\dfrac{1}{6}$

$+1\dfrac{7}{10}$ $+2\dfrac{7}{8}$ $+3\dfrac{5}{6}$ $+ 2\dfrac{4}{5}$

27. $\dfrac{1}{3} \times 42$ 28. $\dfrac{3}{16} \times 48$ 29. $2\dfrac{1}{4} \times 12$ 30. $2\dfrac{2}{5} \times 4\dfrac{1}{6}$

31. $9 \div \dfrac{3}{5}$ 32. $15 \div \dfrac{3}{5}$ 33. $\dfrac{1}{2} \div \dfrac{1}{6}$ 34. $\dfrac{2}{3} \div \dfrac{8}{9}$

35. $1\dfrac{8}{20} \div 1\dfrac{8}{20}$ 36. $1\dfrac{5}{6} \div 1\dfrac{1}{4}$ 37. $5\dfrac{1}{3} \div 2\dfrac{2}{3}$ 38. $2\dfrac{2}{3} \div 1\dfrac{5}{9}$

Solve.

39. Ted and Sheila worked on a science project for $28\dfrac{1}{2}$ hours. If Ted worked $13\dfrac{3}{4}$ hours, how long did Sheila work?

40. The morning session of Day Street School lasts $3\dfrac{1}{2}$ hr. The afternoon session is $2\dfrac{3}{4}$ hr. For how many hours altogether are the students in school?

41. Some students are giving oral book reports. If each report takes about 3 minutes, how many reports can be given in $\dfrac{3}{4}$ of an hour?

42. Thirty-six people were on the beach yesterday. If about $\dfrac{1}{3}$ of those people were swimming, about how many people were doing other things?

43. The pizza shop sold 150 pizzas on Friday. They sold $13\dfrac{5}{8}$ pies before 3:00 P.M. and $102\dfrac{1}{2}$ pies after 4:00 P.M. How many pizzas did they sell between 3:00 P.M. and 4:00 P.M.?

44. A group of mountain climbers took 3 days to reach their goal. They climbed 3240 ft the first day and approximately $\dfrac{2}{3}$ of that distance the next day. What was the approximate total distance climbed in 2 days?

+10 BONUS

45. Imagine that you are a baker known far and wide for your special cake. The recipe calls for $2\dfrac{3}{4}$ cups of flour, and it yields a cake that serves 8. Now suppose that you have received an order for this cake for 60 people. How many cakes do you need and how much flour?

Cumulative Test I

Choose the correct answer.

1. The standard numeral for five billion, four hundred million, three thousand, two is:
   a. 5432   b. 54,302   c. 5,400,003,002   d. 5,403,002   e. none of these

2. The standard numeral for $(6 \times 100) + (8 \times 10) + (7 \times 0.1) + (1 \times 0.01)$ is:
   a. 68.71   b. 68.071   c. 680.071   d. 680.71   e. none of these

3. 5.7416 rounded to the nearest hundredth is:
   a. 5.74   b. 5.7   c. 5.715   d. 6   e. none of these

4. Write an expression for 5 decreased by a number.
   a. $5 - n$ b. $5 + n$ c. $n - 5$ d. $5n$ e. none of these

5. A solution for the inequality $m > 7$ is:
   a. 8, 9, 10 ,11 ...   b. 7, 8, 9, 10 ...   c. 7, 6, 5, 4...   d. 6, 5, 4, 3 ...   e. none of these

6. Which statement is true?
   a. $5,007,219 > 5,070,219$ b. $4630 < 4603$
   c. $10^5 < 10^4$ d. $0.5 \times 10 > 0.005 \times 100$

7. The opposite of $^-6$ is:
   a. $\frac{1}{6}$ b. 6   c. $\frac{^-1}{6}$ d. $^-6$

8. $^+6 + {}^-8 - {}^-9$ equals:
   a. $^+5$ b. $^-11$ c. $^+7$ d. $^+11$

9. Which of the following statements is false?
   a. $^-7 + (^-3 + {}^+8) = (^-3 + 8) + {}^-7$ b. $^-7 + (^-3 + {}^+8) = {}^-7 + (^+8 + {}^-3)$
   c. $(^-7 + {}^-3) - {}^+8 = {}^-7 + (^-3 + {}^+8)$ d. $(^-7 + {}^-3) + {}^+8 = (^+8 + {}^-3) + {}^-7$

10. An example of a composite number is:
    a. 4   b. 29   c. 11   d. none of these

11. An example of a prime number is:
    a. 12   b. 80   c. 24   d. none of these

Compute.

12. $3\frac{4}{10}$
    $+1\frac{7}{10}$

13. $9\frac{5}{6}$
    $+2\frac{7}{8}$

14. $6\frac{2}{3}$
    $-4\frac{1}{3}$

15. $7\frac{1}{6}$
    $-2\frac{4}{5}$

16. 12
    $-4\frac{1}{6}$

17. $10\frac{1}{8}$
    $-3\frac{3}{4}$

18. $\frac{1}{3} \times 36$

19. $4 \div \frac{2}{7}$

20. $\frac{5}{8} \times 40$

21. $\frac{6}{8} \div \frac{1}{9}$

22. $2\frac{3}{5} \times 4\frac{1}{6}$

23. $\frac{3}{5} \div 7$

24. $\dfrac{\frac{1}{3}}{2}$

25. $\dfrac{\frac{1}{4}}{2}$

26. $\dfrac{\frac{4}{7}}{\frac{3}{8}}$

27. $\dfrac{\frac{1}{4}}{6}$

28. $\dfrac{\frac{1}{3}}{6}$

29. $\dfrac{\frac{2}{3}}{\frac{1}{3}}$

30. $\dfrac{\frac{1}{2}}{\frac{1}{6}}$

31. $\frac{9}{16} = \frac{8}{a}$

32. $\frac{1.6}{5} = \frac{b}{14}$

33. $\frac{2}{9} \times n = 8$

34. $2\frac{1}{4} \times 3\frac{1}{8} \div 1\frac{1}{8}$

35. $8\frac{3}{10} \times \frac{2}{3} \div 1\frac{3}{5}$

36. $1\frac{5}{6} \times 9 \div 4\frac{5}{7}$

37. $^-9 \times {}^-8 + {}^+72$

38. $0 \times {}^-6$

39. $7 \times {}^-9 + {}^+63$

40. $^-56 \div {}^+8 - {}^+7$

41. $^+42 \div {}^-7 - {}^-6$

42. $^+12 \div {}^+2 + {}^+6$

Write true (T) or false (F).

43. The greatest common factor of 36 and 45 is 45.
44. An inequality uses the symbol $>$, $<$, or $\neq$.
45. An expression for "15 less than a number" is $15 - n$
46. The quotient when 0.25 is divided by 50 is 200.
47. The identity element for multiplication of integers is $^+1$.
48. The least common multiple of 6 and 15 is 30.
49. Ten to the fifth power written as a standard numeral is 50.
50. The least possible 4-digit number in which no digit repeats is 1230.
51. If $m - 13 = 37$, then $m = 24$
52. The integers $^-1$, $^-2$, $^-3$ . . . are increasing in value.
53. 79,986 rounded to the nearest hundred is 79,000.
54. $\{1, 4, 8\}$ is a subset of $\{0, 1, 2, 3, 4, 5\}$
55. 54,423 is divisible by 3.
56. The product of two negative numbers is always positive.
57. If $\dfrac{m}{4} = 12$, then $m = 48$.
58. 41 is a prime number.
59. $420,000,000 = 4.2 \times 10^8$
60. A reasonable estimate for $81,236 \div 792$ is 200.
61. The difference of two negative integers is always negative.

Solve.

62. John earned $21.78 per day. How much did he earn in 30 days?

63. A watch was marked $80 and sold for $67. What was the discount?

64. On April 1, Mrs. Martinez noted that the odometer on her car read 23460.8 km. On June 1 the odometer read 31786.3 km. How many kilometers had she driven in April and May?

65. The seventh-grade class collected 25,543 wrappers, and the sixth grade collected 24,985. How many more wrappers did the seventh grade collect?

66. Samantha wants a roast to cook for $3\frac{1}{2}$ hours at 320°. She places it in the oven at 2:45 and removes it at 5:10. How long has the roast cooked?

67. An even number greater than 3 can be written as the sum of two primes. Write each even number between 10 and 16 as the sum of two primes.

68. Mr. Lerro shipped 48 steel file cabinets. Each cabinet was valued at $84.74. What was the shipment worth?

69. The scouts had a pancake party. They made 8 batches of pancakes, 36 to a batch. How many scouts were served if each was given 6 pancakes?

70. Rayleen bought 12 meters of a certain fabric at $3.79 per meter. She used 6 meters for a dress and 3.65 meters for a belt and shawl. How much fabric was not used?

71. In a mayoral race Mr. Quinn received 190,654 votes and Ms. Ford 201,963 votes. Through an error in the voting machine, Mr. Quinn had received 1720 additional votes while Ms. Ford had lost 3487 votes. Which candidate won the greater number of votes? How many more?

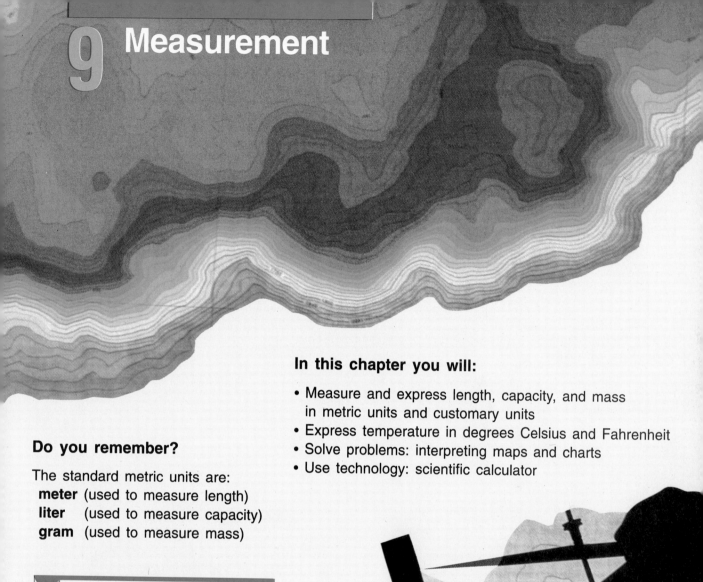

9 Measurement

Do you remember?

The standard metric units are:
- **meter** (used to measure length)
- **liter** (used to measure capacity)
- **gram** (used to measure mass)

In this chapter you will:

- Measure and express length, capacity, and mass in metric units and customary units
- Express temperature in degrees Celsius and Fahrenheit
- Solve problems: interpreting maps and charts
- Use technology: scientific calculator

Light Years Away!

Return to your imaginary planet. Remember when you calculated how much money you had on your planet? (See page 1.) Now can you tell someone how far, in metric measure, it is from Earth to your planet? Hint: your planet is as far away as the planet Pluto — only in a different direction.

Meter (m): the standard unit of length in the metric system.

Metric units of length are related to each other in the same way that place-value positions within the decimal system of numeration are related.

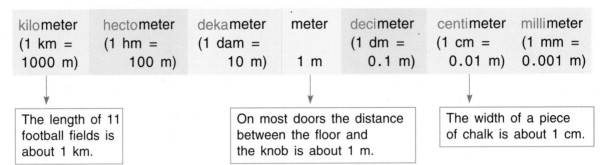

| kilometer (1 km = 1000 m) | hectometer (1 hm = 100 m) | dekameter (1 dam = 10 m) | meter 1 m | decimeter (1 dm = 0.1 m) | centimeter (1 cm = 0.01 m) | millimeter (1 mm = 0.001 m) |

The length of 11 football fields is about 1 km.

On most doors the distance between the floor and the knob is about 1 m.

The width of a piece of chalk is about 1 cm.

Use a metric ruler to find the length of each.

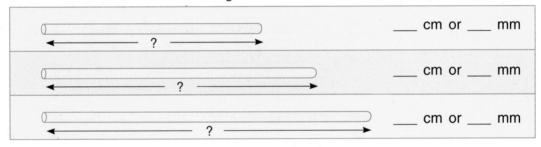

____ cm or ____ mm

____ cm or ____ mm

____ cm or ____ mm

Remember: The *meter, millimeter, centimeter,* and *kilometer* are the most commonly used metric units of length.

Which unit — km, m, cm, or mm — should be used to measure each?

1. drinking straw
2. baking pan
3. hiking trail
4. telephone pole
5. pencil eraser
6. door
7. postage stamp
8. building

Choose the most reasonable measurement.

9. length of a sailboat
   a. 750 mm
   b. 7.5 m
   c. 0.75 km

10. length of a pair of scissors
    a. 20 cm
    b. 2000 mm
    c. 2 m

11. width of your hand
    a. 95 mm
    b. 9.5 dm
    c. 95 cm

12. height of an adult
    a. 175 mm
    b. 1.75 m
    c. 1750 cm

Meanings of Metric Prefixes	Measuring to the Nearest Unit
kilo = 1000 hecto = 100 deka = 10 AND deci = 0.1 centi = 0.01 milli = 0.001	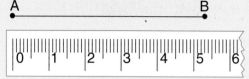 $\overline{AB}$ is closer to 5 cm than it is to 6 cm. So, $\overline{AB}$ measures 5 cm to the nearest centimeter. $\overline{AB}$ is closer to 53 mm than it is to 54 mm. So, $\overline{AB}$ measures 53 mm to the nearest millimeter.

Estimate each. Then use a metric ruler or tape to measure to the nearest unit.

13. your height from the floor to your shoulder

14. the distance around a baseball

15. your wrist measurement

16. the distance around a basketball

17. the length of a strand of your hair

18. the height of a basketball rim

19. the distance around your classroom

20. the length of a chalkboard eraser

Cut a piece of yarn 20 centimeters long.

21. How many millimeters long is the yarn?

22. How many of these pieces would you need to make a meter?

CALCULATOR ACTIVITY

Sometimes it may be necessary to convert from metric units to customary units.
Use this chart and a calculator to convert each measurement. Round to the nearest hundredth. (The first one is done.)

1 cm ≈ 0.394 in.
1 m ≈ 3.28 ft
1 km ≈ 0.621 mi
1 ft ≈ 30.5 cm
1 yd ≈ 0.914 m

23. 2.5 in. = ? cm | 2.5 | ÷ | 0.394 | = | 6.3451776 | , or 6.35 to the nearest hundredth

24. 8 yd = ? m

25. 18.6 mi = ? km

26. 61 cm = ? ft

27. 8.05 km = ? mi

28. 24.84 mi = ? km

29. 4.92 ft = ? m

Draw each of the following freehand, using estimation. Then measure to check your estimating skills.

30. a line 6 cm long

31. a circle with a diameter of 10 cm

Estimate each in centimeters. Then measure to check your estimating skills.

32. the length of your shoe

33. the circumference of your head

34. the distance from your elbow to the tip of your thumb

9-2 | Changing Metric Units

Metric system is based on *multiples of ten*.

So, to change from one unit to another, always *multiply or divide by a multiple of ten*.

Use this metric place-value chart to convert the value of one metric amount to another.

kilometer (1 km = 1000 m)	hectometer (1 hm = 100 m)	dekameter (1 dam = 10 m)	meter 1 m	decimeter (1 dm = 0.1 m)	centimeter (1 cm = 0.01 m)	millimeter (1 mm = 0.001 m)

Each unit is ten times the value of the unit to its *right*.

Each unit is one tenth the value of the unit to its *left*.

$$1 \text{ km} = 10 \times 100 \text{ m} = 1000 \text{ m}$$
$$1 \text{ hm} = 10 \times 10 \text{ m} = 100 \text{ m}$$

$$1 \text{ cm} = \frac{1}{10} \times \frac{1}{10} = \frac{1}{100} \text{ m}$$
$$1 \text{ mm} = \frac{1}{10} \times \frac{1}{100} = \frac{1}{1000} \text{ m}$$

To change larger units to smaller units, *multiply* by 10 for *every* place moved to the *right*.

$$70 \text{ cm} = \underline{?} \text{ mm}$$
$$70 \text{ cm} = (10 \times 70) \text{ mm}$$
$$70 \text{ cm} = 700 \text{ mm}$$

To change smaller units to larger units, *divide* by 10 for *every* place moved to the *left*.

$$80 \text{ mm} = \underline{?} \text{ cm}$$
$$80 \text{ mm} = (80 \div 10) \text{ cm}$$
$$80 \text{ mm} = 8 \text{ cm}$$

Copy and complete the chart so that each row shows equivalent measurements.
Will you multiply or divide?

	km	hm	dam	m	dm	cm	mm
1.	?	?	0.6	6	60	600	?
2.	?	?	9	90	?	?	90 000
3.	?	0.35	3.5	?	?	?	?

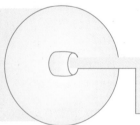

Copy and complete. Think the steps.

4. 7 km = $\underline{?}$ m

5. 20 cm = $\underline{?}$ m

6. 75 mm = $\underline{?}$ dam

7. 9 dm = $\underline{?}$ m

8. 400 hm = $\underline{?}$ m

9. 1875 km = $\underline{?}$ m

10. 12 cm = $\underline{?}$ mm

11. 0.5 m = $\underline{?}$ mm

12. 0.05 m = $\underline{?}$ mm

13. 0.08 cm = $\underline{?}$ mm

14. 3.5 dam = $\underline{?}$ cm

15. 35 460 mm = $\underline{?}$ km

A Shortcut for Converting Metric Units

Move the decimal point on the metric place-value chart the *same number* of places and in the *same direction* as the places between the given unit and the new unit.

600 cm = ? m

Think: To convert from centimeters to meters, move *two places to the left.*

600 cm = ? m

6.00. = 6 m

0.75 km = ? m

Think: To convert from kilometers to meters, move *three places to the right.*

0.75 km = ? m

0.750. = 750 m

Use the shortcut to complete the following:

16. 3 m = ? km

17. 2000 m = ? km

18. 4000 mm = ? m

19. 222 mm = ? m

20. 450 km = ? dm

21. 6.3 km = ? m

22. 1.3 dam = ? dm

23. 1 000 000 mm = ? m

24. 21.6 m = ? mm

25. 8300 m = ? km

26. 29.5 m = ? cm

27. 4 789 000 km = ? m

28. 1155 m = ? km

29. 9.3 hm = ? cm

30. 0.51 cm = ? km

Compare. Write <, =, or >.

31. 49 m ? 49 dm

32. 0.09 m ? 90 cm

33. 4200 km ? 4.2 m

34. 9.4 mm ? 0.0094 m

35. 780 cm ? 78 m

36. 16 dam ? 200 m

37. 335 mm ? 33 cm

38. 47 dm ? 0.05 hm

39. 812 m ? 8.5 km

Solve.

40. Nickels are 2 mm thick. How many nickels are in a stack 4 cm high?

41. Nickels have a diameter of 2 cm. How many nickels are needed to make a row 160 mm long?

42. The students in Ms. Jakob's room used Crunchy Puffs to make a line that stretched for 4.5 m. If each puff was 1 cm wide, how many puffs were used?

43. Ms. Jakob's home is 1.8 km from school. How many Crunchy Puffs would be needed to stretch from school to her home?

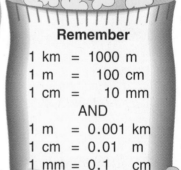

Remember

1 km = 1000 m
1 m   =   100 cm
1 cm =   10 mm

AND

1 m   = 0.001 km
1 cm = 0.01   m
1 mm = 0.1   cm

9-3 | Measuring Capacity in Metric Units

Liter (L): the standard unit of capacity (volume) in the metric system.

Metric units of capacity are related to each other in the same way that place-value positions within the decimal system of numeration are related.

kiloliter (1 kL = 1000 L)	hectoliter (1 hL = 100 L)	dekaliter (1 daL = 10 L)	liter 1 L	deciliter (1 dL = 0.1 L)	centiliter (1 cL = 0.01 L)	milliliter (1 mL = 0.001 L)

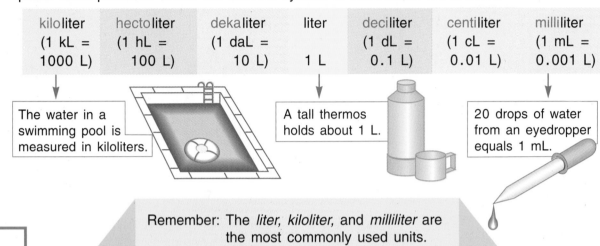

The water in a swimming pool is measured in kiloliters.

A tall thermos holds about 1 L.

20 drops of water from an eyedropper equals 1 mL.

Remember: The *liter, kiloliter,* and *milliliter* are the most commonly used units.

Which unit — kL, L, or mL — should be used to express the capacity of each?

1. teaspoon
2. drop of rain
3. test tube
4. coffee cup
5. lake
6. oil tanker
7. milk carton
8. fish tank
9. car's gas tank

Use the shortcut to complete the following:*

10. 5 L = _?_ kL
11. 20 L = _?_ cL
12. 8000 mL = _?_ L
13. 6.1 kL = _?_ L
14. 140 cL = _?_ daL
15. 2483 mL = _?_ L

16. Use a cylinder marked in milliliters to find the volume of an irregularly shaped object by the *displacement method*. The difference between the two levels of water represents the volume of the object.

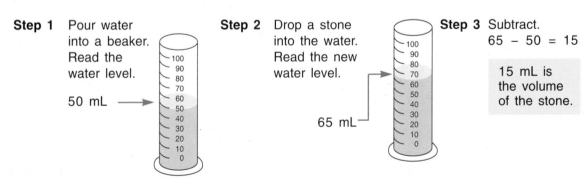

Step 1 Pour water into a beaker. Read the water level.

50 mL

Step 2 Drop a stone into the water. Read the new water level.

65 mL

Step 3 Subtract.
65 − 50 = 15

15 mL is the volume of the stone.

216   *See page 455 for more practice.

9-4 Measuring Mass in Metric Units

Gram (g): the standard unit of mass (weight) in the metric system.

Metric units of mass are related to each other in the same way that place-value positions within the decimal system of numeration are related.

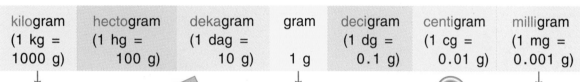

kilogram (1 kg = 1000 g)	hectogram (1 hg = 100 g)	dekagram (1 dag = 10 g)	gram 1 g	decigram (1 dg = 0.1 g)	centigram (1 cg = 0.01 g)	milligram (1 mg = 0.001 g)

A hardcover dictionary has a mass of about 1 kg.

A paper clip has a mass of about 1 g.

A single grain of salt has a mass of about 1 mg.

The mass of extremely heavy things is expressed in metric tons.
1 t = 1000 kg = 1 000 000 g

Remember: The *gram, kilogram, milligram,* and *metric ton* are the most commonly used units of mass.

Which unit — g, mg, kg, or t — should be used to express the mass of each?

1. pair of shoes
2. all your books
3. kernel of corn
4. box of cereal
5. space shuttle
6. feather
7. airmail stamp
8. newborn baby
9. hamster
10. hard-boiled egg
11. sack of potatoes
12. caboose

13. Estimate the mass of four of the items above. Then use a metric balance to test your answers.

Use the shortcut to complete the following:

14. 146 cg = _?_ dg
15. 24 g = _?_ mg
16. 3 mg = _?_ g
17. 16 dag = _?_ mg
18. _?_ t = 5000 kg
19. 5550 cg = _?_ g
20. 4675 g = _?_ kg
21. 0.001 g = _?_ kg
22. 896 dg = _?_ mg

Compare. Write <, =, or >.*

23. 190 g _?_ 1.9 kg
24. 40 000 cg _?_ 4 g
25. 764 kg _?_ 764 000 g
26. 4786 hg _?_ 47.86 kg
27. 4 600 000 g _?_ 4 t
28. 0.0411 g _?_ 0.411 mg
29. 823 mg _?_ 0.0082 kg
30. 280 000 mg _?_ 28 g
31. 896 hg _?_ 8.96 kg

*See page 455 for more practice.

Relating Metric Units

Metric units of length, capacity, and mass are related to one another in this way:

A cube that measures
1 centimeter along
each edge. . . . . . .will hold 1 milliliter of water. . . . . . .which has a mass of 1 gram.

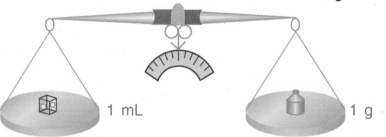

1 cm
1 cm
1 cm

$1\ cm^3$ = 1 cm × 1 cm × 1cm
= 1 cubic centimeter

1 mL   1 g

A cube that measures
1 decimeter along
each edge. . . . . . . will hold 1 liter of water. . . . . . . which has a mass of 1 kilogram.

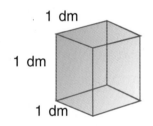
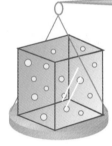
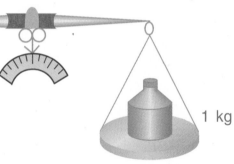

1 dm
1 dm
1 dm

1 L   1 kg

$1000\ cm^3$ = $1\ dm^3$
$1\ dm^3$ = 1 dm × 1 dm × 1 dm
= 1 cubic decimeter

Copy each chart and complete.

	Cube	Capacity	Mass
1.	$2\ cm^3$	2 mL	__?__ g
2.	__?__ cm^3	8 mL	8 g
3.	$15\ cm^3$	__?__ mL	__?__ g
4.	$5\ dm^3$	5 L	?
5.	$27\ dm^3$	?	?
6.	?	?	19 kg

	Cube	Capacity	Mass
7.	$12\ cm^3$	__?__ mL	__?__ g
8.	__?__ dm^3	60 L	__?__ kg
9.	?	?	13 g
10.	$33\ dm^3$	?	?
11.	?	11 mL	?
12.	$10.5\ dm^3$	?	?

9-6 Precision in Measurement

All measurements are approximate.

Greatest possible error (GPE) of a measurement is *one half* of the unit of measure being used.

For example, when something measures 3 cm long, we understand that it is between 2.5 cm and 3.5 cm long.

This may be written as: 3 ± .5 cm

This is read: "three, plus or minus one half centimeter."

GPE: 0.5 cm

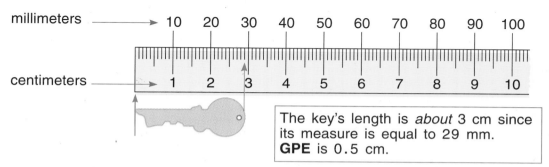

millimeters ⟶ 10 20 30 40 50 60 70 80 90 100

centimeters ⟶ 1 2 3 4 5 6 7 8 9 10

The key's length is *about* 3 cm since its measure is equal to 29 mm. **GPE** is 0.5 cm.

The smaller the unit of measure, the more precise the measurement.

Measurements using millimeters are more precise than measurements using centimeters.

Name and give the GPE of the more precise measurement.*

1. mm or cm
2. ft or yd
3. qt or oz
4. kL or L
5. g or mg
6. in. or ft
7. 12 cm or 12 m
8. 8 g or 8 kg
9. 5 ft or 1 yd
10. 7.2 m or 72.1 cm
11. 9.1 m or 9.1 km
12. 5286 cm or 2 mm
13. 620 km or 6210 m
14. 2 L or 2200 mL
15. 25 kg or 24 900 g

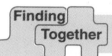

Finding Together

16. List five situations where very precise measurements are required. Then list situations where less precise measurements are acceptable.

9-7 Measuring Length in Customary Units

Customary units of length are the inch, foot, yard, and mile.

This chart shows some of their relationships.

1 foot (ft) = 12 inches (in.)
1 yard (yd) = 3 ft or 36 in.
1 mile (mi) = 5280 ft or 1760 yd

' means *feet*.
" means *inches*.

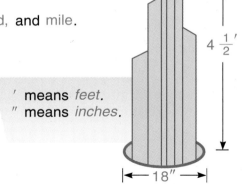

Rhonda builds a model skyscraper $4\frac{1}{2}'$ high.
The front of the model is 18″ wide.

How many inches high is the model?

$$4\frac{1}{2} \text{ ft} = \underline{} \text{ in.}$$

1 ft = 12 in., so *multiply* by 12 in.

$$4\frac{1}{2} \text{ ft} \times 12 \text{ in.} = 54 \text{ in.}$$

How many feet wide is the model?

$$18 \text{ in.} = \underline{} \text{ ft} \underline{} \text{ in.}$$

12 in. = 1 ft, so *divide* by 12 in.

$$18 \text{ in.} \div 12 \text{ in.} = 1 \text{ ft } 6 \text{ in.} = 1\frac{1}{2} \text{ ft}$$

Copy and complete.

1. 1 yd = __?__ in.

2. 1 mi = __?__ in.

3. 16 yd = __?__ ft

4. 2.5 yd = __?__ in.

5. 120 ft = __?__ yd

6. 2640 ft = __?__ mi

7. 30 in. = __?__ ft __?__ in.

8. 52 ft = __?__ yd __?__ ft

9. 118 ft = __?__ yd __?__ ft

10. 161 ft = __?__ yd __?__ ft

11. 3 mi 500 ft = __?__ ft

12. 0.75 mi = __?__ ft

13. 4 ft 5 in. = __?__ in.

14. 1 yd 2 ft 3 in. = __?__ in.

15. 2678 in. = __?__ yd __?__ in.

Select the correct answer.

16. Which is longest?   **a.** 4 ft 9 in.   **b.** 1 yd 19 in.   **c.** 1.5 yd

17. Which is shortest?   **a.** 3 yd 1 ft   **b.** 10.5 yd   **c.** 10 ft 14 in.

18. Which are the same?   **a.** 1.5 miles   **b.** 5280 ft   **c.** 2640 yd

19. Who is taller?   **a.** 6′8″ player   **b.** $6\frac{3}{4}'$ player   **c.** both same height

**Draw each of the following freehand, using estimation.
Then measure to check your estimating skills.**

20. a line 7 in. long

21. a circle with a diameter of 4 in.

Estimate each in inches or feet. Measure to check your estimating skills.

22. the length of your shoe

23. the length of your math book

24. the height of your desk

25. the length of the classroom

220

Computing and Regrouping Measures

To *add* or *subtract* customary units:		
• Add or subtract *like* units. • Regroup to change from one unit to another.	3 ft ⋮ 7 in. +4 ft ⋮ 9 in. —————————— 7 ft ⋮ 16 in. = 1 ft 4 in. ↑—Regroup—↑ 7 ft + 1 ft 4 in. = 8 ft 4 in.	18 1 1̶2̶ + 6̶ 3 yd ⋮ 2̶ ft ⋮ 6̶ in. −1 yd ⋮ 1 ft ⋮ 9 in. —————————— 2 yd ⋮ 0 ft ⋮ 9 in. = 2 yd 9 in.

> Regroup 2 ft 6 in. to 1 ft 12 in.
+ 6 in.
= 1 ft 18 in.

To *multiply* with customary units:	

• Multiply by *each* unit.
• Regroup to change from one unit to another.
        OR
• Change all units to *like* units.
• Multiply and then regroup.
	2 ft ⋮ 8 in.
	×　　　　4
	——————————
	8 ft ⋮ 32 in. =
	8 ft + 2 ft 8 in.
	= 10 ft 8 in.
	        OR
	2 ft 8 in. is same as
	×　　　4
	———————

> Regroup 32 in. to 2 ft 8 in.

> Divide 128 in. by 12.
Regroup to 10 ft 8 in.

24 in. 8 in. = 32 in.
×　4　×　4　　×　4
————————————
128 in. = 10 ft 8 in.

To *divide* with customary units:	
• Change all units to *like* units. • Divide. • Then regroup.	Regroup each to inches. 1 yd 2 ft 3 in. ÷ 3 ↓ ↓ ↓ 36 in. + 24 in. + 3 in. ÷ 3 63 in. ÷ 3 = 21 in. = 1 ft 9 in.

> Regroup to ft and in.

Add or subtract.

26.　8 ft 6 in.　**27.**　4 yd 1 ft　**28.**　4 yd 2 ft 4 in.　**29.**　6 mi 4 yd 2 ft　**30.**　7 mi 17 yd
　　+3 ft 5 in.　　　−2 yd 2 ft　　　+3 yd 2 ft 6 in.　　　−4 mi 2 yd 1 ft　　　−3 mi 23 yd

Multiply.

31.　3 ft 3 in.　**32.**　8 yd 2 ft　**33.**　7 ft 9 in.　　**34.**　8 mi 3 yd　**35.**　2 yd 2 ft 2 in.
　　×　　2　　　×　　4　　　×　　7　　　　×　　7　　　×　　　　8

Divide.

36.　5 yd 1 ft ÷ 4　　**37.**　7 ft 6 in. ÷ 10　**38.**　1 yd 1 ft 8 in. ÷ 7　**39.**　1 mi 220 ft ÷ 100

CHALLENGE　　　　Copy and complete.

40.　$3\frac{1}{3}$ yd = 3 yd __?__ ft　　**41.**　$7\frac{1}{6}$ mi = 7 mi __?__ ft　**42.**　$9\frac{5}{6}$ ft = 9 ft __?__ in.

43.　$\frac{4}{5}$ mi = __?__ yd　　　**44.**　$7\frac{1}{4}$ yd = 7 yd __?__ in.　**45.**　$2\frac{3}{5}$ mi = __?__ ft

46.　6 yd 18 in. = __?__ yd　　**47.**　2 ft 8 in. = __?__ ft　　**48.**　2 mi 3960 ft = __?__ mi

49.　1 mi 660 yd = __?__ mi　　**50.**　4 yd 10 in. = __?__ yd　**51.**　8 yd 2 ft = __?__ yd

9-8 Measuring Capacity in Customary Units

Customary units of capacity are the cup, pint, quart, and gallon.

Smaller units are the teaspoon and the tablespoon.

2 cups (c) = 1 pint (pt)	4 quarts = 1 gallon (gal)
2 pints = 1 quart (qt)	3 teaspoons (tsp) = 1 tablespoon (tbsp)

Which unit — teaspoon, cup, quart, or gallon — should be used to measure each?

1. honey in a cup of tea
2. water for a garden
3. salt for a cake recipe
4. flour to make a cake
5. lemonade for a party
6. water in a lake

Copy and complete.*

7. 1 qt = ___?___ c
8. 1 gal = ___?___ qt
9. 1 gal = ___?___ c
10. 3 pt = ___?___ c
11. 5 qt = ___?___ pt
12. 5 gal = ___?___ qt
13. 6 tbsp = ___?___ tsp
14. 4 pt = ___?___ qt
15. $1\frac{1}{2}$ gal = ___?___ qt
16. 8 qt = ___?___ gal
17. 27 qt = ___?___ gal ___?___ qt
18. 5 pt = ___?___ qt ___?___ c

Compute.*

19.   2 gal 1 qt
    +4 gal 3 qt

20.   4 qt 1 pt 1 c
    +3 qt 0 pt 0 c

21.   3 gal 3 qt 1 pt
    −1 gal 1 qt 2 pt

22.   6 qt 1 pt
    ×   2

23.   4 pt 1 c
    ×   6

24. 2 gal 3 qt 1 pt ÷ 10

25. Which is the least? Why? **a.** 6 qt 1 c **b.** 12 pt 2 c **c.** 2 gal

A NEW LOOK   SUPPOSE THAT...

26. A leaky faucet loses $1\frac{1}{2}$ cups of water an hour. How many gallons are lost in a week?

27. An average shower uses from 25 to 30 gal of water. About how much water is used in a week by a person who takes 1 shower a day?

9-9 Measuring Weight in Customary Units

Customary units of weight are the ounce, pound, and ton.

16 ounces (oz) = 1 pound (lb)   2000 pounds = 1 ton (T)

Which unit — ounce, pound, or ton — should be used to express the weight of each?

1. five raisins
2. your math book
3. a school bus
4. a box of spaghetti
5. a pachyderm
6. your desk
7. a lunch table
8. a moving van
9. a letter

Copy and complete.*

10. 2 lb = ___?___ oz

11. 3 T = ___?___ lb

12. 1 T = ___?___ oz

13. $6\frac{1}{2}$ lb = ___?___ oz

14. $4\frac{1}{4}$ lb = ___?___ oz

15. 44 oz = ___?___ lb

16. 20 oz = ___?___ lb

17. 11,000 lb = ___?___ T

18. 45,000 lb = ___?___ T

Compute.*

19. 5 lb 12 oz
    +4 lb 12 oz

20. 2 lb 10 oz
    +8 lb 8 oz

21. 3 lb 5 oz
    −1 lb 8 oz

22. 6 T 280 lb
    ×   3

23. 4 T 350 lb 10 oz
    ×   10

24. 8 T 1000 lb ÷ 25

25. Which is heavier? Why? **a.** a ton of bricks **b.** a ton of feathers

26. You weigh 82 lb 4 oz. You stand on the scale with your cat. The scale reads 92 lb 1 oz. How much does your cat weigh?

27. A large animal gains 2 ounces of weight a day. How much is that in a year?

SKILLS TO REMEMBER

Solve for *n*.

28. $^+3 + {^-16} + {^+10} = n$

29. $^-7 + {^-3} + {^+4} = n$

30. $^+8 - {^-6} = n$

31. $^-5 \times {^+4} \times {^-3} = n$

32. $^+2 \times {^-7} \times {^+5} = n$

33. $^-7 - {^-8} = n$

34. $^+30 + (^-2 - {^+1}) = n$

35. $(^-6 + {^-4}) \div {^+2} = n$

36. $^-18 \div {^+2} = n$

9-10 | Measuring Temperature

Celsius scale is sometimes used to measure temperature in degrees.

The symbol for Celsius is "C." The symbol for degrees is " ° ."

Temperature: 0°C

Temperature: 32°C

Temperature: 100°C

1. Listen to the weather report. What is the present temperature in degrees Celsius?

2. Estimate the Celsius temperature of cold tap water, ice water, and very hot water. Use a Celsius thermometer to check your estimates.

Write "R" if the statement is *reasonable*.
Write "U" if it is *unreasonable*.
(Use the Celsius thermometer for help.)

3. Your body temperature when you are well is about 37°C.

4. We built a snowman when the temperature was 32°C.

5. Inside a freezer it is 10°C.

6. You can bake a cake at 165°C.

7. The skating pond is frozen when the temperature is ⁻5°C.

8. A cup of hot tea is about 20°C.

9. You need a coat in 25°C weather.

10. When an athlete works hard, his/her body temperature goes up to 100°C.

11. At 45°C, it is swimming weather.

12. It's ⁻12°C. Joe said, "Maybe it will snow!"

Celsius Thermometer

110°
100° ← Water boils at 100°C.
90°
80°
70°
60°
50°
40° → 37°C is Normal body temperature
30° → A warm day
20°
10°
0° ← Water freezes at 0°C.
⁻10° → A cold day
⁻20°
⁻30°
⁻40°
⁻50°

A NEW LOOK SUPPOSE THAT...

13. The temperature falls from 12°C at noon to ⁻14°C at midnight. By how many degrees has the temperature fallen?

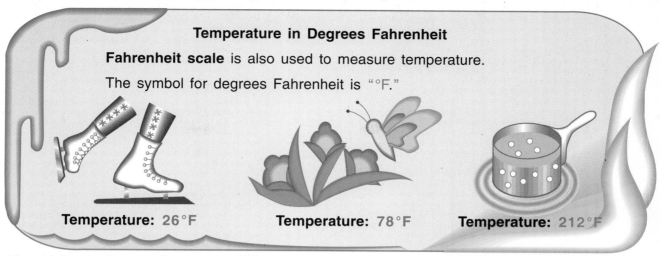

Temperature in Degrees Fahrenheit

Fahrenheit scale is also used to measure temperature.

The symbol for degrees Fahrenheit is "°F."

Temperature: 26°F   Temperature: 78°F   Temperature: 212°F

14. Listen to the weather report. What is the present temperature in degrees Fahrenheit?

15. Estimate the Fahrenheit temperature of cold tap water, ice water, and very hot water. Use a Fahrenheit thermometer to check your estimates.

Choose a reasonable temperature for:

16. the ice in the hockey rink.
    a. 39°F   b. 19°F   c. ⁻10°F

17. the temperature of your classroom during springtime.
    a. 68°F   b. 45°F   c. 80°F

18. the temperature inside a closed car on a hot summer day.
    a. 70°F   b. 80°F   c. 105°F

19. the temperature of a dish of ice cream.
    a. 31°F   b. 0°F   c. ⁻10°F

20. a very cold winter day in Alaska.
    a. 32°F   b. 0°F   c. ⁻25°F

A NEW LOOK SUPPOSE THAT...

21. The pool water temperature at 9 A.M. was 62°F, but by 6 P.M. the temperature was 70°F. How much had the temperature risen?

22. The daytime temperature in the desert was 118.5°F, but during the night the temperature decreased 77°. What was the minimum nighttime temperature?

23. At 11 A.M. the temperature outdoors is 35°F, and at 11 P.M. the temperature is ⁻13°F. By how many degrees has the temperature fallen?

24. When Ted was sick, his temperature was 103°F. After he recovered, his temperature was 98.6°F. How much had his temperature dropped?

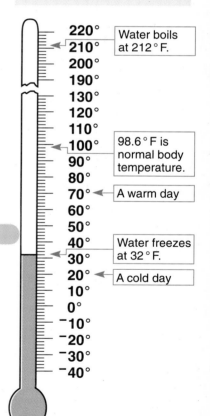

Fahrenheit Thermometer

220°
210° Water boils at 212°F.
200°
190°
130°
120°
110°
100° 98.6°F is normal body temperature.
90°
80°
70° A warm day
60°
50°
40° Water freezes at 32°F.
30°
20° A cold day
10°
0°
⁻10°
⁻20°
⁻30°
⁻40°

Calculator Key-In

There are many keys on a scientific calculator that can simplify your calculations when you understand how to use them.

 This key is called the reciprocal key because it gives the decimal value of the reciprocal of a number.

KEY-IN

 press **DISPLAY READS** `0.2`

 press `1/x` `=` `0.1428571`

The reciprocal of 5 is 0.2.

The reciprocal of 7 is 0.$\overline{142857}$.

 This key is the exponential notation key.
It changes a number from decimal form to scientific notation.

KEY-IN

 press `EE` `=` **DISPLAY READS** `2   03`

`2 1 0 0` press `EE` `=` `2.1   03`

2×10^3

2.1×10^3

 The factorial function key computes a series of multiplications of consecutive integers ($3! = 3 \times 2 \times 1$).

KEY-IN

`4` press `!` `=` **DISPLAY READS** `24`

`5` press `!` `=` `120`

$4! = 24$

$5! = 120$

 The square key and the power key enable you to raise a number to any power.

KEY-IN

`5` press `x²` `=` **DISPLAY READS** `25`

`2` press `yˣ` `6` `=` `64`

$5^2 = 25$

$2^6 = 64$

 This key is called the change-sign key.
It is useful in working with integers.

KEY-IN

 press `=` **DISPLAY READS** `10`

`8` `+/-` `+` `4` `+/-` press `=` `-12`

$6 - {}^-4 = 10$

${}^-8 + {}^-4 = {}^-12$

Find the reciprocal of each number.
Use the reciprocal key on your calculator.

1. 16
2. 125
3. 50
4. 0.5
5. 0.002
6. 0.125
7. 4
8. 0.4

9. For each number in exercises 1–8, press the reciprocal key twice. What is the reciprocal of the reciprocal of a number?

10. Find the reciprocal of 0. What does your calculator display? Why?

Express each number in scientific notation. Use the exponential notation key.

11. 8,000,000
12. 920,000
13. 27,000
14. 1200
15. 46,000,000
16. 900

Use the factorial function key to solve.

17. 8!
18. 10! ÷ 3!
19. 6! + 4!

Complete each statement with <, =, >. Use the square or power key.

20. 3^2 _?_ 2^3
21. 0.5^2 _?_ 0.5^3
22. 1^5 _?_ 1^4
23. 3^3 _?_ 2^5
24. 4^2 _?_ 2^4
25. 1^0 _?_ 0^1

Look for a pattern in these problems.

26. $36^{0.5}$
27. $49^{0.5}$
28. $81^{0.5}$
29. $100^{0.5}$
30. $64^{0.5}$
31. $25^{0.5}$

Copy and complete the chart.

32.

n	1	2	3	4	5	6
$n!$						
n^n						

33. How many factors are multiplied to obtain 2!? to obtain 2^2?

34. How many factors are multiplied to obtain 3!? to obtain 3^2?

35. In general, which is greater, $n!$ or n raised to the n power? Can you explain why?

36. What whole number is an exception to this rule?

CHALLENGE

37. $3^4 - 6 \times (^-4) + (^-8) \times 3 =$ _?_

38. $-(^-2 - 3)^3 \div 5 =$ _?_

Problem: At 1 P.M. Boston time Nadine phoned her grandparents who live in San Diego. What time was it in San Diego?

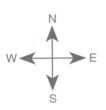

1 IMAGINE Picture the map of the four time zones.

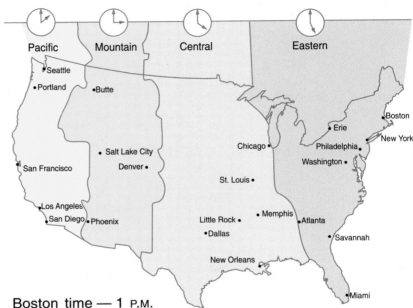

2 NAME *Facts:* Boston time — 1 P.M.

Question: San Diego time — _?_

3 THINK In which time zone is Boston?
In which time zone is San Diego?

> Remember: From time zone to time zone
> it is one hour earlier as you travel west,
> and one hour later as you travel east.

So, subtract one hour for each time zone
as you move from east to west.

4 COMPUTE Boston time − 3 time zones = San Diego time

1 P.M. − 3 hours = 10 A.M.

5 CHECK Begin with the San Diego time and add 1 hour for each
time zone as you move from west to east.

10 A.M. + 3 hours = 1 P.M.

Solve by using a map or chart.

1. Felice uses this chart to record the weather conditions. How many hours and minutes elapsed between the first and second reading on January 18?

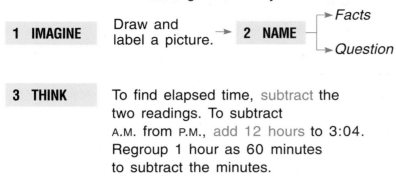

Date	A.M.	°C	P.M.	°C
Jan. 18	9:28	⁻2	3:04	0
Jan. 29	10:06	⁻5	1:59	⁻2
Feb. 7	10:00	⁻3	2:00	⁻1
Feb. 15	9:59	2	1:56	5
Feb. 28	10:05	4	1:55	3
Mar. 10	9:58	7	2:05	9
Mar. 22	10:02	10	2:02	13

1 IMAGINE Draw and label a picture. → **2 NAME** ⌐► *Facts* └► *Question*

3 THINK To find elapsed time, subtract the two readings. To subtract A.M. from P.M., add 12 hours to 3:04. Regroup 1 hour as 60 minutes to subtract the minutes.

4 COMPUTE ——► **5 CHECK**

2. Use the chart to find the elapsed time between the two readings on:

   a. February 7   b. January 29   c. February 15   d. March 10

3. It takes two hours to fly from Philadelphia to St. Louis. If a plane leaves Philadelphia at 8 A.M., what is the local time when it arrives in St. Louis?

4. Elise boarded a plane that took off from Seattle at 10:30 P.M. After 2 hours, the plane landed in Denver, where there was a 1-hour holdover. Then, the plane flew on for 3 hours to Boston. What time was it in Boston when the plane landed?

5. It was 9 P.M. in Memphis when Henri called his parents, who live in Los Angeles. His mother told him that his father was out but was expected back at 9 P.M. Henri said that he would call back later when he could speak to both of them. What time must it be in Memphis before Henri calls back?

6. Brenda got to her office at 8:30 A.M. in New York. If she needs to make a long-distance call to a customer whose business opens at 8:30 A.M. in Seattle, how long must Brenda wait to reach this customer?

7. A deep-sea diver dove underwater at 11:50:32. She resurfaced at 12:03:29. How long was the diver underwater? (Hint: When regrouping time in hours, minutes, and seconds, remember that 60 seconds equals 1 minute.)

8. One year, the winner of the Boston Marathon finished in 2:09:55. The following year, the winner finished in 2:20:19. How much slower was the time of the second year's winner than that of the first?

9. In the 1500-meter men's speed-skating race, the winner finished in 1 minute 55.44 seconds. How much faster was his time than the runner-up's time of 2 minutes 1.07 seconds?

9-13 | Problem Solving: Applications

Name the correct metric measure for each.

1. I am 10 times a centimeter. What am I?

2. I am $\frac{1}{100}$ of a gram. What am I?

3. I am 0.001 of a kiloliter. What am I?

4. I am 1000 times a milligram. What am I?

5. I am $\frac{1}{10}$ of a hectometer. What am I?

6. I am 0.000001 of a kilogram. What am I?

7. I am 0.01 of a decaliter. What am I?

8. I am 10 000 times a centigram. What am I?

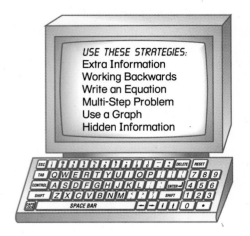

USE THESE STRATEGIES:
Extra Information
Working Backwards
Write an Equation
Multi-Step Problem
Use a Graph
Hidden Information

Solve.

9. Marta's pet weighs 8 kilograms 23 hectograms 57 decigrams. Express the pet's weight in kilograms.

10. A canoe is 1 m 23 dm 15 cm 50 mm in length. Express the canoe's length in meters.

11. Dr. Wing used 1.5 liters 78 centiliters 45 milliliters of a chemical for an experiment. How many milliliters of the chemical did she use?

12. An empty beaker had a mass of 3.8 dag. After 50 mL of a chemical solution had been poured into it, the beaker and its contents weighed 91 g. How much did the solution weigh?

13. A cubic container is 3 cm on each edge. How many milliliters of water will it hold if its capacity is 27 cm³?

14. A freezer is 4 dm on an edge. If it holds 64 dm³, how many liters can it hold?

15. Lori's puppy has grown 2 ± .5 kg this month. If the puppy weighed 3.2 kg last month, between what two weights does the puppy's weight now lie?

16. A carpenter used a saw to cut a piece of wood 23 in. long.

    a. What is the greatest possible error for the length?

    b. Which is more precise, a piece of wood cut $23\frac{1}{2}$ in. long or a piece cut 2 ft long?

17. What is the average of an athlete's three high jumps of 5 ft 10 in., $5\frac{2}{3}$ ft, and 5 ft 9 in.?

18. The hurdles in the women's 100-m race are 2 ft 9 in. high. This is $\frac{1}{4}$ ft higher than those used in the 400-m race. How high are the hurdles in the 400-m race?

19. Adrian, who weighs 112 lb, held his pet and stood on a scale. Find the weight of the pet in ounces if the scale read $113\frac{1}{4}$ lb.

20. A length of ribbon measures 3 yd 1 ft 10 in. If it is cut into five equal pieces to make bows, how long will each piece be?

21. A sailboat of the *Windglider* class is 12 ft 9 in. long. If a boat of the *Tornado* class is $7\frac{1}{4}$ ft longer than a *Windglider*, how long is the *Tornado*?

22. The student council prepared 5 gal 2 qt 1 pt of iced tea for a dance. How many $1\frac{1}{2}$-cup servings are available?

23. The meter on a gas pump reads $10\frac{1}{2}$ gal. How many quarts is that?

24. You weigh 82 lb 4 oz. You stand on a scale with your cat. The scale reads 93 lb 1 oz. How much does your cat weigh?

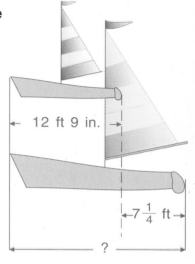

← 12 ft 9 in. →

← $7\frac{1}{4}$ ft →

?

25. Find the distance around a pentagon having sides measuring: 2 ft 5 in., $3\frac{1}{6}$ ft, 2 ft 9 in., $3\frac{1}{3}$ ft, and 2 ft 10 in.

26. The temperature dropped 3° an hour as a cold front moved in. After 4 hours the temperature was ⁻4°C. What was the original temperature?

27. Swimmer A finished a race in 8 min 12 sec. Swimmer B finished in 7 min 58 sec. How much faster was Swimmer B than Swimmer A?

After testing fertilizer and plant growth for 6 weeks, a student graphed the data. Use the graph to complete exercises 28–30.

28. Which plant was $\frac{3}{4}$ ft smaller than plant D?

   a. A   **b.** B   **c.** C

29. Each plant was 2 in. tall at the start of the test. Which plant grew $\frac{2}{3}$ ft during the test?

   a. A   **b.** B   **c.** C

30. How much did plant D grow during the test?

   a. 14 inches   **b.** $2\frac{1}{6}$ ft   **c.** $2\frac{2}{3}$ ft

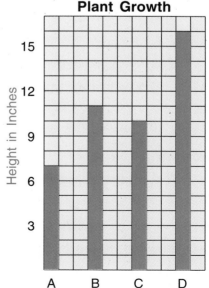

Plant Growth

Height in Inches

15
12
9
6
3

A   B   C   D

More Practice

Match the most reasonable measurement for each using the lettered choices in the box below.

1. capacity of a tablespoon
2. length of a wristwatch
3. mass of a stamp
4. mass of the sandwich
5. width of a pen tip
6. mass of a fruitcake
7. width of the door
8. capacity of the cup
9. capacity of the straw
10. capacity of the swimming pool
11. length of the swimming pool
12. mass of a concert piano
13. temperature of the fire
14. length of the runner's foot
15. temperature of the hot chocolate

a. 100 mg		**b.** 2 mL		**c.** 250 mL		**d.** 5 m		**e.** 1 t
f. 27 cm		**g.** 90 cm		**h.** 1 mm		**i.** 150°C		**j.** 32 cm
k. 50°C		**l.** 1000 kL		**m.** 1 kg		**n.** 125 g		**o.** 1.5 mL

Find the equivalent.

16. 77 g = __?__ kg
17. 5.4 m = __?__ cm
18. 1800 L = __?__ kL
19. 0.98 cg = __?__ dag
20. 307.5 hL = __?__ L
21. 21 999 km = __?__ m

Find the more precise measure.

22. 201 mm or 201 cm
23. 1.5 mm or 15 cm
24. 81 cm or 801 mm

Add or subtract.

25.
```
  5 ft  9 in.
+ 5 ft 11 in.
```
26.
```
  6 ft  4 in.
- 5 ft 11 in.
```
27.
```
  6 gal 3 qt 1 pt
+ 2 gal 1 qt 2 pt
```
28.
```
  6 gal 3 qt 1 pt
- 2 gal 1 qt 2 pt
```

Multiply or divide.

29.
```
  12 ft 8 in.
×          3
```
30.
```
  6 gal 4 qt
×         4
```
31.
```
  3 T 20 lb 8 oz
×             2
```
32.
```
  6 gal 4 qt 2 pt
×              5
```

33. 3 yd 1 ft ÷ 5
34. 4 qt 1 pt ÷ 3
35. 8 T 160 lb ÷ 10
36. 2 yd 3 ft 4 in. ÷ 2

Solve.

37. The distance from Rosedale to Westwood is $2\frac{1}{2}$ mi. If Brad ran that distance twice, how many feet did he run?

38. A vase holds 1 L 250 cL 85 mL of water. How many milliliters does it hold?

Math Probe

NETWORKS

Silvio strings colored lights along each of these diamond designs.

A   B   C   D

He can move through each of these designs without retracing any pathway.

Then Silvio strings colored lights along each of these diamond designs.

E   F   G

He finds that he does have to retrace a pathway in these designs.

The pathways in all these designs are called **networks**.

A network is *traversable* if it can be traveled in one continuous line without retracing a pathway.

1. Copy each of Silvio's designs. Then try to travel through each network without lifting your pencil and without retracing a pathway.

2. How can you make traversable networks from designs E-G? Show two possible solutions.

Another way to tell whether a network is traversable is to discover a pattern.

- Look again at the networks above.

- Each point where the paths meet is a "corner."

- If the number of paths at a corner is odd, the corner is odd.
  If the number of paths at a corner is even, the corner is even.

3. Copy and complete the chart. Then look for a pattern for traversable networks in the number of odd and even corners.

4. Use the pattern you have discovered to draw three more traversable diamond designs.

Network	Number of odd corners	Number of even corners	Traversable?
A	0	4	Yes
B	2	2	Yes
C			

5. Use the same pattern to draw three more diamond designs that are *not* traversable.

6. Draw a square. Then draw additional pathways, keeping the square a traversable network.

Check Your Mastery

Choose the most reasonable measure.   See pp. 212–225

1. the mass of a comb
   a. 10 mg   **b.** 10 g   **c.** 10 kg

2. the mass of a jeep
   a. 1 g   **b.** 1 kg   **c.** 1 t

3. the height of a house
   a. 10 m   **b.** 10 cm   **c.** 10 km

4. the length of a biking trail
   a. 7.5 m   **b.** 7.5 mm   **c.** 7.5 km

5. the temperature of a cup of pasta
   a. 110°C   **b.** 75°C   **c.** 100°C

6. the capacity of a glass pitcher
   a. 1 mL   **b.** 1 cL   **c.** 1 L

Find the equivalent.   See pp. 214–217

7. 9 kg = _?_ g

8. 9.2 g = _?_ mg

9. 2.5 mg = _?_ cg

10. 608 hL = _?_ L

11. 7.6 kL = _?_ L

12. 1800 hL = _?_ L

13. 59 cm = _?_ mm

14. 23 000 cm = _?_ hm

15. 93 hm = _?_ dm

Choose the most precise measurement.   See pp. 219–223

16. **a.** 12.2 cm   **b.** 12 cm

17. **a.** 4 cm   **b.** 39 mm

Find the equivalent.

18. 77 ft = _?_ yd _?_ ft

19. 7 gal. = _?_ qt

20. $3\frac{1}{2}$ lb = _?_ oz

Compute.   See pp. 220–223

21.   6 yd 1 ft 10 in.
    +2 yd 2 ft 2 in.

22.   3 T 1100 lb
    −1 T   853 lb

23. (3 mi 10 yd) × 3

24. (4 lb 11 oz) ÷ 5

Solve.

25. How many 5-mL spoonfuls of honey are in a half-liter jar.

26. Floyd needs 1 kg of flour for making papier-maché. If he has 120 grams, how much more does he need?

27. If a shaker holds 2 oz of salt, how many pounds are needed to fill 25 shakers?

28. Wanda walks along Willow Way every Wednesday. If Willow Way is $\frac{3}{5}$ of a mile long, how far does Wanda walk after four Wednesdays?

29. Jo measured 15 mL of water. How many grams of water was that?

10 Plane Geometry

In this chapter you will:

- Work with basic geometric concepts
- Classify angles
- Use a protractor to measure angles
- Use a compass and straightedge
- Bisect segments and angles
- Construct parallel and perpendicular lines
- Identify line and rotational symmetry
- Solve problems: organized list or chart

Do you remember?

A ray is named by its endpoint and one other point on the ray.

An angle can be named by three points or by its vertex alone.

A line is named by any two points along it.

An arc is part of a circle and is named by two points on the circle.

∥ means: is parallel to

⊥ means: is perpendicular to

≅ means: is congruent to

The Geometric Eye

RESEARCHING TOGETHER

Look at your world through the eye of a camera. Take photographs of at least 5 geometric shapes you can find in the world around you. Be creative — geometry can be found in the most unexpected places!

Share your photos with the group. Can they see what you saw through the camera's eye? Can they name the geometric shapes?

Description	Figure	Symbol	Read
Point: a location or position.	•R	R	*"point R"*
Line: a set of points that extends indefinitely in opposite directions.	B C	$\overleftrightarrow{BC}$	*"line BC"*
Segment: part of a line with two endpoints.	M N	$\overline{MN}$	*"segment MN"*
Ray: part of a line with one endpoint.	G H	$\overrightarrow{GH}$	*"ray GH"*
Angle: formed by two rays with a common endpoint, called the vertex.	vertex P R Q	$\angle PQR$	*"angle PQR"*
Plane: a flat surface that extends indefinitely in all directions.	•R •K •J	RJK	*"plane RJK"*

Use the figure to answer these questions.

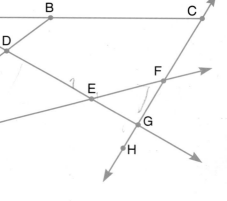

1. Are $\overline{DE}$ and $\overline{DG}$ names for the same segment?
2. Are $\overline{DE}$ and $\overline{ED}$ names for the same segment?
3. Do $\overrightarrow{EF}$ and $\overrightarrow{FE}$ name the same ray?
4. Name all the segments having *B* as an endpoint.
5. Name all the rays having *E* as an endpoint.
6. What rays pass through point *D*?
7. What is another name for $\angle FEG$?

Use *unlined* paper and a ruler.

8. Make a dot anywhere on the paper. Label it point *A*. How many lines can you draw through point *A*?

9. Show a point *B*. How many lines can you draw through points *A* and *B*?

10. How many planes can pass through line *AB*?

11. Show a point *C* that does not lie along the same line as do *A* and *B*. Draw as many lines as possible connecting the three points.

12. How many planes can pass through all three points?

Classifying Angles

Angles are measured in degrees and classified according to their measurements.

An **acute angle** measures less than 90°.  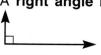	An **obtuse angle** measures between 90° and 180°. 
A **right angle** measures 90°. 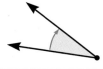 	A **straight angle** measures 180°. 
Two angles are **complementary angles** if the sum of their measures is 90°. 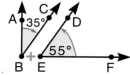 ∠ABC and ∠DEF are *complementary*.	Two angles are **supplementary angles** if the sum of their measures is 180°. 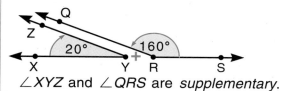 ∠XYZ and ∠QRS are *supplementary*.

Vertical angles are formed when two lines intersect. They share a common endpoint, but no common side. *Vertical angles* have the same measure and are congruent. ∠1 and ∠3 are vertical angles. ∠2 and ∠4 are also vertical angles.

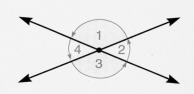

Use the figure at the right for exercises 13–22.

13. ∠PMS is ___?___

 a. obtuse   b. acute   c. right   d. straight

14. ∠QMS has a measure of ___?___

 a. 40°   b. 60°   c. 50°   d. 90°

15. ∠QMS and ___?___ are complementary angles.

 a. ∠RMS   b. ∠QMP   c. ∠SMT   d. ∠RMT

16. Name two different obtuse angles.

17. Name two different acute angles.   18. Name the complement of ∠QMR.

19. Name a right angle.   20. Name a pair of supplementary angles.

21. Name a straight angle.   22. Give the sum of ∠PMW and ∠WMU.

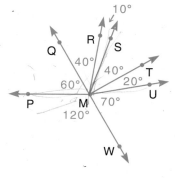

Use the figure at the right for exercises 23–24.

23. Name pairs of congruent angles.

24. Give the measure of each angle.

Protractor: used to measure angles and to draw angles.

To measure an angle:

Place the protractor so that its *base* rests along one ray of the angle and its *center mark* is at the vertex of the angle.

Find the "0" on the scale where the base ray crosses the protractor.

Follow along that scale to the point where the other ray crosses the protractor. The number at that point is the measure of the angle.

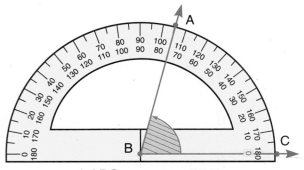

∠*ABC* measures 75°.

To draw an angle:

Mark a point for the vertex of the angle. Draw a ray from that point.

Position the center mark of the protractor so that the base rests along the ray.

To draw a 110° angle, find the "0" on the scale where the ray crosses the protractor. Follow along that scale to "110." Mark another point there. Draw a ray from the vertex to this spot.

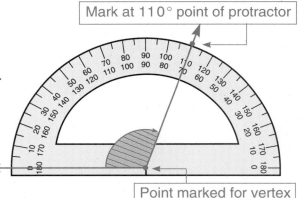

Mark at 110° point of protractor

Point marked for vertex

> **Remember:** An *acute angle* measures less than 90°.
> A *right angle* measures 90°.
> An *obtuse angle* measures between 90° and 180°.
> A *straight angle* measures 180°.

Classify each angle as acute, right, or obtuse.

1. ∠*AOB*
2. ∠*AOC*
3. ∠*DOG*
4. ∠*GOB*
5. ∠*AOD*
6. ∠*AOE*
7. ∠*GOF*
8. ∠*BOE*
9. ∠*COF*
10. The sum of ∠*GOE* and ∠*EOD*
11. The sum of ∠*COD* and ∠*COA*

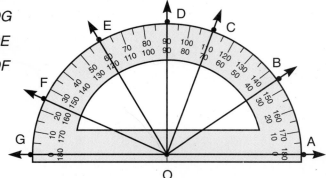

Estimate the measure of each angle. Then use your protractor to find the exact measure. (Within how many degrees of the exact measure was your estimate?)

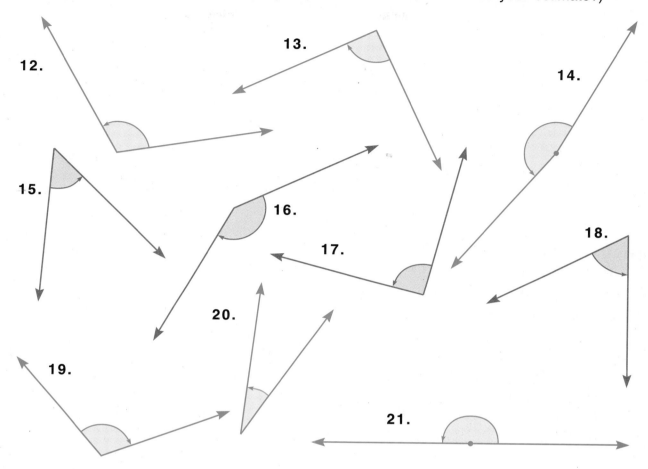

12.

13.

14.

15.

16.

17.

18.

19.

20.

21.

Draw the angle with each measure.

22. 33°
23. 155°
24. 87°
25. 125°
26. 32°
27. 9°

28. 73°
29. 27°
30. 146°
31. 101°
32. 49°
33. 179°

Draw the angle that is the complement of each angle; the supplement of each.

34. 20°
35. 65°
36. 72°
37. 4°
38. 17°
39. 61°

40. 83°
41. 24°
42. 69°
43. 19°
44. 6°
45. 45°

Write in scientific notation.

46. 17,600
47. 305,000,000
48. 2,140,000
49. 89,000,000

50. 60,000
51. 305,000
52. 21,000,000
53. 8,900,000,000

Constructing a Congruent Segment

Congruent figures: figures that have the same size and shape. (They are exact copies of one another.)

Segments of the *same length* are congruent to each other.

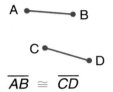

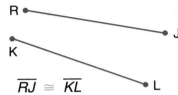

$\overline{AB} \cong \overline{CD}$ $\overline{EF} \cong \overline{GH}$ $\overline{RJ} \cong \overline{KL}$

The symbol ≅ means "is congruent to."

To construct a segment congruent to $\overline{MN}$:

Step 1	Using a straightedge, draw a segment of any length.	Step 3	Keeping the compass opening the same, place the compass point at any point on the segment. (Call it *O*.)
Step 2	Open a compass to match the length of $\overline{MN}$.	Step 4	From point *O*, swing the compass across the segment. (Call the point at which it crosses *P*.)

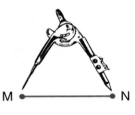

$\overline{OP} \cong \overline{MN}$

For each of the segments below, construct a congruent segment.*

1.

2.

3.

6.

4.

5.   7.

8.

Finding Together   **Use your ruler to find the length of each item.** Draw a line segment to match the length. Then construct another segment congruent to it.

9. your thumb   10. crayon   11. pocket comb   12. key

13. small scissors   14. pencil   15. paper clip   16. eraser

Constructing a Congruent Angle

Angles of the *same measure* are congruent to each other.

∠ABC ≅ ∠DEF   ∠GHR ≅ ∠LKJ   ∠MNO ≅ ∠PQR

To construct an angle congruent to ∠STU:

Step 1 With compass point at vertex *T*, draw an arc that crosses both rays of the angle. Call the crossing points *V* and *W*.	**Step 4** Adjust the compass to measure the opening of arc *VW*.

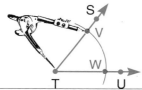

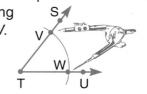

Step 2 Use a straightedge to draw a ray from a point, *Y*.

Step 5 With the compass point on *Z* and the adjusted compass opening, draw another arc intersecting the first one at a point, *X*.

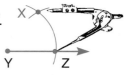

Step 3 With the compass point on *Y* and the same compass opening as in Step 1, draw an arc intersecting the ray at a point, *Z*.

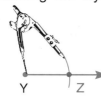

Step 6 Draw a ray from *Y* through *X* to form ∠XYZ.

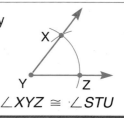

∠XYZ ≅ ∠STU

**Use unlined paper to trace each of the angles below.
Then construct an angle congruent to each.***

1.

2.

3.

4. Draw any acute angle and any obtuse angle. Then use a protractor to draw a right angle. Construct angles congruent to each of the drawn angles. Use a protractor to check the congruency of the pairs of angles.

5. Construct a triangle that has two congruent angles.
    Begin by drawing an acute angle of any size.

10-5 Bisecting a Segment

Bisect a segment by dividing it into two congruent segments.

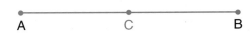

A   C   B

D   F   E

Point *C* is the *midpoint* of $\overline{AB}$.
It divides $\overline{AB}$ into two congruent segments.

$$\overline{AC} \cong \overline{CB}$$

Point *F* is the *midpoint* of $\overline{DE}$.

$$\overline{DF} \cong \overline{FE}$$

To bisect segment $\overline{GH}$:

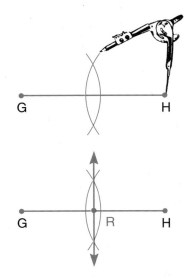

Step 1   Holding a compass point at *G*, swing the compass, making an arc across $\overline{GH}$.

Step 2   Using the same compass opening, place the compass point on *H*. Swing the compass to make another arc crossing $\overline{GH}$.

Step 3   Use a straightedge to draw a line through the two points at which the arcs intersect.

Step 4   Call the point at which the line crosses the segment point *R*.

Point *R* is the midpoint of $\overline{GH}$.

$$\overline{GR} \cong \overline{RH}$$

Trace each segment. Then bisect each.

1.

2.

3.

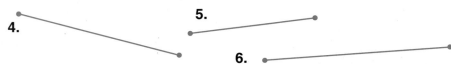

4.

5.

6.

7. Use a centimeter ruler. Draw a rectangle with a length of 10 cm and a width of 6.5 cm. Bisect each of the four sides.

8. Draw a triangle with two sides that measure 5 cm and one side that measures 3 cm. Bisect each of the sides.

10-6 | Bisecting an Angle

Bisect an angle by dividing it into two congruent angles.

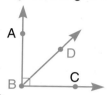

$\overrightarrow{BD}$ bisects $\angle ABC$.
$\angle ABD \cong \angle DBC$

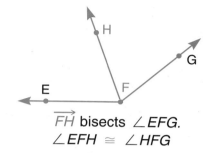

$\overrightarrow{FH}$ bisects $\angle EFG$.
$\angle EFH \cong \angle HFG$

To bisect angle *RJK*:

Step 1 With the compass point at vertex *J*, draw an arc crossing $\overrightarrow{JR}$ and $\overrightarrow{JK}$. (Call the points of intersection *L* and *M*.)

Step 2 With the compass point on *M*, draw an arc.

Step 3 With the compass point on *L* and the same compass opening as in Step 2, draw an arc intersecting the first. (Call the point of intersection *N*.)

Step 4 Use a straightedge to draw a ray from vertex *J* through *N*.

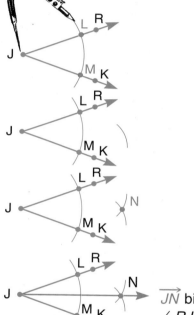

$\overrightarrow{JN}$ bisects $\angle RJK$.
$\angle RJN \cong \angle NJK$

Use unlined paper to trace each of the angles below. Then bisect each.

1. 2. 3. 4.

5. Draw a large triangle with angles of any size. Bisect each angle. (Notice where the ray that bisects each angle crosses the opposite side.)

6. Draw a triangle, each of whose angles measures 60°. Bisect each of the angles. What do you notice about where each of the angle bisectors crosses the opposite sides of this triangle?

243

10-7 Constructing Parallel Lines

Parallel lines: lines that are in the same plane and never intersect.

$\overleftrightarrow{AB} \parallel \overleftrightarrow{CD}$

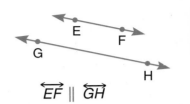

$\overleftrightarrow{EF} \parallel \overleftrightarrow{GH}$

The symbol ∥ means "is parallel to."

To construct a line parallel to $\overleftrightarrow{RJ}$:

Step 1	Draw a point, *K*, anywhere not along $\overleftrightarrow{RJ}$.
Step 2	Use a straightedge to draw a line connecting *R* and *K*.
Step 3	With a compass point on *R*, draw arc *LM*.
Step 4	Using the same compass opening and the compass point on *K*, draw arc *NO*.
Step 5	Adjust the compass opening to match the distance between *L* and *M*. Use this new opening to draw an arc from point *N* intersecting arc *NO* at point *P*.
Step 6	Use a straightedge to draw a line through points *K* and *P*.

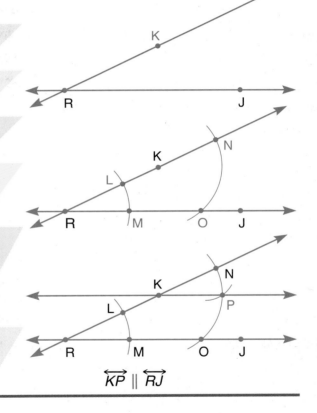

$\overleftrightarrow{KP} \parallel \overleftrightarrow{RJ}$

1. Give examples of parallel lines in your classroom.

2. Trace line *ST* and point *R*.
   Construct a line parallel to line *ST* at point *R*.

3. Place a point, *M*, somewhere below $\overleftrightarrow{ST}$ on your paper.
   Construct a line parallel to line *ST* at point *M*.

4. Are the two lines you constructed parallel to each other?

5. How many lines, do you think, can be parallel to any given line?

Perpendicular lines: lines that cross and form right angles at their intersection.

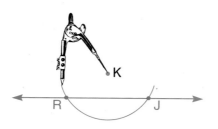

$\overleftrightarrow{AB} \perp \overleftrightarrow{CD}$

The symbol ⊥ means "is perpendicular to."

To construct a perpendicular to a line, *RJ*, from a point not on the line:

Step 1 | With a compass point on a point, *K*, draw an arc that crosses the line at two points. (Call these points *R* and *J*.)

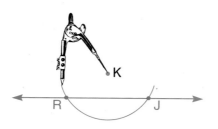

Step 2 | Place the compass point on *R*. Swing the compass to form an arc below $\overleftrightarrow{RJ}$.

Step 3 | Using the same compass opening and the compass point on *J*, draw an arc that intersects the other one below the line. (Call the point of intersection *L*.)

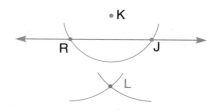

Step 4 | Using a straightedge, draw a line connecting *K* and *L*.

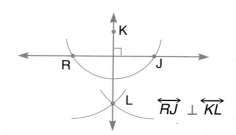

$\overleftrightarrow{RJ} \perp \overleftrightarrow{KL}$

1. Draw any line and a point not on the line. Construct a perpendicular from the point through the line.

2. How many lines, do you think, are perpendicular to a given line?

3. Draw line segment *AB*. Now bisect it. Describe the line that bisects the segment.

CHALLENGE

4. Trace the triangle at the right. Construct a line through vertex *F* that is perpendicular to $\overline{BD}$.

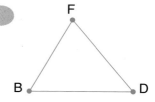

10-9 | Line Symmetry

Line of symmetry: a real or imaginary line that divides something into two congruent parts.

This module has one line of symmetry.

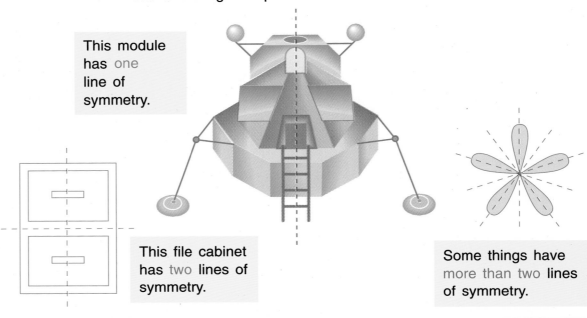

This file cabinet has two lines of symmetry.

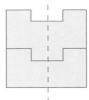

Some things have more than two lines of symmetry.

Tell whether or not the dotted line shows a line of symmetry.

1.
2.
3.
4.
5.

6. Which of the capital letters as shown here have line symmetry?

A B C D E F G H I J K L M N O P Q R S T U V W X Y Z

Trace each figure. Draw every line of symmetry on each.*

7.
8.
9.
10.
11.

12. Cover half of a symmetrical figure. Hold a mirror up along the line of symmetry. Lean the mirror toward the uncovered part. What is reflected in the mirror?

10-10 Rotational Symmetry

Some figures are symmetrical "about a point." If a figure appears the same after rotating through a certain number of degrees, it has **rotational** (or **point**) **symmetry**.

These have 90° rotational symmetry.

These have 180° rotational symmetry.

These have 120° rotational symmetry.

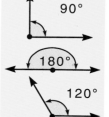

Remember:

90°

180°

120°

Tell whether each figure has 90°, 180°, or 120° rotational symmetry.*
(Hint: A single figure can have more than one kind of symmetry.)

1.   2.   3.    4.    5.

Trace each figure. What is the least number of degrees of rotation through which the figure must be turned to look the same?

6.   7.   8.    9.    10.

11. Which letters of the alphabet shown on the opposite page will look the same when rotated through 180°?

Draw these designs as they would appear when rotated 180° around the given point.

12.    13.    14.

15. Which designs in exercises 12–14 have point symmetry?

MAKE UP YOUR OWN...

16. Draw a figure. Then redraw it to show each kind of symmetry defined above.

Problem: How many line segments connecting five points can be drawn if no three of these points lie on the same straight line?

A

•B

C

E•

•D

1 IMAGINE Draw, label, and count the line segments needed to connect the five points.

2 NAME *Facts:* Five points lie on a plane.
No three of the points lie on the same line.

Question: How many line segments connecting the five points can be drawn?

3 THINK See if there is a pattern or a rule that can be used to solve this problem by setting up an organized list.

Number of points	1	2	3	4	5
Number of segments	0	1	3	6	?

1 point ⟶ 0 segments   The number of segments is one less than the number of points.

2 points ⟶ 1 segment   The number of segments is one less than the number of points, plus the number of segments joined by one additional point (1 + 0 = 1 segment).

Continue to check the computation to determine whether this pattern will solve the problem.

4 COMPUTE For three points the segments should equal the sum of: 2 + 1 + 0 = 3 segments

For four points the segments should equal the sum of: 3 + 2 + 1 + 0 = 6 segments

For five points, compute: 4 + 3 + 2 + 1 + 0 = 10 segments

5 CHECK To check, draw and name the 10 segments that can be used to connect the five points.

$\overline{AB}, \overline{AC}, \overline{AD}, \overline{AE},$ ⟶ 4
$\overline{BC}, \overline{BD}, \overline{BE},$ ⟶ 3
$\overline{CD}, \overline{CE},$ ⟶ 2   ⟶ 4 + 3 + 2 + 1 = 10 segments
$\overline{DE}$ ⟶ 1

Solve by using an organized list or chart.

1. If the sum of the three interior angles of a triangle equals 180° and the sum of the four interior angles of a quadrilateral equals 360°, what is the sum of the six interior angles of a hexagon?

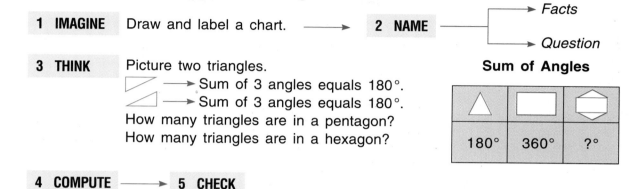

1 IMAGINE Draw and label a chart. ⟶ **2 NAME** ⟶ Facts

⟶ Question

3 THINK Picture two triangles.

⟶ Sum of 3 angles equals 180°.

⟶ Sum of 3 angles equals 180°.

How many triangles are in a pentagon?

How many triangles are in a hexagon?

Sum of Angles

△	▭	⬡
180°	360°	?°

4 COMPUTE ⟶ **5 CHECK**

2. 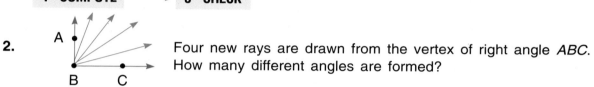 Four new rays are drawn from the vertex of right angle *ABC*. How many different angles are formed?

3. Mr. Zing works in a factory that produces unicycles and bicycles. In one week 60 cycles using 95 wheels are made. How many cycles of each type are produced?

4. How many angles are formed when straight angle *TOP* is divided into six congruent angles?
How many of these angles are straight?
How many are right?
How many are acute?
How many are obtuse?

5. How many different amounts of money can be made with six pennies, two nickels, and one quarter?

6. Mae created a design from this rectangle. If there were eight triangles in the basic design, how many triangles will there be in a column of 4 such rectangles?

7. How many squares can be made from this 7-by-7 unit square?

8. A set of six starships landed on Earth one by one. Each starship had five equal sides. The first starship to land had one light on each of its five sides. Each of the following starships had one more light on each side than the previous one had. How many lights did the sixth starship have all together?

10-12 Problem Solving: Applications

What am I?

USE THESE STRATEGIES:

Use a Model / Drawing
Logical Reasoning
Write an Equation
Guess and Test

1. I am a flat surface extending indefinitely in all directions.

2. I am formed by two intersecting planes.

3. I am the common endpoint of an angle.

4. I am part of a line having two endpoints.

5. I am the vertex of ∠CRT.

6. I am the supplement of a 95° angle.

7. I am the complement of a 42° angle.

8. I am the measure of each angle formed by bisecting a right angle.

9. I am two lines in the same plane that never intersect.

10. I am two lines that form right angles.

11. I am used to measure angles.

12. I am a line that divides something into two congruent parts.

13. I am used with a straightedge to draw congruent angles.

14. I am the point formed when a segment is bisected.

15. I am congruent angles formed when two lines intersect.

16. I am two rays having a common endpoint.

Solve.

17. How many lines of symmetry does a rectangle have?

18. Which of the digits 0, I, 2, 3, 4, 5, 6, 7, 8, 9 has/have:

    a. no lines of symmetry?

    b. only one line of symmetry?

    c. more than one line of symmetry?

19. Which digit(s) in exercise 18 has/have rotational symmetry?

20. One of a pair of complementary angles is half the other. Find the measure of each angle.

21. One of a pair of supplementary angles is half the other. Find the measure of each angle.

22. $\angle XYZ$ is bisected by $\overline{YT}$. $\angle TYZ$ measures $65°$. What is the measure of $\angle XYZ$?

Use figure *DRXT* to complete exercises 23–35.

23. $\overline{DT}$ _?_ $\overline{RX}$ **24.** $\overline{DR}$ _?_ $\overline{RX}$

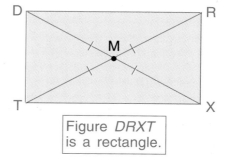

Figure *DRXT* is a rectangle.

25. $m\angle RMX$ is $50°$. Find $m\angle XMT$.

26. $\angle XMT$ and _?_ are vertical angles.

27. $\angle TDM$ is a(n) _?_ angle.

28. Find $m\angle DMR$.

29. $\overline{DM}$ is congruent to $\overline{MX}$, so M is the _?_ of $\overline{DX}$.

30. $\angle DMR$ and $\angle RMX$ are _?_ angles.

31. Find $m\angle TMR$. **32.** $\angle XTM$ and $\angle DTM$ are _?_ angles.

33. Name two straight angles. **34.** Name two obtuse angles.

35. What is the sum of the four consecutive angles ($\angle DMR$, $\angle RMX$, $\angle XMT$, and $\angle TMD$) around point M?

Write *always, sometimes,* or *never*.

36. Two complementary angles are _?_ acute angles.

37. Two supplementary angles are _?_ right angles.

38. Two vertical angles are _?_ congruent.

39. Two vertical angles are _?_ right angles.

40. Two supplementary angles are _?_ acute angles.

> **MAKE UP YOUR OWN...**

41. Use a straightedge, protractor, and compass to draw and measure a figure that includes parallel lines and vertical angles.

42. Draw a figure that has $120°$ rotational symmetry.

More Practice

Choose the correct answer.

1. The complement of an 81° angle is:
   a. 9°   **b.** 90°   **c.** 99°   **d.** none of these

2. The supplement of a right angle is:
   a. an acute angle   **b.** a right angle   **c.** an obtuse angle   **d.** a left angle

3. An angle that measures 137° is:
   a. an acute angle   **b.** a right angle   **c.** an obtuse angle   **d.** a straight angle

4. The supplement of a 40° angle is:
   a. 50°   **b.** 90°   **c.** 104°   **d.** none of these

5. ∠ABC and ∠DBF are vertical angles. ∠ABC measures 75°.
   The measure of ∠DBF is:
   a. 15°   **b.** 75°   **c.** 105°   **d.** none of these

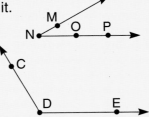

6. Classify the angle at the right. Write three different names for it.

7. Use a protractor to find the measure of ∠CDE.

8. Construct an angle congruent to ∠CDE, using a compass and a straightedge.

9. Bisect the angle you just constructed.

Use your protractor to draw:

10. a 25° angle   11. a 68° angle   12. a 147° angle   13. a 98° angle

14. Bisect each of the angles drawn for exercises 10–13. Use a compass and straightedge.

15. Name the parallel lines in this figure.

16. Name the perpendicular lines.

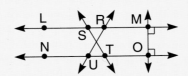

Which figures have at least one line of symmetry?
Which have 180° rotation symmetry?

17.    18.    19.   20.

Math Probe

"SAW" AND "LADDER" ANGLES

Create a graph like this one.

These angles are congruent.
They are called "ladder" angles.

Name the ladder angles in your graph.
Measure them with a protractor.

These angles are congruent.
They are called "saw" angles.
Name the saw angles in your graph.
Measure them with a protractor.

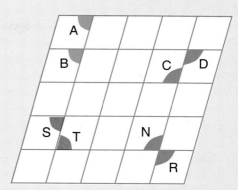

Figure 1

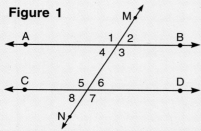

$\overleftrightarrow{AB} \parallel \overleftrightarrow{CD}$

$\overleftrightarrow{MN}$ cuts across the parallel lines.

$\overleftrightarrow{MN}$ is called a *transversal*.

Study the parallel lines above and look for *pairs of congruent angles*.
Then complete exercises 1–3.

1. Why are angles 1 and 3 congruent?

2. Name other angles that are also congruent because they are vertical angles.

3. Which angles are congruent because they are ladder angles? saw angles?

To show that $\angle A \cong \angle T$, use true statements about congruent pairs:

    a. $\angle A \cong \angle B$ because they are ladder angles.

    b. $\angle B \cong \angle S$ because they are ladder angles.

    c. $\angle S \cong \angle T$ because they are saw angles.

    Since $\angle A \cong \angle B \cong \angle S \cong \angle T$, $\angle A \cong \angle T$.

Use congruent pairs of angles to show that:

4. $\angle R \cong \angle C$ 5. $\angle 1 \cong \angle 7$ 6. $\angle 2 \cong \angle 8$

7. Suppose that m $\angle 1 = 140°$. What is the measurement of each angle in Figure 1?

8. Use what you know about congruent pairs of angles to show that the opposite angles of a parallelogram are congruent.

Check Your Mastery

Use the figure at the right.

See pp. 236–237

1. Name a line.
2. Name a segment.
3. Name a ray.
4. Name an obtuse angle.
5. Name a straight angle.
6. Name a pair of supplementary angles.

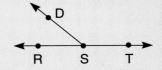

Choose the correct answer.

7. The complement of a 59° angle is __?__
   a. 59° **b.** 31° **c.** 121° **d.** 129°

8. The supplement of a 59° angle is __?__
   a. 59° **b.** 31° **c.** 121° **d.** 129°

Use your protractor. Write the measure of these angles.

See pp. 238–239

9.
10.
11.
12.

With your protractor, draw an angle of each size. Then copy the angle, using a compass and straightedge.

See pp. 238–245

13. a 35° angle
14. a 63° angle
15. a 108° angle

Trace the figure at the right. ($\overline{WX} \parallel \overline{ZY}$)

16. Bisect the segment that is parallel to $\overline{ZY}$.
17. Construct a perpendicular from point P to $\overline{XY}$.
18. Bisect a right angle.
19. Name two congruent angles.
20. Name two pairs of vertical angles.

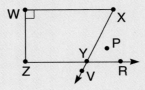

Write the letter of the figure that has:

See pp. 246–247

21. one line of symmetry.
    no symmetry.

22. 180° rotational symmetry.
    120° rotational symmetry.
    90° rotational symmetry.

a.
b.
c.
d.
e.

Cumulative Review

Choose the correct answer.

1. $\overline{AB}$ is 3 m 25 cm long. Which of the following is equal to its length?

   a. 0.325 km
   b. 325 cm
   c. 3.25 cm
   d. 32.5 mm

2. A room is 124 in. long. This is equal to:

   a. 10 ft
   b. 10 ft 4 in.
   c. 3 yd 1 ft
   d. 12.4 ft

3. Name this figure.   F   G

   a. $\overrightarrow{FG}$
   b. $\overleftrightarrow{FG}$
   c. $\overrightarrow{GF}$
   d. none of these

4. $\overrightarrow{NM}$ and $\overrightarrow{NO}$ meet at point *N*. Which figure is formed?

   a. *MNO*
   b. $\angle MNO$
   c. $\angle NOM$
   d. $\overline{MNO}$

5. Which statement is true?

   a. $\angle 1$ measures $90°$
   b. $\angle 2 \cong \angle 3$
   c. $\angle 1 \cong \angle 3$
   d. $\angle 4 \cong \angle 3$

6. The most reasonable Celsius temperature for a room temperature is:

   a. $20°$
   b. $50°$
   c. $68°$
   d. $72°$

7. $\angle XYZ$ measures $80°$. $\overrightarrow{YW}$ bisects $\angle XYZ$. Find the measure of $\angle WYZ$.

   a. $180°$
   b. $160°$
   c. $80°$
   d. $40°$

8. How many lines of symmetry does an equilateral triangle have?

   a. none
   b. six
   c. three
   d. one

9. $\overleftrightarrow{CD}$ and $\overleftrightarrow{EF}$ lie in the same plane and never intersect. Which is true?

   a. $\overleftrightarrow{CD} \perp \overleftrightarrow{EF}$
   b. $\overleftrightarrow{CD} \neq \overleftrightarrow{EF}$
   c. $\overleftrightarrow{CD} \parallel \overleftrightarrow{EF}$
   d. $\overleftrightarrow{CD} \cong \overleftrightarrow{EF}$

10. The supplement of a right angle is:

    a. a right angle
    b. an obtuse angle
    c. an acute angle
    d. a straight angle

11. A board 9 ft 8 in. is cut into 4 equal pieces. Each piece will be:

    a. 116 in.
    b. $8\frac{3}{4}$ in.
    c. 29 in.
    d. $24\frac{1}{2}$ in.

12. $\angle TSU$ and $\angle VWY$ are complementary. If the measure of $\angle VWY$ is $29°$, what is the measure of $\angle TSU$?

    a. $29°$
    b. $151°$
    c. $136°$
    d. $61°$

13. The number of degrees of rotational symmetry for this equilateral triangle is:

    a. $90°$
    b. $60°$
    c. $120°$
    d. $180°$

14. $\angle ABC$ is an obtuse angle. Its supplement could be:

    a. a right angle
    b. an obtuse angle
    c. an acute angle
    d. a straight angle

Complete.

15. 4 kg = __?__ g	**16.** 3.5 km = __?__ m	**17.** 2 gal = __?__ qt	**18.** $10\frac{1}{2}$ ft = __?__ in.
19. 48 oz = __?__ lb	**20.** 4000 mL = __?__ L	**21.** 5 lb = __?__ oz	**22.** 4 qt = __?__ c
23. 66 yd = __?__ ft	**24.** 0.7 m = __?__ mm	**25.** 10 c = __?__ pt	**26.** 0.2 kg = __?__ g

Compute.

27. 2 ft 9 in.
    +4 ft 7 in.

28. 16 yd 7 in.
    – 8 yd 11 in.

29. 4 qt 1 pt
    ×   7

30. 3 pt 1 c
    ×   8

31. 15 yd 1 ft ÷ 2

32. 1 mi 440 ft ÷ 10

33. 16 gal ÷ 32

How many lines of symmetry does each figure have?

34.

35.

36.

37.

Solve.

38. A patient's temperature was 104°F. After he recovered, it was 98.6°F. How much had his temperature dropped?

39. The Appian Way is about 560 km long. If you walked at a rate of 25 km a day, how many days would it take to travel from one end to the other?

40. Sue blends the following to make a health drink: 1 banana, 0.45 L milk, and 0.25 L orange juice. To make a dozen of these drinks, how many milliliters of milk does she need?

41. Two slices of pizza were taken from the center of a circular pie. Each angle at the tip of the slice measured 60°. What is the measurement of the angle of the remaining center part of the pie?

42. After cutting off $5\frac{1}{3}$ yd of fabric from a bolt, $8\frac{2}{3}$ yd were left on the bolt. How many feet of fabric were on the bolt originally?

43. Don needs to triple the following measures of ingredients for a pizza recipe: 8 oz of grated cheese, 1 cup of tomato sauce, 3 tsp of oregano. Write the amount of each ingredient in the greater measure designated. (For example, express ounces as pounds.)

+10 BONUS

44. Imagine that you are going to build a monument to honor your favorite person. Design a model for the monument's foundation, using 5 line segments joined together. One segment measures 3 in., one measures 1 in., and two measure 2 in. Determine for yourself the measurement of the fifth line segment. Label each of the 5 endpoints with a letter. Measure each of the five angles of the foundation. Use the symbols for line segments and angles to make a chart of measurements for each line segment and angle.

11 Polygons and Circles

In this chapter you will:

- Classify polygons
- Identify parts of a circle
- Use perimeter formulas
- Find the circumference
- Identify congruent figures
- Use area formulas for common plane figures
- Solve for missing dimensions
- Solve problems: use simpler numbers
- Use technology: BASIC programming

Do you remember?

Polygons and circles are closed plane figures.

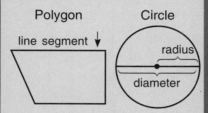

Polygon   Circle

Congruent: same size and same shape

Chord: a line segment whose endpoints are on a circle

RESEARCHING TOGETHER

Pick Pick's Polygon

Pick a polygon. Draw it on dot paper. Now, if you know *Pick's Formula*, you would be fast as lightning in finding its area.

Find out all you can about *Pick's Formula* and test it on several polygons.

257

Classifying Polygons

Polygon: a closed figure all of whose sides are line segments.

Polygons are named by the number of their sides.

triangle
(3 sides)

quadrilateral
(4 sides)

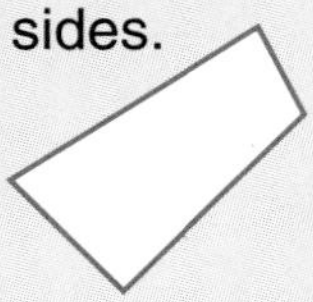

pentagon
(5 sides)

Regular polygons are figures in which all sides are congruent and all angles are congruent.

equilateral triangle
(3 sides and
 3 angles congruent)

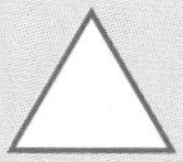

square
(4 sides and
 4 angles congruent)

regular pentagon
(5 sides and
 5 angles congruent)

Some polygons are special in other ways.

isosceles triangle
(2 sides and
 2 angles congruent)

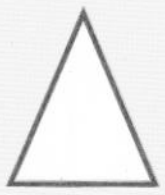

rectangle
(all angles
congruent, opposite
sides congruent)

parallelogram
(2 pairs of
parallel sides)

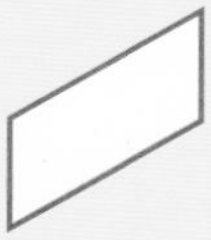

scalene triangle
(no sides or
angles congruent)

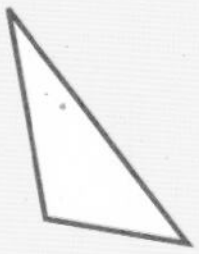

rhombus
(all sides congruent,
opposite angles
congruent)

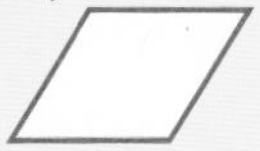

trapezoid
(1 pair of
parallel sides)

Find the meaning of each of these prefixes used to name polygons.

1. *tri-*   2. *penta-*   3. *octa-*   4. *hepta-*

5. *deca-*   6. *nona-*   7. *hexa-*   8. *quadri-*

Copy and complete the charts.

	Figure	Number of Sides	Number of Angles
9.	triangle	?	?
10.	?	4	4
11.	?	5	?
12.	hexagon	?	?

	Figure	Number of Sides	Number of Angles
13.	?	7	?
14.	?	8	?
15.	nonagon	?	?
16.	?	10	?

Solve. (Remember: Use *like* units of measure.)

13. How many meters of brick walkway will surround a 12-meter-square flower bed?

14. A snack area is 10 m long and 7 m wide. What is the perimeter of the snack area?

15. Privet forms a "living hedge" and costs $3.75 a meter. What is the cost of enclosing a garden measuring 280 cm by 220 cm?

16. Edging for a tablecloth costs $2.85 a yard. How much will Mrs. Wynn pay to put edging on a tablecloth $3\frac{3}{4}$ feet on each side?

17. How many centimeters greater is the perimeter of a regular heptagon measuring 3.5 cm on a side than the perimeter of a rectangle that is 6 cm long and 3.9 cm wide?

18. Every morning Hank jogs twice around the school playground, which measures 160 m by 200 m. About how many kilometers does he jog during a 5-day week?

Exercise 15

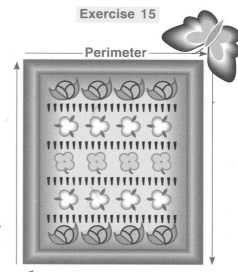

Perimeter

CHALLENGE

Find the perimeter of each figure below.
(Hint: In exercises 21 and 22 the marks show which segments are congruent.)

19.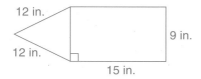
12 in.
12 in.
15 in.
9 in.

20.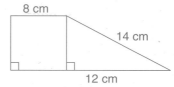
8 cm
14 cm
12 cm

21.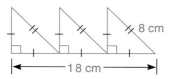
8 cm
18 cm

22.
7.5 dm

Finding Together

23. Draw the letter L on grid paper. Then determine the number of units in the perimeter.

24. Draw your initials on grid paper. Find the perimeter.

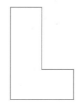

263

11-4 Perimeter and Missing Dimensions

When the perimeter and one side are known, use the perimeter formula to find the missing side.

A rectangle has a perimeter of 70 cm and a length of 21 cm. What is its width?

$P = 70$ cm
$\ell = 21$ cm
$w = \underline{\quad?\quad}$

$P = 2(\ell + w)$
$70 = 2(21) + 2w$
$70 = 42 + 2w$
$28 = 2w$
$14 = w \longrightarrow w = 14$ width

Check: $2(21 + 14) = 70$

- Write the formula.
- Substitute.
- Solve the equation.
- Check.

A triangular sign has three equal sides. Its perimeter is 24 in. Find the length of a side.

$P = 24$ in.
$s = \underline{\quad?\quad}$

$P = 3s$
$24 = 3s$
$8 = s \longrightarrow s = 8$ length of side

Check: $3 \times 8 = 24$

Find the measure of the missing side for each rectangle.*

1. $P = 30$ cm
    $\ell = 9$ cm
    $w = \underline{\ ?\ }$

2. $P = 80$ in.
    $w = 18$ in.
    $\ell = \underline{\ ?\ }$

3. $P = 19.4$ m
    $\ell = 3.5$ m
    $w = \underline{\ ?\ }$

4. $P = 2$ m
    $w = 60$ cm
    $\ell = \underline{\ ?\ }$

Find the measure of a side of a square whose perimeter is:

5. 16 cm   **6.** 24 m   **7.** $2\frac{1}{4}$ ft   **8.** $10\frac{1}{2}$ in.   **9.** 69.2 cm   **10.** 32.35 m

Given the perimeter of the figure, find the measure of a side.*

11. *square* with 16 dm perimeter   **12.** *regular pentagon* with 20 in. perimeter

13. *regular octagon* with 16.8 mm perimeter   **14.** *regular decagon* with 43 cm perimeter

15. *equilateral triangle* with 132 mm perimeter

Solve.

16. The swimming pool at Holiday Haven is in the shape of a regular octagon. If the perimeter is 32 meters, what is the length of a side?

17. A hotel flower bed is surrounded by a fence 10.6 m long. The length of the flower bed is 3.3 m. Find the width.

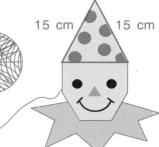
15 cm   15 cm

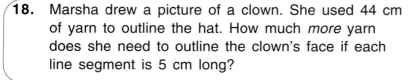

18. Marsha drew a picture of a clown. She used 44 cm of yarn to outline the hat. How much *more* yarn does she need to outline the clown's face if each line segment is 5 cm long?

11-5 Congruent Figures

The **corresponding parts** (or matching angles and matching sides)
of *congruent figures* are congruent.

Compare these triangles:

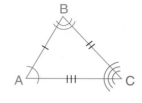

$$\triangle ABC \cong \triangle DEF$$

The triangles are congruent, so
the *corresponding sides* are congruent.

$$\overline{AB} \cong \overline{DE}$$
$$\overline{BC} \cong \overline{EF}$$
$$\overline{CA} \cong \overline{FD}$$

The triangles are congruent, so
the *corresponding angles* are congruent.

$$\angle A \cong \angle D$$
$$\angle B \cong \angle E$$
$$\angle C \cong \angle F$$

Two circles are congruent if their diameters have the same measure.

Triangles *PQR* and *XYZ* are congruent.
Name the part of one that corresponds to this part of the other:

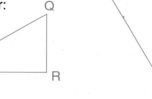

1. $\angle Y$ 2. $\angle R$ 3. $\angle P$ 4. $\angle X$
5. $\overline{PQ}$ 6. $\overline{XY}$ 7. $\overline{ZY}$ 8. $\overline{RP}$

Parallelograms *ABCD* and *GJKH* are congruent.
Name the part of one that corresponds to this part of the other:

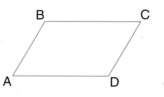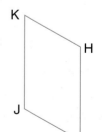

9. $\angle B$ 10. $\angle G$ 11. $\angle C$
12. $\angle J$ 13. $\overline{AB}$ 14. $\overline{KJ}$
15. $\overline{HG}$ 16. $\overline{JG}$ 17. $\overline{CD}$

18. Here are two pairs of congruent triangles.
Name each pair. Then list the pairs of corresponding sides.

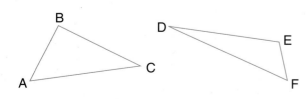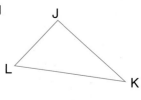

265

11-6 Area of a Rectangle and a Square

Area of a region: the number of square units it contains.

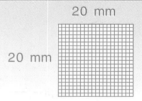

▶ **To find the area of a *rectangle*:**

- Multiply the number of units of length times the number of units of width.

- Label the product in square units.

The formula for the area of a rectangle is:

Area = length × width or
$A = \ell w$

This rectangle contains
2 rows of 5 square centimeters.

$A = \ell w$
$A = 5 \text{ cm} \times 2 \text{ cm}$
$A = 10 \text{ cm}^2$

Read: "square centimeters."

▶ **To find the area of a *square*:**

- Multiply the number of units on one side times the number of units on another side.

- Label the product in square units.

The formula for the area of a square is:

Area = side × side or
$A = s \times s$ or
$A = s^2$

This square contains 20 rows of 20 square millimeters.

$A = s^2$
$A = 20 \text{ mm} \times 20 \text{ mm}$
$A = 400 \text{ mm}^2$

Read: "square millimeters."

Remember: Multiply only *like* units.

Use the area formulas. Copy and complete.

	Rectangles			
	length	width	$A = \ell w$	
1.	7 m	5 m	?	7(5)
2.	9 m	4 m	?	?
3.	10.5 mm	2.5 mm	?	?
4.	18 km	7.65 km	?	?
5.	3 yd	7 ft	?	?
6.	160 cm	2 m	?	?
7.	20 in.	$1\frac{1}{2}$ ft	?	?

	Squares		
	side	$A = s^2$	
8.	7 m	?	7 (7)
9.	12 cm	?	?
10.	8.4 m	?	?
11.	2.1 cm	?	?
12.	0.5 m	?	?
13.	0.3 m	?	?
14.	$2\frac{1}{2}$ yd	?	?

Find the area of each figure.

15.  s = 32 yd

16. 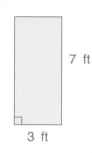 7 ft / 3 ft

17. 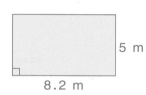 5 m / 8.2 m

18. s = 2.6 cm

19. 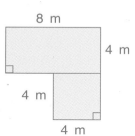 8 m / 4 m / 4 m / 4 m

20. 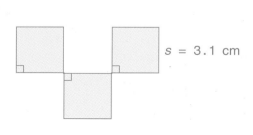 s = 3.1 cm

Solve.

21. How many square meters of contact paper are needed to cover a poster measuring 30 cm by 20 cm?

22. Six square tables are to be refinished with vinyl. How many square meters of vinyl are needed if each table measures 1.2 m on a side?

23. Which has the greater area and by how many units?
    a. a square measuring 85 units on a side
    b. a rectangle measuring 90 by 75 units

24. How many square patches measuring 9 in. on a side can be cut from a piece of rectangular material 72 in. long and 36 in. wide?

25. A rectangular porch is being covered with carpeting. The porch has a length of 6 yd and a width of $2\frac{2}{3}$ yd. Porch carpeting comes in bolts 3 yards wide and is sold only by the yard. How much will it cost to carpet the porch if the carpeting is $6.95 a square yard? How many square yards of carpeting will be left?

Exercise 24
36 in.
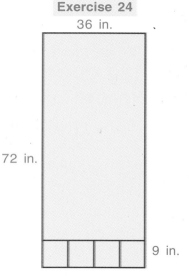
72 in.
9 in.

Finding Together 26. Draw a block letter on grid paper. Begin at the upper left corner and move:

a. right 6 units.   b. down 2 units.   c. left 2 units.   d. down 4 units.
e. left 2 units.   f. up 4 units.   g. left 2 units.   h. up 2 units.

27. What block letter did you draw?

28. What is its area in square units?

29. Choose another block letter. List the steps used to draw it; then find its area.

11-7 Area of a Parallelogram

Area of a parallelogram is the same as the area of a related rectangle.

To find the area of a parallelogram, first imagine a line drawn from a vertex meeting the opposite side at a right angle. This line is the height (h) of the parallelogram.

Now imagine sliding the newly formed right triangle across the figure to form a rectangle.

The length (ℓ) of the rectangle is the same as the base (b) of the parallelogram.

The width (w) of the rectangle is the same as the height (h) of the parallelogram.

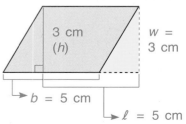

The area of the rectangle is the same as the area of the parallelogram.

So, $A = \ell w$ becomes $A = bh$.

In the parallelogram above:

$A = bh$
$A = 5(3)$
$A = 15 \text{ cm}^2$

> The formula for the area of a parallelogram is:
>
> Area = base × height or
> $A = b \times h$ or
> $A = bh$

Find the area of each figure.

1.
8 in.
3 in.

2. 6 mm
10 mm

3.
1 m
12.5 m

4.
4 cm
1.9 cm

Find the area of each parallelogram. (Compute with like units.)*

5. $b = 8$ m
$h = 5$ m
$A = \underline{?}$

6. $b = 56$ m
$h = 32$ m
$A = \underline{?}$

7. $b = 2.4$ cm
$h = 8.7$ cm
$A = \underline{?}$

8. $b = 11$ km
$h = 3.4$ km
$A = \underline{?}$

9. $b = 80$ cm
$h = 1.2$ m
$A = \underline{?}$

10. $b = 9.8$ cm
$h = 30$ mm
$A = \underline{?}$

11. $b = 2\frac{1}{2}$ in.
$h = 6$ in.
$A = \underline{?}$

12. $b = 12$ ft
$h = 7\frac{1}{4}$ ft
$A = \underline{?}$

Choose the correct answer.

13. The area of a parallelogram with a height of 2.3 cm and a base three times the height is $\underline{?}$.
  a. 9.2 cm^2 **b.** 10.58 cm^2 **c.** 15.87 cm^2 **d.** none of these

14. The area of a rectangle whose base is 285 cm and whose height is 1.35 m is $\underline{?}$.
  a. 3.8475 m^2 **b.** 513 cm^2 **c.** 481 cm^2 **d.** none of these

11-8 Area of a Triangle

Area of a triangle is one half the area of a related parallelogram.

To find the area of a triangle, first imagine flipping it along a side. Trace around it.

The two triangles are congruent. Together they form a parallelogram.

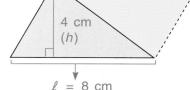

4 cm
(*h*)

ℓ = 8 cm

Since the triangle is one half of the parallelogram, the area of the triangle is one half the area of the parallelogram.

So, the formula for the area of a triangle is one half the area of a parallelogram,

or $A = \frac{1}{2}bh$.

In the triangle above:

$A = \frac{1}{2}bh \longrightarrow \frac{1}{2}(8 \times 4)$

$A = \frac{1}{2}(32) \longrightarrow 16$ cm²

The formula for the area of a triangle is:

Area = $\frac{1}{2}$ base × height or

$A = \frac{1}{2}b \times h$ or

$A = \frac{1}{2}bh$

Find the area of each figure.*

1.

35 cm

30 cm

2.

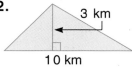

3 km

10 km

3.

8 m

7.5 m

4.

12 cm

2.5 cm

Find the area of each triangle.*

5. b = 6 yd
h = 6 yd
A = ___?___

6. b = 5 ft
h = 8 ft
A = ___?___

7. b = 1 m
h = 45 cm
A = ___?___

8. b = 2 m
h = 96.2 cm
A = ___?___

CHALLENGE

9. Draw a 3 × 4 rectangle on grid paper and determine its area.

10. Multiply each dimension by 2. Draw the new rectangle on grid paper and determine its area.

11. Multiply the original dimensions by 3 and determine the area of the resulting rectangle.

11-9 Area of a Trapezoid

Area of a trapezoid is one half the area of a related parallelogram.

To find the area of a trapezoid, first imagine another trapezoid of the same size turned upside down beside it.

The two trapezoids are congruent. Together they form a parallelogram.

The base of the parallelogram is made up of the lower base (b_1) and the upper base (b_2) of the trapezoid. The height (h) is the same for both.

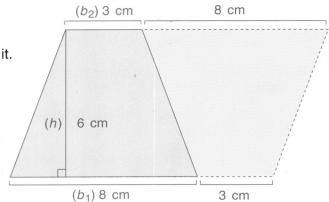

For a parallelogram: $A = bh$
$$A = (b_1 + b_2)h \longrightarrow \text{So, for a trapezoid:} \quad A = \tfrac{1}{2}(b_1 + b_2)h$$

In the trapezoid above:

$A = \tfrac{1}{2}(b_1 + b_2)h$

$A = \tfrac{1}{2}(8 + 3)6 = \tfrac{1}{2}(11)6$

$A = \tfrac{1}{2}(66) = 33 \text{ cm}^2$

The formula for the area of a trapezoid is:

$\text{Area} = \tfrac{1}{2}(\text{base}_1 + \text{base}_2)\,\text{height}$ or

$A = \tfrac{1}{2}(b_1 + b_2)h$

Find the area of each figure.

1.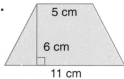
5 cm
6 cm
11 cm

2.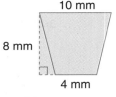
10 mm
8 mm
4 mm

3.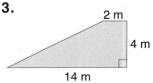
2 m
4 m
14 m

4.
4 dm
20 dm
6 dm

Find the area of each trapezoid.

5. $b_1 = 11$ m
$b_2 = 9$ m
$h = 5$ m
$A = \underline{\quad?\quad}$

6. $b_1 = 23$ mm
$b_2 = 17$ mm
$h = 8$ mm
$A = \underline{\quad?\quad}$

7. $b_1 = 40.6$ m
$b_2 = 36.4$ m
$h = 10$ m
$A = \underline{\quad?\quad}$

8. $b_1 = 12.3$ km
$b_2 = 10.5$ km
$h = 5$ km
$A = \underline{\quad?\quad}$

9. $b_1 = 10$ cm
$b_2 = 8.4$ cm
$h = 5.5$ cm
$A = \underline{\quad?\quad}$

10. $b_1 = 35.6$ mm
$b_2 = 45$ mm
$h = 15.2$ mm
$A = \underline{\quad?\quad}$

11. $b_1 = 3\tfrac{1}{2}$ in.
$b_2 = 5\tfrac{1}{2}$ in.
$h = 4$ in.
$A = \underline{\quad?\quad}$

12. $b_1 = 1\tfrac{3}{4}$ ft
$b_2 = 2\tfrac{1}{4}$ ft
$h = 3\tfrac{1}{2}$ ft
$A = \underline{\quad?\quad}$

Find the area of each figure. (Compute with like units.)

13.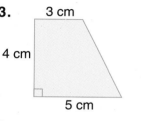
3 cm

4 cm

5 cm

14.
8 ft   12 ft

8 ft

15.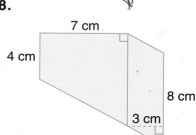
6 ft   10 ft

1 yd

16.
10 cm

12 cm   4 cm

4 cm

10 cm

17.
6 mm

9 mm

6 mm

1.2 cm

18.
7 cm

4 cm

8 cm

3 cm

Solve.

19. Each of the four sides of a pedestal is shaped like a trapezoid. The shorter base measures 1.4 m, the longer base measures 2 m, and the height measures 1.8 m. How many square meters of granite tiling are needed to cover all four sides?

20. One of the outside walls of a house is shaped like a trapezoid with parallel bases of 36 ft and 42 ft and a height of 24 ft between them. If a homeowner can earn $1.20 per square foot by renting this area as billboard space, how much money can the homeowner make?

21. Given these dimensions, which trapezoid has the *greatest* area?
   a. height of 10 cm and parallel bases of 12 cm and 18 cm
   b. height of 18 cm and parallel bases of 10 cm and 12 cm
   c. height of 12 cm and parallel bases of 10 cm and 18 cm

MENTAL ◄▻ MATH

Find the areas of these trapezoids each having a height of 4 cm and parallel bases measuring:

22. 99 cm and 101 cm   **23.** 24 cm and 26 cm   **24.** 16 cm and 14 cm

25. 49 cm and 51 cm   **26.** 35 cm and 45 cm   **27.** 2.1 cm and 1.9 cm

Finding Together   **Look for a pattern in finding the areas of these rectangles:**

28. 15 yd × 15 yd   **29.** 27 cm × 23 cm   **30.** 16 m × 14 m

31. 34 mm × 36 mm   **32.** 48 ft × 42 ft   **33.** 51 in. × 59 in.

11-10 Solving for Missing Dimensions

When the area is known, use the area formula to find missing sides or dimensions.

- Draw a diagram.
- Substitute the known measures into the formula.
- Solve.

A rectangular park has an area of 450 m². It is 30 m long. Find its width.

$A = 450$ m² $w = \underline{\ ?\ }$

$\ell = 30$ m

$A = \ell w$

$450 = 30w \longrightarrow 30w = 450$

$w = 450 \div 30 \longrightarrow w = 15$

The width is 15 m.

A square basement has an area of 900 ft². What is the length of a side?

$A = 900$ ft² $s = \underline{\ ?\ }$

$A = s^2$

$900 = s^2 \longrightarrow s^2 = 900$

Ask: What number multiplied by itself equals 900?

$900 = 30 \times 30 \longrightarrow s = 30$

A side measures 30 ft.

A parallelogram with an area of 78 cm² has a base of 13 cm. What is its height?

$A = bh$

$78 = 13h \longrightarrow 13h = 78$

$h = 78 \div 13 \longrightarrow h = 6$

The height is 6 cm.

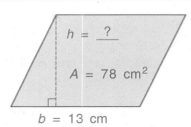

$h = \underline{\ ?\ }$

$A = 78$ cm²

$b = 13$ cm

Find the measure of the missing side for each rectangle.

1. $A = 21$ m²
   $w = 3$ m
   $\ell = \underline{\ ?\ }$

2. $A = 416$ km²
   $\ell = 26$ km
   $w = \underline{\ ?\ }$

3. $A = 2.7$ m²
   $w = 0.9$ m
   $\ell = \underline{\ ?\ }$

4. $A = 6$ cm²
   $w = 1.2$ cm
   $\ell = \underline{\ ?\ }$

Find the measure of a side of a square whose area is:

5. 100 mm²

6. 400 cm²

7. 144 km²

8. 81 m²

9. 64 ft²

10. 169 cm²

11. 225 yd²

12. 16 in.²

Find the missing dimension.
(Which figure is a square?)

	Area	length	width
13.	21 in.²	5 in.	?
14.	121 ft²	11 ft	?
15.	40.5 dm²	9 dm	?
16.	18.24 m²	?	6.08 m

Find the measure of the missing side for each parallelogram.

17. $A = 31.5$ cm^2

$b = 4.5$ cm

$h = $ _?_

18. $A = 41.6$ mm^2

$b = $ _?_

$h = 8$ mm

19. $A = 14$ ft^2

$b = 5\frac{1}{4}$ ft

$h = $ _?_

20. $A = 28$ in.^{2}

$b = $ _?_

$h = 3\frac{1}{2}$ in.

Triangles and Missing Dimensions

To find a missing dimension of a triangle when the area is known, use the area formula.

Draw a diagram, substitute the known measures into the formula, then solve.

Find the height of a triangle with an area of 21 in.^{2} and a base of 10 in.

$A = \frac{1}{2} bh$

$21 = \frac{1}{2} \times \frac{10}{1} \times h$

$21 = 5h \longrightarrow 5h = 21$

$h = 21 \div 5 \longrightarrow h = 4\frac{1}{5}$ or 4.2

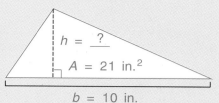

$h = $ _?_

$A = 21$ in.^{2}

$b = 10$ in.

The height is 4.2 in.

Find the missing dimension of each triangle.

21. $A = 85$ cm^2

$b = 17$ cm

$h = $ _?_

22. $A = 18$ ft^2

$b = $ _?_

$h = 4$ ft

23. $A = 350$ mm^2

$b = $ _?_

$h = 20$ mm

24. $A = 17.5$ m^2

$b = 5$ m

$h = $ _?_

Solve.

25. A triangular pennant has an area of 364 cm^2. If the base of the pennant is 14 cm long, what is the height?

14 cm

Go Team!

$h = $ _?_

26. What is the width of a billboard that is 4 m long and has an area of 12.8 m^2?

27. The area of a large platform is 56 m^2. Its length is 14 m. What is its width?

28. The base of a parallelogram is 4.5 cm. Its area is 13.05 cm^2. What is the height of the parallelogram?

29. $\triangle ABC$ is an isosceles right triangle. $\overline{DE}$ is parallel to $\overline{CB}$. $AD = 4$ cm. $DE = 4$ cm. Find the area of:

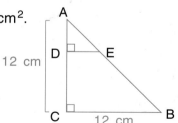

A

D E

12 cm

C 12 cm B

  a. $\triangle ADE$ **b.** $DCBE$ **c.** $\triangle ACB$

11-11 Circumference of a Circle

Circumference of a circle: the distance around it.

The symbol π (pi) stands for the ratio of any circle's circumference to its diameter.

$$\pi = \frac{C}{d}$$

The *approximate* value of π is expressed as the decimal 3.14 or as the fraction $\frac{22}{7}$.

$$\pi \approx \frac{22}{7}$$

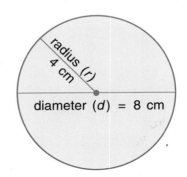

radius (r)
4 cm

diameter (d) = 8 cm

To find the circumference of a circle:

If the **diameter** is given, multiply the diameter by π.

$$C = \pi d$$

In the circle above:

$C = \pi d$
$C = \pi 8$
$C \approx 3.14 \times 8$
$C \approx 25.12$ cm

The symbol $\approx$ means "is almost or approximately equal to."

If the **radius** is given, multiply two radii by π.

$$C = 2\pi r$$

In the circle above:

$C = 2\pi r$
$C = 2\pi 4$
$C \approx 2(3.14) \times 4$
$C \approx 6.28 \times 4$
$C \approx 25.12$ cm

The formulas for the circumference of a circle are: $C = \pi d$ and $C = 2\pi r$

Given the diameter, find the circumference. (Use 3.14 for π.)

1. $d = 4$ cm
   $C \approx$ ___
2. $d = 5$ dm
   $C \approx$ ___
3. $d = 16$ m
   $C \approx$ ___
4. $d = 20$ ft
   $C \approx$ ___
5. $d = 111$ ft
   $C \approx$ ___

6. $d = 112$ m
   $C \approx$ ___
7. $d = 1.6$ cm
   $C \approx$ ___
8. $d = 7.1$ cm
   $C \approx$ ___
9. $d = 100$ ft
   $C \approx$ ___
10. $d = 400$ mm
    $C \approx$ ___

Given the radius, find the circumference. (Use 3.14 for π.)

11. $r = 10$ mm
    $C \approx$ ___
12. $r = 15$ in.
    $C \approx$ ___
13. $r = 40$ mm
    $C \approx$ ___
14. $r = 35$ km
    $C \approx$ ___
15. $r = 3.5$ m
    $C \approx$ ___

16. $r = 2.5$ mm
    $C \approx$ ___
17. $r = 7.5$ mm
    $C \approx$ ___
18. $r = 12.5$ cm
    $C \approx$ ___
19. $r = 4.6$ in.
    $C \approx$ ___
20. $r = 3.1$ ft
    $C \approx$ ___

Find the circumference. (Use $\frac{22}{7}$ for π.)

21. $r = 14$ in.
$C \approx$ _?_

22. $r = 126$ cm
$C \approx$ _?_

23. $d = 210$ ft
$C \approx$ _?_

24. $d = 35$ m
$C \approx$ _?_

25. $r = 17.5$ mm
$C \approx$ _?_

26. $r = 10.5$ ft
$C \approx$ _?_

27. $r = 3.5$ ft
$C \approx$ _?_

28. $d = 7$ yd
$C \approx$ _?_

29. $d = 56$ km
$C \approx$ _?_

30. $d = 28$ m
$C \approx$ _?_

Expressing Circumference in π Units

Sometimes circumference may be expressed in terms of π.
Find the circumference of a circle having a radius of 3 cm.

$$C = 2\pi r$$
$$C = 2 \times \pi 3$$
$$C = 6\pi \text{ cm} \longleftarrow \text{ Exact Circumference}$$

To express circumference in more familiar units:

- Use 3.14 for π.
  $$C \approx 6 \times 3.14 \longrightarrow C \approx 18.84 \text{ cm} \qquad \text{Approximate Circumference}$$
- Use the key on a calculator.
  $$C \approx \boxed{6} \times \boxed{\pi} \boxed{=} \boxed{18.849556} \qquad \text{Approximate Circumference}$$

Find the circumference in terms of π.

31. $r = 12$ mm
$C =$ _?_

32. $r = 8.3$ in.
$C =$ _?_

33. $d = 30$ m
$C =$ _?_

34. $d = 15$ ft
$C =$ _?_

35. $d = 7\frac{1}{2}$ ft
$C =$ _?_

36. $d = 2\frac{3}{4}$ in.
$C =$ _?_

37. $r = 200$ yd
$C =$ _?_

38. $r = 450$ yd
$C =$ _?_

39. $d = 0.3$ cm
$C =$ _?_

40. $d = 0.8$ cm
$C =$ _?_

Solve. (Use 3.14 or $\frac{22}{7}$ for π.)

Exercise 41

41. How many meters of colored lights will be needed to string around the outside rim of a Ferris wheel if the diameter of the wheel is 9 meters?

42. A carousel needs a new steel frame around the outside of the base. If the diameter of the base is 14 meters, and the steel costs $14.50 a meter, how much will it cost to repair the carousel?

43. Which has a smaller perimeter: a circle with a radius of 10.5 cm or a square with a side of 16 cm?

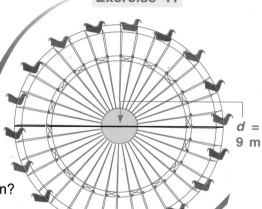

$d = 9$ m

275

Find the **area of a circle** by multiplying π (pi) by the square of the radius.

Circle A

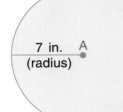

7 in. A
(radius)

'The formula for
the area of
a circle is: $A = \pi r^2$

Circle B

B
12 mm
(diameter)

Remember:

$r = \dfrac{d}{2}$

$r = \dfrac{12}{2} = 6$

In circle A (using $\frac{22}{7}$ for π):

$A = \pi r^2$

$A = \pi 7^2$

$A = \pi \times 7 \times 7$

$A \approx \dfrac{22}{\overset{}{\underset{1}{7}}} \times \overset{1}{7} \times 7$

$A \approx 154 \text{ in.}^2$

In circle B (using 3.14 for π):

$A = \pi r^2$

$A = \pi \left(\dfrac{12}{2}\right)^2$

$A = \pi (6)^2$

$A = \pi \times (6 \times 6)$

$A \approx 3.14 \times 6 \times 6$

$A \approx 113.04 \text{ mm}^2$

Find the area to the nearest tenth. (Use 3.14 for π.)

1.

1 in.

2.

2 mm

3.

20 m

4.

2 yd

5.

4 km

6.

2 km

7.
2.5 m

8.

1.5 cm

Find the area of each. (Use $\frac{22}{7}$ for π.)

9. a coin with a diameter of 28 mm

10. a mirror with a diameter of 70 mm

11. a cake plate with a radius of 21 cm

12. a cookie cutter with a radius of 2.8 cm

13. a table cover with a diameter of 5 ft 3 in.

14. a porthole with a diameter of 2 ft 4 in.

15. a button with a diameter of 21 mm

Exercise 14

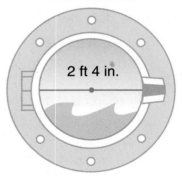

2 ft 4 in.

Use a metric ruler to find the radius or diameter of each to the nearest millimeter. Then find the area.

16.

17.

18.

19.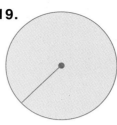

Expressing Area in π Units

Sometimes area may be expressed in terms of π.
Find the area of a circle having a radius of 3 cm.

$$A = \pi r^2$$
$$A = \pi \times 3^2$$
$$A = \pi \times 3 \times 3$$
$$A = 9\pi \text{ cm}^2 \quad \longleftarrow \quad \boxed{\text{Exact Area}}$$

To express area in more familiar units:

- Use 3.14 for π.

  $A \approx 9 \times 3.14$

  $A \approx 28.26 \text{ cm}^2$ **Approximate Area**

- Use the **π** key on a calculator.

  $A \approx$ **9** × **π** = `28.274334`

  $A \approx 28.274334 \text{ cm}^2$ **Approximate Area**

Find the area in terms of π.

20. $r = 1$ m
$A = \underline{?}$

21. $r = 2$ m
$A = \underline{?}$

22. $r = 4$ ft
$A = \underline{?}$

23. $d = 6$ in.
$A = \underline{?}$

24. $d = 10$ in.
$A = \underline{?}$

Solve. (Use 3.14 or $\frac{22}{7}$ for π.)

25. A dog tied to a post by a 10-ft rope has how much area in which to run?

26. A garden sprinkler spins around a point and waters the lawn in a circular region. If the farthest the water reaches is 3.8 m, what is the area of the lawn that is watered?

27. The public address loudspeaker will be mounted on a pole at the center of the carnival grounds. A voice on the speaker can be heard over a circular area of 5024 m². What is the radius of the area?

Exercise 27

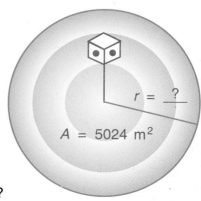

$r = \underline{?}$

$A = 5024 \text{ m}^2$

TECHNOLOGY

Welcome to BASIC Training!

You can have fun using your BASIC programming skills to solve problems like this.

Your friend Evan entered a science competition in which he had to solve 20 problems. Five points were given for each correct answer, and two points were deducted for each incorrect answer. If Evan's score was 79, how many correct answers did he have?

BASIC training will help you to write a computer program to GUESS and TEST solutions. Each guess (line 40) is computed (line 60) and tested (line 90) until a solution is found.

```
10  REM   GUESS  AND  TEST  SOLUTION
          FOR  PROBLEM  #1
20  CLS
30  PRINT  "HOW  MANY  ANSWERS  DOES
           EVAN  HAVE  CORRECT?"
40  INPUT  C
50  LET  I = 20 - C
60  LET  S = (C*5) - (I*2)
70  PRINT  "IF  EVAN  HAS  ";C;"  PROBLEMS
           CORRECT  HIS  SCORE  IS  ";S
80  PRINT
90  IF  S <> 79  THEN  PRINT
    "PLEASE  TRY  AGAIN."  :  GOTO  30
100 PRINT  "CONGRATULATIONS!"
110 END
```

REM: gives a message or a remark in a program.

INPUT: stops the program, displays a "?," and waits for information to be entered.

IF...THEN: tests for a condition and then directs the continued flow of the program.

How easy! Now see what BASIC can do with the following problem.

Two cubes, each numbered 1-6, are thrown. The resulting numbers from each throw represent the length and width of a rectangle. If perimeter and area are computed, for what combinations of length and width will the number result be the same?

An ORGANIZED LIST will help you to solve this problem.
Use FOR-NEXT loops in your program to create the organized list.

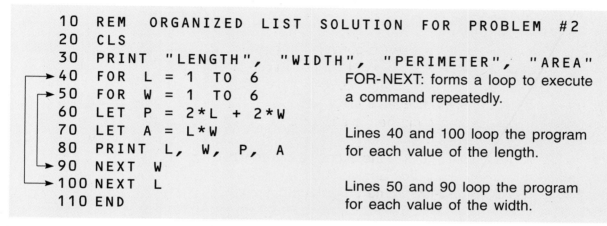

```
10  REM   ORGANIZED  LIST  SOLUTION  FOR  PROBLEM  #2
20  CLS
30  PRINT  "LENGTH",  "WIDTH",  "PERIMETER",  "AREA"
40  FOR  L = 1  TO  6                FOR-NEXT: forms a loop to execute
50  FOR  W = 1  TO  6                a command repeatedly.
60  LET  P = 2*L + 2*W
70  LET  A = L*W                     Lines 40 and 100 loop the program
80  PRINT  L,  W,  P,  A             for each value of the length.
90  NEXT  W
100 NEXT  L                          Lines 50 and 90 loop the program
110 END                             for each value of the width.
```

See If You Can Get Through This BASIC Training!

Tell how you would change the computer program for problem #1 in each case.

1. Evan's score was 86.

2. Thirty problems were given.

3. Seven points were given for each correct answer.

4. Three points were deducted for each incorrect answer.

5. Suppose that spinners, numbered from 1 to 10, were used instead of numbered cubes in problem #2. Explain how you would change the program.

Write a computer program in BASIC to help you solve each of these problems.

6. Twice a number increased by the square of the number equals 288. The number is between 10 and 20. What is the number?

7. A number cubed divided by two less than the number equals 54. What is the number?

8. Find all the 2-digit numbers that divide 360 with no remainders.

9. If the sum of 5 consecutive even numbers is 320, what is the least of the numbers?

10. Jaime is 4 years younger than his sister Inez. Fourteen years ago Inez was twice as old as Jaime. How old are they now?

STRATEGY
Problem Solving: Use Simpler Numbers

Problem: Edwina is making a square frame measuring 29.7 cm
on each side. She has a strip of wood 1.25 m long.
Estimate whether she will have enough wood to make the frame.

1 IMAGINE   Draw and label
a picture of this frame.

←——————— 1.25 m ———————→

2 NAME   *Facts:*   29.7   cm — length of each side
of square frame
1.25 m — length of wood
from which frame is cut

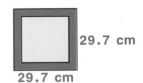

29.7 cm

29.7 cm

Question: Does Edwina have enough wood?

3 THINK   Estimate: 29.7 cm is about 30 cm.

To make the square frame, Edwina needs 4 strips.

4 × 30 cm = 120 cm ←——— Estimated perimeter
of the frame

The actual amount of wood Edwina needs must be
less than 120 cm, because 29.7 cm was rounded up to 30 cm.

Is 1.25 m > 120 cm?

To answer, change 1.25 meters to centimeters.

1.25 m = __?__ cm

4 COMPUTE   1.25 m × 100 = 125 cm
125 cm > 120 cm

So, Edwina has enough wood.

5 CHECK   Find the actual perimeter of the frame to check your estimate.

29.7 × 4 = 118.8 Perimeter of Frame

Edwina needs 118.8 cm of wood.

Since she has 125 cm (1.25 m) of wood,
she will have enough to make the frame.

Solve by using estimation.

1. One pasta salad recipe calls for $\frac{3}{4}$ teaspoon of spices. Another pasta salad recipe calls for $\frac{1}{2}$ teaspoon of spices. If Bob wants to make two batches of both recipes, will one tablespoon of spices be enough?

Recipe 1 $\frac{3}{4}$ tsp $\frac{3}{4}$ tsp

Recipe 2 $\frac{1}{2}$ tsp $\frac{1}{2}$ tsp

3 tsp = 1 tbsp

1 IMAGINE	Draw and label a picture.
2 NAME	→ *Facts* → *Question*

3 THINK   If the amount of spices needed for the first batch of each recipe is doubled, then that will be the number of teaspoons of spices needed to make two batches of each recipe.

Then change the total number of teaspoons of spice needed for both recipes to tablespoons, using the hidden fact (3 teaspoons = 1 tablespoon).

4 COMPUTE ⟶ **5 CHECK**

2. To make table trimmings for the spring dance, the decoration committee decided they needed 150 pieces of ribbon measuring $5\frac{7}{9}$ inches each. Estimate how many yards of ribbon they must purchase.

3. Some backpackers have been hiking for $4\frac{1}{2}$ hours. They have completed $\frac{7}{16}$ of their hike. About how many hours have they planned to hike?

4. A manufacturing plant can make 482 cars in one day. If almost $\frac{1}{16}$ of them are white, estimate the number of white cars made in one day.

5. Alyssa removed the trim from two circular pieces of cloth, each having a radius of $4\frac{5}{11}$ in. She plans to attach this trim to two rectangular pieces of cloth, each measuring $6\frac{1}{4}$ in. by $6\frac{2}{7}$ in. Does she have enough trim?

6. Ronnie bought a coat on sale for nearly $\frac{2}{3}$ of the original price. If the original price was $35.80, about how much did Ronnie pay for the coat?

11-15 Problem Solving: Applications

What am I?

1. I am a triangle having a 90° angle.

2. I am a triangle having three congruent angles.

3. I am a triangle that has angles measuring 55°, 45°, and 80°.

4. I am a four-sided polygon having *no* congruent or parallel sides.

5. I am a quadrilateral with only one pair of parallel sides.

6. I am a rhombus having four congruent angles.

7. I am a triangle having one right angle and two congruent sides.

8. I am a triangle having an angle measuring 110° and two congruent sides.

9. I am a polygon having six congruent sides.

10. I am a triangle having a right angle and no congruent sides.

USE THESE STRATEGIES:
Write an Equation
Hidden Information
Use a Model/Drawing
Multi-Step Problem
Use a Formula

Complete exercises 11–21, given: $\overline{RP}$ = 6 cm, m∠SPT = 76°, and △RPS ≅ △TPU.

11. $\overline{SP}$ = __?__ cm

12. $\overline{RT}$ = __?__ cm

13. m∠RPS = __?__°

14. m$\overparen{SRT}$ = __?__°

Use 3.14 for π in exercises 15 and 16.

15. Circumference = __?__ cm

16. Area of circle P = __?__ cm²

17. If the perimeter of △TPU is 22 cm, what is the perimeter of △RPS?

18. What is the length of $\overline{UT}$?

19. The height of △TPU is 4 cm. What is its area?

20. What is the area of △RPS?

21. What is the area of the colored region of circle P ?

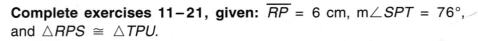

Solve.

22. A museum display case is covered by a piece of plexiglass measuring 3.2 m by 2.8 m. What is the area of the cover?

23. What is the width of a mural having an area of 6272 cm² and a length of 1.12 m?

24. What is the area of a triangle that has a height of 47.5 mm and a base of 1 m?

25. How many feet of portable fencing are needed to enclose a circular garden having a 21-foot diameter?

26. A circle has a circumference of 44 ft. Find the diameter. (Use $\frac{22}{7}$ for π.) $C = \pi d \longrightarrow 44 = \pi d \longrightarrow 44 \div \pi = d$

27. What is the diameter of a circular wading pool if its circumference is 25.12 ft? (Use 3.14 for π.)

28. Hector and Gil are testing a ring-toss game for the school carnival. The bottle over which the rings must fit is 2.6 cm in diameter. The rings are 90 mm in circumference. Will they fit over the bottle?

29. A tree trunk has a circumference of $7\frac{1}{3}$ feet. If a "slice" of the trunk is used as a table top, what will its area be?

30. Find the area of a frame 7 cm wide that fits around the rim of a circular mirror having a 14-cm radius.

31. One angle of an isosceles triangle is 100°. What is the measure of the other two angles?

32. What is the measure of the third angle of a right triangle if one angle is 35°?

33. **Find the perimeter and area of each figure.**

   a.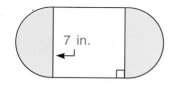
      7 in.

   b.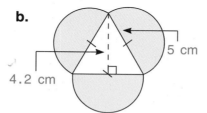
      5 cm
      4.2 cm

   c.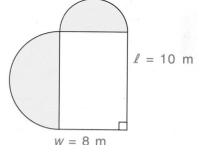
      ℓ = 10 m
      w = 8 m

34. **Find the area of the shaded region in each figure.**

   a.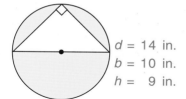
      d = 14 in.
      b = 10 in.
      h = 9 in.

   b.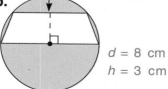
      6 cm
      d = 8 cm
      h = 3 cm

   c.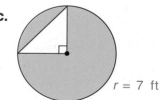
      r = 7 ft

More Practice

Choose the correct answer or answers.

1. Which figures are *regular polygons*?
   a. equilateral triangle   **b.** square   **c.** parallelogram   **d.** heptagon

2. Which figures are *quadrilaterals*?
   a. equilateral triangle   **b.** square   **c.** parallelogram   **d.** heptagon

3. The perimeter of a rectangle whose length is 14.2 cm and whose width is 8.5 cm is __?__
   a. 22.7 cm   **b.** 45.4 cm   **c.** 90.8 cm   **d.** none of these

4. The area of a square with a side that measures 9 mm is __?__
   a. 18 m²   **b.** 48 mm²   **c.** 81 mm²   **d.** none of these

5. A parallelogram has an area of 600 m². If its base is 30 m, its height is __?__
   a. 20 m   **b.** 30 m   **c.** 200 m   **d.** 300 m

6. A trapezoid has bases of 90 cm and 50 cm and a height of 52 cm. Its area is __?__
   a. 3640 cm²   **b.** 7280 cm²   **c.** 10 920 cm²   **d.** 14 560 cm²

7. A central angle measures 240°. The measure of its arc is:
   a. 120°   **b.** 240°   **c.** 60°   **d.** none of these

8. △JYM and △LNR are congruent equilateral triangles. The perimeter of △JYM is 36 in. The length of $\overline{LN}$ is:
   a. 12 in.   **b.** 36 in.   **c.** 9 in.   **d.** none of these

Find the perimeter and area.

9.
   4.4 cm

10.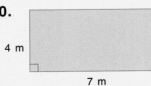
    4 m
    7 m

11.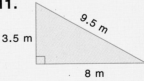
    3.5 m   9.5 m
    8 m

12.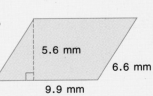
    5.6 mm   6.6 mm
    9.9 mm

Find the circumference and area.

13.
    3.5 cm

14. 12.6 cm

15. 18 m

16. 48 mm

Math Probe

TANGRAMS OR POLYGONAL PUZZLES

Some polygons can be combined with others to form
still other polygons and interesting figures.

A tangram puzzle can be constructed from a square divided into seven parts:

> 2 large triangles
> 1 medium triangle
> 2 small triangles
> 1 square
> 1 parallelogram

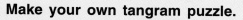

Make your own tangram puzzle.

- Draw a 4-inch square and cut it out.
- Use your ruler to measure and draw the lines that divide the puzzle into seven pieces.
- Number the pieces as shown.
- Cut the seven pieces apart.

Now use the tangram pieces to make these polygons:

1. a triangle using both of the small triangles and the parallelogram;

2. a square using all the pieces except the two large triangles;

3. a rectangle using all the pieces.

Use the numbered tangram pieces to answer exercises 4–13.

4. Name two pairs of congruent triangles.

5. The trapezoid made from pieces 3 and 4 is congruent to what other trapezoid?

6. Triangle 1 is what part of the whole tangram?

7. Triangle 7 is what part of the whole tangram? of triangle 2?

8. Triangle 4 is what part of the whole tangram?

9. Triangle 4 is what part of square 5?

10. Triangle 4 is what part of triangle 1?

11. Triangle 4 is what part of triangle 7?

12. Triangle 6 is what part of parallelogram 3?

13. Parallelogram 3 is what part of triangle 1?

14. Make up your own part-whole relationship using the tangram pieces.

Check Your Mastery

Choose the correct answer.

See pp. 258–265

1. Which figures are regular polygons?
   a.   b.   c.   d.

2. The perimeter of a regular pentagon can be found with this formula.
   a. $P = s \times s$ b. $P = s + s$ c. $P = 5s$ d. $P = 5 + s$

3. The perimeter of a rectangle is 20 cm and its length is 4 cm. Which equation should be used to find the width?
   a. $20 = 8 + w$ b. $20 = 4w$ c. $20 = 8 + 2w$ d. $40 = \frac{1}{2}w$

4. The measure of an arc is equal to:
   a. its central angle   b. radius   c. 360°   d. 90°

5. Figures $ABCD$ and $EFGH$ are congruent. Which pairs of line segments are congruent?
   a. $\overline{AB}$ and $\overline{EH}$ b. $\overline{CD}$ and $\overline{GH}$ c. $\overline{BC}$ and $\overline{EF}$ d. $\overline{AB}$ and $\overline{FG}$

Find the perimeter and area of each figure.

See pp. 262–277

6.

7.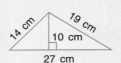

8.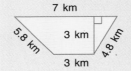

9.

10. (Use $\frac{22}{7}$ for π.)

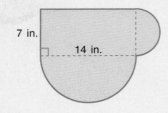

11. In $\triangle ABC$, $\overline{AB} \cong \overline{BC} \cong \overline{CA}$
    (Use 3.14 for π.)

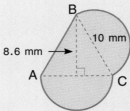

Solve.

12. A square field measures 45 m on a side. What is its area?

13. What is the circumference of a paddle wheel with a diameter of 1.25 m?

14. What is the length of a poster that is 28 in. wide and has a perimeter of 136 in.?

12 | Ratio and Proportion

In this chapter you will:

- Identify ratios
- Find the missing term in a pair of equal ratios
- Read, write, and solve proportions
- Use proportions to make and read scale drawings and models
- Identify similar figures
- Write proportions for similar triangles
- Solve problems: constructing models

Do you remember?

You can find equivalent fractions by multiplying or dividing the numerator and denominator by the same number.

$$\frac{3}{5} = \frac{6}{10} = \frac{9}{15} = \frac{12}{20}$$

$$\frac{3 \times 1}{5 \times 1} = \frac{3 \times 2}{5 \times 2} = \frac{3 \times 3}{5 \times 3} = \frac{3 \times 4}{5 \times 4}$$

$$\frac{16}{20} = \frac{12}{15} = \frac{8}{10} = \frac{4}{5}$$

$$\frac{16 \div 4}{20 \div 4} = \frac{12 \div 3}{15 \div 3} = \frac{8 \div 2}{10 \div 2} = \frac{4 \div 1}{5 \div 1}$$

R E S E A R C H I N G   T O G E T H E R

Golden Ratios

The ancient Greeks loved the beauty of mathematics. They used what they called the "golden rectangle." Find out what this meant to the Greeks and construct some "golden rectangles." Show their respective ratios of length to width.

Perhaps you have a different idea about what shapes are "golden" or beautiful. Draw your own figure and tell why you think it is mathematically "golden" to you.

12-1 Ratios and Rate Pairs

Ratio: a comparison of two numbers by division.
A *ratio* is used to compare two *like* quantities.

Out of 10 runners, 7 runners finished the race.

Here, we are comparing runners to runners.

We compare by saying 7 out of 10 runners finished.

We can write this three ways.

7 to 10   or 7 : 10   or $\dfrac{7}{10}$

Each ratio is read: "7 to 10."

Rates are ratios that compare *different* quantities.

A runner ran 1 kilometer in 3 minutes.

The rate, 1 kilometer to 3 minutes, compares two *different* quantities, distance to time. It can be written three ways:

1 to 3   or   1 : 3   or $\dfrac{1}{3}$

Each rate is read: "1 kilometer to 3 minutes."

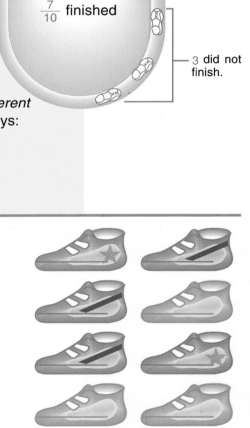

We say:

7 out of 10 finished

OR

7 to 10 finished

OR

7 : 10 finished

OR

$\dfrac{7}{10}$ finished

— 7 finished.

Finish Line

3 did not finish.

Compare the sneakers. Write each ratio three ways.

1. The number of sneakers that are orange or green to the total number of sneakers.

2. The number of sneakers with stripes to the number with stars.

3. The total number of sneakers to the number with stripes.

4. The number of green sneakers to the number of orange sneakers.

5. The number of sneakers in all to the number with stars.

For each ratio, write a statement of comparison about the sneakers.

6. 3 : 1   7. 3 to 9   8. $\dfrac{4}{9}$ 9. 5 : 9   10. $\dfrac{1}{9}$

Use the drawing to write a ratio, or comparison, of:

11. the height to the width

12. the length to the width

13. the length to the height

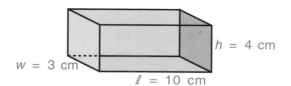

$h = 4$ cm
$w = 3$ cm
$\ell = 10$ cm

Express each rate as a fraction. Simplify. (The first one is done.)

14. 6 cans of juice for 96¢ $\longrightarrow$ $\dfrac{6}{\$.96} = \dfrac{1}{\$.16}$

15. 3 plants for $6

16. 165 words typed in 3 minutes

17. 2 books read every 10 days

18. 180 kilometers traveled in 2 hours

19. 150 miles in 3 hours

20. 30 tumbles in 5 minutes

21. 50 balloons inflated in 10 minutes

22. $8.00 earned for 2 hours' work

23. 55 miles per hour

24. 5 copies per second

Express each ratio as a fraction. To express the ratio of two measures as a fraction, both units must be the same. (The first one is done.)

25. ⌐3 weeks⌐   to   ⌐1 week 1 day⌐ $\longrightarrow$ change all units to days

⌐(3 × 7 = 21 days)⌐ to ⌐[(1 × 7) + 1 = 8 days]⌐ $\longrightarrow$ 21 days to 8 days $\longrightarrow$ $\dfrac{21}{8}$

26. 4 nickels to 3 dimes

27. 2 gallons to 5 quarts

28. 4 days to 2 weeks

29. 1 quarter to 3 cents

30. 1 quarter 1 dime to 1 dollar

31. 1 quart 3 pints to 4 pints

32. 8 weeks 3 days to 1 year

33. 2 kilometers to 800 meters

MAKE UP YOUR OWN...

Express each ratio as a fraction.

34. hand width across the knuckles to hand length

35. hand length to arm length

36. foot length to arm length

37. hand length to foot length

38. leg length to total height

39. Compare your answers to exercises 34–38 with someone else. Are any of your ratios almost the same?

Equal ratios, written as *equivalent* fractions, express the same comparisons or rates. Equal ratios have the same value.

For every turn of the large gear the small gear turns three times.

The ratio of large-gear turns to small-gear turns is *always* the same, no matter how many turns are made.

Look at the table.

Turns of large gear	1	2	3	4	5	?	24
Turns of small gear	3	6	9	12	?	18	?
Ratios of turns: large : small	$\frac{1}{3}$	$\frac{2}{6}$	$\frac{3}{9}$	?	$\frac{5}{15}$	?	?

$$\frac{1}{3} = \frac{2}{6} = \frac{3}{9}$$

Notice that these ratios, written as equivalent fractions, are equal.

▶ **To find equal ratios for a given ratio:**

- Express the ratio as a fraction.
- Find equivalent fractions by multiplying or dividing *both* the numerator and the denominator by the same number.

Multiplying	*Dividing*
$\frac{3}{7} = \frac{3 \times 1}{7 \times 1} = \frac{3 \times 2}{7 \times 2} = \frac{3 \times 3}{7 \times 3}$	$\frac{18}{24} = \frac{18 \div 1}{24 \div 1} = \frac{18 \div 2}{24 \div 2} = \frac{18 \div 3}{24 \div 3}$
Equal ratios: $\frac{3}{7} = \frac{6}{14} = \frac{9}{21}$	Equal ratios: $\frac{18}{24} = \frac{9}{12} = \frac{6}{8}$

▶ **To find if two ratios are equal:**

- Express each ratio as a fraction in simplest form.
- Compare the two fractions.

$$1.2 : 1.8 \overset{?}{=} 2.4 : 3.6 \longrightarrow \frac{1.2}{1.8} \overset{?}{=} \frac{2.4}{3.6}$$

$$\frac{1.2 \times 10}{1.8 \times 10} = \frac{12}{18} = \frac{12 \div 6}{18 \div 6} = \frac{2}{3} \quad \text{and} \quad \frac{2.4 \times 10}{3.6 \times 10} = \frac{24}{36} = \frac{24 \div 12}{36 \div 12} = \frac{2}{3}$$

$$\text{So, } \frac{1.2}{1.8} = \frac{2.4}{3.6}$$

1. Copy and complete the table above.

2. Express each ratio in the table as a fraction in simplest form. Why are all the fractions the same?

Find the missing number in each set of equal ratios.

3. $\dfrac{4}{5} = \dfrac{?}{10} = \dfrac{16}{?} = \dfrac{?}{40} = \dfrac{?}{80} = \dfrac{96}{?}$ 4. $\dfrac{5}{8} = \dfrac{15}{?} = \dfrac{?}{32} = \dfrac{30}{?} = \dfrac{?}{72} = \dfrac{75}{?}$

Write each ratio as a fraction in lowest terms or simplest form.

5. 12 : 15   6. 14 : 16   7. 36 to 9   8. 72 to 6

9. 30 : 35   10. 17 : 34   11. 64 : 48   12. 625 : 125

13. 1.5 : 2.4   14. 1.3 : 3.9   15. 1.2 : 0.8   16. 2.7 : 0.6

Write three equal ratios for each.

17. $\dfrac{4}{9}$ 18. $\dfrac{5}{8}$ 19. $\dfrac{27}{9}$ 20. $\dfrac{32}{16}$ 21. $\dfrac{0.3}{0.12}$ 22. $\dfrac{0.5}{0.25}$

Write each ratio as a fraction in simplest form. (The first one is done.)

23. $\dfrac{3}{4} : 6 = \dfrac{3}{4} \div 6 = \dfrac{3}{4} \div \dfrac{6}{1} = \dfrac{\overset{1}{\cancel{3}}}{4} \times \dfrac{1}{\underset{2}{\cancel{6}}} = \dfrac{1}{8}$ 24. $\dfrac{2}{5} : 10$

25. $6 : \dfrac{2}{3}$ 26. $8 : \dfrac{4}{5}$ 27. $\dfrac{1}{2} : \dfrac{2}{3}$ 28. $\dfrac{1}{8} : \dfrac{2}{5}$

29. $\dfrac{5}{8} : 2\dfrac{1}{2}$ 30. $\dfrac{4}{9} : 1\dfrac{1}{3}$ 31. $1\dfrac{1}{4} : 2\dfrac{1}{2}$ 32. $2\dfrac{2}{5} : 2\dfrac{2}{3}$

Which ratios do not belong with the first one?

33. 12 : 21   a.   8 : 14   b. 23 : 32   c. 28 : 49   d. none of these

34. 4 : 34   a.   6 : 51   b. 12 : 102   c.   1 : 17   d. none of these

35. 72 : 6   a. 34 : 2   b. 24 : 2   c. 84 : 7   d. none of these

These problems compare *different* quantities. Write each rate as an equivalent fraction. Simplify. (The first one is done.)

36. Sena uses 2 tomatoes for 4 salads.
    Rate of tomatoes to salads is 2 : 4 or $\dfrac{2}{4} = \dfrac{1}{2}$ (1 tomato to 2 salads).

37. 3 cucumbers cost 99¢.

Exercise 36

38. 4 pounds of tomatoes cost $5.60.

39. Jay helps 15 customers in 5 minutes.

40. The train travels 130 miles in 2 hours.

41. 45 balloons are inflated in 9 minutes.

42. Dawn reads 2 books every ten days.

MAKE UP YOUR OWN...

Make up a comparison and a rate situation for each.

43. $\dfrac{2}{3}$ 44. $\dfrac{9}{12}$ 45. $\dfrac{7}{6}$ 46. $\dfrac{10}{9}$ 47. $\dfrac{9}{15}$

12-3 Proportions

Proportion: an equation stating that two ratios are equal.

A proportion can be written as:

means

$a : b = c : d$ or $\dfrac{a}{b} = \dfrac{c}{d}$

extremes

The two middle terms are called the means.

The two end terms are called the extremes.

A photographer can enlarge a 5 × 7 photograph (A) to a 10 × 14 photograph (B). The two ratios, 5 : 7 and 10 : 14, form a proportion because the two ratios are equal.

To find out if two ratios form a proportion, use *one* of these ways:

- Use the "cross-products" rule:

  Cross-Products Rule: **Product of Means = Product of Extremes**

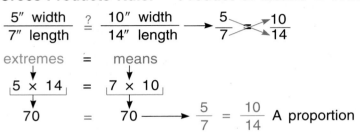

  $$\dfrac{5'' \text{ width}}{7'' \text{ length}} \overset{?}{=} \dfrac{10'' \text{ width}}{14'' \text{ length}} \longrightarrow \dfrac{5}{7} \times \dfrac{10}{14}$$

  extremes = means

  $5 \times 14 = 7 \times 10$

  $70 = 70 \longrightarrow \dfrac{5}{7} = \dfrac{10}{14}$ A proportion

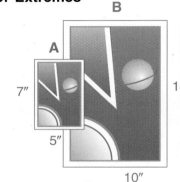

- Write each ratio as a fraction in simplest form.

  Compare the width to the length of the two photographs:

  Photo A: 5" width to 7" length $\longrightarrow$ 5 : 7 $= \dfrac{5}{7}$

  Photo B: 10" width to 14" length $\longrightarrow$ 10 : 14 $\longrightarrow \dfrac{10 \div 2}{14 \div 2} = \dfrac{5}{7}$

  So, 5 : 7 = 10 : 14 or $\dfrac{5}{7} = \dfrac{10}{14}$ forms a proportion.

  Remember: When writing proportions, the *order* of the labels used in the ratios must be the same.

  $$\dfrac{5 \text{ width A}}{7 \text{ length A}} = \dfrac{10 \text{ width B}}{14 \text{ length B}} \quad \text{or} \quad \dfrac{5 \text{ width A}}{10 \text{ width B}} = \dfrac{7 \text{ length A}}{14 \text{ length B}}$$

Write a proportion for each statement in two ways. (The first one is done.)

1. 30 miles in 15 minutes is the same as 10 miles in 5 minutes.   Incorrect. Why?

   $$\dfrac{30 \text{ mi}}{15 \text{ min}} = \dfrac{10 \text{ mi}}{5 \text{ min}} \quad \text{or} \quad \dfrac{30 \text{ mi}}{10 \text{ mi}} = \dfrac{15 \text{ min}}{5 \text{ min}} \qquad \dfrac{30 \text{ mi}}{10 \text{ mi}} \overset{?}{=} \dfrac{5 \text{ min}}{15 \text{ min}}$$

2. 2 melons for $.98 is the same as 10 melons for $4.90.

3. 5 hours to go 250 miles is the same as 2 hours to go 100 miles.

The cross-products rule states:

If two ratios form a proportion, then the cross products are equal.

If the cross products are equal, then the two ratios form a proportion.

Use the cross-products rule to find out which of these form a proportion.
(Hint: First express each ratio in simplest form.)

4. $\dfrac{2}{10} \overset{?}{=} \dfrac{5}{25}$ 5. $\dfrac{6}{8} \overset{?}{=} \dfrac{24}{32}$ 6. $\dfrac{18}{27} \overset{?}{=} \dfrac{16}{24}$ 7. $\dfrac{24}{4} \overset{?}{=} \dfrac{60}{5}$

8. $\dfrac{72}{24} \overset{?}{=} \dfrac{8}{3}$ 9. $\dfrac{18}{1} \overset{?}{=} \dfrac{36}{3}$ 10. $\dfrac{6}{8} \overset{?}{=} \dfrac{4}{3}$ 11. $\dfrac{1}{9} \overset{?}{=} \dfrac{81}{9}$

12. $\dfrac{8}{24} \overset{?}{=} \dfrac{12}{36}$ 13. $\dfrac{30}{36} \overset{?}{=} \dfrac{5}{6}$ 14. $\dfrac{10}{5} \overset{?}{=} \dfrac{2}{1}$ 15. $\dfrac{25}{25} \overset{?}{=} \dfrac{5}{1}$

Write a proportion for each word problem.
Use the cross-products rule to check the proportions.

16. Four wallet-size photos cost $1.60, so 8 photos cost $3.20.

17. Two rolls of film cost $12.50, so 6 rolls will cost $37.50.

18. Five students can be photographed in 20 minutes, so 15 students can be photographed in one hour.

19. One out of every 35 students is pleased with his or her pictures, so 12 out of 420 are pleased with theirs.

20. For 60 passengers there are 3 flight attendants, so there are 6 flight attendants for 120 passengers.

21. There are 3 yards of red material for every 2 yards of green material, so there are 4 yards of green material for 6 yards of red material.

22. In 45 minutes the plane travels 300 kilometers, so in 30 minutes it travels 200 kilometers.

> ### MAKE UP YOUR OWN...

Write a proportion, using each set of numbers. The order of the numbers may be changed. Check by using either "equal ratios" or the "cross-products rule."

23. 2, 7, 14, 4   24. 12, 36, 3, 1   25. 84, 72, 28, 24

26. $1.80, $1.20, 14, 21   27. 5, 20, 10, 40   28. 2 hr, 3 hr, 500, 750

Use the cross-products rule to write a proportion for each given cross product.

29. 24 = 24   30. 42 = 42   31. 63 = 63   32. 54 = 54

12-4 Solving Proportions

Find the cost of 9 grams of gold if 5 grams of gold cost $400.

5 grams gold	=	$400
9 grams gold	=	$?

To find a missing term, n, in a proportion, use the cross-products rule.

- Set up a proportion:

$$\frac{\text{grams}}{\text{cost}} = \frac{\text{grams}}{\text{unknown cost, } n}$$

- Substitute:

$$\frac{5}{400} = \frac{9}{n}$$

- Cross multiply:

$$\frac{5}{400} \diagup\!\!\!\diagdown \frac{9}{n}$$

$$5n = 3600$$

- Solve for n:

$$5n = 3600$$

Divide both sides by 5. $\longrightarrow 5n \div 5 = 3600 \div 5$

$$n = 720$$

Check

$$\frac{5}{400} \diagup\!\!\!\diagdown \frac{9}{720}$$

$$3600 = 3600$$

The cost of 9 grams of gold is $720.

The missing term can be anyplace in the proportion.

$$\frac{3}{8} \diagup\!\!\!\diagdown \frac{c}{32} \longrightarrow 3 \times 32 = 8c \longrightarrow 8c = 3 \times 32$$

Check

$$\frac{3}{8} \diagup\!\!\!\diagdown \frac{12}{32}$$

$$96 = 96$$

Solve for c: $\quad 8c = 96 \longrightarrow c = \frac{96}{8} \longrightarrow c = 12$

Copy and complete.

1. $\dfrac{2}{x} = \dfrac{10}{50}$

   $10x = \underline{\ ?\ }$

   $x = \underline{\ ?\ }$

2. $\dfrac{25}{4} = \dfrac{100}{r}$

   $25r = \underline{\ ?\ }$

   $r = \underline{\ ?\ }$

3. $\dfrac{y}{7} = \dfrac{7}{49}$

   $49y = \underline{\ ?\ }$

   $y = \underline{\ ?\ }$

4. $\dfrac{1}{b} = \dfrac{3}{51}$

   $3b = \underline{\ ?\ }$

   $b = \underline{\ ?\ }$

5. $\dfrac{3}{8} = \dfrac{t}{14}$

   $\underline{\ ?\ } = 42$

   $t = \underline{\ ?\ }$

6. $\dfrac{8}{56} = \dfrac{n}{14}$

   $\underline{\ ?\ } = 112$

   $n = \underline{\ ?\ }$

7. $\dfrac{3}{7} = \dfrac{15}{d}$

   $\underline{\ ?\ } = \underline{\ ?\ }$

   $d = \underline{\ ?\ }$

8. $\dfrac{a}{24} = \dfrac{12}{48}$

   $\underline{\ ?\ } = \underline{\ ?\ }$

   $a = \underline{\ ?\ }$

9. $\dfrac{d}{49} = \dfrac{6}{14}$

10. $\dfrac{4}{5} = \dfrac{x}{75}$

11. $\dfrac{7}{8} = \dfrac{56}{n}$

12. $\dfrac{10}{15} = \dfrac{r}{45}$

What is the next step in each solution?

13. $\dfrac{d}{49} = \dfrac{6}{14}$ **a.** $14d = 294$ **b.** $6d = 686$ **c.** $14d = 84$ **d.** none of these

14. $\dfrac{4}{5} = \dfrac{x}{75}$ **a.** $4x = 365$ **b.** $5x = 300$ **c.** $75x = 20$ **d.** none of these

15. $\dfrac{7}{8} = \dfrac{56}{n}$ **a.** $8n = 56$ **b.** $7n = 56$ **c.** $7n = 448$ **d.** none of these

16. $\dfrac{10}{15} = \dfrac{r}{45}$ **a.** $15r = 45$ **b.** $10r = 675$ **c.** $10r = 450$ **d.** none of these

Solve each proportion.

17. $\dfrac{9}{t} = \dfrac{3}{6}$ 18. $\dfrac{3}{4} = \dfrac{c}{24}$ 19. $\dfrac{b}{10} = \dfrac{15}{50}$ 20. $\dfrac{a}{25} = \dfrac{400}{100}$

21. $\dfrac{a}{4} = \dfrac{20}{16}$ 22. $\dfrac{r}{7} = \dfrac{98}{49}$ 23. $\dfrac{x}{15} = \dfrac{18}{45}$ 24. $\dfrac{15}{b} = \dfrac{6}{14}$

25. $\dfrac{12}{7} = \dfrac{48}{y}$ 26. $\dfrac{3}{12} = \dfrac{18}{h}$ 27. $\dfrac{66}{121} = \dfrac{s}{11}$ 28. $\dfrac{27}{63} = \dfrac{12}{d}$

29. $\dfrac{x}{18} = \dfrac{27}{54}$ 30. $\dfrac{125}{15} = \dfrac{25}{y}$ 31. $\dfrac{27}{51} = \dfrac{s}{17}$ 32. $\dfrac{32}{t} = \dfrac{64}{18}$

33. $9 : 16 = b : 48$ 34. $n : 55 = 12 : 15$ 35. $11 : b = 22 : 121$

36. $20 : 5 = 6 : n$ 37. $\dfrac{2}{3} : \dfrac{5}{6} = n : 10$ 38. $1\dfrac{1}{2} : \dfrac{3}{4} = \dfrac{1}{3} : n$

39. $1.2 : n = 1.6 : 4$ 40. $1 : 6 = y : 1.5$ 41. $4 : 5.6 = 3.7 : x$

Distance, Rate, and Time   Rate can be expressed as the comparison of distance to time. Proportions can be used to solve problems involving *distance, rate,* and *time.*

How long will it take a car traveling at a rate of 50 mph (miles per hour) to go 450 miles?

• Write a proportion.

$$\dfrac{distance}{time} = \dfrac{miles}{hour}$$

• Solve, using the cross-products rule.

$$\dfrac{450}{t} = \dfrac{50}{1}$$

$$50t = 450$$

$$t = 9 \longrightarrow \text{Car travels 9 hours.}$$

See pages 306–307 for more practice.

Use proportions to find the answers.

42. How far will a motor boat travel in 7 hours if it travels at a rate of 35 miles an hour?

43. How long will it take an airplane to cover 3750 miles if it travels 300 miles an hour?

44. How far will a bus go in 90 minutes if it travels 50 miles an hour?

45. How long will it take a train to travel 360 miles if it travels 60 miles an hour?

46. What is the rate per hour for traveling 270 miles in 90 minutes?

47. What is the rate *per hour* for traveling 6.3 miles in 70 minutes?

12-5 | Scale Drawings

Scale drawing: an accurate picture of something, but *different* in size.

Scale: the ratio of the *pictured* measure to the *actual* measure.

$$\text{Scale ratio} = \frac{\text{Scale measure}}{\text{Actual measure}}$$

This is a scale drawing of the Ward's apartment. In the drawing the apartment length is 3 inches.

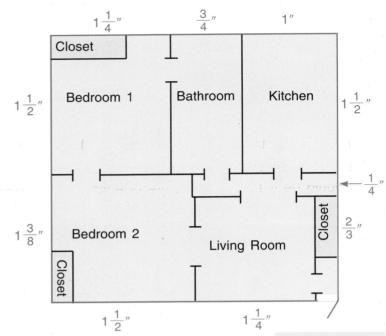

Scale:
1 in. = 12 ft

To solve problems involving scale and actual measures, form a proportion.

To find the actual length of the Ward's apartment:

- Set up a proportion.

$$\frac{\text{Scale measure}}{\text{Actual measure}} = \frac{\text{Scale length}}{\text{Actual length}}$$

- Substitute.

$$\frac{1 \text{ in.}}{12 \text{ ft}} = \frac{3 \text{ in.}}{\ell \text{ ft}}$$

- Use the cross-products rule to solve.

$$1\ell = 12 \times 3 \longrightarrow \ell = 36 \text{ ft} \quad \text{Actual length}$$

To find the actual width of this apartment, follow the above.

$$\frac{1 \text{ in.}}{12 \text{ ft}} = \frac{2\frac{7}{8} \text{ in.}}{w \text{ ft}} \longrightarrow \frac{1}{12} = \frac{2\frac{7}{8}}{w} \longrightarrow w = 12 \times 2\frac{7}{8} = 34\frac{1}{2} \text{ ft Actual width}$$

Find the actual measurements for the:

1. width of the bathroom.
2. length of bedroom 1.
3. width of the kitchen.
4. width of bedroom 2.
5. length and width of the entry closet.
6. length and width of the living room.

Find the actual measurements. (The first one is done.)

If the scale measure of the house floor plan was 1 cm = 3 m,
what would be the *metric* scale measure of each of these?

7. a picture window 2.4 m wide

$$\frac{1 \text{ cm}}{3 \text{ m}} = \frac{w \text{ cm wide}}{2.4 \text{ m wide}} = \longrightarrow \frac{1}{3} = \frac{w}{2.4} \longrightarrow 3w = 2.4 \longrightarrow w = 0.8 \text{ cm}$$

8. a garage 9 m by 12 m   9. a garden 2.1 m by 3.3 m

10. a wood stove 1.5 m by 1.5 m   11. a playground 7.2 m by 9 m

Measure the three segments of the bee drawn below.

If the scale measure of the bee is 1 mm = $\frac{1}{2}$ mm,
what would be the actual measure of each of these parts?

12. Head   13. Thorax   14. Abdomen

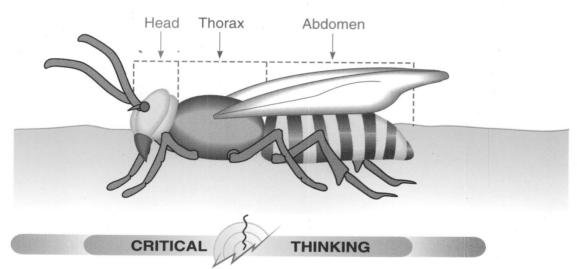

Head   Thorax   Abdomen

CRITICAL   THINKING

Look at the proportion below.

15. What effect will inverting the proportion have
    on the length of the Ward's apartment?

$$\frac{\text{Actual measure}}{\text{Scale measure}} = \frac{\text{Actual length}}{\text{Scale length}}$$

MAKE UP YOUR OWN...

Draw a floor plan of your apartment or the first floor of your home.

16. Choose a convenient scale in customary measure.
    Fill in the scale measures on your floor plan.

17. Choose a convenient scale in metric measure. Next to the customary
    scale measures on your floor plan, write metric scale measures.

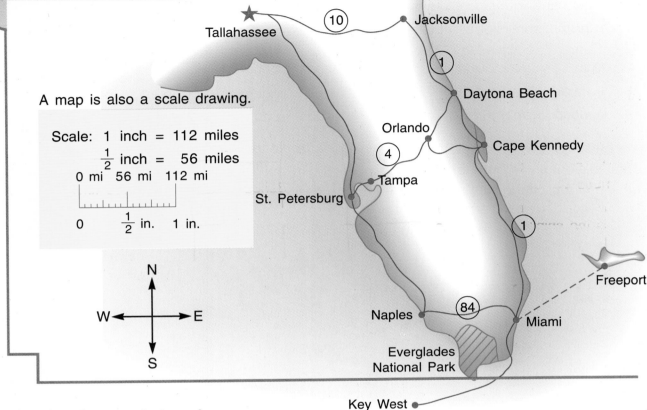

A map is also a scale drawing.

Scale: 1 inch = 112 miles
$\frac{1}{2}$ inch = 56 miles
0 mi 56 mi 112 mi

0    $\frac{1}{2}$ in.   1 in.

Use the given scale to solve.

1. What distance would 2 inches on this map represent?

2. What distance on the map represents 336 miles?

**Estimate the distance between the cities
in exercises 3–8.** (Hint: Use a piece of paper to measure.)

3. Naples and Miami

4. Orlando and Cape Kennedy

5. Jacksonville and Tallahassee

6. St. Petersburg and Tampa

7. Key West and Miami

8. Daytona Beach and Cape Kennedy

9. A car travels from Miami through the Everglades National Park
to the coast of Florida. If the average speed of the car
is 40 miles per hour, how long will the trip take?

10. If the car gets 20 miles to a gallon of gasoline,
how much gasoline will be used on the trip?
How much will the gasoline cost at $1.24 a gallon?

11. The Lane family flies from Miami to Freeport in the Bahamas
for a vacation. How many miles is the flight?

Use this metric scale for the map of Florida for exercises 12–17.

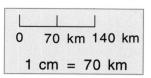

1 cm = 70 km

12. What distance on the map represents 350 km?

13. What distance would 2.5 cm on the map represent?

Estimate the distance in kilometers between the cities.

14. St. Petersburg and Naples

15. Cape Kennedy and Tampa

16. Tallahassee and St. Petersburg

17. Jacksonville and Daytona Beach

Use the map of the island to answer exercises 18–21.

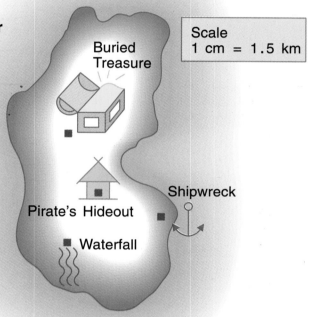

Scale
1 cm = 1.5 km

18. About how many kilometers is the shipwreck from the buried treasure?

19. About how many kilometers is the pirate's hideout from the buried treasure?

20. About how many kilometers is the shipwreck from the waterfall?

21. About how many kilometers long is the island?

Make a map of the streets in your neighborhood.

Mark at least four different locations on your map; for example, school, home, athletic field, and so on. Determine an appropriate scale in miles or kilometers for your map.

22. Use the scale to determine the distances between each of the four locations.

23. Write a word problem that uses your map to determine total distance traveled in moving from home to three different locations.

Find the solution set for each inequality. Let n = the set of integers.

24. $^-6 + n > {}^-2$

25. $n - {}^+1 < {}^+7$

26. $n + {}^-4 \geq {}^+6$

27. $n - {}^-7 \leq {}^+8$

28. $n + {}^+4 > 0 - {}^-1$

29. $n - {}^-3 < {}^+2 - {}^+7$

12-7 Similar Figures

Similar figures: figures that have the same shape, but may be different in size.

Corresponding sides of similar figures are in proportion.

Compare these triangles:

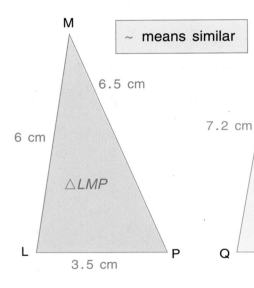

M
6.5 cm
6 cm
△LMP
L
3.5 cm
P

R
7.8 cm
7.2 cm
△QRS
Q
4.2 cm
S

~ means similar

Set up a proportion
of corresponding sides.
If the sides are in proportion,
the triangles are similar.

Do this for △LMP and △QRS:

$$\frac{LM}{QR} = \frac{LP}{QS}$$

$$\frac{6}{7.2} \times \frac{3.5}{4.2}$$

Use cross
products.

$$6 \times 4.2 = 7.2 \times 3.5$$

$$25.2 = 25.2$$

△LMP ~ △QRS

Similar figures have corresponding angles that are congruent.

See why the corresponding angles in the similar triangles, △XNR and △BFS, are congruent.

∠X ≅ ∠B   because   120° = 120°
∠N ≅ ∠F   because   23° = 23°
∠R ≅ ∠S   because   37° = 37°

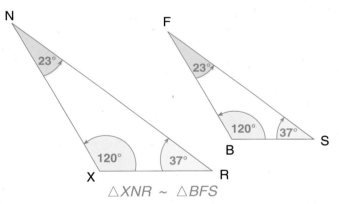

△XNR ~ △BFS

Use △LMP and △QRS above. Show that these sides are proportional.

1. $\dfrac{LM}{QR} = \dfrac{MP}{RS}$ 2. $\dfrac{MP}{RS} = \dfrac{LP}{QS}$

3. Set up a proportion of corresponding sides. Use cross products to find whether △FGH and △LKJ are similar.

4. Name the corresponding angles in △FGH and △LKJ. What can you say about these corresponding angles? Measure them to see if you are correct.

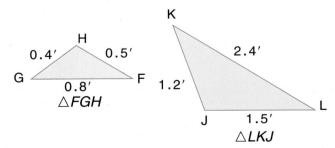

H
0.4'   0.5'
G
0.8'   F
△FGH

K
2.4'
1.2'
J   1.5'   L
△LKJ

Missing Dimensions

You can use the cross-products rule to find
the missing dimensions of similar figures.

Figure *ABCD* is similar to figure *UTSR*.

To find the length of $\overline{TS}$:

- Name the corresponding sides.
- Set up a proportion and solve.

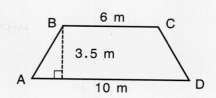

$$\frac{\overline{AD}}{\overline{UR}} = \frac{\overline{BC}}{\overline{TS}} \longrightarrow \frac{10 \text{ m}}{8 \text{ m}} \underset{x}{\times} \frac{6 \text{ m}}{}$$

$$10x = 6(8)$$

$$10x = 48$$

$$x = 4.8$$

$$\overline{TS} = 4.8 \text{ m}$$

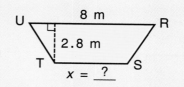

5. Find the length of $\overline{NM}$.
Figure *MNOP* is similar to figure *ZYWX*.

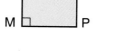

6. Find the length of $\overline{HF}$.
△*GFH* is similar to △*BAC*.

CHALLENGE

Use the formulas you have learned to determine the area and
perimeter or circumference for the pairs of figures below.

7. For each pair of similar figures what is the ratio
of their areas? of their perimeters?

8. Compare the area ratios with the perimeter ratios.
What pattern do you find?

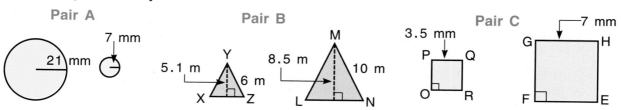

12-8 Using Similar Right Triangles

To find missing dimensions for similar right triangles,
write proportions using the corresponding sides.

A person 2 m tall casts a shadow 3 m long,
while a tree casts a shadow 60 m long.
How tall is the tree?

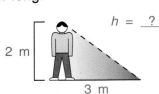
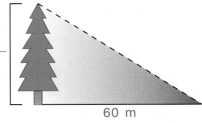

To find the height of the tree:

- Draw a picture.
  (The right triangles
  formed are similar.)

  2 m — 3 m person's shadow

  $h = $? — 60 m tree's shadow

- Set up a proportion.

$$\frac{\text{person's height}}{\text{tree's height}} = \frac{\text{person's shadow}}{\text{tree's shadow}}$$

- Substitute. $\dfrac{2}{h} = \dfrac{3}{60}$

- Cross multiply. $3h = 120$

- Solve. $h = 40$ The tree is 40 m tall.

Draw a picture. Set up a proportion. Then solve.

1. A flagpole casts a shadow 14 m long at the
    same time a boy 1.2 m tall casts a shadow
    2 m long. How tall is the flagpole?

    - Draw a picture.

    - Set up a proportion.

    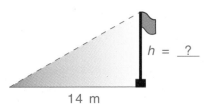

    1.2 m (boy's height) — 2 m

    $h = $? — 14 m

    $$\frac{\text{boy's height}}{\text{flagpole's height}} = \frac{\text{boy's shadow}}{\text{flagpole's shadow}}$$

    - Substitute. $\dfrac{1.2}{h} = \dfrac{2}{14}$

    - Now cross multiply and solve.

2. How long is the shadow cast by a 50.5-meter-tall building if,
    at the same time, the shadow cast by a 1.5-meter-tall passerby
    is 3 meters long?

3. △*ABC* and △*EDC* are similar.
Find the distance across the lake along $\overline{AB}$.

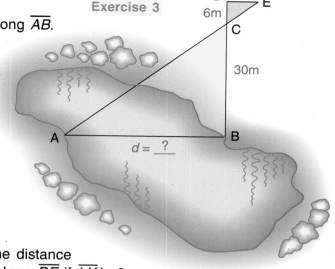

Exercise 3

4. A lamp post is 6.5 meters high.
Next to it, a 1.2-meter-high mailbox
casts a shadow 4.8 meters long.
How long is the shadow of the
lamp post?

5. John casts a shadow 5.4 m long
at the same time that his little
brother casts a shadow 3.3 m long.
If John is 1.8 m tall, how tall
is his little brother?

6. △*DEL* and △*JKL* are similar. Find the distance
of the track span across the gorge along $\overline{DE}$ if $\overline{LK}$ is 8 m.

Exercise 6

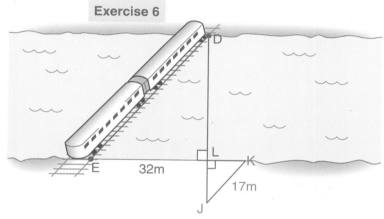

7. Similar triangular sails are raised on two sailboats. If the smaller sail
is 1.5 m wide and 4.5 m high and the larger sail is 6 m wide, how
high is the larger sail?

8. Two powerboats leave the same dock and travel north at the same
speed. The *Misty* travels for 20 miles and the *Surfer* travels for 32
miles. Then each boat turns east. The *Misty* travels for 15 miles.
How far east does the *Surfer* travel?

9. Use the information from exercise 8. How far is the *Surfer* from the
starting point if the *Misty* is 25 miles from the starting point?

MAKE UP YOUR OWN...

10. Write a word problem using these numbers to form
a true proportion: *x*, 7, 21, and 24. Find the missing term.
Compare your problem with problems that others made up.

12-9 Problem Solving: Constructing Models

Problem: Minshi designs a circular pillow. She places 6 sequins evenly around the edge of the circle. Then she connects each sequin to the others by a piece of ribbon. How many pieces of ribbon does she need?

1 IMAGINE Draw and label how you will connect 6 sequins: *A, B, C, D, E, F.*

2 NAME *Facts:* 6 sequins total

Each sequin connected once to each of the other 5 sequins

Question: How many pieces of ribbon are used?

3 THINK Construct a model.

- Use the letters *A* to *F* to stand for the 6 sequins. Arrange the letters in a circle.

- Starting at *A*, draw lines connecting letters to indicate ribbon (between *A* and *B*, *A* and *C*, *A* and *D*, and so on).

- As you connect pairs of letters, keep a record of each new set of pieces of ribbon.

  A joins with *B, C, D, E, F.*

  Already having been joined to *A*, *B* now joins with *C, D, E, F.*

  Already having been joined to *A* and *B*, *C* now joins with *D, E, F.*

  Already having been joined to *A, B,* and *C*, *D* now joins with *E, F.*

  Already having been joined to *A, B, C,* and *D*, *E* now joins with *F.*

4 COMPUTE The number of pieces of ribbon ⟶ 5 + 4 + 3 + 2 + 1 = 15 pieces of ribbon

5 CHECK Does your answer make sense? Yes.

Solve each problem by constructing a model.

1. What is the least number of equilateral triangles that can be added to this pattern to form a figure having a perimeter of 15″? What is the greatest number that can be added and still have a perimeter of 15″?

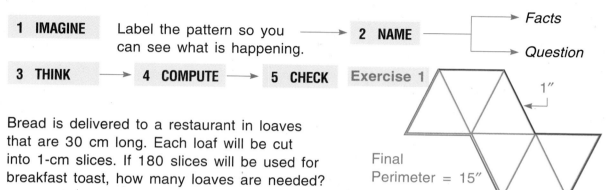

1 **IMAGINE**   Label the pattern so you can see what is happening.   →   2 **NAME** → Facts / Question

3 **THINK** → 4 **COMPUTE** → 5 **CHECK**   Exercise 1

Final Perimeter = 15″

Least number = _?_
Greatest number = _?_

2. Bread is delivered to a restaurant in loaves that are 30 cm long. Each loaf will be cut into 1-cm slices. If 180 slices will be used for breakfast toast, how many loaves are needed?

3. If the cook burns every fourth slice, how many loaves are needed to get 180 edible slices?

4. Jane has 3 skirts, a brown one, a white one, and a blue one. She has five blouses. The blouses are yellow, green, blue, red, and white. How many different skirt-and-blouse outfits can she make?

5. The first five runners to finish a marathon were Steve, Kathy, Maryanne, Ralph, and Bill. Steve finished 1 meter ahead of Bill. Kathy finished ahead of Steve but behind Ralph. Maryanne finished 4 meters ahead of Bill and 1 meter behind Kathy. Who won? Who came in second, third, fourth, and fifth?

6. Tsing used 4 chords to cut her "pie" into 5 pieces, but Mike could cut his "pie" into 8 pieces also using 4 chords. Can you use 4 chords and make more pieces than Mike?

7. The ratio of the length to the width of a garden is 7 : 2. If the garden has a perimeter of 45 feet, what are its length and width?

8. A total of 37 students participated in these sports: football, baseball, and basketball. Draw three overlapping circles, each labeled for one of the sports. Within the circles, write the number of students that played as follows:

    22 play baseball   8 play all three sports
    16 play football   25 play basketball
    14 play basketball and baseball
     9 play football and basketball

How many played football and baseball? How many played football only, baseball only, basketball only?

12-10 | Problem Solving: Applications

Use the distance formula to find:

1. the rate per hour for traveling 220 miles in 2.5 hours.

2. the rate per minute for traveling 78 miles in 20 minutes.

3. the rate per minute for traveling 1260 miles in 1 hour. (Hint: 1 hr = 60 min.)

4. the rate per hour for traveling 25.4 miles in 30 minutes.

5. the rate for traveling 237.5 miles in 4 hours 45 minutes. (Hint: Assume the rate to be "per hour.")

6. In 45 minutes a pilot can fly a jet plane 702 miles. How far can she fly in $2\frac{1}{2}$ hours?

7. How long will it take a blimp to travel 20 miles if it travels 12 miles in $\frac{1}{2}$ hour?

8. How far will a cyclist go in 45 minutes if he travels 12 mph?

9. Rita can swim 84 yards in 42 seconds. At that rate, how many yards can she cover in 880 seconds? (Hint: Find the unit rate, or rate per second.)

10. A motorist travels 342 miles in 6 hours. At that rate, how many hours will it take to travel 627 miles?

USE THESE STRATEGIES:
Use a Model/Drawing
Hidden Information
Write an Equation
Use Simpler Numbers
Multi-Step Problem

Use this chart to complete exercises 11–14. Write the ratio of states that:

11. border the Pacific Ocean to states that border the Atlantic Ocean.

12. border the Gulf of Mexico to states that border the Great Lakes.

13. border the Arctic Ocean to states that border the listed bodies of water.

14. Write a statement of comparison about states listed that form equal ratios.

No. of States	Body of Water
15	Atlantic Ocean
5	Pacific Ocean
1	Arctic Ocean
5	Gulf of Mexico
8	Great Lakes

Solve.

15. Mary Lou paid $23.75 for five tickets.
    What ratio represents the cost per ticket?

16. If 4 cans of soup cost $2.28, what will 8 cans cost?

17. Ms. Swartz earns $45 for 6 hours of work.
    How long must she work to earn $67.50?

18. If Kara can read 90 pages of a book in $1\frac{1}{2}$ hours,
    how many pages of the the book can she read in $3\frac{1}{4}$ hours?

Use the scale drawing for exercises 19–22.

19. How tall is the actual bookcase?

20. How wide is the actual bookcase?

21. How high is the bottom shelf?

22. Will the bookcase fit in a space
    1 yard wide? Explain.

23. The ratio of quarters to dimes
    in Ben's bank is $4:3$.
    If Ben has $1.20 in dimes,
    how much money has he in all?

24. The ratio of nickels to pennies in April's change purse
    is $3:8$. If she has $.45 in nickels, how many pennies
    are there in her change purse?

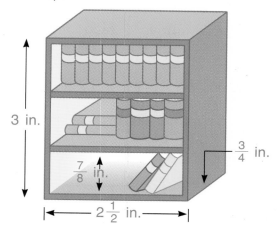

3 in.

$\frac{7}{8}$ in.

$2\frac{1}{2}$ in.

$\frac{3}{4}$ in.

Scale: 1 in. = $1\frac{1}{3}$ ft

**Marsha measured the distance between various locations
on her map in centimeters. Use the scale to find the
actual distances between each location in exercises 25–28.**

25. campground to waterfall = 1.4 cm

26. home to campground = 11.2 cm

27. school to campground = 13.9 cm

28. campground to lake = 0.8 cm

Scale:
1 cm = 2.5 km

Polygon *ABCD* is similar to polygon *RSTU*.

29. $\overline{AB}$ is 6 in., $\overline{BC}$ is 8 in., and $\overline{ST}$ is 4 in.
    What is the length of $\overline{RS}$?

30. $\overline{TU}$ is 11 in. What is the length of $\overline{CD}$?

31. $\overline{AD}$ is 9 in. What is the length of $\overline{RU}$?

32. What is the perimeter of each polygon?

33. Dee casts a shadow 2.1 m long and Matt casts a
    shadow 2.4 m long. If Dee is 1.4 m tall, how tall is Matt?

More Practice

Use the chart for exercises 1–2.

Teams	Actual	Goal
Team A	18 mi	24 mi
Team B	21 mi	27 mi
Team C	24 mi	32 mi

1. Write the ratio of actual distances to goal distances for each team.

2. Which teams had equal ratios for actual distances to goal distances?

Write each ratio as a fraction. (Remember: Both units must be the same.)

3. 2 years : 4 years 6 months

4. 2 feet : 1 yard 4 inches

5. 9 gallons : 3 quarts 1 pint

6. 3 km : 5 km 2 m

Which are proportions? Write = or ≠.

7. $\dfrac{1}{4}$? $\dfrac{20}{100}$

8. $\dfrac{2}{7}$? $\dfrac{26}{90}$

9. $\dfrac{10}{16}$? $\dfrac{45}{72}$

10. $\dfrac{7}{8}$? $\dfrac{64}{49}$

Solve these proportions. Use the cross-products rule.

11. $\dfrac{2}{3} = \dfrac{x}{15}$

12. $\dfrac{8}{n} = \dfrac{2}{5}$

13. $\dfrac{a}{9} = \dfrac{9}{27}$

14. $\dfrac{x}{6} = \dfrac{36}{12}$

Assume you could cycle at a rate of 0.4 mile per minute.

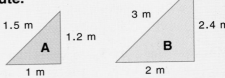

15. How far would you go in 20 minutes?

16. How long would it take you to go 27 miles?

Solve.

17. Determine whether triangles A and B are similar.

18. Triangle C is similar to triangle A above. Find the height of triangle C.

19. If a map-printing machine prints 2400 maps in one hour, how many maps can it print in 35 minutes? in 1 hour and 35 minutes?

20. If a typist types 120 words per minute, how many words can be typed in $5\frac{1}{2}$ minutes?

21. A person 5 feet tall casts a shadow 3 feet long. A nearby building casts a shadow 18 feet long. How tall is the building?

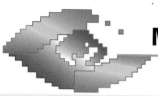

Math Probe

PROPORTIONAL PUZZLERS

Choose the correct answer to each of these problems from those given in the "Answer Chart."

1. ▢ is to ◸ as 4 is to __?__ .

2. XXX is to X as ●●● is to __?__ .

3. 3 is to 9 as ▱ is to __?__ .

4. △ is to ◺ as 5 is to __?__ .

5. 12 is to 3 as ◯ is to __?__ .

6. 1 lb is to 4 oz as 1 ft is to __?__ .

7. 7 is to 1 as ◖◖◖◗ is to __?__ .

Answer Chart		
1	2	$2\frac{1}{2}$
X	●●	⊓
●●	▽	✶hexagon
$\frac{1}{2}$	$3\frac{1}{2}$	10
◗	$\frac{3}{4}$	◿
4 ft	3 in.	8 in.

Find two answers from the "Answer Chart" for each of these problems.

8. ●●●●●●●● is to ●● as __?__ is to __?__ .

9. ⧖ is to ◖ as __?__ is to __?__ .

10. XX is to XXX as __?__ is to __?__ .

11. 6 is to 1 as __?__ is to __?__ .

12. Use the chart to create your own ratio and proportion puzzlers.

309

Check Your Mastery

1. For every 5 contestants, there are 2 prizes. Write the ratio of contestants to prizes in two ways.

See pp. 288–293

Name	Number correct	Total
Lynne	16	24
Mandy	10	12
Steve	12	18

2.
3.
4.

For exercises 2–4 write the ratio of the number of correct answers to the total number of questions. Determine which ratios are equal.

Write = or ≠ for each pair of ratios.

5. $\dfrac{3}{8}$? $\dfrac{18}{48}$ 6. $\dfrac{22}{17}$? $\dfrac{66}{53}$ 7. $\dfrac{4}{9}$? $\dfrac{32}{81}$ 8. $\dfrac{8.4}{4.6}$? $\dfrac{4.2}{1.8}$

See pp. 294–295

Solve. Use the cross-products rule.

9. $\dfrac{14}{16} = \dfrac{t}{24}$

10. $\dfrac{12}{10} = \dfrac{36}{n}$

11. $\dfrac{n}{8} = \dfrac{32}{64}$

12. $\dfrac{r}{7} = \dfrac{42}{49}$

13. $\dfrac{3}{s} = \dfrac{9}{53}$

14. $\dfrac{m}{9} = \dfrac{28}{81}$

15. If a jogger jogs 10 miles in 50 minutes, what is the jogger's rate per minute?

See pp. 296–297

16. If a car is drawn to a scale of 1 inch = 5 feet, how wide will a 4-ft windshield be on the drawing?

See pp. 300–303

17. Determine whether these parallelograms are similar.

3.5 cm 2 cm 8.75 cm 5 cm

18. Find the height of the telephone pole.

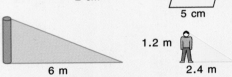

h 1.2 m 6 m 2.4 m

Solve.

19. If $2\frac{1}{2}$ inches on a map represents 750 miles, what does one inch represent?

20. If Dave drives 1200 miles in 5 days, how many miles does he drive in 3 days? Choose one and solve:

a. $\dfrac{1200 \text{ miles}}{m \text{ miles}} = \dfrac{5 \text{ days}}{3 \text{ days}}$

b. $\dfrac{1200 \text{ miles}}{3 \text{ days}} = \dfrac{5 \text{ days}}{m \text{ miles}}$

c. $\dfrac{1200 \text{ miles}}{m \text{ miles}} = \dfrac{3 \text{ days}}{5 \text{ days}}$

d. $\dfrac{1200 \text{ miles}}{5 \text{ days}} = \dfrac{m \text{ miles}}{3 \text{ days}}$

21. How long will it take an airplane to travel 1675 mi if it travels 335 mph?

Cumulative Review

Choose the correct answer.

1. A polygon that has two pairs of parallel sides:

   a. pentagon   b. trapezoid

   c. rhombus   d. triangle

2. What is the sum of the angles of any triangle?

   a. $90°$ b. $180°$

   c. $120°$ d. $360°$

3. At a speed of 55 mph, how long will it take to travel 660 miles?

   a. 14 hours   b. 12 hours

   c. 15.5 hours   d. 33 hours

4. In the proportion $2 : 9 = 8 : 36$ the means are:

   a. 2, 9
   b. 9, 8
   c. 9, 36
   d. 18, 288

5. If 1 cm represents 10 km on the map, how many centimeters would represent 40 km?

   a. 4   b. 6

   c. 8   d. 10

6. What is the perimeter of a yard 320 ft long and 500 ft wide?

   a. 820 ft   b. 1140 ft

   c. 3280 ft   d. 1640 ft

7. The area of a square garden plot that has a side measuring 20 ft is:

   a. 80 ft^2 b. 240 ft

   c. 400 ft   d. 400 ft^2

8. The approximate circumference of a circle having a 7.1 cm radius is:

   a. 58.28 cm   b. 44.588 cm

   c. 22.294 cm   d. 1139.6 cm^2

9. If 350 miles were traveled in 7 hours, how long will it take to travel 1500 miles?

   a. 30 hours
   b. 50 hours
   c. 2 days
   d. $4\frac{1}{2}$ hr

10. The area of a trapezoid having a height of 2.5 and parallel bases of 4 cm and 7 cm is:

    a. 27.5 cm^2 b. 13.75 cm^2

    c. 55 cm^2 d. 105 cm^2

11. A triangular banner has an area of 440 cm^2. If the base is 22 cm, what is the height?

    a. 44 cm   b. 60 cm

    c. 40 cm   d. 64 cm

12. In $\dfrac{2.1}{2.9} = \dfrac{8}{c}$ the best estimate for the missing term is:

    a. 20   b. 12

    c. 16   d. 9

13. On a map the distance from one point to another is 3.5 cm. What is the actual distance between points if 1 cm = 8 km?

    a. 2.8 km   b. 280 km
    c. 28 km   d. 80 km

14. How much greater is the area of a 28-cm square than the area of a circle with a diameter of 28 cm?

    a. 784 cm^2 b. 1677 cm^2
    c. 168 cm^2 d. 616 cm^2

Find the area of each figure.

15.
2.7 cm

16.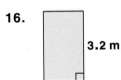
3.2 m
1.8 m

17.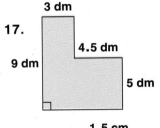
3 dm
4.5 dm
9 dm
5 dm

18.
3 m
4 m

19.
3.5 cm
2.5 cm

20.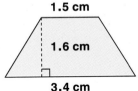
1.5 cm
1.6 cm
3.4 cm

Solve.

21. Thirteen of the first 39 presidents had previously been vice-presidents. Express this fact as a ratio.

22. The sail on a rental boat is triangularly shaped. If the base is 2.6 m and the height is 3.6 m, what is the area of the sail?

23. Sandy gets paid $.03 for each envelope she stuffs for a business firm. She gets a bonus of $.05 for each two dozen she fills. If Sandy stuffed 1296 envelopes, how much did she earn?

24. The pieces of glass in a small window are in the shape of small parallelograms. Each has a base of 10.5 cm and a height of 5.6 cm. What is the total area of the window if there are 150 parallelogram pieces in the window?

25. Justin's microscope kit has glass slides that measure 20 mm wide and 75 mm long. After he puts an 18 mm square coverslip on top of the slide, what area is not covered by the coverslip?

26. Mr. Alonso uses a rotating sprinkler to irrigate the corn field. If the sprinkler is 120 m long from center to edge, what is the area of the circular plot he can irrigate?

+10 BONUS

27. Draw a mosaic using the shapes of at least one of each of the following polygons: triangle, square, rectangle, rhombus, parallelogram, and trapezoid. Label each figure.

Let the finished mosaic be a scale drawing for a mural six times its size. Record what the measurements for each full-size figure would be. Then find the area of each figure that would be on the mural.

13 Percent

In this chapter you will:

- Rename decimals and fractions as percents
- Find fraction equivalents for decimals
- Use percents greater than 100%
- Find the percentage
- Find the percent one number is of another
- Find the original number
- Solve problems: hidden questions
- Use technology: patterns and conjectures

Do you remember?

A ratio is a comparison of two numbers by division:

$\frac{1}{4}$ means $1 \div 4$

A proportion is an equation that states that two ratios are equal:

$\frac{1}{4} = \frac{25}{100}$

RESEARCHING TOGETHER

Animal Habitat—Today *and* Tomorrow?

Your architecture firm, Stone, Sand, and C. Ment, has just been asked to build the Safari Habitat. Here is your information.

Total size of area:
12,390,400 sq yd
  65% for safari land
  35% for park area
    40% for an Environmental Learning Lab
    36% for food and rest areas
    24% for parking

How many square yards will be set aside for each section? Use graph paper to create a scale drawing. Why have an Environmental Learning Lab?

13-1 Changing Decimals and Fractions to Percents

Percent: a ratio or comparison of a number to 100.
The ratio of the number of shaded boxes to
the total number of boxes in the grid is 30 : 100.

This ratio may be expressed as a:

Fraction	Decimal	Percent
$\frac{30}{100}$	0.30	30%

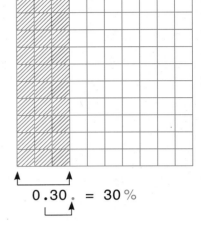

To change a decimal to a percent:

- Move the decimal point two places to the right.
- Write the percent symbol (%).

$$0.15 = 15\% \qquad 0.02\,4 = 2.4\%$$

$$0.30 = 30\%$$

To change a fraction to a percent:

- Change the fraction to a decimal.
- Then change the decimal to a percent.

$$\frac{3}{5} \rightarrow 5\overline{)3.00}^{\,0.60} = 60\% \qquad \frac{1}{8} \rightarrow 8\overline{)1.000}^{\,0.125} = 0.12\,5 = 12.5\%$$

$$\frac{20}{100} = \underline{\ ?\ }\%$$

$$\frac{20}{100} = 0.20 = 20\%$$

Shortcut — When the denominator is a factor of 100, change the denominator to hundreds.

$$\frac{3}{20} = \underline{\ ?\ }\%$$

20 is a factor of 100:
$20 \times 5 = 100$

$$\frac{3 \times 5}{20 \times 5} = \frac{15}{100} = 0.15 = 15\%$$

Copy and complete this chart about the grid. (The first one is done.)

	Color	Ratio	Fraction	Decimal	Percent
1.	▨	15 : 100	$\frac{15}{100}$	0.15	15%
2.	▨	10 : 100	$\frac{10}{100}$	0.10	?
3.	▨	5 : 100	$\frac{5}{100}$	?	5%

Write each ratio as a percent.

4. 32 : 100   **5.** 41 : 100   **6.** 27 : 100   **7.** 98 : 100   **8.** 3 : 100   **9.** 5 : 100

Write each decimal as a percent.

10. 0.75   **11.** 0.25   **12.** 0.07   **13.** 0.05   **14.** 0.99   **15.** 0.29

16. 0.175   **17.** 0.375   **18.** 0.035   **19.** 0.015   **20.** 0.022   **21.** 0.011

This square has 100 parts. Write a ratio, a fraction, a decimal, and a percent to describe the parts that are:

22. blue.

23. red.

24. white.

25. green.

26. blue and red.

27. blue and green.

28. red and green.

29. not green.

30. red, white, and blue.

31. not red.

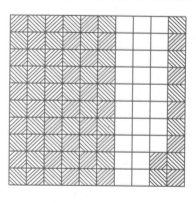

Write each fraction as a percent.

32. $\frac{3}{10}$ **33.** $\frac{8}{10}$ **34.** $\frac{2}{5}$ **35.** $\frac{1}{5}$ **36.** $\frac{1}{25}$ **37.** $\frac{2}{25}$ **38.** $\frac{5}{8}$

39. $\frac{7}{8}$ **40.** $\frac{3}{4}$ **41.** $\frac{1}{50}$ **42.** $\frac{3}{25}$ **43.** $\frac{3}{20}$ **44.** $\frac{3}{8}$ **45.** $\frac{9}{10}$

46. $\frac{3}{5}$ **47.** $\frac{9}{25}$ **48.** $\frac{19}{20}$ **49.** $\frac{14}{50}$ **50.** $\frac{4}{5}$ **51.** $\frac{1}{10}$ **52.** $\frac{12}{50}$

Solve.

53. This diagram shows a fitness center. What percent of the center is used for:

     a. track?   **b.** equipment?

     c. racquetball?   **d.** weight room?

     e. pool?

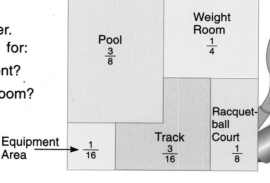

54. Color a 10-by-10 grid to show:

     a. 12% yellow.   **b.** 20% red.   **c.** 25% purple.

     d. 18% green.   **e.** 25% blue.

55. Express these fractions as percents. Then illustrate them.

     a. $\frac{1}{2}$, $\frac{1}{4}$, $\frac{1}{5}$, $\frac{1}{10}$, $\frac{2}{5}$, $\frac{3}{4}$ **b.** $\frac{2}{3}$, $\frac{3}{5}$, $\frac{3}{8}$, $\frac{3}{10}$, $\frac{4}{5}$, $\frac{5}{8}$

SKILLS TO REMEMBER

Order each set of fractions from *least to greatest*. (Hint: Use percent equivalents.)

56. $\frac{2}{5}$, $\frac{1}{3}$, $\frac{3}{8}$ **57.** $\frac{3}{10}$, $\frac{1}{3}$, $\frac{1}{4}$ **58.** $\frac{1}{2}$, $\frac{3}{5}$, $\frac{2}{3}$

59. $\frac{3}{10}$, $\frac{2}{5}$, $\frac{1}{4}$ **60.** $\frac{3}{8}$, $\frac{2}{3}$, $\frac{1}{2}$ **61.** $\frac{5}{8}$, $\frac{1}{2}$, $\frac{2}{5}$

13-2 Changing Percents to Decimals and Fractions

To change a percent to a decimal:

- Drop the percent symbol, %.

$60\% = \underline{} \text{ decimal}$

- Move the decimal point two places to the *left*.

$60\,\%\, = 0\cdot 60$

- Prefix zeros if necessary.

$1\cdot 5\% = 01\cdot 5\% = 0\cdot 015$

To change a percent to a fraction:

- Drop the percent symbol, %.

$12\frac{1}{2}\% = \underline{} \text{ fraction}$

- Write the percent as a fraction with a denominator of 100.

$$\frac{12\frac{1}{2}\%}{100} = 12\frac{1}{2} \div 100$$

- Express the fraction in lowest terms.

$$= \frac{25}{2} \div 100 = \frac{25}{2} \times \frac{1}{100} = \frac{1}{8}$$

Change each percent to a decimal.*

1. 32%	**2.** 27%	**3.** 77%	**4.** 44%	**5.** 85%	**6.** 15%	**7.** 7%
8. 5%	**9.** 4%	**10.** 1%	**11.** 6%	**12.** 8%	**13.** 0.9%	**14.** 0.4%

Change each percent to a fraction.*

15. 45%	**16.** 35%	**17.** 20%	**18.** 70%	**19.** 21%	**20.** 33%
21. 7%	**22.** 2%	**23.** 55%	**24.** 65%	**25.** 82%	**26.** 28%
27. 15%	**28.** 18%	**29.** $33\frac{1}{3}\%$	**30.** $16\frac{2}{3}\%$	**31.** $6\frac{1}{4}\%$	**32.** $7\frac{1}{2}\%$

Solve.

33. The girls' basketball team made 0.78 of their foul shots. The boys' basketball team made 69% of their foul shots. What is the percent of difference in foul shots made between the girls' and boys' teams?

34. The I. M. Patient family used this graph to show the distribution of the family medical expenses for the year. Express each percent as a fraction and as a decimal.

- **a.** pediatrician
- **b.** orthodontist
- **c.** allergist
- **d.** optician
- **e.** dermatologist

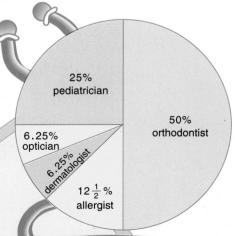

25% pediatrician

50% orthodontist

6.25% optician

6.25% dermatologist

$12\frac{1}{2}\%$ allergist

*See page 459 for more practice.

13-3 Changing Percents Greater than 100%

Change a percent greater than 100% to a decimal or fraction the same way any percent is changed. The tables show how these percents are changed.

Percent to Decimal

Percent		Decimal
225%	2.25	2.25
510%	5.10	5.10
119%	1.19	1.19
120%	1.20	1.20

Percent to Fraction

Percent	Fraction Ratio to 100	Fraction expressed in lowest terms
225%	$\frac{225}{100}$	$\frac{9}{4}$ or $2\frac{1}{4}$
510%	$\frac{510}{100}$	$\frac{51}{10}$ or $5\frac{1}{10}$
119%	$\frac{119}{100}$	$\frac{119}{100}$ or $1\frac{19}{100}$
120%	$\frac{120}{100}$	$\frac{6}{5}$ or $1\frac{1}{5}$

Remember The ratio of a number greater than 100 to 100 is equivalent to:
- a percent greater than 100 (101%, 102%, . . .).
- a decimal greater than 1 (1.1, 1.2, 1.3, . . .).
- a fraction greater than 1 ($\frac{3}{2}$, $\frac{4}{2}$, $\frac{5}{2}$, $\cdots$).

Write each percent as a decimal.

1. 300% 2. 250% 3. 248% 4. 101% 5. 313% 6. 500% 7. 208%

8. 106% 9. 102% 10. 250% 11. 518% 12. 495% 13. 289% 14. 177%

Write each percent as a fraction.

15. 208% 16. 105% 17. 300% 18. 125% 19. 180% 20. 340%

21. 116% 22. 134% 23. 225% 24. 175% 25. 310% 26. 600%

Change to a percent. * (Hint: Reverse the rules.)

27. $\frac{400}{100}$ 28. $\frac{201}{100}$ 29. $\frac{120}{100}$ 30. $\frac{199}{100}$ 31. 1.58

32. 1.23 33. 1.08 34. 4.01 35. 1.30 36. 1.90

37. $\frac{7}{2}$ 38. $\frac{9}{4}$ 39. $\frac{25}{20}$ 40. $\frac{75}{25}$ 41. $\frac{25}{5}$

13-4 Changing More Difficult Decimals and Fractions to Percents

Percent means per 100 or $\frac{1}{100}$ or 0.01

Decimals to Percents

To change a decimal to a percent, move 2 places *to the right*

1.5 = __?__ %

1.5 = 1.50 = 150% ←

Move 2 places to right. | Write % symbol.

0.005 = __?__ %

0.005 = 0.005 = 0.5% ←

Move 2 places to right. | Write % symbol.

Fractions to Percents

- Change to a decimal by dividing to the hundredths place.

$\frac{1}{8}$ = __?__ %

$\frac{1}{8}$ = $8\overline{)1.00}$ → 0.12 R4

$\frac{1}{3}$ = __?__ %

$\frac{1}{3}$ = $3\overline{)1.00}$ → 0.33 R1

- Express the remainder as a fraction.

0.12 R4 = $0.12\frac{4}{8}$
= $0.12\frac{1}{2}$

0.33 R1 = $0.33\frac{1}{3}$

- Express the decimal as a percent.

$0.12\frac{1}{2}$ = $12\frac{1}{2}$ %

$\frac{1}{8}$ = $12\frac{1}{2}$ %

$0.33\frac{1}{3}$ = $33\frac{1}{3}$ %

$\frac{1}{3}$ = $33\frac{1}{3}$ %

Change each decimal to a percent.*

1. 1.25
2. 1.75
3. 0.003
4. 0.006
5. 3.333
6. 2.125
7. 0.013
8. 0.025
9. 0.3
10. 0.2
11. 0.005
12. 0.001
13. 0.009
14. 0.008
15. 8.375
16. 1.255
17. 1.10
18. 7.60
19. 3.35
20. 1.55
21. 4.5
22. 1.1
23. 0.04
24. 0.08

Change each fraction to a percent.*

25. $\frac{3}{8}$
26. $\frac{5}{8}$
27. $\frac{1}{9}$
28. $\frac{2}{9}$
29. $\frac{1}{7}$
30. $\frac{3}{7}$
31. $\frac{5}{6}$
32. $\frac{1}{6}$
33. $\frac{1}{25}$
34. $\frac{12}{25}$
35. $\frac{2}{11}$
36. $\frac{5}{11}$
37. $\frac{5}{7}$
38. $\frac{11}{14}$
39. $\frac{1}{30}$
40. $\frac{1}{32}$
41. $\frac{1}{40}$
42. $\frac{7}{30}$
43. $\frac{1}{15}$
44. $\frac{1}{12}$
45. $\frac{3}{16}$
46. $\frac{5}{16}$
47. $\frac{9}{32}$
48. $\frac{3}{64}$
49. $\frac{7}{8}$
50. $\frac{8}{9}$
51. $\frac{7}{15}$
52. $\frac{11}{12}$

13-5 Changing More Difficult Percents to Decimals and Fractions

Percents are either:

Whole numbers ⟶ 46%

Not whole numbers ⟶ 12.5% or $12\frac{1}{2}$%

Less than 1 ⟶ $\frac{3}{5}$%

To change any percent to a decimal and a fraction, follow the four steps:

Step 1 Percent (%) means per 100 or $\frac{1}{100}$. Put the number over 100.

Step 2 To change to a decimal, divide the number by 100. Shortcut to divide by 100: Move the decimal point 2 places to the left.

Step 3 Now the decimal may be written as a fraction.

$$7.5\% = \frac{7.5}{100} = 0.075 = \frac{75}{1000} = \frac{3}{40}$$ ⟵ **Step 4** Express in lowest terms.

Study the table to see how each percent was changed to a decimal and a fraction.

%	Step 1 Ratio to 100	Step 2 Decimal	Step 3 Fraction	Step 4 Fraction expressed in lowest terms
41.3%	$\frac{41.3}{100}$	0.413	$\frac{413}{1000}$	fraction is in lowest terms
2.5%	$\frac{2.5}{100}$	0.025	$\frac{25}{1000}$	$\frac{1}{40}$
12.5%	$\frac{12.5}{100}$	0.125	$\frac{125}{1000}$	$\frac{1}{8}$
0.40%	$\frac{0.40}{100}$	0.004	$\frac{4}{1000}$	$\frac{1}{250}$
$8\frac{1}{3}$%	$\frac{8\frac{1}{3}}{100}$	$0.08\frac{1}{3}$	$\frac{8\frac{1}{3}}{100} = \frac{\frac{25}{3}}{100}$	$\frac{1}{12}$

Change to a decimal, then to a fraction in lowest terms.*

1. 80.9%   2. 10.1%   3. 3.75%   4. 6.25%   5. 0.3%   6. 0.50%

7. 18.1%   8. 21.3%   9. 0.75%   10. 0.45%   11. 12.5%   12. 65.2%

13. 20.6%   14. 12.2%   15. 8.9%   16. 16.25%   17. 0.05%   18. 0.04%

19. 4.8%   20. 6.5%   21. $9\frac{1}{3}$%   22. $3\frac{4}{5}$%   23. $8\frac{1}{8}$%   24. $6\frac{1}{4}$%

13-6 Using Percents

To use percents easily, memorize the common equivalent fractions.
Look for patterns.

Fraction	$\frac{1}{2}$
Percent	50%

Fraction	$\frac{1}{3}$	$\frac{2}{3}$
Percent	$33\frac{1}{3}$%	$66\frac{2}{3}$%

Fraction	$\frac{1}{4}$	$\frac{2}{4}$	$\frac{3}{4}$
Percent	25%		75%

Fraction	$\frac{1}{5}$	$\frac{2}{5}$	$\frac{3}{5}$	$\frac{4}{5}$
Percent	20%	40%	60%	80%

Fraction	$\frac{1}{6}$	$\frac{2}{6}$	$\frac{3}{6}$	$\frac{4}{6}$	$\frac{5}{6}$
Percent	$16\frac{2}{3}$%				$83\frac{1}{3}$%

Fraction	$\frac{1}{7}$	$\frac{2}{7}$	$\frac{3}{7}$	$\frac{4}{7}$	$\frac{5}{7}$	$\frac{6}{7}$
Percent	$14\frac{2}{7}$%	$28\frac{4}{7}$%	$42\frac{6}{7}$%	$57\frac{1}{7}$%	$71\frac{3}{7}$%	$85\frac{5}{7}$%

Fraction	$\frac{1}{8}$	$\frac{2}{8}$	$\frac{3}{8}$	$\frac{4}{8}$	$\frac{5}{8}$	$\frac{6}{8}$	$\frac{7}{8}$
Percent	$12\frac{1}{2}$%		$37\frac{1}{2}$%		$62\frac{1}{2}$%		$87\frac{1}{2}$%

Fraction	$\frac{1}{9}$	$\frac{2}{9}$	$\frac{3}{9}$	$\frac{4}{9}$	$\frac{5}{9}$	$\frac{6}{9}$	$\frac{7}{9}$	$\frac{8}{9}$
Percent	$11\frac{1}{9}$%	$22\frac{2}{9}$%		$44\frac{4}{9}$%	$55\frac{5}{9}$%		$77\frac{7}{9}$%	$88\frac{8}{9}$%

1. Why are percents not given for all of the fractions in the tables above?

2. Copy each table. What pattern do you see?

Make similar tables for the following.

3. $\frac{1}{10}$, $\frac{2}{10}$, $\frac{3}{10}$, ..., $\frac{9}{10}$
4. $\frac{1}{25}$, $\frac{2}{25}$, $\frac{3}{25}$, ..., $\frac{24}{25}$
5. $\frac{1}{11}$, $\frac{2}{11}$, $\frac{3}{11}$, ..., $\frac{10}{11}$

Write <, =, or >.

6. $\frac{1}{2}$? 50%
7. $\frac{3}{4}$? 75%
8. 20% ? $\frac{1}{8}$
9. $\frac{1}{12}$? $12\frac{1}{2}$%

10. 30% ? $\frac{3}{5}$
11. $16\frac{2}{3}$% ? $\frac{1}{6}$
12. $\frac{4}{9}$? $42\frac{6}{7}$%
13. $\frac{1}{7}$? 15%

14. $\frac{3}{9}$? $33\frac{1}{3}$%
15. $16\frac{2}{3}$% ? $\frac{1}{5}$
16. 60% ? $\frac{15}{25}$
17. $88\frac{8}{9}$% ? $\frac{6}{7}$

Study the design.
Write both a fraction and a
percent that represent the ratio
of each color to the whole mosaic.

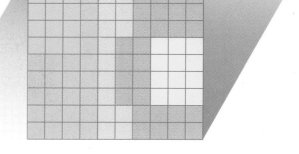

18. pink   19. yellow

20. blue   21. purple

22. green

Write <, =, or >.

23. 1.08 __?__ $1\frac{3}{4}$ %

24. 6.17 __?__ 6%

25. $5\frac{1}{2}$ % __?__ 5.0175

26. 43% __?__ 0.43

27. $20\frac{1}{2}$ % __?__ 0.205

28. 6.068 __?__ $6\frac{4}{5}$ %

29. 0.05 __?__ 0.5%

30. 10.2% __?__ 0.102

31. $30\frac{1}{4}$ % __?__ 3.025

Solve.

32. A department store sold 2 of its 3 remaining microwave ovens. What percent was that?

33. Mr. Link uses a microwave oven to cook 8 out of 24 meals. What percent of the meals does he cook without a microwave oven?

34. A certain discount shop found that $\frac{1}{2}$ % of the microwave ovens it sold were defective. Write $\frac{1}{2}$ % as a fraction.

35. The families of three fourths of the students have microwave ovens. What percent of the families do not have microwave ovens?

36. 25% of the students in the same school prefer microwave popcorn; $\frac{1}{3}$ of the students prefer prepackaged popcorn. Which type of popcorn do most of the students prefer?

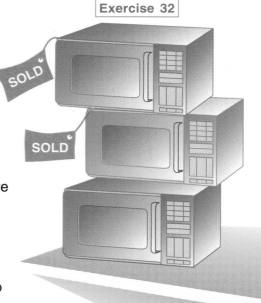

Exercise 32

CALCULATOR  **ACTIVITY**

37. Use a calculator to change the fractions in your tables to equivalent decimals.

Fraction	$\frac{1}{2}$
Percent	50%
Decimal	?

`1` `÷` `2` `=` `0.5`

13-7 Finding the Percentage

Myrna sold $160 worth of stamps to a stamp collector. She kept 15% of the money for herself. How much money did she keep for herself?

To find the percentage of a number, use *one* of these two ways:

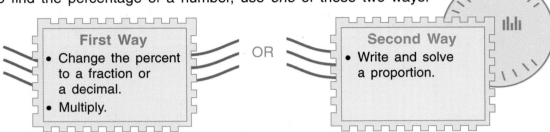

First Way
- Change the percent to a fraction or a decimal.
- Multiply.

OR

Second Way
- Write and solve a proportion.

$15\% = \dfrac{15}{100} = \dfrac{3}{20}$

$\dfrac{\overset{}{3}}{\underset{1}{20}} \times \dfrac{\overset{8}{\$160}}{1} = \$24$

Myrna kept $24.

$15\% = 0.15$

$$\begin{array}{r} \$1\ 6\ 0 \\ \times\quad 0\ .\ 1\ 5 \\ \hline 8\ 0\ 0 \\ 1\ 6\ 0 \\ \hline \$2\ 4\ .\ 0\ 0 \end{array}$$

15% of $160 = $24

$\dfrac{\text{(part kept)}}{\text{(whole)}}\ \dfrac{\$n}{\$160} = 15\% \text{ kept}$

$\dfrac{n}{160} = \dfrac{15}{100}$

$100n = 160 \times 15$

$100n = 2400$

$n = \dfrac{2400}{100} = \$24$

Memorizing fraction equivalents sometimes makes it easier to change a percent to a fraction.

How much is $66\frac{2}{3}\%$ of 30? $\longrightarrow$ $66\frac{2}{3}\% = \dfrac{2}{3}$

$\dfrac{2}{3}$ of $30 = \dfrac{2}{\underset{1}{3}} \times \dfrac{\overset{10}{30}}{1} = 20$

$66\frac{2}{3}\%$ of 30 is 20.

Find the percentage by changing the percent to a fraction.

1. 25% of 172
2. 50% of 842
3. $62\frac{1}{2}\%$ of 656
4. $87\frac{1}{2}\%$ of 176

5. 80% of 240
6. 40% of 600
7. $37\frac{1}{2}\%$ of 80
8. $12\frac{1}{2}\%$ of 464

9. 20% of 240
10. 50% of 848
11. $33\frac{1}{3}\%$ of 330
12. $66\frac{2}{3}\%$ of 330

Find the percentage by changing the percent to a decimal.

13. 30% of 120
14. 40% of 45
15. 16% of 20
16. 15% of 40

17. 90% of $5.00
18. 80% of $1.20
19. 11% of $47
20. 30% of $108

Find the percentage by using proportion.

21. 60% of 44   **22.** 5% of 1.2   **23.** 20% of 5   **24.** $\frac{1}{2}$% of 15

25. 40% of 60   **26.** 3% of 80   **27.** 15% of 90   **28.** 8% of 260

29. 18% of 50   **30.** 22% of 200   **31.** $2\frac{1}{2}$% of 300   **32.** $10\frac{1}{4}$% of 400

Estimating to Find the Percentage

In a rain forest 32% of the 1796 species of fauna and flora are endangered by deforestation. About how many species are endangered?

To estimate to find the percentage:

- Round the percent
  to a common percent. ⟶ $32\% \approx 33\frac{1}{3}\% = \frac{1}{3}$

- Round the other factor
  to a compatible number. ⟶ $\frac{1}{3} \times 1796 \approx \frac{1}{3} \times 1800$ ⟵ 3 and 1800 are compatible.

- Compute mentally. ⟶ $\frac{1}{3} \times 1800 = 600$

About 600 species are endangered.

To check the accuracy of your estimate, use a calculator.

| 1796 | × | 32 | % | = | 574.72 |

Estimate the answer. Then find the percentage.

33. 51% of 650   **34.** 12.5% of 71   **35.** $1\frac{3}{4}$% of 800   **36.** 21% of 80

37. 63% of 101   **38.** 76% of 29   **39.** 15.2% of 81   **40.** 24% of 250

41. 30% of 89   **42.** 12% of 160   **43.** 59% of 150   **44.** $62\frac{1}{2}$% of 319

45. 11% of 40   **46.** 20% of 44   **47.** 4% of 74   **48.** 11% of 32

MENTAL ◄ ► MATH

Compare by writing <, =, or >.

49. 25% of 50 _?_ 30% of 50   **50.** 20% of 200 _?_ 40% of 100

51. 15% of 100 _?_ 15% of 200   **52.** 75% of 8 _?_ 75% of 4

53. $12\frac{1}{2}$% of 80 _?_ $12\frac{1}{2}$% of 40   **54.** 25% of 96 _?_ $\frac{1}{4}$% of 96

55. 50% of 250 _?_ 90% of 250   **56.** $2\frac{1}{2}$% of 100 _?_ $0.2\frac{1}{2}$% of 100

13-8　Finding the Percent

The TV weather man said that it had rained 21 days out of the last 30. What percent of 30 is 21?

To find the percent one number is of the other, use *one* of these two ways:

First Way	OR	**Second Way**

First Way
- Form a ratio.
- Divide to change the ratio to a decimal.
- Change the decimal to a percent.

Ratio ⟶ 21 out of 30 = $\frac{21}{30}$ = $\frac{7}{10}$

Divide ⟶ $\frac{7}{10}$ ⟶ 0.70

Change to % ⟶ 0.70 ⟶ 70%

Second Way
- Write and solve a proportion.

$$21 \text{ out of } 30 = \frac{21}{30}$$

$$\frac{21}{30} = p\%$$

$$\frac{\text{part} \longrightarrow 21}{\text{whole} \longrightarrow 30} = \frac{p}{100}$$

$$30p = 21 \times 100$$

$$30p = 2100$$

$$p = \frac{2100}{30} = 70\%$$

Find each answer by forming a ratio and changing it to a percent.

1. What percent of 30 is 6?
2. What percent of 200 is 10?
3. What percent of 40 is 30?
4. What percent of 50 is 20?
5. What percent of 40 is 15?
6. What percent of 90 is 60?
7. What percent of 63 is 9?
8. What percent of 18 is 6?
9. What percent of 240 is 120?
10. What percent of 240 is 240?

Find each answer by writing and solving a proportion.

11. What percent of 442 is 221?
12. What percent of 100 is 4?
13. What percent of 160 is 4?
14. What percent of 80 is 20?
15. What percent of 1.8 is 0.6?
16. What percent of 1.2 is 0.2?
17. What percent of $\frac{1}{2}$ is $\frac{1}{8}$?
18. What percent of $\frac{1}{2}$ is $\frac{1}{4}$?
19. What percent of $\frac{3}{4}$ is $\frac{1}{4}$?
20. What percent of $\frac{3}{8}$ is $\frac{1}{8}$?

(Hint:　For exercises 17–20, change to decimals.)

Estimating to Find the Percent

In a sample study of a rain forest 118 of the 476 types of insects are listed as endangered. About what percent of insects are endangered?

To estimate to find the percent:

- Round the numerator and denominator to compatible numbers. $\longrightarrow$ $\dfrac{118}{476} \approx \dfrac{120}{480}$ $\longleftarrow$ Compatible Numbers

- Express the fraction in lowest terms and compute the % mentally. $\longrightarrow$ $\dfrac{120}{480} = \dfrac{1}{4} = 25\%$

About 25% of insects are endangered.

Use a calculator to check: 118 ÷ 476 % = 24.789915

Estimate. Use a calculator to check.

21. 20 is what percent of 79?　　　　　　**22.** 6 is what percent of 31?

23. 4 is what percent of 13?　　　　　　**24.** 18 is what percent of 19?

25. 0.9 is what percent of 50?　　　　　　**26.** 0.5 is what percent of 8?

27. 7 is what percent of 40?　　　　　　**28.** 115 is what percent of 231?

Find each percentage.

29. 20% of 60　　**30.** 20% of 75　　**31.** 0.2% of 6　　**32.** 0.5% of 4

33. 18% of 36　　**34.** 27% of 300　　**35.** 60% of 90　　**36.** 60% of 400

37. 100% of 450　　**38.** 200% of 1200　　**39.** 120% of 60　　**40.** 140% of 300

41. 600% of 1800　　**42.** 200% of 750　　**43.** 800% of 1800　　**44.** 70% of 50

Solve.

45. The Ecology Club raised $320 to divide among various concerns. Find the percent of the total allotted for each of these donations:
  a. $76 for City Garden.　　**b.** $102 for wildlife preservation.
  c. $74 for fuel conservation.　　**d.** $68 for wetland preservation.

46. A local weather reporter asked 560 people what temperature range they preferred. Find the percent of each group's preference:
  a. 84 people — 80°F to 90°F.　**b.** 224 people — 70°F to 80°F.
  c. 210 people — 60°F to 70°F. **d.** 42 people — 50°F to 60°F.

> ### MAKE UP YOUR OWN...

47. Conduct a survey of 20 people. Ask them which of the four seasons they like best and which they like least. Express the results in terms of percents.

13-9 Finding the Original Number

Alex is responsible for washing 25% of the windows of a 20-story skyscraper in a week. If he washes 40 windows a week, how many windows are there all together?

To find the original number for a given percent, use *one* of these two ways. (Here, we are looking for the original, or total, number of windows in the building.)

25% of **unknown number of *original* windows** is 40 windows.

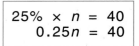

$$25\% \times n = 40$$
$$0.25n = 40$$

Let n be the unknown number of original windows.

First Way

- Change percent to a decimal or fraction.
- Divide to solve the equation.

OR

Second Way

- Write and solve a proportion.

$$25\% \times n = 40 \qquad 25\% = 0.25$$

$$40 \div 0.25 \rightarrow 0.25\overline{)40.00} = 160$$

or

$$25\% = \frac{1}{4}$$

$$40 \div \frac{1}{4} \rightarrow 40 \times \frac{4}{1}$$

$$40 \times \frac{4}{1} = 160$$

(part) $\longrightarrow$ $\dfrac{40 \text{ windows}}{n \text{ windows all together}} = \dfrac{25}{100}$ $\longleftarrow$ (whole)

$$\frac{40}{n} = \frac{25}{100}$$
$$25n = 40 \times 100$$
$$25n = 4000$$
$$n = \frac{4000}{25}$$
$$n = 160$$

There are 160 windows all together.

Complete the solution.

1. 12% of x is 48.

$48 \div 0.12$

2. 30% of a is 15.

$15 \div 0.30$

3. 10% of m is 6.

$\dfrac{6}{m} = \dfrac{10}{100}$

4. 60% of r is 12.

$\dfrac{12}{r} = \dfrac{60}{100}$

5. 80% of $t = 25$.

$25 \div \dfrac{4}{5}$

6. 50% of $s = 5$.

$5 \div \dfrac{1}{2}$

7. $66\frac{2}{3}$% of $p = 48$.

$\dfrac{48}{p} = \dfrac{2}{3}$

8. $12\frac{1}{2}$% of $m = 40$.

$\dfrac{40}{m} = \dfrac{1}{8}$

Solve.

9. 75% of what number is 30?

10. 25% of what number is 40?

11. 40% of what number is 32?

12. 60% of what number is 24?

13. $16\frac{2}{3}$% of what number is 18?

14. $66\frac{2}{3}$% of what number is 30?

15. $240 = 120\%$ of r

16. $300 = 125\%$ of s

17. $18 = 120\%$ of s

18. $44 = 110\%$ of m

19. $26 = 130\%$ of y

20. $90 = 150\%$ of r

21. $320 = 160\%$ of b

22. $230 = 115\%$ of n

23. $28 = 133\frac{1}{3}\%$ of p

24. Pedro spends $1\frac{1}{2}$ hours each day setting up and taking down his safety equipment. If this is 15% of his work time, how many hours a day does he work?

25. Forty-eight customers have their windows washed more than once a year. If this number represents 40% of Alex's customers, how many customers does he have in all?

26. Alex used 40% of the money in his wallet to buy lunch for himself and his girlfriend. If the lunch cost $25, how much money did Alex have before lunch?

27. Today 72 people were served lunch at a restaurant. If this was 60% of the number served for the entire day, how many ate there today?

28. Twelve of the workers brought lunch from home. If this is 75% of all the workers, how many workers are there?

29. If 80% of the enrollment in the school is 416, what is the total enrollment?

30. If $8\frac{1}{3}$% of all the books in the library are reference books and there are 250 reference books, how many books are in the library?

CHALLENGE

31. Joyce is 25% as old as Mary. Carla is 50% as old as Mary. If Joyce is 12, how old is Carla?

Just Imagine: Patterns, Conjectures, and Calculations

Calculators can reveal fascinating things about mathematics if you
use your mathematical knowledge and a little imagination.
One class explored some patterns in mathematical computations.

Joy discovered a pattern with squares.

$$(11)^2 = \quad 11 \times 11 \quad = 121$$

$$(111)^2 = \quad 111 \times 111 \quad = 12321$$

$$(1111)^2 = 1111 \times 1111 = 1234321$$

Matt remembered a pattern
in fraction/decimal conversions.

$$\frac{1}{11} = 0.090909\ldots = 0.\overline{09}$$

$$\frac{2}{11} = 0.181818\ldots = 0.\overline{18}$$

$$\frac{3}{11} = 0.272727\ldots = 0.\overline{27}$$

Now Joy and Matt guess the answers to
other examples that followed in these patterns.

Joy demonstrated:

$$(111,111)^2 = 12,345,654,321$$

6 digits

The number of digits (1's) becomes the middle
digit of the square. The ascending and
descending pattern centers on that middle digit.

Matt explained:

$$7 \times 9 = 63$$

$$\frac{7}{11} = 0.636363\ldots = 0.\overline{63}$$

The repeating pattern of the
decimal (63) is 9 times the
numerator of the fraction.

Joy and Matt had each made a **conjecture**.
A **conjecture** is a conclusion reached by guessing. In mathematics, conjectures
are often made after observing patterns with numbers or figures.

Use a calculator to compute. Look for a pattern and make a conjecture.
Compute mentally as soon as you are able.

1.

$(1 \times 9) + 1 =$
$(12 \times 9) + 2 =$
$(123 \times 9) + 3 =$

2.

$3 \times 37037 =$
$6 \times 37037 =$
$9 \times 37037 =$

3.

$9 \times \quad 3 \ =$
$9 \times \quad 33 \ =$
$9 \times \quad 333 \ =$

4.

$9 \times 44 =$	$9 \times 444 =$	$9 \times 77 =$	$9 \times 777 =$
$9 \times 666 =$	$9 \times 22,222 =$	$9 \times 555 =$	$9 \times 888,888 =$

5.

99^2	999^2	9999^2	$99{,}999^2$	$999{,}999^2$

6.

11^2	111^2	1111^2	$11{,}111^2$	$111{,}111^2$

The patterns in the next two exercises are more difficult to observe.
Study the first several examples carefully. Then make a conjecture.

7.

$15^2 =$
$25^2 =$
$35^2 =$
$45^2 =$
$55^2 =$

8.

$105^2 =$
$205^2 =$
$305^2 =$
$405^2 =$
$505^2 =$

9.

$123 \times 9 + 3 =$
$234 \times 9 + 4 =$
$345 \times 9 + 5 =$
$456 \times 9 + 6 =$
$678 \times 9 + 8 =$

**Use a calculator to change each fraction to a decimal. Watch for a pattern
and make a conjecture.** Compute mentally as soon as you are able.

10.

fraction	$\dfrac{1}{3}$	$\dfrac{1}{30}$	$\dfrac{1}{300}$	$\dfrac{1}{3000}$	$\dfrac{1}{30{,}000}$
decimal	0.333...				
	$0.\overline{3}$				

11.

fraction	$\dfrac{1}{22}$	$\dfrac{3}{22}$	$\dfrac{5}{22}$	$\dfrac{7}{22}$	$\dfrac{9}{22}$
decimal	0.04545...				
	$0.0\overline{45}$				

12.

fraction	$\dfrac{27}{99}$	$\dfrac{82}{99}$	$\dfrac{61}{99}$	$\dfrac{19}{99}$	$\dfrac{42}{99}$
decimal	$0.\overline{27}$				

13.

fraction	$\dfrac{188}{999}$	$\dfrac{127}{999}$	$\dfrac{24}{999}$	$\dfrac{5296}{9999}$	$\dfrac{137}{9999}$
decimal					

STRATEGY
Problem Solving: Hidden Information

Problem: Chef Pierre spends 40% of his workday creating desserts. If yesterday he worked $10\frac{1}{2}$ hours, how many minutes did he spend creating desserts?

1 IMAGINE Draw and label a picture of the time spent creating these desserts.

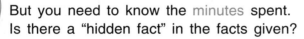

40% of time on desserts
$10\frac{1}{2}$ hours total time

? minutes
on desserts

2 NAME *Facts:* 40% of the time worked on desserts
$10\frac{1}{2}$ hours total time worked

Question: __?__ minutes on desserts

3 THINK Write a word sentence to explain time spent on desserts.
40% of $10\frac{1}{2}$ hr = Time (T) spent on desserts

$$T = 40\% \times 10\frac{1}{2}$$
$$T = \frac{4}{10} \times \frac{21}{2}$$
$$T = \frac{84}{20} = 4\frac{4}{20} = 4\frac{1}{5}$$

hours spent on desserts

But you need to know the minutes spent.
Is there a "hidden fact" in the facts given?

1 hour = 60 minutes
So, replace $T = 4\frac{1}{5}$ hours with
$$T = 4\frac{1}{5} \times 60 = \text{minutes spent}$$

4 COMPUTE $T = 4\frac{1}{5} \times 60 = \dfrac{21}{\cancel{5}_1} \times \dfrac{\overset{12}{\cancel{60}}}{1} = 252$ minutes spent
making desserts

5 CHECK Use estimation to check the reasonableness of your answer.

40% of $10\frac{1}{2}$ hr $\approx$ 40% of 10 hours = 4 hours

4 hr $\times$ 60 minutes = 240 minutes, which is close to computed answer of 252 minutes

Solve each problem. What is the hidden fact in each?

1. Samantha's 206 bones make up 25% of her body weight. How many ounces do her bones weigh if she weighs 110 pounds?

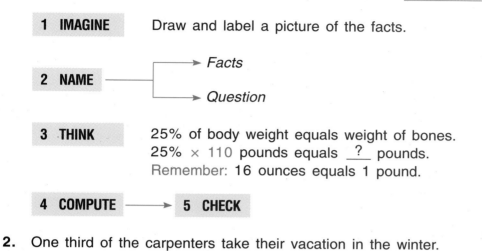

206 Bones = 25% Body Weight
Weight   = 110 pounds
          = __?__ ounces

1 IMAGINE   Draw and label a picture of the facts.

2 NAME ——————→ Facts
                 └——→ Question

3 THINK   25% of body weight equals weight of bones.
                25% × 110 pounds equals __?__ pounds.
                Remember: 16 ounces equals 1 pound.

4 COMPUTE ——→ **5 CHECK**

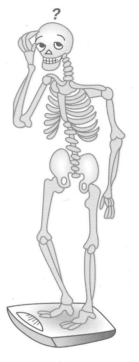

2. One third of the carpenters take their vacation in the winter. The rest of them take it in the spring or fall. If 38 take their vacation in the spring or fall, how many carpenters are there?

3. A carpenter is fencing a square park having an area of 625 m². How many meters of wooden fencing does he need?

4. 70% of a 5-gallon can of paint was used to paint the exterior of a house. How many quarts of paint were left over?

5. The Stitch Factory workers spend 15% of their workday sewing on zippers. If the workday is 8 hours, how many minutes are spent sewing zippers?

6. Each hair on your head can support a little more than 3 ounces. If you have about 100,000 hairs, how many 150-pound people could your hair support?

7. Jane is 5 years younger than Sue. Jane is 6 years older than Maria, who is 12. How old is Sue?

8. A family spent 37.5% of its weekly income on housing, 25% on food, 12.5% on clothing, and put the rest in a savings account. If it saved $190, what is the weekly income of the family?

9. If a human heart beats 70 times a minute, how many times does it beat in one day?

10. A rectangular field is 120 yards long and has a perimeter of 380 yards. If the cost of seeding is $1.25 per 200 square yards, how much will it cost to seed this field?

13-12 Problem Solving: Applications

Solve. You may combine strategies.

USE THESE STRATEGIES:
Write an Equation
Multi-Step Problem
Hidden Information
Use a Graph
Extra Information
Use Simpler Numbers

1. Only $\frac{1}{4}$% of the televisions sold by an appliance outlet were defective. What fraction was that?

2. Four out of 12 televisions on display at the store have wide screens. What percent do not have wide screens?

3. Between 6 and 7 P.M., half of the television viewers in Videoville watch the news, $\frac{1}{4}$ watch game shows, and $16\frac{2}{3}$% watch reruns. What percent watch other shows?

4. Because of an ice storm, $37\frac{1}{2}$% of the homes in Videoville were without power. What fractional part of the homes were without power?

5. A truck delivered a shipment of 140 televisions. If this number represents 70% of the truck's capacity, what fractional part of the truck was empty? How many more televisions might have been shipped in the truck?

6. Dani has saved 75% of the cost of a new stereo system. If the system costs $150, how much more money does she need?

7. Lynn received $180 for graduation. She spent $60 of it on jewelry. What percent did she spend?

8. A jeweler sold 18 of the 24 gold watches she had in stock. What percent did the jeweler *not* sell?

9. Of the pieces in a jewelry collection 0.2 are rings, $\frac{1}{8}$ are pendants, 16% are gold chains, $\frac{1}{3}$ are bracelets, and the rest are pins. What part of the collection is the smallest?

10. Milo bought a $375 diamond ring. He paid a 20% down payment. How much money does he still owe for the ring?

11. "Green" gold is 75% gold; the rest is silver. Ms. Watz has a 2.4-oz chain made of green gold. How many ounces of gold and of silver are in the chain?

12. Ms. Watz has a "yellow" gold pendant that weighs 200 g. The pendant contains 106 g of gold and 44 g of copper. The rest is silver. What is the percent of silver in yellow gold?

13. If 12-karat gold is 50% gold, how many karats are in something pure gold?

14. Sterling silver is 7.5% copper; the rest is silver. How many pounds of silver are in a sterling silver tea set weighing 4 lb?

15. Ms. Robins spent $3\frac{1}{2}$%, or $27.30, of her weekly earnings on jewelry. What were her weekly earnings?

16. Celine engraved only $8\frac{1}{3}$% of the silver rings she had made. If she engraved five rings, how many rings did she *not* engrave?

17. José designed 24, or 15%, of the gold charms in the craft shop display. How many of the charms were *not* designed by José?

18. Ms. Murtha sold $46\frac{2}{3}$% of the watches in her store for half off and the rest at $\frac{1}{4}$ off. How many watches were sold at $\frac{1}{4}$ off if she sold 14 watches at half off?

19. Forty percent, or 10.89 g, of a silver dollar is silver. How many grams of silver are in $125 in silver dollars?

20. Aluminum gold is 22% aluminum; the rest is gold. If an aluminum gold bar contains 55 g of aluminum, how many grams of gold does it contain?

21. A sheet of gold foil 0.0025 mm thick is $1\frac{2}{3}$% of the thickness of a sheet of aluminum foil. How thick is a sheet of aluminum foil?

22. Marsha has two magnets and a 120-g pile of iron filings. One magnet picks up $\frac{7}{8}$ of the filings; the other picks up $62\frac{1}{2}$% of the filings. How many more grams does the stronger magnet pick up than the weaker?

The graph shows the content of an alnico magnet. Use the graph to complete exercises 23–27.

23. What fractional part of the magnet is cobalt?

24. One third of the magnet is made of what two metals?

25. What percent of the magnet is iron?

26. If the magnet contains 24 g of nickel, what is the total weight of the magnet?

27. How many grams of each metal are in the magnet in exercise 26?

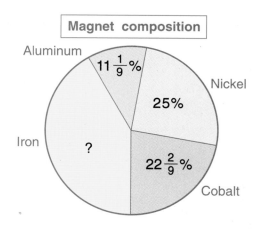

Magnet composition

Aluminum $11\frac{1}{9}$%

Nickel 25%

Iron ?

$22\frac{2}{9}$% Cobalt

More Practice

Write as a percent.

1. $\frac{42}{100}$ 2. $\frac{124}{100}$ 3. 0.26   4. 0.03   5. $\frac{1}{5}$

6. $\frac{7}{10}$ 7. $\frac{1}{6}$ 8. $\frac{3}{8}$ 9. 1.0   10. 0.004

Write the fraction equivalents.

11. 40%   12. 80%   13. 85%   14. 120%   15. $66\frac{2}{3}$%   16. $2\frac{1}{2}$%

Write as a decimal.

17. 8.5%   18. $10\frac{1}{5}$%   19. 260%   20. 8%   21. 27%   22. $15\frac{4}{5}$%

Complete.

23.

Fraction	$\frac{1}{2}$	$\frac{1}{3}$	$\frac{1}{4}$	$\frac{1}{5}$	$\frac{1}{8}$	$\frac{1}{10}$
Decimal						
Percent						

24.

Fraction	$\frac{2}{3}$	$\frac{3}{4}$	$\frac{2}{5}$	$\frac{3}{5}$	$\frac{4}{5}$	$\frac{3}{8}$	$\frac{5}{8}$	$\frac{7}{8}$
Decimal								
Percent								

Solve.

25. Sean works after school. If 25% of his salary is $20, what is his salary?

26. The physical education department owns 14 basketballs. $14\frac{2}{7}$% of them need air. How many basketballs need air?

27. The temperature reached above 80°F for 21 days in June. What percent of days in the month has it reached above 80°F? (Hint: Check a calendar for the total number of days in June.)

28. The art show had 36 scene paintings displayed. 18 paintings were landscapes. What percent were not landscapes?

29. If $\frac{3}{4}$ of Carmen's money is $24, how much money does Carmen have?

30. Miss Paxson earns $2200 a month. If she spends 25% of her salary on rent, how much money does she have left for everything else?

Math Probe

LOGIC

In logic we are interested only in sentences that state a fact and can be judged true or false.

Sentence
(complete thought)

Statement
(States a fact.)

Disneyland is in California. (**T**)
Five is an odd number. (**T**)
Picasso was a great general. (**F**)
George Washington died in 1903. (**F**)

Not a statement
(Does not state a fact.)

Open the window.
Clean out your desk.
Oh, no!
Is there a test tomorrow?

Open statement (**O**)
(Contains a variable or an unknown.)

His favorite color is red. (**O**)
They love to do math. (**O**)
7 – x = 4 (**O**)
You have a cold. (**O**)

Closed statement
(Can be judged true (**T**) or false (**F**).)

Most plants are green. (**T**)
Dallas is in Texas. (**T**)
All polygons have four sides. (**F**)
All roses are red. (**F**)

Tell whether each sentence is open, true, or false.

1. Tallahassee is the capital of Florida.

2. It is located in the northwest.

3. Shakespeare wrote *Romeo and Juliet*.

4. He has a big appetite.

5. $2 \times 2 \times 2 = 8$

6. Mozart was a famous painter.

7. $10 + x = 12$

8. Nine is a prime number.

9. She lives in San Francisco.

10. Theodore Roosevelt became president in 1987.

11. $36 - n = 14$

12. The South won the War Between the States.

Check Your Mastery

Write as a percent.

See pp. 314–319

1. 0.36   2. 0.8   3. $\frac{2}{5}$ 4. $\frac{3}{50}$

5. $\frac{5}{12}$ 6. $\frac{3}{11}$ 7. 0.075   8. 1.002

Write as a decimal.

9. 25%   10. 5%   11. 6.3%   12. 118%

13. 220%   14. 47.9%   15. 0.30%   16. $9\frac{2}{3}\%$

Write as a fraction.

17. 80%   18. 8%   19. 210%   20. 250%

21. $1\frac{1}{2}\%$ 22. $55\frac{5}{9}\%$ 23. 1.9%   24. 0.1%

See pp. 320–321

Write <, =, or >.

25. $\frac{2}{3}$ _?_ 65%   26. $\frac{5}{6}$ _?_ 80%   27. $\frac{3}{8}$ _?_ 39%   28. $\frac{3}{5}$ _?_ 65%

29. $\frac{2}{25}$ _?_ 8%   30. $\frac{4}{9}$ _?_ 46%   31. 0.43 _?_ 43%   32. 0.75 _?_ 75%

33. 0.6% _?_ 60.0   34. $\frac{1}{4}\%$ _?_ 0.4   35. 12.2% _?_ 0.102   36. $25\frac{1}{5}\%$ _?_ 25.2

See pp. 322–325

Solve.

37. 60% of 35
38. 150% of 1.6
39. What percent of 800 is 40?
40. What percent of 20 is 35?
41. 20 is 4% of what number?
42. 9 is 180% of what number?
43. $66\frac{2}{3}\%$ of 720
44. 80% of $10.50
45. What percent of $\frac{1}{4}$ is $\frac{1}{8}$?
46. What percent of 2.5 is 0.25?
47. 75% of what number is 30?
48. 65% of what number is 585?

49. The music department has 1 piano, 3 trumpets, 2 electric guitars, 4 violins, and 1 set of drums. What percent of the instruments are the electric guitars?

50. A school has 900 students. About 11% of the students ride bicycles to school. How many of the students do not ride bicycles to school?

51. This weekend, 15 of the seventh graders went to the game. This is 37.5% of all the seventh graders. How many students are in the seventh grade?

14 Consumer Mathematics

In this chapter you will:

- Find the discount and sale price
- Compute sales tax and commissions
- Learn about using the interest formula
- Examine installment buying
- Solve multi-step problems

Do you remember?

5 % = half of 10 %

20 % = double 10 %

30 % = three times 10 %

To multiply a decimal by 10 %, move the decimal point one place *to the left.*

$$10\% \text{ of } \$416.70$$

$$\$41.67$$

RESEARCHING TOGETHER

No Place Like Home?

You have a pen pal living in another state. Show some of the advantages and disadvantages of where you each live. Use graphs to compare data such as sales tax, cost of living, unemployment rate, and average income.

14-1 Discount

Discount (D): a reduction on the original or **list price (LP)** of an item.

Rate of discount (R of D): the percent taken off the price.

Sale price (SP): the difference between the
list (original, or regular) price (*LP*) and the discount (*D*).

A beach umbrella regularly sells for $49.
It is being sold at a 30% discount.
What is the discount? What is the sale price?

Sale
30% off

To find the discount (D), use *one* of these two ways:

First Way		Second Way
• Use a formula.	**OR**	• Write a proportion.
• Solve it.		• Solve it.

First Way

$D = LP \times R \text{ of } D$

$D = \$49 \times 30\%$

$D = \$49 \times 0.30$

$$D = \begin{array}{r} \$49 \\ \times 0.30 \\ \hline \$14.70 \end{array} \quad \text{Discount}$$

The discount is $14.70.

Second Way

$$\frac{\text{Part}}{\text{Whole}} \longrightarrow \frac{\text{Discount}}{\text{List Price}} = \frac{\text{Rate of Discount}}{\text{Rate of List Price}}$$

$$\frac{D}{\$49} \times \frac{30}{100}$$

$100D = \$1470$

$D = \dfrac{\$1470}{100}$

$= \$14.70 \quad \text{Discount}$

To find the sale price (SP), use the formula: **SP = LP – D.**

$SP = LP - D$

$SP = \$49 - \$14.70$

$SP = \$34.30$ The sale price is $34.30.

Find the discount and sale price. (Hint: Decide if it is easier to do the problem if the percent is a fraction or if it is a decimal. In exercise 1 the fraction is easier. Why?)

1. Outdoor Grill — Regular price: $360; discount: $33\frac{1}{3}\%$

2. Lawn Sprinkler — List price: $12.60; discount: $66\frac{2}{3}\%$

3. Bags of Charcoal — Original price: $6.00; discount: 15%

4. Radio and Headset — List price: $25.00; discount: 20%

5. Beach Towels — Regular price: $15.00; discount: 50%

6. Picnic Cooler — List price: $22.50; discount: 60%

Finding the Rate of Discount

At a sale Jane paid $5.10 for a canvas bag that had a list price of $6.00. What was the rate of discount?

SP = $5.10

LP = $6.00

D = ?

To find the **rate of discount (R of D)**, one must know *both* the discount and the list price.

- Subtract the selling price from the list price to find the discount.

$D = SP - LP$
$D = \$6.00 - \$5.10 = \$.90$ Discount

- Solve by proportion.

$$\frac{\text{Discount}}{\text{List Price}} = \frac{\text{Rate of Discount}}{\text{Rate of List Price}}$$

Use a calculator:

| .9 | ÷ | | 6 |
| × | 100 | = | 15 |

$$\frac{\$.90}{\$6.00} \; \diagdown \; \frac{R \text{ of } D}{100}$$

R of $D = 15\%$ Rate of Discount

Copy and complete.

	List Price	Sale Price	Discount	Rate of Discount
7.	$660.00	?	$33.00	?
8.	$ 28.00	$ 21.00	?	?
9.	$ 6.40	?	$ .96	?
10.	?	$119.97	$79.98	?
11.	$850.00	?	$76.50	?
12.	$ 15.40	?	$ 3.08	?
13.	?	$ 12.98	?	$33\frac{1}{3}\%$
14.	$200.00	?	?	12%

Solve. You can use a calculator to find discount and rate of discount.

Discount

| LP | × | R OF D | % | = |

Rate of Discount

| D | ÷ | LP | × | 100 | = |

15. A family beach pass costs $150.00 for the season. The town offers a $33\frac{1}{3}$ % discount on any that are purchased before June 1. What will a beach pass cost if purchased before June 1?

16. A tube of sun screen costs $4.50. A 10% discount is offered when 2 or more are purchased. What will be the sale price for 3 tubes?

17. Batteries selling for $3 are reduced one dollar. What is the rate of discount?

18. Jessica bought a tote bag that usually sold for $14.95 at a 5% reduction. How much money did she save buying the tote bag at the sale price?

19. Paul bought two beach chairs for $35. Each chair was originally listed at $25. What was the rate of discount?

$150.00
BEACH PASS
$33\frac{1}{3}\%$ Discount
Before June 1

339

Sales tax (*T*): the amount added to the marked price (*MP*) of an item and collected by the local and state governments that levied the tax.

Rate of sales tax (*R* of *T*): the percent of the marked price (*MP*) levied as a tax.

Total cost (*TC*): the sum of the marked price (*MP*) and the tax (*T*).

A calculator is marked $29.95. There is a 6% state sales tax.
What is the sales tax? What is the total cost of the calculator?

To find the sales tax (*T*), use *one* of these two ways:

First Way		Second Way
• Use a formula.	**OR**	• Write a proportion.
• Solve it.		• Solve it.

First Way:

$T = MP \times R \text{ of } T$
$T = \$29.95 \times 6\%$
$T = \$29.95 \times 0.06$
$T = \$29.95$
$\underline{\times \quad 0.06}$
$\$1.7970$ Sales Tax

Second Way:

$$\frac{Part}{Whole} \rightarrow \frac{Sales\ Tax}{Marked\ Price} = \frac{Rate\ of\ Sales\ Tax}{Rate\ of\ Marked\ Price}$$

$$\frac{T}{\$29.95} \times \frac{6}{100}$$

$100T = \$29.95 \times 6$
$100T = \$179.70$
$T = \dfrac{\$179.70}{100}$
$= \$1.797$ Sales Tax

Round sales tax to the nearest cent: $\$1.797 \approx \$1.80$.

To find the total cost (*TC*), use the formula: ***TC* = *MP* + *T*.**

$TC = MP + T$
$TC = \$29.95 + \$1.80 = \$31.75$ Total Cost

Find the sales tax and the total cost of:

1. a $450 rug with a 5% sales tax.
2. a $245.00 VCR with a 7% sales tax.
3. a $40.00 skateboard with a 4% sales tax.
4. a $69.50 answering machine with a $6\frac{1}{2}$% sales tax.
5. a $926.00 bedroom set with a $3\frac{1}{2}$% sales tax.
6. three videotapes priced at $9.95 each with a sales tax of 8%.

Taken from a 6% Sales Tax Table:

Sale	Tax		Sale	Tax
$.00-$ .10	$ .00		$20.11-$20.22	$1.21
.11- .22	.01		20.23- 20.38	1.22
.23- .38	.02		20.39- 20.56	1.23
.39- .56	.03		20.57- 20.72	1.24
.57- .72	.04		20.73- 20.88	1.25
.73- .88	.05		20.89- 21.10	1.26
.89- 1.10	.06		21.11- 21.22	1.27

Use the table above to find the sales tax and the total cost of:

7. a roll of recycled paper towels costing $.75.
(Hint: $.75 falls between $.73-$.88 on the table.)

8. a calculator marked $14.95 and four pads of paper at $1.50 each.

9. a camera listed at $25.50 and sold at a 20% discount.

10. six 5-pound bags of grass seed regularly costing $6.90 and sold at $\frac{1}{2}$ off.

11. two posters usually listed at $12.75 apiece and selling at a 20% discount.

MENTAL MATH If 10% of $4.20 is $.42 ──→ 5% of $4.20 is $\frac{1}{2} \times$ $.42 = $.21.

──→ 20% of $4.20 is 2 × $.42 = $.84.

Use this pattern to find the sales tax mentally.

12. 10% of $3.40   13. 5% of $3.40   14. 20% of $3.40

15. 10% of $26   16. 5% of $26   17. 20% of $26

18. 10% of $10.22   19. 5% of $10.22   20. 20% of $10.22

21. 10% of $1.53   22. 5% of $140   23. 20% of $3.60

24. 10% of $98   25. 5% of $5.80   26. 20% of $4.19

Estimate the total cost of:

27. A $200 suit sold at a 20% discount, with a 5% sales tax.
Find the discount: 10% of $200.00 is $20.00; so, 20% is 2 × $20 = $40
Find the selling price: $200 − $40 = $160
Find the sales tax: 10% of $160 is $16.00; so, 5% is $16 ÷ 2 = $8
Total Cost = ___?___

28. A $200 air fare sold at a 5% discount, with a 10% sales tax.

29. A $150 car rental at a 20% discount, with a 5% sales tax.

30. A $60 raincoat that was sold at a 10% discount, with a 5% sales tax.

14-3 | Commission

Commission (C): the amount of money earned for selling goods or services.

Rate of commission (R of C): the percent of the total amount of goods or services sold that is earned by the seller.

Total sales (TS): the total amount of goods or services sold.

Mr. Grayhawk sold $1500 worth of office supplies. If his rate of commission is 7%, how much money did he earn on the sale?

To find the commission (C), use *one* of these two ways:

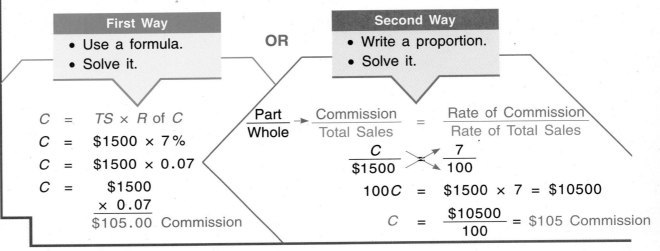

First Way		Second Way
• Use a formula.	**OR**	• Write a proportion.
• Solve it.		• Solve it.

First Way:

$C = TS \times R$ of C

$C = \$1500 \times 7\%$

$C = \$1500 \times 0.07$

$C = \begin{array}{r} \$1500 \\ \times\ 0.07 \\ \hline \$105.00 \end{array}$ Commission

Second Way:

$$\frac{\text{Part}}{\text{Whole}} \rightarrow \frac{\text{Commission}}{\text{Total Sales}} = \frac{\text{Rate of Commission}}{\text{Rate of Total Sales}}$$

$$\frac{C}{\$1500} \times \frac{7}{100}$$

$100C = \$1500 \times 7 = \$10500$

$C = \dfrac{\$10500}{100} = \$105$ Commission

Find the commission:

1. on a $975 wide-screen television at a 6% rate of commission.

2. on a $12,000 truck at a 5% rate of commission.

3. on a $400 sale of magazine subscriptions at a 5% rate of commission.

4. on a $56 radio at a 7.5% rate of commission.

5. on a $1180 microcomputer at a 3% rate of commission.

6. on a $175,000 house at an 8% rate of commission.

7. on total sales of $9000 at a 20% rate of commission.

8. on total sales of $445 at a 5% rate of commission.

9. on total sales of $245 at an 11% rate of commission.

10. on total sales of $2500 at an 8% rate of commission.

11. on three $750 refrigerators at an $11\frac{1}{9}$% rate of commission.

12. on two $8740 pianos at a 6.5% rate of commission.

Finding Rate of Commission

Lloyd earned $57 when he sold an electronic keyboard. The total sale was $1140. Find Lloyd's rate of commission.

To find the **rate of commission (R of C)**, one must know both the commission and the total sales.

TS	=	$1140
C	=	$57
R of C	=	?

Solve by proportion:

Use a calculator:

$$\frac{\text{Commission}}{\text{Total Sales}} = \frac{\text{Rate of Commission}}{\text{Rate of Total Sales}}$$

 57 ÷ 1140 × 100 = 5

$$\frac{\$57}{\$1140} \diagup\!\!\!\diagdown \frac{R \text{ of } C}{100}$$

R of C = 5% Rate of Commission

Copy and complete.

	13.	14.	15.	16.	17.	18.
Total Sales	$1520	$370.40	$437	$868	$243	$621
Rate of Commission	15%	6%	?	?	?	8%
Commission	?	?	$17.48	$30.38	$21.87	?

Solve. You can use a calculator to find commission and rate of commission.

Commission ▷ TS × R ᴏꜰ C % =

Rate of Commission ▷ C ÷ TS % =

19. Mrs. Martone, a real estate agent, sold a luxury condominium for $190,000 at a commission rate of 9%. Mrs. Ewing sold an apartment building for $220,000 at a rate of 7.5%. Who earned more money? How much more?

20. Bella sold an alarm system to the Clark family for $3800. Her commission for the sale was $323. What was the rate of commission?

21. Kevin receives 4.5% commission on his sales of insurance policies. During the month of February, he sold $5120 worth of insurance. How much did he earn in commissions that month?

SOLD $190,000

SOLD $220,000

22. Peg sells Rudy's paintings at an 8% commission. Rudy's latest painting sold for $12,000. How much money did Peg make on the sale?

23. Find the selling price or total sales on an entertainment center if the rate of commission is 15% and the commission is $180. (Hint: If $C = TS \times R$ of C, $TS = C \div R$ of C.)

14-4 Earning Interest

Principal (P): the amount of money deposited in a bank.

Simple Interest (I): the amount paid by the bank to the depositor *only* on the principal for a stated period of time.

Rate of Interest (R): the percent of interest paid to the depositor on the principal.

Time (T): represents how long the principal is left on deposit. When calculating, time must be in years or a fractional part of a year.

Gail has $300. How much interest will she earn in a year if the bank pays 6% interest annually?

To find simple interest (I):

Interest = Principal × Rate of Interest × Time

$I = P \times R \times T$

$I = \$300 \times 6\% \times 1 \text{ year}$

$I = \$300 \times 0.06 \times 1 = \$18.00 \text{ Interest}$

To find the new total principal in the bank after the interest is paid:

Total = $P + I \longrightarrow$ Total = $\$300 + \$18.00 = \$318$ New Total

Suppose Gail left her money in the bank for 33 months. How much would she have?

$I = P \times R \times T$

$I = \$300 \times 0.06 \times \dfrac{33}{12}$

Express time as a fraction of a year.

$I = \dfrac{3\cancel{00}}{1} \times \dfrac{\overset{1}{\cancel{6}}}{1\cancel{00}} \times \dfrac{33}{\underset{2}{\cancel{12}}} = \dfrac{99}{2} = \$49.50 \text{ Interest}$

Total = $\$300 + \$49.50 = \$349.50$ New Total

Find the simple interest and the new total in the bank.

1. Janet deposits $500 at 5% for 1 year.
2. Darrin deposits $1100 at 6% for 1 year.
3. Kate deposits $425 at 8% for 12 months.
4. Beth deposits $365 at 5% for 24 months.

Write as a fractional part of a year. Express in simplest form.

5. 7 months
6. 6 months
7. 8 months
8. 30 months
9. 42 months
10. 3 years 6 months
11. 2 years 9 months
12. 5 years 8 months
13. 2 years 2 months

Find the simple interest.

	Principal	Rate	Time
14.	$ 100	$6\frac{1}{2}$ %	6 mo
15.	$ 200	4%	9 mo
16.	$ 810	5%	3 yr
17.	$ 550	$4\frac{1}{2}$ %	4 yr
18.	$1880	5%	5 yr 3 mo

	Principal	Rate	Time
19.	$ 440	7%	5 yr 6 mo
20.	$ 3000	3%	18 mo
21.	$ 4850	$6\frac{1}{2}$ %	$4\frac{1}{2}$ yr
22.	$10,000	5.8%	$1\frac{1}{2}$ yr
23.	$ 5200	8.2%	$1\frac{1}{4}$ yr

Find the simple interest and the new total in the bank.
You can use a calculator to find the interest and the total amount.

Interest P × R % × T = Total Amount P + I =

24. Meg deposits $865 at 6% for 6 months.

25. Kadim deposits $5000 at $7\frac{1}{4}$ % for 2 years.

26. Nikki deposits $650 at $5\frac{1}{2}$ % for $1\frac{1}{2}$ years.

27. Frank deposits $200 at 6.8% for 1 year.

28. Taylor deposits $440 at 8% for $1\frac{1}{2}$ years.

29. Jesse deposits $380 at $7\frac{1}{2}$ % for 24 months.

CHALLENGE

30. Santos invested $1000 in a savings certificate. At the end of the year, it had earned $68 in interest. What was the rate of interest?

CRITICAL  THINKING

Molly, Ollie, and Polly each invested $1000 for 3 years at a different bank — Safe Savings, True Trust, Big Bucks. One of the banks offered a 6% interest rate, another offered 7%, and the third offered 8%.

31. Use the clues and the chart to match each person with his/her bank.

32. Find each person's total earnings after three years.

Clue 1: Molly did not bank at Big Bucks.
Clue 2: Ollie earned $210 interest.
Clue 3: Polly earned the least amount of interest.
Clue 4: True Trust offered the most interest.
Clue 5: Big Bucks offered 6 % interest.

	Molly	Ollie	Polly
Safe Savings			
True Trust			
Big Bucks			

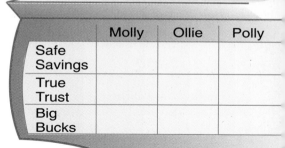

14-5 Borrowing Money

Principal (P): the amount of money borrowed from a bank.

Interest (I): the amount paid by the borrower for the use of the principal for a stated period of time.

Rate of Interest (R): the percent of interest paid to the bank.

Time (T): represents how long the principal is borrowed.

The Matthews' want to borrow $2000 from Long Branch Bank to pay the orthodontist for Linda's braces. The interest rate is 9.5% a year. How much interest will they pay in a year? What is the total the Matthews' will have to repay the bank in a year?

To find the interest, use the simple interest formula:

$$I = P \times R \times T$$
$$I = \$2000 \times 9.5\% \times 1 \text{ year}$$
$$I = \frac{200\cancel{0}}{1} \times \frac{9.5}{10\cancel{0}} \times \frac{1}{1} = \$190 \text{ Interest}$$

To find the total to be repaid to the bank:

Total $= P + I \longrightarrow$ $2000 + $190 $=$ $2190 Total to be repaid

If the Matthews' repaid the money in 6 months instead of a year, how much would they have to pay back?

$$I = P \times R \times T$$
$$I = \frac{\overset{10}{200\cancel{0}}}{1} \times \frac{9.5}{10\cancel{0}} \times \frac{\overset{1}{\cancel{6}}}{\underset{\underset{1}{\cancel{2}}}{\cancel{12}}} = \$95$$

Total $=$ $2000 + $95 $=$ $2095 Total to be repaid to the bank

Find the interest and the total amount paid to the bank. (Hint: Use 360 days for a year.)

	Principal	Rate	Time		Principal	Rate	Time
1.	$ 750	8%	1 yr	8.	$4000	12.5%	2 yr
2.	$ 600	10%	2 yr	9.	$ 950	10%	36 mo
3.	$2500	12%	12 mo	10.	$1000	18%	24 mo
4.	$9000	12%	24 mo	11.	$1200	12%	180 days
5.	$3800	20%	90 days	12.	$ 500	22%	90 days
6.	$ 400	15%	180 days	13.	$ 375	$8\frac{1}{2}$%	1 yr
7.	$1620	9.5%	1 yr	14.	$5500	$9\frac{1}{2}$%	2 yr

Installment Buying and Finance Charges

Installment buying: paying part (down payment) of the total price at the time that the item is purchased. The rest (unpaid balance) is paid in specified monthly payments.

Finance charge: the interest paid to the store for delaying total payment of the item purchased.

Geofrey's portable disc player costs $210. He will pay $30 down and $16.50 a month for 12 months. How much will he pay in all? What will be the finance charge?

To find the total cost of installment buying:

Total Cost = Down Payment + Total Monthly Payments

Down Payment	⟶	$ 30.00
Monthly Payments	⟶ 12 × $16.50 =	+$198.00
		$228.00 Total Cost

To find the finance charge:

Finance Charge = Total Cost − Original Price

Total Cost	⟶	$228.00
Original Price	⟶	−$210.00
		$ 18.00 Finance Charge

Find the finance charge.

	Price	Down Payment	Monthly Payment	Number of Months
15.	$ 200	$ 20	$10	20
16.	$ 50	$ 5	$ 5	12
17.	$ 135	0	$ 9	18
18.	$1000	$200	$25	36
19.	$1500	10%	$90	20

Solve.

20. Which is cheaper and by how much: buying a set of speakers for $1200 on the installment plan, paying 25% down and $85.50 a month for a year, or borrowing $900 from the bank at 15% simple interest for a year?

21. Who should not sign an installment plan agreement because the monthly payments are more than 20% of his/her monthly earnings?

	Earnings	Monthly Payment
Jamal:	$ 500 mo	$100
Vladimir:	$ 1800 mo	$320
Sue Li:	$12,000 yr	$ 65
Caitlin:	$30,744 yr	$ 40
Joyce:	$ 9780 yr	$440

14-6 Comparing Unit Prices

Unit price: the cost per unit of measure. Knowing the unit price enables you to determine the best buy per unit of measure.

59¢ an ounce ⟶ 59¢ : 1 ⟶ $\dfrac{\$.59}{1}$

$1.19 a package ⟶ $1.19 : 1 ⟶ $\dfrac{\$1.19}{1}$

> A unit price is a ratio.

Which is less expensive:

Happy Hound Hash at 5 cans for $2 or
Diggidy Dog Dinner at 4 cans for $1.64?

To compare the cost of two items:

- Find the unit price of each item by setting up and solving a proportion.
- Then compare the two unit prices.

$\dfrac{5 \text{ cans}}{\$2 \text{ cost}} \times \dfrac{1 \text{ can}}{x \text{ cost}}$

$5x = 2$

$x = \dfrac{2}{5}$ or $.40 per can

Happy Hound Hash

$\dfrac{4 \text{ cans}}{\$1.64 \text{ cost}} \times \dfrac{1 \text{ can}}{y \text{ cost}}$

$4y = 1.64$

$y = \dfrac{1.64}{4}$ or $.41 per can

Diggidy Dog Dinner

Happy Hound Hash at $.40 per can is less expensive by one cent per can.

Which item is less expensive?

1. Party Favors: 10 for $1.50 or 30 for $4.45

2. Tomato Sauce: 8 oz for 69¢ or 16 oz for $1.39

3. Fruit Punch: 2 liters for 79¢ or 3 liters for $1.19

4. Toothbrushes: package of 6 for $7.98 or package of 8 for $10

5. Fresh Cherries: 1 pint for 99¢ or 1 quart for $1.99

6. Detergent: 12 oz for $1.75 or 16 oz for $2.29

7. Napkins: 100 count for 99¢ or 145 count for $1.10

8. Pork Chops: package of 8 for $9.99 or package of 4 for $4.99

9. Chicken Broth: 1 16-oz can for 99¢ or 3 6-oz cans for $1

10. Mushrooms: 3 10-oz packages for $4.15 or 2 16-oz packages for $4.49

11. Mouthwash: 2 12-oz bottles for $3.98 or 1 48-oz bottle for $7

12. Apple Cider: 3 15-oz bottles for $3.05 or 1 48-oz bottle for $3.35

13. Crew Socks: 4 pairs for $4.79 or 6 pairs for $6.99

14. Wrapping Paper: 60 sq yd for 99¢ or 200 sq yd for $3.19

**Use a proportion to find the price per pound
to the nearest cent.** (Remember: 1 pound = 16 ounces.)

15. 12-oz jar of peanuts for $2.88

16. 5.4-oz box of fruit snacks for $1.89

17. 2-oz package of cookies for $.75

18. 6-oz can of tuna for $1.09

19. 4.25-oz box of crackers for $1.70

Finding Together

20. Plan a birthday party for 24. Use the chart to help you be a smart shopper. Use a newspaper to find two different prices for each item. Then determine the unit price and compare. Find out how much money you can save by buying a less expensive item.

Birthday Party Expense Sheet			
Item	**Brand/Differences**	**Price**	**Savings (if any)**
Paper Plates	plain		
	decorated		
Napkins	180 count		
	300 count		
Balloons	package of 25		
	25 helium-filled		
Streamers	plain		
	with glitter		
Pizza	plain		
	vegetable		
Snack Bags	8 oz		
	14 oz		
Frozen Yogurt	quart		
	half gallon		
Fruit Juice	24 oz		
	32 oz		
Birthday Cake	1 sheet cake		
	2 dozen cupcakes		

14-7 Checking Accounts

Checking account: one type of bank account.

Check: tells the bank to take the stated amount of money from the account and deposit it into the account of the person or company named on the check (payee).

A check is safer than cash because the money is withdrawn from the account only when the payee has endorsed or signed the back of the check and the bank verifies the signature.

Janice Jones bought a gift for her mother at Fifi's Flower Shop. She paid $30.03, using check No. 1248. This is her check.

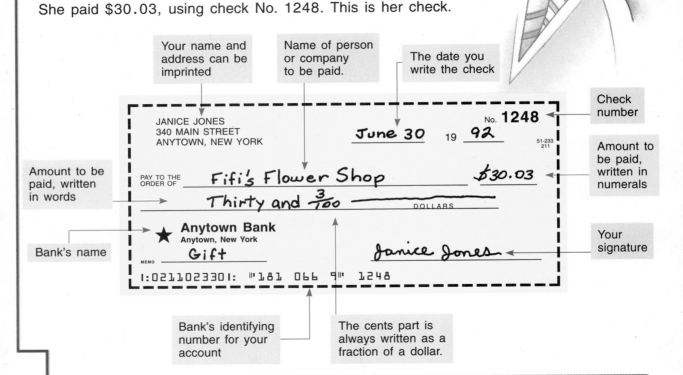

Your name and address can be imprinted

Name of person or company to be paid.

The date you write the check

Check number

Amount to be paid, written in words

Amount to be paid, written in numerals

Bank's name

Your signature

Bank's identifying number for your account

The cents part is always written as a fraction of a dollar.

JANICE JONES
340 MAIN STREET
ANYTOWN, NEW YORK

No. **1248**

June 30   19 92

51-233
211

PAY TO THE ORDER OF   Fifi's Flower Shop $30.03

Thirty and ³⁄₁₀₀   DOLLARS

★ **Anytown Bank**
Anytown, New York

Gift

Janice Jones

MEMO

I:0211023301: "181 066 9" 1248

Write each in proper word form for a check.

1. $56.85
2. $7.75
3. $33.00
4. $129.03

Write out a check for each of the following. Sign your own name.

5. $502.41 for a sofabed purchased from Miller's Department Store
6. $32.67 for jeans from John's Sportswear Store
7. $12.00 for baseball tickets from your school
8. $46.20 for monthly phone bill from your phone company
9. $72.90 for food from your local supermarket

Every checkbook has an **account record**. This is used to keep track of all money taken out of and put into the checking account.

This is a sample account record for Janice Jones.

		PLEASE BE SURE TO DEDUCT CHARGES THAT AFFECT YOUR ACCOUNT						BALANCE FORWARD	
NO.	DATE	ISSUED TO OR DESCRIPTION OF DEPOSIT	PAYMENT (-)				DEPOSIT (+)	158	95
1248	6-30-92	TO Fifi's Flower Shop	30	03				-30	03
		FOR Gift						128	92
	7-6-92	TO Deposit-salary					55 00	+55	00
		FOR June babysitting						183	92
1249	7-17-92	TO Town Telephone Co.	30	91				-30	91
		FOR telephone bill						153	01
	7-23-92	TO Deposit					100 00	+100	00
		FOR birthday money						253	01

Study Janice Jones' checking account record to answer these.

10. How much did she have left in her account after she wrote check no. 1248?

11. When did she deposit $55?

12. What was her balance after depositing $55?

13. To whom was check no. 1249 written?

14. What was the balance after check no. 1249 was paid?

15. **Copy and complete.** Replace the question marks with the correct answers.

		PLEASE BE SURE TO DEDUCT CHARGES THAT AFFECT YOUR ACCOUNT						BALANCE FORWARD	
NO.	DATE	ISSUED TO OR DESCRIPTION OF DEPOSIT	PAYMENT (-)				DEPOSIT (+)	253	01
1250	7-26-92	TO Music Time	29	49					
		FOR CD's						BAL ?	
1251	7-25-92	TO Sound Shoe Store	45	95					
		FOR sneakers						BAL 177	57
	7-27-92	TO Deposit					30 00		
		FOR odd jobs						BAL ?	
1252	8-10-92	TO Town Telephone Co.	?						
		FOR telephone bill						BAL 179	41
1253	8-17-92	TO Party Palace	19	69					
		FOR decorations						BAL ?	

16. You have won a raffle for $500 worth of merchandise that can be spent at: Sam's Record Shop, Sneaker Palace, Jim's Jean Store, or Allen's Department Store. The $500 is deposited in your checking account and must be spent in 90 days. Write checks and a record of your withdrawals and balance. Do not overdraw your account.

14-8 Budgets and Circle Graphs

Budget: a plan to spend money wisely.

One way to display a budget is with a circle graph.

This circle graph shows how a $4000 monthly income is to be spent.

To find the amount of money spent on an item, **find the percentage**.

How much is spent on education?

6% of $4000 = ?

$$\frac{6}{100} \times \frac{4000}{1} = ?$$

6 × $40 = $240

$240 is spent on education.

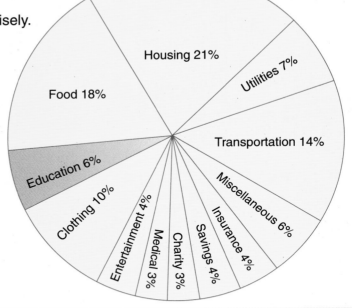

$4000 Monthly Income
The sum of all the percents in the graph is 100%.

Food 18% Housing 21% Utilities 7% Transportation 14% Education 6% Clothing 10% Entertainment 4% Medical 3% Charity 3% Savings 4% Insurance 4% Miscellaneous 6%

How much of the monthly income of $4000 is spent on each item?

1. Charity
2. Utilities
3. Transportation
4. Housing
5. Savings
6. Food
7. Clothing and Entertainment
8. Miscellaneous and Transportation

Copy and complete. (The first one is done.)

Budget of Birthday Money

	Use	Amount Spent	Percent
9.	Food	$28 → $\frac{28}{200}$ = $\frac{14}{100}$	14%
10.	Savings	$90	?
11.	Clothing	$30	?
12.	Jewelry	$40	?
13.	Miscellaneous	$12	?
	Total	$200	100%

Solve.

King Mall has an operating budget of $5,000,000 per year. Use the circle graph to find how much money is spent on each item.

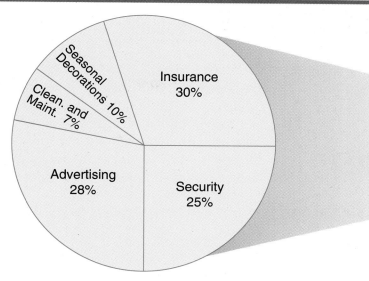

14. Seasonal Decorations

15. Cleaning and Maintenance

16. Advertising

17. Insurance

18. Security

The Art Department at the Bell School budgeted $3000 to buy supplies for all its classes. How much money was spent on each class?

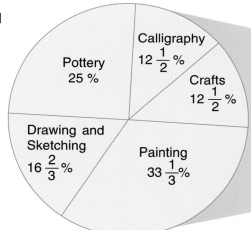

19. Pottery

20. Crafts

21. Drawing and Sketching

22. Painting

23. Calligraphy

You are given $1,000,000 to spend. Use the circle graph to determine how much money is to be spent on an item.

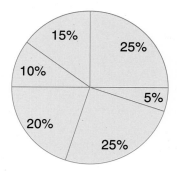

24. List 6 possible ways you might spend the money.

SKILLS TO REMEMBER

Solve for v.

25. $3v + 5 = 70$

26. $7v - 10 = 32$

27. $\dfrac{v}{5} - 3 = 3$

28. $\dfrac{2}{3}v + 1 = 9$

29. $^-2v + {}^+3 = {}^-1$

30. $\dfrac{v}{-4} + {}^-2 = {}^+1$

14-9 Problem Solving: Multi-Step Problems Using Formulas and Proportion

Problem: If Moya orders 3 or more books from a paperback book club, she can take either a 5% discount or $5 off the total cost of the order. Which money-saving option should she choose if she decides to order 2 books at $5.95 each, 2 books at $6.95 each, and one book for $7.50? Why?

5% discount or $5 off the total cost

1 IMAGINE Draw and label your options.

	Option One	Option Two
	5% Discount	$5 Discount
2@5.95 =		
2@6.95 =		
1@7.50 =		

2 NAME

Facts: Two Options: 5% or $5 off on 3 or more books

Prices: 2 books at $5.95 each
2 books at $6.95 each
1 book at $7.50

Question: Which is the better option? Why?

3 THINK Compute the total cost of the books for each option.

Option One: Use the formula to find the amount of money that will be saved using the 5% discount.

Option Two: Now compute the cost with the $5 discount.

Compare the two totals to find which is the better offer.

4 COMPUTE To find the total list price of the five books:

$$\underbrace{(2 \times \$5.95)}_{\$11.90} + \underbrace{(2 \times \$6.95)}_{\$13.90} + \underbrace{(1 \times \$7.50)}_{\$7.50} = \underline{\quad ? \quad}$$

$11.90 + $13.90 + $7.50 · = $33.30 Total List Price

Option One: To find the discount:

$D = LP \times R$ of $D \longrightarrow$ $\underline{\quad ? \quad} = \$33.30 \times 5\%$
 $\underline{\quad ? \quad} = \$33.30 \times 0.05$
Discount $= \$1.665 = \$1.67$ ·

$TC = LP - D \longrightarrow \$33.30 - \$1.67 = \$31.63 \longleftarrow$ Total Cost of Option One

Option Two: Total List Price $- \$5.00 = \underline{\quad ? \quad}$
$\$33.30 - \$5.00 = \$28.30 \longleftarrow$ Total Cost of Option Two

Clearly, Option Two, the $5 off, is the better buy.

5 CHECK Check your computations. You can also solve this problem using proportion. Compare the solution you find using proportion with the solution you found by using a formula.

Solve using formulas or proportions.

1. Emma Jiles sells magazines. She receives an 8.5% commission on sales, plus an additional $50 for any month in which her sales exceed $400. If last month she had sales totaling $420, how much money did she make?

1 IMAGINE	Draw and label a picture of these sales.

2 NAME	*Facts:*	Commission:	8.5%
		Bonus:	$50 when sales > $400
		Last month:	$420

Question: Total earnings: ?

3 THINK	Find the commission on $420 in sales, using the formula $C = TS \times R$ of C

Then add the $50 bonus to the commission to find her total earnings.

4 COMPUTE ⟶ **5 CHECK**

Total Sales: $420

2. Lori bought 6 blankets and 4 pillows. The blankets, regularly $30 each, were marked down 40%. The pillows, regularly $12 each, were discounted 50%. There was also a 6% sales tax. If Lori paid with three 50-dollar bills, how much change did she receive?

3. A Milwaukee agent bought 48,000 bushels of wheat for an eastern firm at $2.10 a bushel. What was the entire cost to the eastern firm if the agent charged 3% of the cost of the wheat for his services?

4. Mrs. O'Lin began a job at a starting salary of $15,500. Melody's job started at $12,800. Melody also received a 10% increase each year. Mrs. O'Lin received a raise of $870 each year. In 5 years' time who made more money? How much more?

5. Divida, the 2 and 5 divisor gremlin, flits about dividing numbers. How many numbers between 1 and 200 inclusively are not divisible by 2 and/or by 5?

6. Mr. Tanner spent $28.50 for movie tickets for his family. The tickets cost $5.50 for each adult and $3.00 for each child. If three adults went to the movie, how many children went?

7. Art-O Stationery sells calligraphy sets listed at $2.40 for $2.25, while Rainbow Art Supply sells sets listed at $2.50 for $2.30. Which store has the better rate of discount?

8. Pianoville has a 6% sales tax. In that city a store lists a stereo at $375 with a 20% discount. In a store in Cleftown, which has a 7% sales tax, the same stereo is listed at $369 with an 18% discount. In which town does the stereo cost less?

14-10 Problem Solving: Applications

Solve. Round each to the nearest cent. (The first one is done.)

1. Gus bought a VCR at $\frac{1}{4}$ off. If the sale price was $269.70, what was the original price?

Subtract to find the original sale price.

Original $-$ R of D = R of SP

$$100\% - \frac{1}{4} = \frac{3}{4}$$

Solve, using this formula:

$$SP \div R \text{ of } SP = LP$$

$$\$269.70 \div \frac{3}{4} = LP$$

$$\$269.70 \times \frac{4}{3} = \$359.60 \text{ List Price}$$

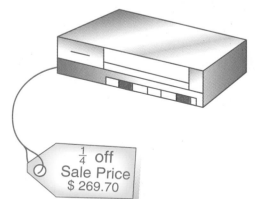

$\frac{1}{4}$ off
Sale Price
$ 269.70

2. Ms. Gomez bought a cordless phone on sale for only $87.98. If this was $\frac{1}{3}$ off the list price, what was the list price of the phone?

USE THESE STRATEGIES:
Use Simpler Numbers
Use a Formula
Use a Graph
Hidden Information
Multi-Step Problem

3. What percent did Arlene save when she paid $6.21 for cosmetics regularly priced at $6.75?

4. Pablo saved $9.99 when he bought an aquarium regularly listed at $29.97. What percent did he save?

5. The coach bought 3 sets of weights and saved a total of $24.30, or 27%. How much had each set of weights originally cost?

6. Todd bought 2 videos for 15% off. If he saved $1.98 and each video cost the same amount, what was the sale price of each video?

7. Laura bought a set of baseball cards regularly priced at $24.45 for $1.25 off. What was the total cost if she paid an 8% sales tax?

8. A clock radio listed at $24.45 is sold for $2.75 less. If a 6% sales tax is added to the discounted price, what would be the total cost of the radio?

9. A real estate agent charges a 7% commission for the sale of a $115,000 town house. What is the total cost of the town house?

10. An agent adds an $8\frac{1}{3}$% commission on a $25,000 shipment of corn. What is the total cost of the shipment to the buyer?

11. A supervisor receives a $4\frac{1}{2}$% commission for selling office products. What is the commission from selling a word processor for $449.99, an answering machine for $159.99, a speaker phone for $199.99, a typewriter for $349.95, and a fax machine for $429.95?

12. Holly sells computers and software and receives a 6.25% commission on all sales above $250. She sold a $1299 computer, 10 boxes of computer disks for $9.95 each, a printer for $195, and 3 software programs for $49.50 each. What commission did she earn?

13. Clarence's $1600 certificate will earn $6\frac{1}{4}$% interest. How much interest will he earn in 2 years 6 months?

14. Phil deposited $1800 for 1 year 4 months at $5\frac{1}{3}$% interest. What is his total in the bank?

15. Troy borrowed $1200 to buy an electric keyboard. If the bank charged $9\frac{1}{3}$% interest, how much money had he paid back after 2 years 3 months?

16. Ms. Link earned $72 on a sale of carpeting. If her rate of commission is 3%, how much was the sale? (Hint: Use $C \div R$ of $C = TS$.)

17. A travel agent earned $133 in commission on a trip she arranged. If the agent makes $4\frac{3}{4}$% commission, what was the cost of the trip?

18. Howard paid a $29 down payment on a 10-speed bicycle that cost $89.75. He then paid three monthly payments of $25. How much could he have saved by paying cash for the bicycle?

19. Jan bought a 5-piece set of luggage that cost $189.20. She made a 20% down payment and paid 8 monthly payments of $20 each. How much more did Jan spend on the luggage by paying in installments than by paying the full price upon purchase?

Sumita collected $150. The circle graph shows the sources of her income. How much did she receive for the following?

20. Babysitting

21. Errands

22. Birthday

23. Allowance

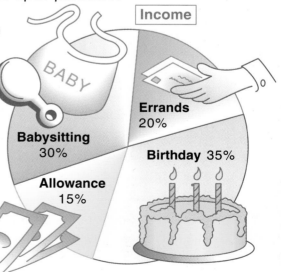

Income

Errands 20%

Babysitting 30%

Birthday 35%

Allowance 15%

24. Which is the better buy: a 6-oz box of rice for $1.68 or a 5-lb bag for $4.68?

25. Which is the better buy: a 6-pack of yogurt weighing 24.6 oz for $1.99 or four 6-oz containers of yogurt for $1.89?

More Practice

Copy and complete.

1. List Price: $104
    Rate of Discount: $12\frac{1}{2}$ %
    Discount: _?_
    Sale Price: _?_

2. List Price: $330
    Rate of Discount: _?_
    Discount: $110
    Sale Price: _?_

3. Total Sales: $540
    Rate of Commission: $5\frac{1}{2}$ %
    Commission: _?_

4. Total Sales: $120
    Rate of Commission: _?_
    Commission: $9.30

Find the sales tax and the total cost. (Round to the nearest cent.)

5. 9% sales tax on a $850 video camera

6. $7\frac{1}{2}$ % sales tax on a $1100 sailboat

7. 5% sales tax on a $20.00 clock radio discounted at 10%

8. 6% sales tax on a $450 washing machine discounted at 15%

Solve.

9. The Lesters bought kitchen appliances for $4600. They paid
    $600 cash and borrowed the rest at $12\frac{3}{4}$ % interest for 3 years.
    At the end of the 3 years how much had they paid?

10. On his seventh birthday Brad's grandmother put $1500 for him
    in a savings account at 6% simple interest. How much was in
    the account if he withdrew it on the following birthdays:
    Twelfth? fifteenth? eigtheenth?

The circle graph shows how Sarah budgets her money
each week. If she spends $36 per week, how much
does she spend on each item?

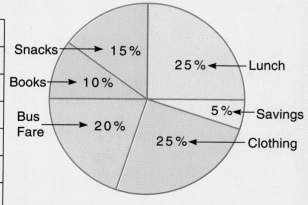

	Percent	Use	Weekly Expenses
11.	20%	Bus Fare	$
12.	25%	Lunch	
13.	15%	Snacks	
14.	25%	Clothing	
15.	5%	Savings	
16.	10%	Books	

Math Probe

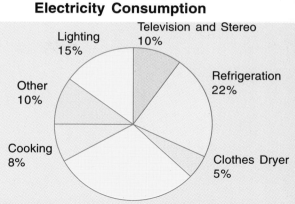

Electricity Consumption

Lighting 15%
Television and Stereo 10%
Other 10%
Refrigeration 22%
Cooking 8%
Clothes Dryer 5%
Heat: Space and Water 30%

BECOME AN ENERGY EXPERT!

This circle graph shows how the Watts family consumed electricity in their home last month.

At the beginning of the month the electric meter read 72 911 kilowatt hours. At the end of the month it read 74 969 kilowatt hours.

The cost of electricity is 8.5¢ per kilowatt hour.

Use the circle graph and the information from the meter to complete exercises 1–9.

1. How many kilowatt hours of electricity did the Watts family consume last month?

2. At 8.5¢ per kilowatt hour, what was the cost of their electricity consumption for the month?

3. What item used the most electricity? How much money did that item alone cost?

4. How might the Watts family save money in the category that consumes the most electricity?

5. What was the cost for the month of each of the following?
   a. refrigeration   b. television and stereo   c. lighting
   d. cooking   e. clothes dryer

6. What electrical appliances might fit under the category "Other"?

7. If the Watts family decides to convert to fluorescent lighting, which reduces energy consumption in that category by about 2%, how much money can they expect to save each month?

8. By turning down the temperature of their hot-water heater, the Watts family saved $9.01. How many kilowatt hours does this savings represent?

9. If the Watts family had turned back their thermostat and saved 148 kilowatt hours, how much money would they have saved last month?

Study your electric or gas meter. Find out how many kilowatt hours of electricity your family consumes. Does the consumption of electricity remain constant or vary from month to month? List five things your family can do to reduce your electricity consumption. Write to the gas and electric company that serves your home for information on energy use.

Check Your Mastery

Find the missing information.

See pp. 338–339

1.	Original price	$80		2.	Original price	$70
	Rate of discount	4%			Rate of discount	?
	Discount	?			Discount	$56
	Sale price	?			Sale price	?

Find the total cost if a 5% sales tax is added to each.

See pp. 340–341

3. $8   4. $114   5. $2450

Find the missing information.

See pp. 342–343

6.	Total sales	$1170		7.	Total sales	?
	Rate of commission	8%			Rate of commission	15%
	Commission	?			Commission	$225

Find the simple interest and the new total in the bank.

See pp. 344–345

8. Deposit of $800 at $5\frac{1}{4}$% for 1 year

9. Deposit of $1100 at 8% for 2 years, 6 months

Find the amount to be paid back on each loan.

See pp. 346–347

10.	Principal	$2000		11.	Principal	$550
	Rate	9%			Rate	7%
	Time	3 yr			Time	6 months

12. **Which is a better buy:**
8 apples for $1.89 or a dozen apples for $2.80?

See pp. 348–349

Write each in proper word form for a check.

See pp. 350–353

13. $30.29   14. $209.08   15. $540   16. $2010.01

Solve.

17. The balance in a checkbook is $1067.12. After a deposit of $345.00 and a withdrawal of $189.90, what should the new balance read?

18. The Fox family budgets 9% of their monthly income for savings and 12% for utilities. Their monthly income is $6100. How much more money is spent on utilities than is saved?

Cumulative Review

Choose the correct answer.

1. Write $\frac{3}{5}$ as a percent.
   - **a.** 20%
   - **b.** 55%
   - **c.** 60%
   - **d.** 75%

2. What is 0.45 as a percent?
   - **a.** $4\frac{1}{2}$%
   - **b.** 45%
   - **c.** 4.5%
   - **d.** 450%

3. What is the amount of sales tax on a $30 jacket if the sales tax is 5%?
   - **a.** $1.50
   - **b.** $1.75
   - **c.** $2.50
   - **d.** $3.05

4. A set of dishes marked for a 10% discount originally cost $45. What was the sale price?
   - **a.** $43.50
   - **b.** $35.00
   - **c.** $49.50
   - **d.** $40.50

5. $2\frac{1}{2}$ can be expressed as a decimal (2.5) and a percent. What is the percent?
   - **a.** $2\frac{1}{2}$%
   - **b.** 25%
   - **c.** 250%
   - **d.** 225%

6. How much would Gail save by buying a dozen pencils for $1.25 rather than 12 pencils at $.15 each?
   - **a.** $1.80
   - **b.** $.75
   - **c.** $.65
   - **d.** $.55

7. If 80% of the visitors to the art museum are 420 people, how many people visited the museum?
   - **a.** 420
   - **b.** 525
   - **c.** 600
   - **d.** 625

8. How much would you save by buying a lunch special for $3.75 rather than ordering three items separately for $2.49, $1.10, $.89?
   - **a.** $1.05
   - **b.** $.73
   - **c.** $.80
   - **d.** $.85

9. 80% of the choral group went on a field trip. This was $12\frac{1}{2}$% more than was expected. What percent shows how many were expected?
   - **a.** 67.5%
   - **b.** 0.675%
   - **c.** 6.75%
   - **d.** 675%

10. Estimate to find the answer. At a baseball game, about 43% of 26,400 attending were under 18 years of age. About how many were in this age group?
    - **a.** 10,000
    - **b.** 15,000
    - **c.** 20,000
    - **d.** 30,000

11. 0.02 is what percent of 0.0001?
    - **a.** 200%
    - **b.** 500%
    - **c.** 2000%
    - **d.** 20,000%

12. 6.003 ÷ 2.9 equals what percent?
    - **a.** 20.7%
    - **b.** 2.07%
    - **c.** 207%
    - **d.** 0.207%

13. Find the simple interest for a principal of $250 for a 2-year period at a 6% rate of interest.
    - **a.** $50
    - **b.** $25
    - **c.** $30
    - **d.** $40

14. Find the rate of commission if the total was $926 and the commission was $55.56.
    - **a.** 8%
    - **b.** 4%
    - **c.** 6%
    - **d.** 9%

Complete.

	Percent	Ratio to 100	Fraction in Lowest Terms	Decimal
15.	30%	?	?	?
16.	120%	?	?	?
17.	?	?	?	0.025
18.	?	?	?	$0.08\frac{1}{3}$

	Cost	Discount	Sales Tax	Total Cost
19.	$3736.00	No	4%	?
20.	$200.00	10%	$7\frac{1}{2}$%	?
21.	$512.50	30%	No	?
22.	$115.00	25%	6%	?

	Principal	Rate of Interest	Time	Interest
23.	$4000	6%	36 months	?
24.	$5150	$8\frac{1}{8}$%	9 months	?

Solve.

25. Of the 250 teachers at a conference, 40 were middle school teachers. What percent taught middle schools?

26. A car in a dealer's showroom was marked $13,590. When the car was moved to the lot, it was marked for a 15% discount. What was the cost of the car on the lot?

27. Mr. Hernandez sells educational software. His commission for last month's sales was $630. If he received a $14\frac{1}{2}$ % commission, what was the total amount of his sales last month?

28. Carla's monthly income is $3600. If she spends 15% for fuel and 35% for rent, how much money remains for other costs?

29. Ms. Antonio put $3000 in a savings account. If the principal remains in the bank for 2 years at a rate of 5.5%, how much will she have?

30. The owners of Aqua Park bought a new boat for $35,000. They paid 30% down and financed the balance at a rate of 15% for 36 months. What was the monthly finance charge?

+10 BONUS

31. Imagine that you have raised $550,000 to produce a low-budget movie. Make a circle graph to show what percentage of your total budget you will allow for various costs; for example, director's and actors' fees, props and scenery, and so on. Now imagine that once the movie is released, it brings in profits triple the total production costs. If you had decided that your salary would be 35% of the profits, what will your salary be?

15
Surface Area and Volume

In this chapter you will:

Identify solid figures having flat and curved surfaces

Compute the *surface area* of: a rectangular prism, cube, triangular prism, cylinder

Compute the *volume* of: a rectangular prism, cube, triangular prism, cylinder

Solve problems: combining strategies

Use technology: spreadsheets

Do you remember?

A solid figure can have *faces*, *bases*, *edges*, and *vertices*.

face any of the polygons that make up the figure

vertex a point of intersection of three or more edges

edge a line segment common to two faces

base the face on which the figure rests

Architecturally Speaking

Every architect tries to be an artist creating "a thing of beauty" as well as places that make our work, play, and daily living as easy as possible. Study the buildings in your neighborhood or a picture of a famous building. Choose one in which the architect used many different types of space figures. Use models or nets of space figures to recreate the building. Then write about the geometric figures you used.

15-1 Space Figures

Polyhedrons: solid figures whose faces are polygons.

Prism: a polyhedron with two parallel and congruent bases. The shape of the base names the prism. The other faces are rectangles.

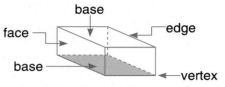

Rectangular Prism

Square Prism (Cube)

Triangular Prism

Pyramid: a polyhedron whose faces are triangles. It has a single base, which may be a triangle or other polygon.

Rectangular Pyramid

Square Pyramid

Triangular Pyramid

Some solid figures have *curved* surfaces.

Cylinder

Cone

Sphere

Find examples of items in school and at home that are shaped like these polyhedrons.

1. rectangular prism 2. cube 3. triangular prism 4. pyramid

Name items that are shaped like each of these curved figures.

5. cylinder 6. cone 7. sphere

Copy and complete the table. Look for patterns.

	Prism	Number of Faces	Number of Vertices	Number of Edges
8.	triangular	5		
9.	square		8	
10.	rectangular			12
11.	pentagonal	7		
12.	hexagonal		12	

13. How are prisms like pyramids? How are they different?
14. What is the difference between a rectangular prism and a triangular prism?
15. How many faces does a rectangular pyramid have?
16. How many faces does a pentagonal prism have?
17. Why is a cone *not* called a polyhedron?
18. How does a sphere differ from other solid figures?
19. Imagine a line dividing a sphere in half. What would be a name for each half?

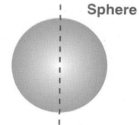
Sphere

20. Use this formula to show that the figures in exercises 8–12 are polyhedrons.
Faces + Vertices = Edges + 2 ⟶ (F + V = E + 2)

Nets

A **net** is a flat pattern for a solid figure. When folded, a *net* forms a space figure. Here are nets for a square prism and cylinder.

Square Prism Net

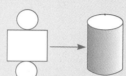

Cylinder Net

Make nets like these. Into what figure would each of these nets fold?

21.

22.

23.

CHALLENGE

Imagine how each of these figures would look. Use a ruler to draw them.

24. pentagonal prism
25. hexagonal prism
26. pentagonal pyramid
27. hexagonal pyramid

SKILLS TO REMEMBER

Compute.

28. 20^2
29. $(1.2)^2$
30. $(3\frac{1}{3})^2$
31. $(1.05)^2$
32. 7^3
33. 30^3
34. $(1.5)^3$
35. $(2\frac{1}{2})^3$

365

15-2 Surface Area of a Rectangular Prism and a Cube

Surface area (S) of a rectangular prism or **a cube:** the **sum** of the areas of all **six** faces.

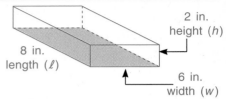

8 in. length (ℓ)
2 in. height (h)
6 in. width (w)

▶ **To find the surface area (S) of a *rectangular prism*:**

- Find the area of one of each of the parallel faces.

$A = \ell w$	$A = \ell h$	$A = wh$
$A = 8 \times 6$	$A = 8 \times 2$	$A = 6 \times 2$
$A = 48$ in.^{2}	$A = 16$ in.^{2}	$A = 12$ in.^{2}

- Double each area. (Each face has a congruent face opposite it.)

  $2 \times 48 = 96$ in.^{2} $2 \times 16 = 32$ in.^{2} $2 \times 12 = 24$ in.^{2}

- Find the sum of the areas. The sum is the surface area (S).

  $S = 96 + 32 + 24$

  $S = 152$ in.^{2}

The formula for the surface area of a rectangular prism is:

$$S = 2(\ell w + \ell h + wh)$$

▶ **To find the surface area (S) of a cube** that measures 9 cm on an edge (e):

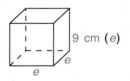

9 cm (e)
e
e

- Find the area of one of the faces.

  $A = e \times e = e^2$
  $A = 9 \times 9 = 81$ cm^2

- Multiply the area by 6 since all six faces are congruent.

  $S = 6 \times 81$
  $S = 486$ cm^2

The formula for the surface area of a cube is:

$$S = 6e^2$$

Use this net to find the surface area of this prism.

1. Name the parallel faces.
2. Find the area of the face labeled 1.
3. Find the area of the face labeled 2.

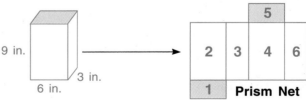

9 in.
3 in.
6 in.

5
2 3 4 6
1   **Prism Net**

4. Find the area of the face labeled 3.
5. What is the sum of the areas of faces 1, 2, and 3?
6. What is the sum of the areas of faces 4, 5, and 6?
7. What is the surface area of all six faces?
8. Draw a net for a cube that measures 3 units on an edge. Then find its surface area.

Find the surface area of each rectangular prism.

9.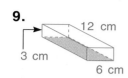
12 cm
3 cm
6 cm

10.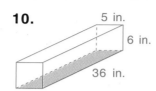
5 in.
6 in.
36 in.

11.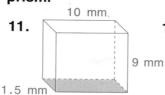
10 mm.
9 mm
1.5 mm

12.
4 ft
20 ft
12 ft

Find the surface area of each prism with the following measurements:

13. $\ell = 2.2$ m
$w = 1$ m
$h = 4.5$ m

14. $\ell = 10$ cm
$w = 2.5$ cm
$h = 14$ cm

15. $\ell = 3\frac{1}{2}$ in.
$w = 2$ in.
$h = 5$ in.

16. $\ell = 30$ in.
$w = 14$ in.
$h = 2\frac{1}{2}$ in.

Find the surface area of a cube with an edge of:

17. 16 in. **18.** 8 ft **19.** 6.5 m **20.** 3.2 cm **21.** $9\frac{1}{2}$ yd **22.** $10\frac{1}{2}$ ft

Solve. (Hint: The area of all 6 faces is not always needed.)

23. Find the surface area of the sides and bottom of a rectangular prism having a length of 7 inches, a width of 4 inches, and a height of 15 inches.

24. Mr. Santos is paneling the four walls and ceiling of a closet with cedar. How many square feet of cedar will he use if the closet has a length of 5 feet, a width of 3 feet, and a height of 10 feet?

25. Yolanda is lining the bottom and sides of a drawer with paper. How many square inches of paper does she need if the drawer is 12 inches wide, $3\frac{1}{2}$ inches deep, and 30 inches long?

26. Brenda covers a photo cube measuring 3.5 cm on an edge with contact paper. How many square centimeters of contact paper does she use?

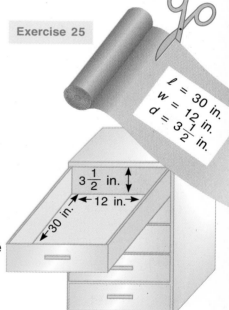

Exercise 25

$\ell = 30$ in.
$w = 12$ in.
$d = 3\frac{1}{2}$ in.

$3\frac{1}{2}$ in.
12 in.
30 in.

 CALCULATOR **ACTIVITY**

To find the surface area of a cube measuring 10 cm on an edge, use this key sequence:

 × × = OR   × x² =

Determine the key sequence needed to find the edge of a cube when the surface area is given. Tell the order of the key sequence that you can use to find the edge. Then find the edge for each of the following cubes:

27. $S = 96$ cm^2 **28.** $S = 150$ ft^2 **29.** $S = 384$ in.^{2} **30.** $S = 486$ mm^2

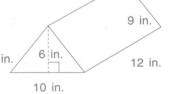

Surface area (S) of a triangular prism: the sum of the areas of its two triangular and three rectangular faces.

To find the surface area (S) of a triangular prism:

- Find the area of one of the triangular bases.

$$A = \frac{1}{2} bh$$

$$A = \frac{1}{2} (10) (6) \longrightarrow A = \frac{1}{2} (60) \longrightarrow A = 30 \text{ in.}^2$$

- Double the area because there are two bases.

$$2 \times 30 = 60 \text{ in.}^2$$

- Find the area of each of the rectangular faces.

$A = \ell w$		$A = \ell w$		$A = \ell w$
$A = 12 \times 9$	+	$A = 12 \times 8$	+	$A = 12 \times 10$
$A = 108 \text{ in.}^2$		$A = 96 \text{ in.}^2$		$A = 120 \text{ in.}^2$

- Find the sum of the areas. The sum is the surface area (S).

$$S = 60 + 108 + 96 + 120$$

$$S = 384 \text{ in.}^2$$

To find the surface area (S) of a right triangular prism:

Follow the same steps. (Remember that since the bases are right triangles, a side of the triangle serves as the *height*.)

Remember: Use these formulas to find the *surface area of a triangular prism:*

$$S = \left\{ \begin{array}{c} A = \frac{1}{2} bh \\ \text{(for the area of each of the two triangular faces)} \end{array} + \begin{array}{c} A = \ell w \\ \text{(for the area of each of the three rectangular faces)} \end{array} \right.$$

Then **add** to find the sum of the areas of all five faces.

Use this net to find the surface area of this prism.

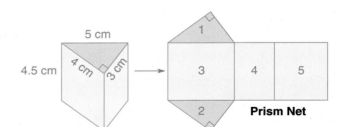

Prism Net

1. How many rectangular faces are there? Name them.

2. How many triangular faces are there? Name them.

3. Name any congruent faces.

4. Find the area of each rectangle.

5. Find the area of the triangles.

6. Find the sum of the areas of all five faces.

7. What is the total surface area of the prism?

Find the surface area of each triangular prism.

8.

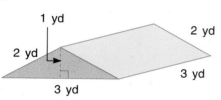
1 yd
2 yd
2 yd
3 yd
3 yd

9.

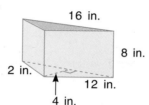
16 in.
8 in.
2 in.
12 in.
4 in.

10.

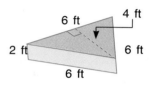
6 ft
4 ft
2 ft
6 ft
6 ft

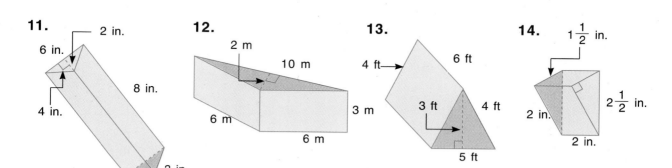

11.
2 in.
6 in.
8 in.
4 in.
3 in.

12.
2 m
10 m
6 m
6 m
3 m

13.
4 ft
6 ft
3 ft
4 ft
5 ft

14.
$1\frac{1}{2}$ in.
2 in.
$2\frac{1}{2}$ in.
2 in.

Solve. (Hint: The area of all 5 faces is not always needed.)

15. The club made a tent shaped like a triangular prism. It was 8 ft long, 3 ft wide, and 6 ft tall at its highest point. If the sloping sides measured $6\frac{1}{2}$ ft from peak to ground and the tent did not have a floor, how many square feet of material did the club use? (Draw a picture.)

16. Find the surface area of the walls and ceiling of a greenhouse that is shaped like a triangular prism on top of a rectangular prism.

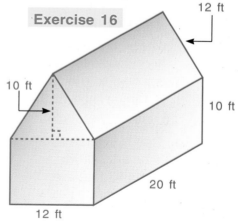
Exercise 16
12 ft
10 ft
10 ft
20 ft
12 ft

CRITICAL THINKING

17. Double the dimensions of the prism in exercise 14. Predict its surface area. Then find the surface area.

18. Double the dimensions of the prism in exercise 17. Predict its surface area. Then find the surface area.

19. What is the effect of doubling the dimensions of a triangular prism on its surface area?

20. Does doubling the dimensions of a rectangular prism have the same effect on its surface area?

Surface area (S) of a cylinder: the sum of the areas of all its surfaces.

To find the surface area (S) of a cylinder:

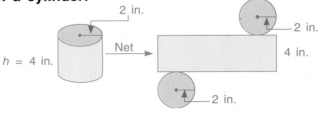

- Imagine how the net of the cylinder would look.

- Find the area of one of the circular bases.

$A = \pi r^2$
$A = \pi \times 2 \times 2$
$A \approx 3.14 \times 2 \times 2$
$A \approx 12.56$ in.^{2}

- Double the area because there are two bases.

$2 \times 12.56 = 25.12$ in.^{2}

- Find the area of the rectangular surface whose height is given and whose length is equal to the circumference of the base of the circle.

First find length of the rectangle.

$C = 2\pi r$
$C = 2 \times \pi \times 2$
$C \approx 2 \times 3.14 \times 2$
$C \approx 12.56$ in.
(length of rectangle)

Now find area of rectangular surface.

$A = \ell w$
$A \approx 12.56 \times 4$
$A \approx 50.24$ in.^{2}

- Now add all the areas to find the surface area (S).

$S \approx \underbrace{\qquad 25.12 \qquad}_{\text{Area of 2 bases}} + \underbrace{\qquad 50.24 \qquad}_{\text{Area of rectangular surface}}$

$S \approx 75.36$ in.^{2}

So, the surface area (S) of the cylinder is approximately equal to 75.36 in.^{2}.

Remember: Use these formulas when finding surface area (S) of a cylinder.

| $A = \pi r^2$ (for the area of each circular base) | Plus | $C = 2\pi r$ (for the length of the rectangle) | $A = \ell w$ (for the area of the rectangle) |

Find the surface area. (Hint: Draw and label a net for each cylinder.)

1. 3 ft, 5 ft

2. 1 ft, 8 ft

3. 25 m, 10 m

4. 6 m, 4 m

Find the surface area. (Remember: The diameter is twice the radius.)

5.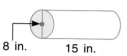
8 in.   15 in.

6.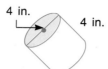
4 in.   4 in.

7. 6 cm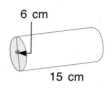
15 cm

8.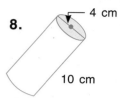
4 cm
10 cm

Expressing Surface Area in π Units

Surface area may be expressed in terms of π units.
Find the surface area of this cylinder.

3 cm
4 cm

Length of rectangle:
$C = 2\pi r$
$C = 2 \times 3 \times \pi$
$C = 6\pi$ (cm)

Area of *TWO* bases:
$A = 2\pi r^2$
$A = 2\pi \times 3 \times 3$
$A = 2 \times 9\pi$
$A = 18\pi$ (cm^2)

$+$

Area of rectangle:
$A = \ell w$
$A = 6\pi \times 4$
$A = 24\pi$ (cm^2)

$=$

Surface area **(S)**:
$S = 18\pi + 24\pi$
$S = 42\pi$ (cm^2)

You can use the key to find the approximate surface area:

$S \approx$ 42 $\times$ π = 131.94689

Find the surface area for each cylinder in terms of π.

9. $d = 2$ ft
$h = 9$ ft

10. $d = 2$ ft
$h = 11$ ft

11. $r = 7$ yd
$h = 7$ yd

12. $r = 3$ yd
$h = 10$ yd

13. $r = 10$ in.
$h = 30.5$ in.

Exercise 14

Solve. Round answers to the nearest tenth.

14. A marble sculpture stand is $4\frac{1}{2}$ feet tall with a
base whose area is 64π in.^{2}. What is the surface
area of the stand? (Hint: Draw a net.)

15. How many square centimeters of aluminum are
needed to make 6 cans, each having a height of
10 cm and a base whose diameter is 4 cm?

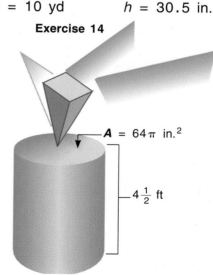
$A = 64\pi$ in.^{2}
$4\frac{1}{2}$ ft

15-5 Volume of a Rectangular Prism and a Cube

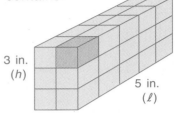

Volume (V) of a solid figure: the number of cubic units it contains.

To find the volume (V) of any prism, multiply the area of the base by the height.

▶ **To find the volume of a rectangular prism:**

- Find the area of the base (B). $B = \ell w$
    $B = 5$ in. $\times$ 2 in.
    $B = 10$ in.^{2}

- Multiply B by the height (h). $V = Bh$
    $V = 10$ in.$^2 \times 3$ in.
    This product is the volume. $V = 30$ in.^{3}

    The formula for the volume of a rectangular prism (where B is the area of the base) is:
    $$V = Bh$$

▶ **To find the volume of a cube** that measures 3 ft on an edge:

- Use the edge (e) to find the area of the base (B). (The edges of a cube are congruent.)

    $B = e^2$
    $B = 3$ ft $\times$ 3 ft
    $B = 9$ ft^2

- Multiply the area (e^2) by the height (e). This is the same as cubing e:

    $V = e^2 \times e$
    $V = 9$ ft$^2 \times 3$ ft
    $V = 27$ ft^3

$$e^2 \times e = (e \times e) \times e = e^3$$

The formula for the volume of a cube is:
$$V = e^3$$

Find the volume of each rectangular prism.*

1.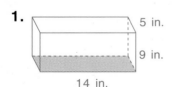
5 in.
9 in.
14 in.

2.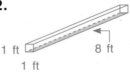
1 ft
8 ft
1 ft

3.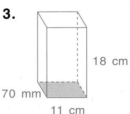
18 cm
70 mm
11 cm

4.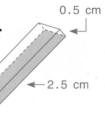
0.5 cm
2.5 cm
14 mm

Find the volume of a cube with an edge of:*

5. 3 in.   **6.** 18 ft   **7.** 9.5 ft   **8.** 4.5 yd   **9.** 13 in.   **10.** 100 in.

 SUPPOSE THAT...

A rectangular prism has a volume of 36 cubic units.

11. What are some possible dimensions for this rectangular prism?

15-6 | Volume of a Triangular Prism

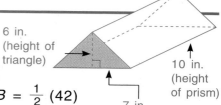

To find the volume (V) of a triangular prism:

- Find the area of the base (B). $B = \frac{1}{2} bh$
  (Use the formula for the area of a triangle.)

  6 in. (height of triangle)
  7 in. (base of triangle)
  10 in. (height of prism)

  $B = \frac{1}{2} (7 \times 6) \rightarrow B = \frac{1}{2} (42)$
  $B = 21$ in.^{2}

- Multiply B by the height of the prism. This product is the volume (V).

  $V = Bh$
  $V = (21$ in.$^2) (10$ in.$) = 210$ in.^{3}

To find the volume (V) of a triangular prism:	First, find B, area of the triangular base: $\longrightarrow$	Then, find V:
	$B = \frac{1}{2} bh$	$V = Bh$

Given the area of the base and height of the prism, find the volume.*

1. $B = 7.4$ cm^2
   $h = 12$ cm
   $V = \underline{\ ?\ }$

2. $B = 2.5$ m^2
   $h = 1.4$ m
   $V = \underline{\ ?\ }$

3. $B = 64$ in.^{2}
   $h = 1\frac{1}{2}$ in.
   $V = \underline{\ ?\ }$

4. $B = 80$ mm^2
   $h = 2.25$ mm
   $V = \underline{\ ?\ }$

Find the volume of each triangular prism.

5.
   40 ft
   30 ft
   18 ft

6.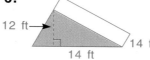
   12 ft
   14 ft
   14 ft

7.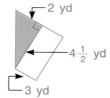
   2 yd
   $4\frac{1}{2}$ yd
   3 yd

8.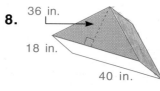
   36 in.
   18 in.
   40 in.

CRITICAL THINKING This rectangular prism can be cut into many different triangular prisms that have the same base as it has.

9. Find the volume of the rectangular prism.

10. Find the volume of the triangular prism.

11. Copy the rectangular prism and draw a different triangular prism with the same base as the rectangular prism. Then determine the volume of the triangular prism.

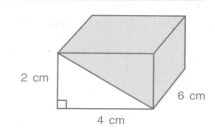
2 cm
6 cm
4 cm

12. What is the relationship between the volume of a rectangular prism and the volume of its related triangular prism?

13. Is this true for every triangular prism cut from a rectangular prism when the bases are the same?

To find the volume (V) of a cylinder:

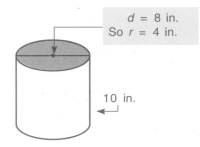

$$d = 8 \text{ in.}$$
$$\text{So } r = 4 \text{ in.}$$

- Find the area (B) of the (circular) base. (Use the formula for area of a circle.)

$$B = \pi r^2$$
$$B = \pi (4)^2$$
$$B = \pi 16$$
$$B \approx 50.24 \text{ in.}^2$$

10 in.

- Multiply this area (B) by the prism's height (h). The product is the volume (**V**).

$$V = Bh$$
$$V \approx 50.24 \text{ in.}^2 \times 10 \text{ in.}$$
$$V \approx 502.4 \text{ in.}^3$$

Find the volume. (Use 3.14 for π.)

1.

r = 2 cm

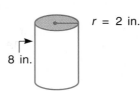

8 cm

2.

r = 2 ft

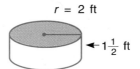

$1\frac{1}{2}$ ft

3.

6 yd

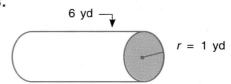

r = 1 yd

4.

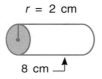

r = 2 in.

8 in.

5.

6 cm

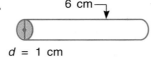

d = 1 cm

6.

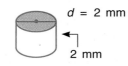

d = 2 mm

2 mm

Solve.

Exercise 9

7. What is the capacity in cubic inches of a thermos which is 5 in. high and has a 3-in. diameter?

8. Which glass holds the most liquid?
  a. one having a 7.5-cm diameter and standing 12 cm high
  b. one having a 4-cm radius and standing 11.5 cm high

2 ft

7 ft

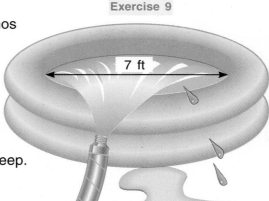

9. A circular wading pool is 7 ft across and 2 ft deep. How many cubic feet of water are needed to fill the pool?

10. What is the volume of the largest cylinder that can be packed into a cube measuring 10 ft on an edge?

Expressing Volume in π Units

Although 3.14 and $\frac{22}{7}$ are often used as approximations for π, it is more exact to express volume in terms of π units.

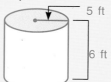

5 ft

6 ft

$V = Bh$
$V = (\pi r^2) h$
$V = (\pi \times 5 \times 5) \times 6 = \pi (5 \times 5 \times 6)$
$V = 150\pi$ (ft³) *Exact* Volume

You can use the key to find the approximate volume.

     471.2389 *Approximate* Volume

Find the volume of each cylinder in terms of π.

11. $r = 10$ mm
    $h = 7$ mm

12. $r = 11$ mm
    $h = 3$ mm

13. $r = 21$ cm
    $h = 4$ cm

14. $r = 8$ in.
    $h = 3$ in.

15. $d = 10$ ft
    $h = 6$ ft

16. $d = 14$ ft
    $h = 8$ ft

17. $d = 20$ cm
    $h = 5$ cm

18. $d = 8$ in.
    $h = 3$ in.

 SUPPOSE THAT...

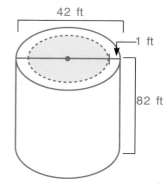

42 ft

1 ft

82 ft

19. The outside diameter of a closed cylindrical tank measures 42 ft. The tank is 82 ft tall. If the tank has an inner lining of insulation that is 1 ft thick, what is its capacity?

CHALLENGE

20. What is the volume of the largest cylinder that can be packed into these cubes? Express the volume in π units.

21. Can you find a pattern to explain what happens to the volume of the cylinder as the edge of the cube doubles?

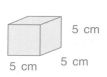

5 cm
5 cm
5 cm

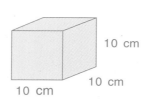

10 cm
10 cm
10 cm

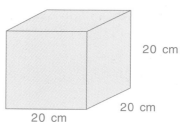

20 cm
20 cm
20 cm

TECHNOLOGY

Shortcuts with Spreadsheets

How "coordinated" are you? How many ways have you learned to use *coordinates*? Can you top these student "whizzes"?

Betty, a math whiz, remembers how to graph in the coordinate plane. She uses the term "coordinates" to refer to the ordered pair associated with a point on a graph.

The coordinates of point *A* are (0,0).

Matthew, a history whiz, says that he has used the term "coordinates" in map activities. Maps sometimes use grid sections marked by numbers and letters.

City Hall is located at D2.

Janet, a computer whiz, points out that an electronic spreadsheet uses "coordinates" to name cells, or locations for information. Cell D6 is formed by the intersection of column D with row 6.

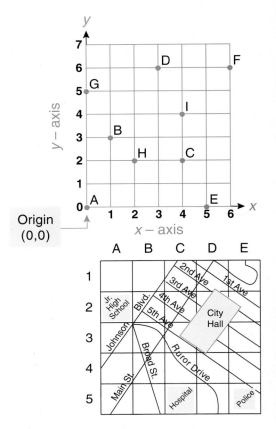

```
          A         B      C      D      E        F
1   Mr. Anthony DiCarlo: Math 7-C3
2   NAME        TEST1  TEST2  TEST3  TEST4    AVERAGE
3   Broadax,M.    80     92     90     97    +(B3+C3+D3+E3)/4
4   Hart,J.       75     96     90     91    @AVG(B4..E4)
5   Levan,R.      95     88     80     85    @SUM(B5..E5)/4
6   Nguyen,T.     85    100     80     82
7   Nunez,J.     100     96     90     88

                              Cell D6
```

Once information is entered into a spreadsheet, formulas and functions allow you to manipulate the information to discover new information.

The formula +(B3 + C3 + D3 + E3)/4 may be placed in cell F3 to compute the average of the first student's marks.

The function @AVG(B4..E4) will accomplish the same result for the second student.

The combination of function and formula @SUM(B5..E5)/4 is another method for computing the average for the third student.

376

Name the coordinates of the points on the graph on page 376.

1. *B* 2. *C* 3. *D* 4. *E*
5. *F* 6. *G* 7. *H* 8. *I*

9. Which points lie on the vertical axis? Give their coordinates.
10. Which points lie on the horizontal axis? Give their coordinates.

Use the map on page 376 to answer these questions.
11. Name the location of the Junior High School.
12. Where do 5th Avenue and Johnson Boulevard intersect?
13. What building is located at C5?
14. Where do Broad Street and Main Street intersect?
15. Where is the Police Station?

Use Mr. DiCarlo's spreadsheet on page 376 to answer these questions.
16. What information is entered in cell A7?
17. What row gives the marks for J. Hart?
18. What entry in cell F6 would compute the average grade?

A spreadsheet is used to compute batting averages for the Wickstown Warriors. Enter the information into a spreadsheet program and find the averages.

	A	B	C	D
1	BATTING AVERAGE	AB–AT BAT	H–HITS	AVERAGE
2	PLAYER			
3	Bannas,J.	23	6	0.261
4	Cassady,I.	15	4	
5	Compton,F.	14	3	
6	Cronin,S.	12	3	
7	Lawler,G.	12	3	
8	Rodriguez,E.	11	3	
9	Windle,A.	12	5	
10	McCullough,S.	10	4	
11	Hayes,T.	6	2	
18	TEAM	115	33	

19. What formula should be entered into cell D3 to compute the batting average?

20. What formula should be entered into cell B18 to compute the total number of AT BAT's for the team?

21. At the next game the HITS and AT BAT's of the team are entered into the spreadsheet:

Bannas(4AB,2H),   Windle(3AB,3H),   McCullough(3AB, 2H)
Lawler(3AB,1H),   Hayes(3AB,0H),   Cronin(3AB,1H)
Cassady(4AB,3H),   Compton(3AB,2H),   Rodriguez(3AB,1H)

Compute the new individual and team averages.

Which strategy is best for solving each problem? Use one of the strategies used in this book, or make up one of your own.

Problem A: A *dodecahedron* is a polyhedron with 12 faces, each of which is a regular pentagon. How many edges does it have? How many vertices?

Vertex (*V*)
Edge (*E*)
Face (*F*)

1 IMAGINE   Draw and label a 12-sided polyhedron.

2 NAME   *Facts:*   A regular polyhedron with 12 pentagonal faces is a dodecahedron.

Question: How many edges does it have? How many vertices?

3 THINK   Should you make a model?
Should you use a formula? ($F + V = E + 2$)
Should you look for patterns?

Dodecahedron

Then → **4 COMPUTE** — and → **5 CHECK**

Problem B: A wedge was cut from a large block of wood 40 in. by 16 in. by 20 in. The wedge was a triangular prism having a height of 4 inches, a base of 12 inches, and it is 8 inches long. Find the volume of the wooden block after the wedge was removed.

8 in.   4 in.
40 in.
12 in.
20 in.
16 in.

1 IMAGINE   Draw and label a large block of wood with a wedge-shaped figure cut out of it.

2 NAME   *Facts:*   Block of wood: $40'' \times 16'' \times 20''$
Wedge cut from block: $4'' \times 12'' \times 8''$

Question: What is the volume of the block of wood?
What is the volume of the wedge?
What is the difference between the two volumes?

3 THINK   Volume of block − Volume of wedge = Volume of block
without the wedge

Then → **4 COMPUTE** — and → **5 CHECK**

Solve by choosing a strategy.

1. A sign-painting company is painting an advertisement on a cylindrical tank that is 60 feet high and has a base 28 feet in diameter. The ad, which is rectangular, will be 12 feet high; in length it will cover 60% of the tank's circumference. How much of the tank will be covered by the advertisement?

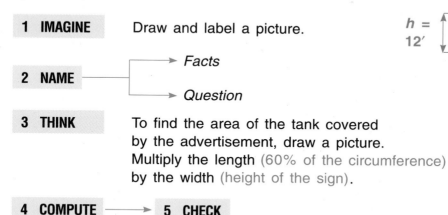

| 1 IMAGINE | Draw and label a picture. |

2 NAME ⟶ Facts
           ⟶ Question

| 3 THINK | To find the area of the tank covered by the advertisement, draw a picture. Multiply the length (60% of the circumference) by the width (height of the sign). |

| 4 COMPUTE | ⟶ | 5 CHECK |

2. A scout leader is planning to make a tent. It will be in the shape of a triangular prism, with a base measuring 12.5 ft by 9.5 ft and an altitude of 9 ft. There will be closing flaps on each end and a window measuring 3 ft by 8 ft on one side. How many square yards of canvas must the scout leader purchase to make the tent?

3. Robbie drew 5 squares. Inside each square he drew 3 circles. Inside each circle he drew 1 triangle. If none of the figures overlapped, how many plane figures did Robbie draw?

4. The height of the TEXAM building is 636 feet. A flagpole extends from the building at a point one-sixth of the distance from the ground to the roof. How many yards from the ground is the flagpole mounted?

5. A wall in a kindergarten classroom is to be covered with cork. If the wall measures $18\frac{3}{4}$ ft by $10\frac{1}{2}$ ft, what will it cost to cover it with cork that sells at $12 a square yard?

6. The volume of a certain cube equals the volume of a rectangular prism. The rectangular prism has a length of 9 m. Its width is $\frac{2}{3}$ of the length and the height is 5 cm less than the length. What is the volume of the cube?

7. Which has the largest area: a square 7 cm on a side, a circle with a diameter of 7 cm, or an isosceles triangle with a base of 7 cm and a height of 7 cm?

What Am I?

1. I am a space figure whose faces are polygons.

2. I am a space figure whose net has two circular bases and a rectangular face.

3. I am a polyhedron whose net is six congruent squares.

4. I am a line segment formed by the intersection of two faces of a polyhedron.

5. I am the common endpoint of three or more edges of a polyhedron.

6. I am the shape of the base of a cylinder.

7. I am the polygon that makes up the faces, excluding the base, of any pyramid.

8. I am a general formula that can be used to find the volume of any prism.

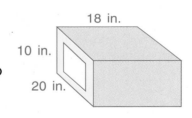

USE THESE STRATEGIES:
Multi-Step Problem
Use a Model / Drawing
Hidden Information
Write an Equation
Use Simpler Numbers
Use a Formula

Solve. You may combine strategies.

9. An opening 2 in. from each edge is cut into one face of the carton shown. What is the surface area that remains on the carton?

18 in.

10 in.

20 in.

A piece of marble is carved in the shape of a triangular prism.

10. If all five faces are polished, how many square feet of marble are polished?

11. How many cubic feet of marble does the prism contain?

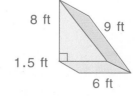

8 ft

9 ft

1.5 ft

6 ft

The storage area of this truck is a prism whose base is a trapezoid.

12. Find the volume of gravel it can hold.
    (Hint: Use $V = Bh$ where B is the area of the trapezoid.)

13. At $5.50 a cubic foot, what is the cost of a truckload of gravel?

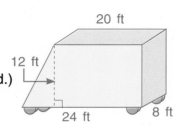

20 ft

12 ft

24 ft

8 ft

14. How many square meters of wrapping paper are needed to wrap a gift box 60 cm on each edge?

15. The surface area of a 20-cm cube is how many times greater than the surface area of a 10-cm cube?

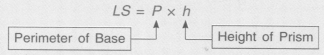

Lateral Surface Area

The *lateral surface area* of a space figure is the surface area excluding the area of the two parallel bases.

For a prism the general formula for lateral surface area is:

$$LS = P \times h$$

| Perimeter of Base | | Height of Prism |

What is the lateral surface of a cylinder having a 7-in. radius and a 10-in. height?

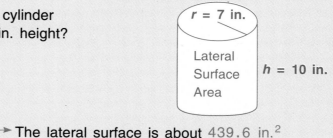

$LS = P \times h$
$LS = \pi d \times h$
$LS = \pi 14 \times 10$
$LS = \pi 140$
$LS \approx 439.6$

The lateral surface is about 439.6 in.^{2}

---

16. How many square feet of cedar are needed to line the four walls of a closet that is a square prism $5\frac{1}{2}$ ft on each edge?

17. How much paper is used to make labels for 100 juice cans 3.5 cm in diameter and 8 cm tall?

18. A building has four columns, each of which is 15 ft high and has a circumference of $1\frac{2}{3}$ yards.

    • Find the lateral surface area of the four columns.

    • How much will it cost to paint the four columns if paint costs $1.10 a square yard?

19. A heating pipe that has a 6-in. diameter is to be covered with insulation. If the pipe is 15 feet long, how many square feet of insulation are needed?

20. How many square feet of plastic will be used to make 10 containers (without lids) if each has a square base 4 inches on each edge and a height of 8 inches?

21. Which is larger and by how many cubic meters?
    a. a rectangular prism 8 m by 9 m by 7.5 m
    b. a cube 8.2 m on an edge

22. Melissa is painting the walls and ceiling of a room. How many square meters will be painted if the room is 15 m long, 10 m wide, and 4 m high?

23. How many cubic yards of soil will be needed to fill a circular flower bed that has a radius of 2 yards and is 1 yard deep?

24. What is the capacity of a cylindrical water tank that is 40 ft high and has a 7-yd diameter?

More Practice

Write the letter of the formula or formulas to be used in finding each.

1. volume of a cube
2. surface area of a rectangular prism
3. surface area of a triangular prism
4. surface area of a cube
5. volume of a triangular prism
6. surface area of a cylinder
7. volume of a rectangular prism

---

a. $V = e^3$

b. $A = \frac{1}{2}bh$, $V = Bh$

c. $S = 6e^2$

d. $A = \frac{1}{2}bh$, $A = \ell w$

e. $V = Bh$

f. $A = \pi r^2$, $C = 2\pi r$, $A = \ell w$

g. $S = 2(\ell w + \ell h + wh)$

Choose the correct answer.

8. The number of edges of a cube is __?__
   a. 5   **b.** 6   **c.** 7   **d.** 12

9. The number of faces in a square pyramid is __?__
   a. 3   **b.** 4   **c.** 5   **d.** 6

10. The number of faces in a pentagonal prism is __?__
    a. 5   **b.** 6   **c.** 7   **d.** 10

11. If a cube measures 10 in. on a side, its volume is __?__
    a. 10 in.^{3} **b.** 600 in.^{3} **c.** 100 in.^{3} **d.** 1000 in.^{3}

12. If a cube measures 10 in. on a side, its surface area is __?__
    a. 10 in.^{2} **b.** 600 in.^{2} **c.** 1000 in.^{2} **d.** 1000 in.^{3}

Find the surface area.

13.
    $d = 8$ in.
    15 in.

14.
    15 cm

15.
    2 ft
    $5\frac{1}{2}$ ft
    3 yd

16.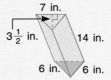
    7 in.
    $3\frac{1}{2}$ in.
    14 in.
    6 in.   6 in.

Find the volume.

17.
    6.5 ft

18.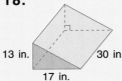
    13 in.   30 in.
    17 in.

19.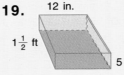
    12 in.
    $1\frac{1}{2}$ ft
    5 in.

20.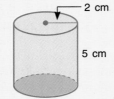
    2 cm
    5 cm

See *Still More Practice*, pp. 462–463.

Math Probe

SQUARE FACES

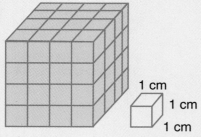

1 cm
1 cm
1 cm

This cube was constructed from small cubes, each 1 cm on an edge.

Then the six faces of the large cube were painted.

Six faces

Can you visualize what the smaller cubes would look like if the larger cube were now taken apart? Use your mental picture to answer these questions. Then copy and complete the chart below.

1. How many 1 cm by 1 cm by 1 cm cubes are in the larger cube?

2. Are any cubes painted on four faces? Why?

3. How many cubes are painted on three faces? Where are these smaller cubes located in the larger cube?

4. How many cubes are painted on two faces? Where are these smaller cubes located in the larger cube?

5. How many cubes are painted on only one face? Where are these smaller cubes located in the larger cube?

6. How many unpainted cubes are there?

7. Use a calculator to change each fraction to a percent.

Number of painted faces	Number of cubes	Fraction	Percent
4	0	$\frac{0}{64}$	
3			
2			
1			
0			

8. Suppose that you have used 1-cm cubes to construct a larger cube measuring 5 cm on each edge and that you have painted all six faces of the larger cube. Create a chart for it like the one above.

Check Your Mastery

 a.

 b.

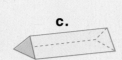 c.

 d.

 e.

 f.

 g.

 h.

Write the letter of the correct figure or figures.

See pp. 364–365

1. Which figure is a pyramid?
2. Which figure is a right triangular prism?
3. Name figures e, f, and g.
4. How are e, f, and g different from the other figures?
5. Two of the figures have 8 vertices each. Name them.

Find the *surface area* of each rectangular prism.

See pp. 366–367

6. length = 8 in.
   width = 6 in.
   height = 3 in.

7. length = 6 ft
   width = $2\frac{1}{2}$ ft
   height = 4 ft

8. length = 8 ft
   width = 5 ft
   height = $2\frac{1}{2}$ ft

9. length = 8 ft
   width = 2 ft
   height = $1\frac{1}{2}$ ft

Find the *surface area* of each cube.

10. edge = 3 in.
11. edge = 10 in.
12. edge = 1.9 m
13. edge = 30 ft

Find the *surface area* of each triangular prism.

See pp. 368–369

14.
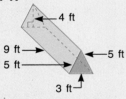
9 ft
5 ft
4 ft
5 ft
3 ft

15.
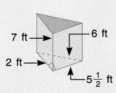
7 ft
2 ft
6 ft
$5\frac{1}{2}$ ft

16.
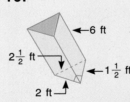
6 ft
$2\frac{1}{2}$ ft
$1\frac{1}{2}$ ft
2 ft

17.
15 in.   33 in.
22 in.
22 in.
42 in.

Find the *volume* of each prism.

See p. 372

18. length = 7 in.
    width = 3 in.
    height = $\frac{1}{2}$ ft

19. length = 10 yd
    width = 9 yd
    height = 2 ft

20. length = 24 ft
    width = 3 yd
    height = 2 yd

21. length = 1.5 cm
    width = 1.5 cm
    height = 1.5 cm

Find the *surface area* and *volume* of each cylinder.

See pp. 370–375

22. area of base = 9π in.^{2}
    height = 12 in.

23. area of base = 16π cm^2
    height = 7 cm

24. radius = 2 ft
    height = 3 ft

384

16 Probability, Statistics, and Graphing

In this chapter you will:

- Express probabilities as ratios
- Read and interpret tree diagrams
- Find the probability of one event; more than one event
- Use the counting principle
- Read and make pictographs; bar, line, and circle graphs
- Plot points on a coordinate grid
- Solve problems: organized list

Do you remember?

A ratio is a comparison of two numbers by division.

2 : 3 or $\frac{2}{3}$, 2 out of 3

If ◯ represents 60 tons of corn, then ◠ represents 30 tons of corn and ◿ represents 15 tons of corn.

To find the percentage of a number, change the percent to a decimal and multiply.

RESEARCHING TOGETHER

Amazing Headlines

Find out the stories behind these amazing headlines.

- **MATHEMATICS SAVES A LIFE!** (Paul Wolfskehl)

- **KARL GAUSS STUNS TEACHER!** (Karl Gauss)

- **PAPYRUS WIZARDRY!** (Ancient Egyptians solve multiplication)

Choose one and write the newspaper article as if "you are there!"

16-1 Identifying Outcomes

Equally Likely

Experiment 1
Toss a marker on this board.

1	2
3	4

Possible outcomes: 1, 2, 3, 4
There are 4 possible outcomes.

The outcomes from *Experiment 1* are **equally likely** because the chances of getting *any one* of them are the same.

Sample Space
ALL the outcomes that are possible

Not Equally Likely

Experiment 2
Toss a marker on this board.

4	1	2
	3	

Possible outcomes: 1, 2, 3, 4
There are 4 possible outcomes.

The outcomes from *Experiment 2* are **not equally likely** because the chances of getting *any one* of them are *not* the same.

It is more likely that the marker will land on 4 than on 1, 2, or 3.

Certainty

Experiment 3
Toss a marker on this board.

1	2	3

Can a marker land on a number less than 4?
Yes, the outcome is *certain* because *all* of the numbers are less than 4.

Can a marker land on a number greater than 3?
No, the outcome is *impossible* because *none* of the numbers are greater than 3.

Answer the questions about each experiment.

Experiment A: Find the ring hidden under one of these cups.

1. How many possible outcomes are there?

2. Are the outcomes equally likely?

Experiment B: Spin the dial.

3. How many separate regions are on the spinner?

4. How many different outcomes are possible?

5. Are the outcomes equally likely?

Experiment C: A box contains 3 red marbles, 4 yellow marbles, and 5 blue marbles. One is picked without looking.

6. How many marbles are there?
7. How many different outcomes are possible?
8. Are the outcomes equally likely?
9. Which outcome is the least likely?

Experiment D: Each letter in the word APPLE is written on a card. Then a card is chosen without looking.

10. How many cards are there?
11. How many different outcomes are possible?
12. Are the outcomes equally likely?
13. Which is the most likely outcome?

Experiment E: Roll this number cube.

14. What are the different possible outcomes?
15. Are the outcomes equally likely?

Experiment F: Roll this number cube.

16. What are the different possible outcomes?
17. Are the outcomes equally likely?

Tree Diagrams

Tree diagrams show *all possible outcomes* of an event or of more than one event.

Toss a coin and, without looking, draw a chip from a bag that contains 2 green chips and 1 red chip. What are all the possible outcomes?

See how the tree diagram gives all possible outcomes.

Event 1 Coin	Event 2 Chip	Write	Event 1 Coin	Event 2 Chip	Write
Heads (H)	Green (G)	(H, G)	Tails (T)	Green (G)	(T, G)
	Green (G)	(H, G)		Green (G)	(T, G)
	Red (R)	(H, R)		Red (R)	(T, R)

There are 6 possible outcomes.

Draw a tree diagram and list all possible outcomes.

18. Toss a coin and roll the cube from experiment E.

    T ?

19. Select a card and spin the dial.

20. Pick a card and roll the cube.

Finding Probability

Probability (*P*): the number representing the chance that a given event **E** will occur.

***P*(E), the probability of one event:** the ratio of the number of *favorable* outcomes to the *total number* of possible outcomes *if each is equally likely.*

$$P(E) = \frac{\text{Number of Favorable Outcomes}}{\text{Total Number of Possible Outcomes}}$$

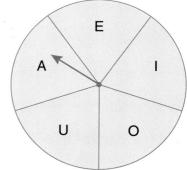

What is the probability of spinning an A on this dial?

$$P(A) = \frac{\text{There is } 1 \text{ chance of spinning an A}}{\text{There are } 5 \text{ possible outcomes}} \longrightarrow \frac{1}{5}$$

If $\frac{1}{5}$, or 1 out of 5, is the probability of spinning an A, what is the probability of *not* spinning an A?

$$P(not \text{ A}) = \frac{4 \text{ chances of } \underline{\text{not}} \text{ spinning A}}{5 \text{ possible outcomes}} \longrightarrow \frac{4}{5}$$

> 5 equally likely outcomes

or

If $P(A) = \frac{1}{5}$, then $P(not \text{ A}) = \frac{4}{5}$, because $P(\text{all}) = \frac{5}{5}$, and $\frac{5}{5} - \frac{1}{5} = \frac{4}{5}$

On this dial, what is the probability of spinning . . .

the letter Z?

$$P(Z) = \frac{\text{no chance}}{5 \text{ possible outcomes}}$$

$$P(Z) = \frac{0}{5}$$

$$P(Z) = 0$$

All *impossible* events have a probability of 0.

a vowel?

$$P(\text{a vowel}) = \frac{5 \text{ chances}}{5 \text{ possible outcomes}}$$

$$P(\text{a vowel}) = \frac{5}{5}$$

$$P(\text{a vowel}) = 1$$

All *certain* events have a probability of 1.

Find the probability for this spinner.

1. $P(3)$ **2.** $P(1)$ **3.** $P(0)$ **4.** $P(5)$

5. $P(not \text{ } 5)$ **6.** $P(<5)$ **7.** $P(>4)$ **8.** $P(<4)$

9. $P(\text{even})$ **10.** $P(\text{odd})$ **11.** $P(\text{even or odd})$ **12.** $P(not \text{ } 3)$

A jar contains 4 red marbles, 2 blue marbles, and 1 white marble.
A marble is picked without looking.

13. Find P(blue).

14. Find P(red).

15. Find P(white or red).

16. Find P(white or blue).

17. Find P(green or yellow).

18. Find P(red or green).

19. Find P(*not* blue).

20. Find P(*not* white).

21. Find P(*not* orange).

A coin is tossed. (H = Heads; T = Tails)

22. Find P(H).

23. Find P(T).

24. Find P(H or T).

25. Find P(*not* T).

More Than One Event

The dial has numbers 1, 2, and 3. There are white chips and black chips.
Spin the dial once and draw a chip.
Find the probability of spinning
the dial and getting a 1 *and* at
the same time drawing a white chip.

   This is written as: P(1, W).

This problem deals with the probability of *more* than one event.

The outcomes shown on a tree diagram
are often used to determine the probability
of more than one event.

There are 6 possible outcomes.

There is 1 favorable outcome.

$P(1, W) = \dfrac{1}{6}$

Event 1 Dial	Event 2 Chips
Spin (1)	white (W) / black (B)
Spin (2)	white (W) / black (B)
Spin (3)	white (W) / black (B)

Use the dial and chips above. Find the following probabilities.

26. P(2, W)

27. P(3, B)

28. P(1, W)

29. P(2, B)

30. P(1, B)

31. P(odd, W)

32. P(even, B)

33. P(odd, B)

34. P(5, W)

35. P(8, B)

Toss 2 coins at the same time. (Hint: Make a tree diagram.)

36. Find P(H, H).

37. Find P(T, T).

38. Find P(H, T).

39. Find P(T, H).

40. Toss a pair of dice 50 times. Find the sum of the dice for each toss and keep a tally of the number of times each sum occurs. Use a chart like the one below for the tally.

2	3	4	5	6	7	8	9	10	11	12

16-3 Probability of Compound Events

Compound event: two or more simple events considered as a single event.

Experiment 1: Draw a card. Replace it. Draw a second card. How many outcomes are there? What is the probability of drawing a circle and then a triangle?

There are 16 outcomes.

$P(\bullet, \blacktriangle) = \dfrac{2}{16}$ or $\dfrac{1}{8}$

Since the card is replaced after the first draw, the outcome of the second event is *independent* of the outcome of the first event.

Experiment 2: Draw a card. Do NOT replace it. Draw a second card. What is the probability of drawing a circle and then a triangle?

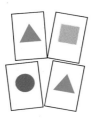

There are 12 outcomes.

$P(\bullet, \blacktriangle) = \dfrac{2}{12}$ or $\dfrac{1}{6}$

Since the card is not replaced after the first draw, the outcome of the second event is *dependent* on the outcome of the first event.

Use this shortcut for finding the probability of a compound event:
Find the product of the probability of each simple event. For example:

Experiment 1	*Experiment 2*
$P(\bullet) = \dfrac{1}{4}$ $P(\blacktriangle) = \dfrac{2}{4}$	$P(\bullet) = \dfrac{1}{4}$ $P(\blacktriangle) = \dfrac{2}{3}$
$P(\bullet, \blacktriangle) = \dfrac{1}{4} \times \dfrac{2}{4} = \dfrac{2}{16}$ or $\dfrac{1}{8}$	$P(\bullet, \blacktriangle) = \dfrac{1}{4} \times \dfrac{2}{3} = \dfrac{2}{12}$ or $\dfrac{1}{6}$

Using experiment 1, compute each probability. Verify on the tree diagram.

1. $P(\blacktriangle, \blacktriangle)$
2. $P(\bullet, \blacksquare)$
3. $P(\bullet, \blacktriangle \text{ or } \blacksquare)$
4. $P(\bullet, \bullet)$

Using experiment 2, compute each probability. Verify on the tree diagram.

5. $P(\blacktriangle, \blacktriangle)$
6. $P(\bullet, \blacksquare)$
7. $P(\bullet, \blacktriangle \text{ or } \blacksquare)$
8. $P(\bullet, \bullet)$

Spin the dial and toss a coin.

9. What are all of the possible outcomes?

10. Find each probability.

    a. P(3, H)   **b.** P(prime number, T)   **c.** P(odd number, T)

Find each probability. (Hint: Use a tree diagram to list all possible outcomes.)

Experiment A:

Toss a coin and choose one of 6 cards without looking. There are 4 red cards and 2 blue cards.

11. P(H, red)   **12.** P(H or T, red)

13. P(T, blue)   **14.** P(T, red or blue)

Experiment B:

Toss a coin twice.

15. P(H, H)   **16.** P(T, T)

17. P(T, H)   **18.** P(T, T or H, H)

Experiment C:

Roll a number cube, labeled 1–6, and spin the dial.

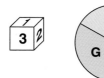

19. P(2, B)   **20.** P(1, *not* R)

21. P(1 or 6, G)   **22.** P(even, G)

Experiment D:

Select a card and toss a coin.

23. P(A, H)   **24.** P(B, T)

25. P(A, H or T)   **26.** P(A or B, T)

Experiment E:

A box contains 2 striped blocks and 1 checkered block. A block is chosen without looking and replaced. Then another block is chosen.

27. P(S, S)   **28.** P(S, C)   **29.** P(one of each)

Experiment F:

A box contains 2 red blocks, 1 yellow block, and 1 green block. A block is chosen without looking and *not* replaced. Then another block is chosen.

30. P(R, Y)   **31.** P(R, R)   **32.** P(G, G)

Experiment G:

An envelope contains 5 cards numbered 1 to 5. A card is chosen without looking and *not* replaced. Then another card is chosen.

33. P(1, 5)   **34.** P(2, 2)   **35.** P(two prime numbers)

16-4 Counting Principle

How many different ways can these three cards, X, Y, and Z, be arranged?

A tree diagram can be used to show the possible arrangements.

First	Second	Third

X — Y — Z
X — Z — Y

If X is placed first, there are only 2 possible positions left for Y and Z. Then, if Y is placed second, there is only 1 position left for Z.

Y — Z — X
Y — X — Z

Similarly, if Y is placed first, there are only 2 possible positions left for X and Z.

Z — X — Y
Z — Y — X

The tree diagram shows that there are *only* 6 possible arrangements.

The counting principle also can be used to determine the number of possible arrangements.

Counting Principle

If one event occurs in m possible ways, and a second event occurs in n possible ways, the total number of ways in which both events can occur is the product of $m \times n \times 1$.

3	×	2	×	1	=	6
possible first cards		possible second cards		possible third card		

The product $3 \times 2 \times 1$ can be shown using this special mathematical symbol (!).

$3! = 3 \times 2 \times 1 = 6$

Read: "3 factorial."

Remember these two special definitions:
$$1! = 1 \quad \text{and} \quad 0! = 1$$

When combining expressions having the factorial symbol, use the proper order of operations.

$$\frac{4! - 2!}{2!} = \frac{(4 \times 3 \times 2 \times 1) - (2 \times 1)}{(2 \times 1)} = \frac{24 - 2}{2} = \frac{22}{2} = 11$$

Find the value of each expression.

1. $5!$ **2.** $3!$ **3.** $6!$ **4.** $8!$

5. $2!$ **6.** $7!$ **7.** $0!$ **8.** $1!$

9. $4! + 1!$ **10.** $3! + 2!$ **11.** $7! - 2!$ **12.** $4! + 4!$

Solve. (The first one is done.)

13. $\dfrac{5! \times 2!}{4!} = \dfrac{(5 \times 4 \times 3 \times 2 \times 1) \times (2 \times 1)}{(4 \times 3 \times 2 \times 1)} = 5 \times 2 = 10$

14. $\dfrac{7! \times 3!}{5!}$

15. $\dfrac{6! \times 3!}{8!}$

16. $\dfrac{4! \times 2!}{6!}$

17. $\dfrac{3! \times 4!}{5! \times 2!}$

18. $\dfrac{4! \times 5!}{3! \times 2!}$

19. $\dfrac{5! \times 3!}{6! \times 2!}$

True or false? Explain.

20. $\dfrac{8!}{4!} = 2!$

21. $(2 + 3)! = 5!$

22. $6! = 120$

23. $1! + 3! = 4!$

24. $9! \times 0! = 0$

25. $1! = 1$

Complete.

26. $5! = 120$, then $6! = $ __?__

27. $7! = 5040$, then $6! = $ __?__

28. $8! = 40,320$, then $9! = $ __?__

29. $5! = 120$, then $4! = $ __?__

30. Explain the pattern that can be used to answer exercises 26–29.

Solve by applying the counting principle.

31. In how many ways can 5 cars be parked in 5 parking places?

32. Mark's new radio will program 6 of his favorite radio stations. In how many different ways can he preset the stations?

33. Ten horses are led into 10 stalls at Red Barn Stables. In how many different ways can the horses be led in?

34. Manuel buys the digits 4, 5, and 6 at the hardware store to mark his address on the door of his house. How many different addresses can be made using all three digits?

35. In how many different ways can four junior-high math teachers line up in the graduation procession?

36. A sandwich shop sells 4 kinds of hot sandwiches, 5 kinds of submarine sandwiches, and 3 kinds of club sandwiches. How many different sandwich assortments can be bought?

SKILLS TO REMEMBER

Solve for x.

37. $x = 5\%$ of 360

38. $8\frac{1}{3}\%$ of $60 = x$

39. 7.5% of $2000 = x$

40. $45 = x\%$ of 180

41. $\$75 = x\%$ of $200

42. $\$12.50 = x\%$ of $50

16-5 Graphing Sense

Graphs: pictures used to communicate ideas.

Graphs present data so that the information being illustrated may be quickly understood.

There are many kinds of graphs.

- **Pictographs** use pictures or symbols to represent numerical quantities. The key for a pictograph tells the number that each picture or symbol represents.

- **Line graphs** present information on one subject so that trends can be identified and comparisons can be made. The scale on a line graph is divided into equal intervals.

- **Bar graphs** present information so that comparisons of two or more given subjects can be made. The scale on a bar graph is similar to the scale on a line graph. When the range of the data is not convenient, the scale can be broken.

- **Circle graphs** present information as percents or percentages, and show how a whole is divided into fractional parts. The sum of the data on the circle graph must total 100%.

Pictograph

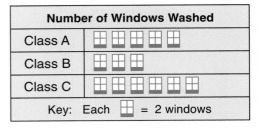

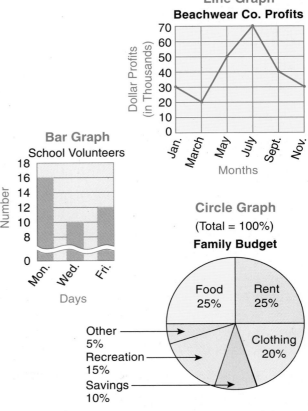

Use the graphs above to answer questions 1–5.

1. Which graphs show information that was recorded at regular intervals?

2. Which graphs compare three sets of data?

3. Which graph depends on the data adding up to 100%?

4. Which graphs use vertical and horizontal axes?

5. Why is it essential that each graph has a title?

Pictographs

Survey of Junior-High Students' Film Preferences

The pictograph below illustrates the data. (Let any symbol represent the amount in a pictograph. Here, ☻ = 2 people.)

Kind	No. of Students
Comedy	17
Nature	11
Horror	8
Space	25
Musical	12
Other	5

Film Preferences of Junior-High Students

Kind	Number of Students
Comedy	
Nature	
Horror	
Space	
Musical	
Other	

Key:
☻ = 2 people
☺ = 1 person

8 full symbols + 1 half symbol = $8\frac{1}{2}$ symbols ⟶ 17 people who like comedy

---

SUPPOSE ▽ THAT...

You want to display the following situations on a graph. For each situation: collect or make up data; decide which type of graph is best; draw such a graph.

6. The number of foreign cars imported from four different countries

7. The bowling average for each member of a team for a season

8. A school's budget for a year

9. The maximum and minimum temperature each day for a week

10. A population growth of 145% over a 5-year period

11. The percent of oil imported from foreign countries for a month

Finding Together

12. Find in newspapers examples of pictographs and bar, line, and circle graphs. Choose 3 different examples and explain why each was used to illustrate the data.

Single Bar Graphs

Bar graphs may be horizontal or vertical.
Below is a horizontal bar graph.

To make a bar graph:

- Decide on a title for the graph.
- Round all numerical information.
- Draw a vertical and a horizontal axis.
- List the items along one axis.
- Choose an interval that will fit the data.
- Write a scale, beginning at 0, along the other axis. Choose convenient units for the scale.
- Draw the bars on the graph.

Data on Length of Ocean Liners	
Name	Length
Lusitania	790 ft
Titanic	892 ft
Queen Mary	1019 ft
Queen Elizabeth	1031 ft
Queen Elizabeth II	963 ft

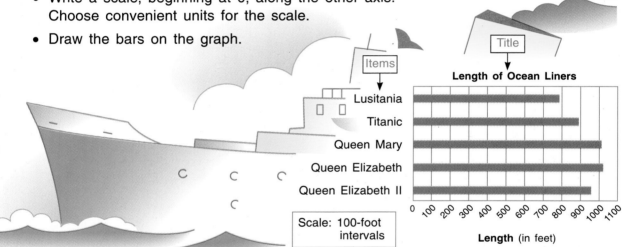

Scale: 100-foot intervals

1. Make a horizontal bar graph to show:

Results of Saturday's Game	
Player	Number of Hits
Jean	3
Ann	2
Rose	1
Mary	4
Jill	3

2. Make a vertical bar graph to show:

Scores on Math Test	
Score	Number of Students
51-60%	2
61-70%	4
71-80%	8
81-90%	14
91-100%	6

3. Make a vertical bar graph to display John's earnings for last week: Monday, $15; Tuesday, $17; Wednesday, $0; Thursday, $12; Friday, $9.

4. Make a horizontal bar graph to show rainfall in your area in the last 8 weeks or months.

16-7 Double Bar Graphs

Double bar graphs are used to show *two* sets of data.
Usually two different colors are used to show each set.

This table shows the results of a survey of 100 seventh graders and 100 eighth graders who were invited to work in the Environmental Learning Lab. They were asked, "Which topic would you like to explore?"

Topics	Biosphere	Acid Rain	Conservation	Recycling	Energy	Pollution
7th Graders	8	7	6	30	12	37
8th Graders	4	22	5	17	32	20

Key: 🔲 7th Graders
     🔲 8th Graders

The key identifies to which group each set of data belongs.

To make a double bar graph:
Make a bar graph for each set of data. (See page 396.)

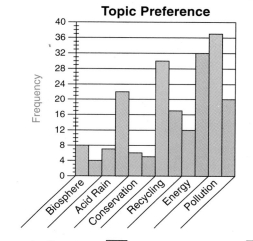

Topic Preference

Key: 7th Graders 🔲   8th Graders 🔲

---

1. Make a vertical double bar graph for the data in this table.

Number of Books Read		
Name	**March**	**April**
Beverly	1	5
Nadia	2	4
Chris	0	3
Edgar	5	2
Sam	8	4

2. Make a horizontal double bar graph for the data in this table.

Campground	Number of Sites	Sites with Electricity
Hartland	20	10
Navajo	40	40
Frost Ridge	50	35
Niagara	100	95
Beach	75	75

3. Make a horizontal double bar graph to compare rainfall for these two cities.

City	June	July	August
Dry Gulch	5 cm	6 cm	8 cm
Seaside	15 cm	10 cm	6 cm

4. Make a vertical double bar graph to compare these test scores.

	Test 1	Test 2	Test 3
1st month	85	90	92
2nd month	75	80	88

16-8 Single Line Graphs

Line graphs are another way to display data. They show a pattern or trend.

This chart gives the rainfall in Central Park for April through September.

Month	Apr.	May	June	July	Aug.	Sept.
Rain	5.12	9.1	2.5	3.51	12.66	2.24
Rounded Data	5	9	3	4	13	2

To make a line graph:

- Decide on a title for the graph.
- Round the numerical information.
- Draw a vertical and a horizontal axis.
- List the items along one axis.
- Choose an interval that will fit the data.
- Write a scale, starting at 0, along the other axis. Choose convenient units for the scale.
- Place points on the graph and connect them with line segments.

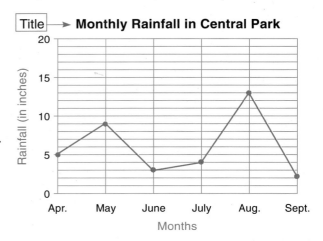

Construct a line graph to display each set of data.

1.

Paper-Route Customers			
1989	1990	1991	1992
186	205	200	220

2.

Math Test Scores				
Mar. 4	Mar. 11	Mar. 18	Mar. 25	Apr. 1
80	82	95	87	90

3.

Monthly Student Attendance									
Sept.	Oct.	Nov.	Dec.	Jan.	Feb.	Mar.	Apr.	May	June
220	210	215	190	188	190	210	205	212	218

4.

7th Grader's Daily Spending						
Mon.	Tues.	Wed.	Thurs.	Fri.	Sat.	Sun.
$2.25	$3.15	$1.85	$.00	$4.50	$5.50	$4.00

Answer these questions. Use the graphs you made for exercises 1–4.

5. Describe the paper-route trend in exercise 1.

6. During which month(s) was the attendance in exercise 3 the highest? the lowest? the same?

7. What might account for a drop in student attendance in exercise 3?

8. Give an explanation for Thursday's spending in exercise 4.

16-9 Double Line Graphs

Double line graphs compare multiple sets of data.

To make a double line graph:
Make a line graph for each set of data.
(See page 398.)

This graph shows the monthly attendance, reported in thousands, at two different resorts.

	SplashDown!	Rides-A-Plenty
May	28	45
June	58	73
July	91	89
Aug.	106	82
Sept.	31	44
(Numbers reported in thousands)		

The **key** below the graph identifies the two items.

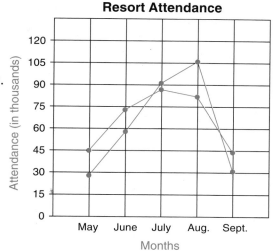

Resort Attendance

Key: SplashDown! —   Rides-A-Plenty —

Make a double line graph to illustrate each set of data.*

1. Compare Heidi's math and science grades for each quarter.

	First	Second	Third	Fourth
Math	88	91	87	90
Science	86	88	86	89

2. Compare Mrs. Vega's and Mrs. Ti's commissions for 1 year.

	Jan.	Feb.	Mar.	Apr.	May	June	July	Aug.	Sept.	Oct.	Nov.	Dec.
Vega	$190	$140	$165	$240	$290	$290	$375	$390	$365	$240	$190	$290
Ti	$265	$240	$240	$310	$290	$240	$310	$265	$340	$240	$265	$365

CHALLENGE

Make a triple line graph using the data from exercise 3.

3.

Automobile Stopping Distances			
Speed (mph)	Reaction Distance (ft)	Breaking Distance (ft)	Stopping Distance (ft)
10	10	5	15
20	20	20	40
30	30	45	75
40	40	80	120
50	50	125	175
60	60	180	240

Circle graphs are a common way of showing how to represent a whole.
In a circle graph the whole represents 100% or 360°.

This chart and the circle graph below show how the *Never-Lose Team* spends its budget of $900.

Items	Amount Spent	Percent of Total	Degrees in Central Angle
Transportation	$180	20%	72°
Uniforms	$225	25%	90°
Equipment	$360	40%	144°
Snacks	$135	15%	54°
Totals	$900	100%	360°

To make, or construct, a circle graph:

- Find the percent that each item represents.

$$\frac{\$180}{\$900} \diagdown \frac{P}{100} \longrightarrow 900P = 18{,}000$$
$$P = 20\%$$

- Find the number of degrees that the percent represents.

$$20\% \text{ of } 360° = \underline{\quad?\quad}$$
$$0.20 \times 360° = 72°$$

- Use a compass to draw a circle. Mark the center of the circle and draw a radius.

- Use your protractor to draw the central angles found for each percent.

- Label each sector of the graph and the graph itself.

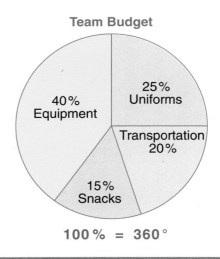

Team Budget

40% Equipment

25% Uniforms

Transportation 20%

15% Snacks

100% = 360°

Give the size of the central angle needed to show these percents.

1. 10%
2. 50%
3. 40%
4. 75%
5. 60%

6. $33\frac{1}{3}\%$
7. $12\frac{1}{2}\%$
8. $37\frac{1}{2}\%$
9. $66\frac{2}{3}\%$
10. 16%

11. Complete this chart showing the languages spoken by the families of the students in Super-Student School. Make a circle graph to show this data.

Language	Number of Families	Percent of Total	Angle Measure
English	200		
Spanish	125		
Chinese	50		
French	75		
German	25		
Other	25		
Totals	500	100%	360°

12. Draw a circle graph to show Raoul's daily activities.

Activity	Time Spent	Percent of Total	Angle Measure
Sleep	8 hours		
Meals	3 hours		
School	6 hours		
Study	3 hours		
Sports	4 hours		
Totals	24 hours	100%	360°

13. Make a circle graph to illustrate the types of animals on MacDonald's Farm: horses, $16\frac{2}{3}$%; cows, $33\frac{1}{3}$%; hogs, 25%; sheep, 25%.

14. Make a circle graph to illustrate a family's monthly budget.

Rent: $ 775   Clothing: $465   Entertainment: $310
Food: $1240   Medicine: $155   Miscellaneous: $155

CALCULATOR ACTIVITY

15. Use a calculator in completing this chart. (Hint: Round each percent to the nearest whole number.) Make a circle graph of the data.

Endangered Species

Species	Number	Percent of Total	Angle Measure
Mammals	555		
Birds	1044		
Reptiles	186		
Amphibians	54		
Fishes	625		
Invertebrates	2125		
Totals	4589	100%	360°

To make it easy to interpret and discuss a set of data, look for *one* number that will represent all the data in the set.

Noontime Temperatures March 26 – 31	
Mar. 26	49
Mar. 27	55
Mar. 28	45
Mar. 29	57
Mar. 30	55
Mar. 31	33

▶ **Measures of Central Tendency** are such numbers.

Here are three measures of central tendency.

Mean: the average of the data.

Median: the middle number of the data, found by arranging the numbers in order.

Mode: the most frequently used number among the data.

Find the mean, median, and mode for the chart above.

Mean: Find the average of the 6 temperatures.

$$\frac{49 + 55 + 45 + 57 + 55 + 33}{6} = \frac{294}{6} = 49° \text{ Mean}$$

Median: Arrange the temperatures in order to find the middle temperature. For an even set, average the two middle numbers.

33, 45, 49, 55, 55, 57

2 middle numbers $\longrightarrow \dfrac{49 + 55}{2} = 52°$ **Median**

Mode: most frequently used number.

33, 45, 49, 55, 55, 57

used most $\longrightarrow 55°$ **Mode**

Range: another statistical measure that gives the *difference* between the highest and the lowest numbers in a set of data.

The range of temperatures is the difference between 57° and 33°:

57° − 33° = 24° **Range**

Find the mean, median, mode, and range for each set of data.

1. 85, 90, 75, 85, 95

2. 40°, 38°, 26°, 36°, 30°

3. $24, $17, $26, $32, $26, $19

4. 120, 96, 104, 84, 104

5. 1017, 1000, 1001, 1006

6. $\frac{3}{4}$, 0.25, 0.50, $\frac{2}{5}$

Box-and-Whisker Plots

Box-and-whisker plots are helpful in determining the central tendency of collected data. This chart shows the speeds of the world's fastest animals.

World's Fastest Animals			
Animal	Speed (mph)	Animal	Speed (mph)
Cheetah	70	Hyena	40
Pronghorn antelope	61	Zebra	40
Wildebeest	50	Mongolian wild donkey	40
Lion	50	Greyhound	39.35
Thomson's gazelle	50	Whippet	35
Quarter horse	47.5	Rabbit (domestic)	35
Elk	45	Mule deer	35
Cape hunting dog	45	Jackal	35
Coyote	43	Reindeer	32
Gray fox	42	Giraffe	30

Median of Lower Half ← (35, Rabbit/Whippet)

To construct a box-and-whisker plot from this data:

- Draw a number line, using appropriate numbers.

  20 25 30 35 40 45 50 55 60 65 70

- Determine the median of the given data. ⟶ $\dfrac{(40 + 42)}{2} = 41$ Median of Whole

- Determine the median of the upper half of the data. ⟶ $\dfrac{(50 + 47.5)}{2} = 48.75$ Median of Upper Half

- Determine the median of the lower half of the data. ⟶ 35 Median of Lower Half

- Locate these three points on the number line and draw the BOX.

  50% of the data lies in the box.

- The whiskers are lines that are drawn to the extremes.

  20 25 30 35 40 45 50 55 60 65 70

Below is the box-and-whisker plot for the slowest moving animals. Compare it with the plot above and answer the questions.

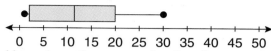

0 5 10 15 20 25 30 35 40 45 50

7. How does the range for the fastest animals compare with that of the slowest?

8. What is the mean for the speed of the fastest animals?

9. Can the mean for the speed of the slowest animals be computed? Why or why not?

10. How does the median for the slowest animals compare with that of the fastest?

11. Which group of animals has a wider variety in speeds? How can you tell?

12. Choose a topic and collect data that you can analyze using a box-and-whisker plot.

Coordinate axes: formed in the coordinate plane by the intersection of a horizontal number line, called the **x-axis**, and a vertical number line, called the **y-axis**.

Ordered pair: the name given to the **x-value**
(x, y) and the **y-value** of each point in the coordinate plane.

Coordinates of a point: the **x-** and **y-** values of the point.

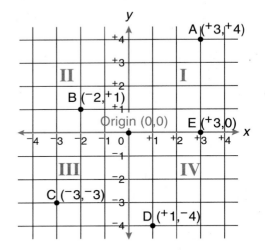

The coordinates of point A are ($^+3$, $^+4$).

The coordinates of point B are ($^-2$, $^+1$).

The coordinates of point C are ($^-3$, $^-3$).

The coordinates of point D are ($^+1$, $^-4$).

The coordinates of point E are ($^+3$, 0).

Origin: the name given to the point **(0, 0)**.

Quadrants: the 4 equal parts into which the coordinate plane is divided. These are named: Quadrant I, Quadrant II, Quadrant III, and Quadrant IV.

Quadrant II	Quadrant I
(−, +)	(+, +)
x — negative	x — positive
y — positive	y — positive
Quadrant III	**Quadrant IV**
(−, −)	(+, −)
x — negative	x — positive
y — negative	y — negative

Give the ordered pair for each point on the graph.

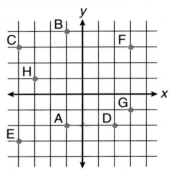

1. A
2. B
3. C
4. D
5. E
6. F
7. G
8. H
9. Which points are in Quadrant I?
10. Which points are in Quadrant II?
11. Which points are in Quadrant III?
12. Which points are in Quadrant IV?

Draw a pair of coordinate axes on graph paper. Then graph these points.

13. I ($^-2$, $^-1$)
14. J (0, $^-4$)
15. K ($^+1$, $^-3$)
16. L ($^-4$, $^+4$)

Use this graph for exercises 17–21.

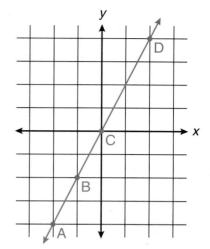

Name the coordinates of each point.

17. *A*   **18.** *B*   **19.** *C*   **20.** *D*

When points *A*, *B*, *C*, and *D* are connected, a straight line is formed.

21. Name another point on $\overleftrightarrow{AD}$ and give its coordinates.

Draw a pair of coordinate axes on graph paper.

22. Graph these points and connect them.

M ($^-$2, $^+$3); N (0, $^-$1); P ($^+$1, $^-$3)

23. Which of these points lies on $\overleftrightarrow{MP}$?

R ($^+$2, $^+$5); S ($^-$1, $^+$2); T ($^-$3, $^-$5)

24. Graph the coordinates of lines 1 and 2.
Describe the relationship between the two lines.

line 1: G ($^-$2, $^-$2); H (0, $^-$1); K ($^+$4, $^+$1)

line 2: R (2, 0); T ($^+$1, $^+$2); S ($^-$1, 6)

25. Graph the coordinates of lines 3 and 4.
Describe the relationship between the two lines.

line 3: X ($^-$2, $^+$5); Y ($^+$1, $^-$1); Z ($^+$2, $^-$3)

line 4: J ($^-$3, $^-$2); K ($^+$3, $^+$1); L ($^+$5, $^+$2)

26. Draw a line parallel to the *y*-axis. Locate three points on the line. Name the coordinates of these points. What do all of these coordinates have in common?

27. Draw a line parallel to the *x*-axis. Locate three points on the line. Name the coordinates of these points. What do all of these coordinates have in common?

CHALLENGE   Draw a pair of coordinate axes on graph paper. Then draw a rectangle that has an area of 12 square units and lies in all four quadrants. Label the vertices of the rectangle *Q, U, A, D*.

28. Name the coordinates of each point. Then find the perimeter of the rectangle.

29. Repeat these steps, using a different rectangle still having an area of 12 square units.

16-13 Graphing Equations

To graph equations on a coordinate grid:

- Choose *at least* 3 different values for x.
- Substitute these values for x in the equation.
- Find the corresponding y-values.
- Write an ordered pair for each x- and y-value.
- Graph each ordered pair.

To graph $y = 2x + 2$, make a table of x- and y-values.
Then list the ordered pairs.

x	$2x + 2$	y	ordered pair
$^-2$	$2(^-2) + 2$	$^-2$	$(^-2, {}^-2)$
$^-1$	$2(^-1) + 2$	0	$(^-1, 0)$
0	$2(0) + 2$	$^+2$	$(0, {}^+2)$
$^+1$	$2(^+1) + 2$	$^+4$	$(^+1, {}^+4)$
$^+2$	$2(^+2) + 2$	$^+6$	$(^+2, {}^+6)$

Here is the graph of these points.
Notice that the points lie in a straight line.

Any point obtained from the
equation, $y = 2x + 2$, will lie on
the line indicated by these points.

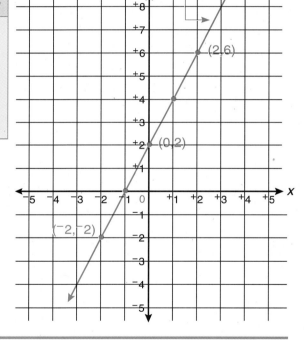

Graph of:
$y = 2x + 2$

Copy and complete the table of values and write the ordered pairs.

1.

x	$x + 2$	y	ordered pair
$^-1$			
0			
$^+1$			
$^+2$			

2.

x	$3x - 2$	y	ordered pair
0			
$^+1$			
$^+2$			
$^+3$			

Graph four points for each equation.

3. $y = x - 3$ 4. $y = x - 2$ 5. $y = 3x + 1$ 6. $y = 2x$

Graph three points for each equation. Then describe the relationship between these two lines.

7. $y = x + 1$; $y = x - 3$

8. $y = 2x - 1$; $y = \dfrac{x}{^-2} - 1$

9. Graph $y = x$. Describe what the line does to Quadrants I and III.

10. Graph $y = {}^-x$. Describe what the line does to Quadrants II and IV.

11. Graph square *SQUA*: *S* $({}^-2, 0)$; *Q* $({}^-2, {}^+4)$; *U* $({}^+2, {}^+4)$; *A* $({}^+2, 0)$. Then graph the equation $y = 2 - x$. What does the line form?

Lines Parallel to Axes

Look at the graph of $y = {}^+2$.
It is the graph of a line parallel to and 2 units above the *x*-axis.

Now look at the graph of $x = {}^+2$.
It can be described as the graph of a line parallel to and 2 units to the right of the *y*-axis.

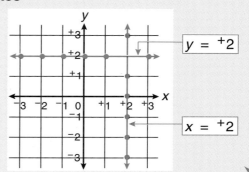

Graph these equations and describe the relationship between the lines.

12. $y = {}^-1$ **13.** $y = {}^+3$ **14.** $y = {}^-4$

15. $x = {}^+1$ **16.** $x = {}^-2$ **17.** $x = {}^+4$

18. Graph triangle *TRI*: *T* $({}^-3, 0)$; *R* $({}^+1, {}^+6)$; *I* $({}^+3, 0)$.
Then graph the equation $y = {}^+1$.
What is the relationship between $y = {}^+1$ and $\overline{TI}$?

CHALLENGE

19. Use a coordinate graph to find the area of trapezoid *ZOID* formed by joining these points:

Z $({}^-3, {}^+3)$; *O* $({}^+4, {}^+3)$; *I* $({}^-1, {}^-1)$; *D* $({}^+2, {}^-1)$.

[Hint: $A = \dfrac{1}{2}(b_1 + b_2)h$]

Problem: Captain Short Jack Bronze has been shipwrecked on an island. He has managed to salvage two sacks of food from the debris that has washed ashore. One sack is labeled to show it contains one can each of hash, stew, and tuna. The other sack is labeled to show it contains one can each of potatoes, rice, beans, corn, and yams. The cans themselves, however, have no labels. If Captain Bronze picks one can from each sack, what is the probability that he will eat tuna and rice for a meal?

1 IMAGINE Draw and label the sets of cans without labels.

2 NAME *Facts:* Sack One – 3 unlabeled cans of hash, stew, tuna

Sack Two – 5 unlabeled cans of potatoes, rice, beans, corn, yams

Question: What is the probability that the captain will have tuna and rice for a meal?

3 THINK Probability = $\dfrac{\text{Number of Favorable Outcomes}}{\text{Total Number of Possible Outcomes}}$

A tree diagram, like the one shown, will help you find all possible outcomes. Count them.

4 COMPUTE There are 15 possible outcomes.

Favorable outcomes that would give the captain tuna and rice for a meal are: (tuna, rice).

So, P (tuna, rice) = $\dfrac{1}{15}$

5 CHECK Look back at the tree diagram. Was there any other combination besides this that would give the captain tuna and rice for a meal?

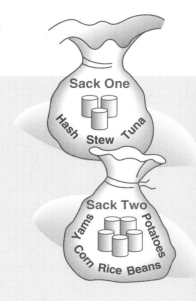

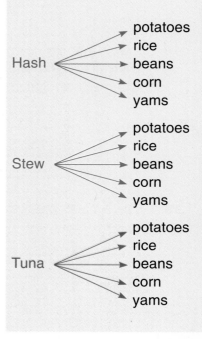

Solve by using an organized list.

1. An envelope contains 5 cards that are numbered 6, 8, 10, 12, and 14. A bag contains 3 balls, numbered 1, 2, and 3. Without looking, Jake chooses a card from the envelope and a ball from the bag. He adds the number on the card to the number on the ball. What is the probability that the sum is greater than 13?

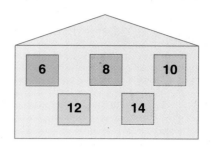

| 1 IMAGINE | Draw and label a picture showing your choices. |

| 2 NAME | → Facts
→ Question |

| 3 THINK | Use a tree diagram to find the possible outcomes. Then find the probability that the sum of the two numbers is greater than 13. |

| 4 COMPUTE | → | 5 CHECK |

2. Joanna spun the dial on this spinner twice. She took the first number she got and multiplied it by the second number. What is the probability that the product was an even number?

3. Juan spun the same dial twice and then tossed a coin. If the coin came up heads, he added the two numbers. If the coin came up tails, he multiplied the two numbers. What is the probability that the result was less than 10?

Exercises 2 and 3

4. How many ways can 4 differently colored sailboats be docked in adjacent slots at a pier?

5. How many different combinations do you have to choose from if a frozen yogurt shop offers 2 flavors of yogurt and 3 toppings?

6. Conchetta has a choice. She can spin this dial twice or she can choose a ball without looking, replace it, and choose again. Which option gives her the best chance of getting the same number twice?

Exercise 6

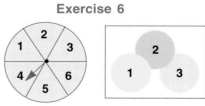

7. Marvin has spun the dial eight times and tallied his results. Where must the dial land on the next two spins so that the mean is the same as the mode?

TALLY

Number	1	2	3	4	5	6
Frequency	I	I	I	III	I	I

16-15 Problem Solving: Applications

Choose the correct answer.

1. Which of the following is an equally likely outcome?
   a. drawing a red card from the color cards
   b. drawing a 1 from the number cards
   c. spinning a 6 on the dial

2. The most likely outcome for the number cards is:
   a. 1   b. 2   c. 3

3. The number of possible outcomes for the number cards is:
   a. 3   b. 4   c. 5

4. How many possible outcomes are there on the dial?
   a. 4   b. 2   c. 6

5. In drawing a color card the P (*not* blue) is:
   a. $\frac{1}{5}$ b. $\frac{4}{5}$ c. $\frac{1}{4}$

6. In drawing a number card the P (2) is:
   a. $\frac{1}{4}$ b. $\frac{1}{2}$ c. 1

7. In spinning the pointer the P (odd) is:
   a. 1   b. 0   c. $\frac{1}{4}$

8. In spinning the pointer the P (even) is:
   a. 1   b. 0   c. $\frac{1}{4}$

> USE THESE STRATEGIES:
> Use a Model
> Make an Organized List
> Use a Graph
> Write an Equation

9. Make a tree diagram to show all possible outcomes for choosing a green card and spinning an 8 on the dial. There are how many possible outcomes?

   a. 9 outcomes   b. 11 outcomes   c. 30 outcomes

10. Make a tree diagram to show all outcomes for getting any number from the cards and from the dial. There are how many possible outcomes?

    a. 8 outcomes   b. 24 outcomes   c. 10 outcomes

11. The P (red) and P (spinning 6) is:
    a. $\frac{1}{5}$ b. $\frac{1}{2}$ c. $\frac{1}{10}$

12. The P (*not* blue) and P (spinning a composite number) is:
    a. $\frac{2}{3}$ b. $\frac{4}{5}$ c. $\frac{1}{6}$

13. Draw a number card. Do *not* replace it. Draw again. What is P (odd, even)?
    a. $\frac{1}{4}$ b. $\frac{1}{2}$ c. $\frac{1}{6}$

14. $\dfrac{9!}{3! \times 5!} = \dfrac{?}{}$ a. 63   b. 504   c. 126

Solve.

15. In how many ways can the four members of the relay team be arranged for the swim meet?

16. How many ways can the postal service arrange the digits 0 through 4 in a zip code so that no digits repeat?

17. A cafeteria offers two types of hot dogs (chicken or turkey) and three toppings (cheese, sauerkraut, or onions) for lunch. How many different combinations of hot dogs and toppings can you buy?

The chart shows the foul shots made by members of two opposing teams.

School	Shots Made					
Radelyn	6	3	5	7	8	1
Brenside	8	2	4	4	10	2

18. What is the median and range of the shots made by each school?

19. What is the mode and mean of the shots made by each?

20. Make a double bar graph to display this data.

Ned made a double line graph to compare his weight with his brother's on each of the nine planets. Use it to complete exercises 21–24.

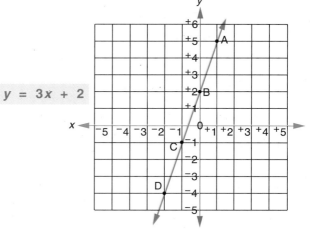

21. How much would Ned weigh on Jupiter?

22. How much would Ned's brother weigh on Jupiter?

23. Write a ratio to compare these two weights.

24. How much more would Ned weigh on Mercury than his brother?

25. Make a circle graph to display these estimated expenses of a company: taxes, $80,000; wages, $160,000; overhead, $100,000; maintenance, $60,000.

Use the graph of $y = 3x + 2$ to complete exercises 26–30.

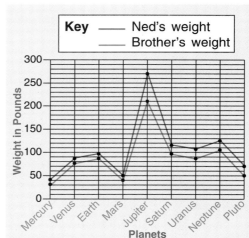

$y = 3x + 2$

26. Which point lies at (1, 5)?

27. What are the coordinates of point B? point C? point D?

28. In what quadrants do the points lie?

29. Make a table for the line $y = 6 - x$. Then graph the line.

30. Does the line $y = 6 - x$ cross the graph of $y = 3x + 2$ at any point?

411

More Practice

List the different possible outcomes.

A container holds 4 red pens, 4 blue pens, and 4 yellow pens.
Select one without looking. Find:

1. P(red) **2.** P(*not* yellow) **3.** P(white) **4.** P(*not* black)

A box contains 7 orange blocks, 1 green block, and 6 purple blocks.
Select one without looking. Find:

5. P(green) **6.** P(orange or purple) **7.** P(*not* orange) **8.** P(*not* green)

9. Luis rolls a number cube, labeled 1–6, once. What is the probability that he will get a number greater than 3?

10. An envelope contains a blue card, a yellow card, and a red card. One card is chosen without looking. What is the probability that it is *not* red?

11. From cards numbered 1 to 6, Leo chooses a card and does not replace it. He chooses a second card. What is the probability both cards will be even numbered? What is the probability that both cards will be less than 3?

The chart shows the number of people who volunteered to help at the school's fairs.

For this chart, what is:

Season	Volunteers
Winter	30
Spring	58
Summer	65
Fall	44

12. the range? **13.** the median?

14. the mode? **15.** the mean?

16. Make a line graph for this chart.

Give the coordinates of each point.

17. H **18.** J **19.** N

20. P **21.** S **22.** T

Draw a set of coordinate axes and graph the following:

23. $J(^-3, ^+2)$ **24.** $K(0, ^-2)$

25. $L(^+3, ^-2)$ **26.** $M(^-5, ^-1)$

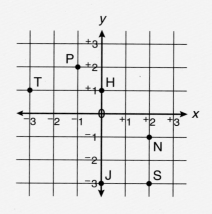

Math Probe

SLIDES AND FLIPS

Each set of points below forms a polygon.

Connect the points and name the polygon for each set.

1. (6, 2) (9, 2) (11, 1) (4, 1) (6, 2)

2. (4, 3) (5, 6) (1, 6) (0, 3) (4, 3)

3. (3, 0) (2, 4) (5, 5) (3, 0)

Write a set of coordinates that will form:

4. a square having a perimeter of 16 units.

5. a rectangle having an area of 15 square units.

> **Slide**
> Move a figure along a line to a new position.
> **Flip**
> Move a figure about a line.

Copy the graph below on grid paper. Answer these questions.

6. Starting from point *A*, move clockwise and list the vertices of the figure graphed.

7. Make a new set of ordered pairs by adding 3 to each coordinate. Graph the new points. Connect them. What happened to the figure?

8. Multiply each original coordinate by 2.
   A (1, 5) → (2, 10)
   Make a new graph. What happened?

9. Multiply each original coordinate by ⁻1. Graph and describe the result.

10. Divide each original coordinate by ⁻1. Graph and describe the result.

11. Multiply the second number in each pair by 3. Graph and describe the result.

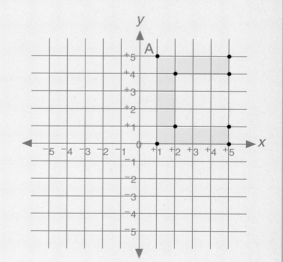

Describe the rule that has this effect on a graph:

12. changes its shape.   13. slides it over.   14. flips it over.

15. Graph this polygon: (2, 2) (6, 2) (4, 6) (2, 2)
    Then predict what happens when each original coordinate is divided by ⁻2 and graphed. Check your prediction.

Check Your Mastery

Spin the dial.

See pp. 386–393

1. Find $P(A)$
2. Find $P(E)$
3. Find $P(C$ or $D)$
4. Find $P(A$ or B or C or $D)$

5. An envelope contains 5 cards numbered 1 through 5. A card is chosen without looking, then put back. Another card is chosen. What is the probability that both cards are even? that both cards are odd?

6. In how many ways can 4 posters be arranged side by side on a cork board?

7. Make a circle graph to show the Grant family's telephone use. See pp. 400–401

Type of Call	Business	Family	Friends	Miscellaneous
Percent	15%	45%	35%	5%

This chart shows the number of pizzas delivered to homes for Pan Fresh Pizza last week. See pp. 396, 402–403

Days	Mon.	Tues.	Wed.	Thurs.	Fri.	Sat.	Sun.
Number of Pizzas	28	16	23	20	52	37	26

8. What is the range?
9. What is the mode?
10. What is the median?
11. What is the mean?
12. Make a bar graph for this data.

Give the coordinates of each point. See pp. 404–407

13. A
14. B
15. C
16. G
17. D
18. E
19. F
20. H

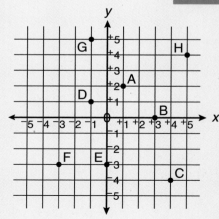

Complete the table for the equation $y = 3x - 2$. Draw a set of coordinate axes and graph.

21. 22. 23. 24. 25.

x	$^-2$	$^-1$	0	$^+2$	$^+3$
y	?				

414

Cumulative Review

Choose the correct answer.

1. A circle graph is used to represent the whole. What percent of the whole is X?

   a. 25%
   b. $33\frac{1}{3}$ %
   c. 35%
   d. 50%

2. Point A is in what quadrant?

   a. I
   b. II
   c. III
   d. IV

3. Tammy received the following scores on her Spanish tests: 86, 85, 83, 85, 92. What is the range?

   a. 10   **b.** 9
   c. 8   **d.** 7

4. In which region is the spinner most likely to stop?

   a. 1
   b. 2
   c. 3
   d. none of these

5. Into what figure would this net fold?

   a. triangle
   b. triangular prism
   c. triangular pyramid
   d. rectangular prism

6. How much greater is the volume of the rectangular prism than the cube?

   a. 160 cm^3
   b. 64 cm^3
   c. 224 cm^3
   d. 96 cm^3

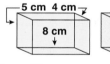

7. To show a trend, comparison, or a growth pattern, which graph would be most useful?

   a. circle   **b.** line
   c. bar   **d.** pictograph

8. The expression 5! – 3! means:

   a. 5 – 3
   b. 5 × 5 × 5 – 3 × 3 × 3
   c. (5 × 4 × 3 × 2 × 1) – (3 × 2 × 1)
   d. $5^2 - 3^2$

9. The surface area of $ABDC$ is how many times larger than the surface area of $ACFE$?

   a. $3\frac{1}{3}$
   b. $2\frac{1}{2}$
   c. 28
   d. 40

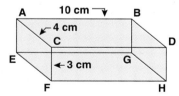

10. How many square pyramids 5 in. high can fit into this cube?

   a. 3
   b. 2
   c. 4
   d. 6

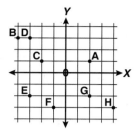

11. How many 0.5-cm cubes will fit into this box?

   a. 200
   b. 1000
   c. 140
   d. 875

12. How many cu ft of sand will occupy this box if it is $\frac{3}{4}$ full?

   a. 12 ft^3
   b. 147 ft^3
   c. 110 ft^3
   d. $9\frac{3}{16}$ ft^3

13. In spinning the wheels, for which are outcomes equally likely?

   a. 1 and 2
   b. 2
   c. 3
   d. 1

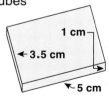

14. The volume of cylinder A is about how much greater than cylinder B?

   a. 109.9 cm^3
   b. 392.5 cm^3
   c. 580.9 cm^3
   d. 1428.7 cm^3

415

Find the volume of each.

15.
17 dm
4 dm
6 dm

16.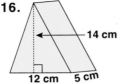
14 cm
12 cm   5 cm

17.
3 m
7 m
1 m

18.
2.4 cm
2.4 cm
2.4 cm

Here is a set of test grades: 97, 82, 75, 78, 75, 91

19. What is the *median* for these grades?
    a. 80   **b.** 83   **c.** 75   **d.** 76.5   **e.** none of these

20. What is the *mean* for these grades?
    a. 80   **b.** 83   **c.** 75   **d.** 76.5   **e.** none of these

21. What is the *mode* for these grades?
    a. 97   **b.** 80   **c.** 75   **d.** 83   **e.** none of these

22. What is the *range* for these grades?
    a. 23   **b.** 17   **c.** 75   **d.** 97   **e.** none of these

23. The total surface area of this rectangular prism is:
    a. 480 cm^2 **b.** 384 cm^2 **c.** 96 cm^2 **d.** none of these

12 cm
6 cm   8 cm

24. The total surface area of a cylindrical pipe 12 m long with a diameter of 2 m and open on both ends is:
    a. 6.28 m^2 **b.** 37.67 m^2 **c.** 75.36 m^2 **d.** none of these

Give coordinates on the points on the graph.

25. *A*   **26.** *B*   **27.** *C*

28. *D*   **29.** *E*   **30.** *F*

31. *G*   **32.** *H*

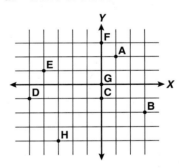

Solve.

33. A house trailer is 65 ft long. It is 24 ft wide and $7\frac{1}{2}$ ft in height. What is the total surface area of the trailer?

34. Three different letters are picked at random from the word BEAM. What is the probability of picking at least one vowel?

35. A cylindrical water tank has a radius of 15 ft and is 20 ft high. What is the surface area of the water tank?

36. Joyce has an aquarium 80.6 cm long. It is 45.5 cm wide and 40 cm deep. How many cubic centimeters of water are needed to fill the aquarium?

37. Lois asked 250 visitors leaving a zoo which exhibits they would like to revisit. 25 visitors said the reptile house, 75 said the petting zoo, and 150 said the aquarium. On a circle graph, what would the angle measures for this data be?

38. Joe has 5 blue socks and 3 white socks in a drawer. He takes out one sock and then another without replacing the first. What is the probability that Joe will remove 2 blue socks?

Cumulative Test II

Choose the correct answer.

1. The length of a classroom
   a. 5 km   b. 5 m   c. 5 cm   d. none of these

2. What is the missing number? 0.15 cm = __?__
   a. 15 mm   b. 150 mm   c. 1.5 mm   d. none of these

3. The ratio used to compare one dime to 4 cents is:
   a. 1 : 4   b. 1 : 10   c. 10 : 1   d. none of these

4. Mary reads 6 pages in 10 minutes. How many pages can she read in 25 minutes?
   a. 12   b. 15   c. 25   d. none of these

5. 0.72 written as a percent is:
   a. 0.72%   b. 7.2%   c. 7200%   d. none of these

6. $33\frac{1}{3}$ % of x = 1150
   a. x = 3450   b. x = 10,350 c. x = 2300   d. none of these

7. What percent of 75 is 15?
   a. 15%   b. 5%   c. 20%   d. none of these

8. At a rate of 8% the commission on $700 is:
   a. $56   b. $5.60   c. $65   d. none of these

9. The interest on $65,000 at 15% for 3 years is:
   a. $9750   b. $29,250   c. $292.50   d. none of these

10. The triangle in which no sides are congruent is called:
    a. equilateral b. isosceles   c. scalene   d. none of these

11. The probability of rolling an even number on a die is:
    a. $\frac{1}{6}$ b. 1   c. $\frac{1}{2}$ d. $\frac{1}{3}$ e. none of these

12. Which outcome has the greatest probability when rolling 1 die?
    a. $P(1)$ b. $P(7)$ c. P(multiple of 3)   d. P(even)   e. none of these

13. On a circle graph a sample of 75% would be represented by:
    a. 100°   b. 75°   c. 270°   d. 360°   e. none of these

Use the drawing to answer the questions.

14. $\angle ABC \cong \angle$ __?__   15. $\overleftrightarrow{AF} \parallel$ __?__   16. $\overleftrightarrow{GL} \perp$ __?__

17. An acute angle is $\angle$ __?__

18. An obtuse angle is $\angle$ __?__

19. $\angle DJL$ is supplementary to $\angle$ __?__

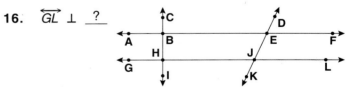

Supply the missing information.

20. Commission: __?__
    Rate of com.: 14%
    Amount sold: $3500

21. Principal: $6000
    Rate of int.: $5\frac{1}{2}$ %
    Time: 4 yr
    Interest: __?__
    Total amount: __?__

22. Principal: $15,000
    Rate of int.: 11.2%
    Time: 18 mo
    Interest: __?__
    Total amount: __?__

Find the area:

23.
    2.7 cm

24.
    3.2 m
    1.8 m

25.
    3 dm
    4.5 dm
    9 dm
    5 dm

417

Write true (T) or false (F). Then rewrite each false statement to make it true.

26. $\frac{4}{7} = 47\%$

27. 7 is $14\frac{2}{7}\%$ of 49

28. 240% of $180 = 192$

29. 18% of 48 is 8.64

30. An hourly wage always gives a greater salary than commission.

31. Polyhedrons have polygons as faces.

32. The five faces of a triangular prism are triangles.

33. A rectangular prism is made up of six rectangles.

34. The total surface area of this polyhedron is 438 cm².

35. The volume of a cube measuring 8 dm on an edge is 64 dm³.

36. ($^+7$, $^-2$) and ($^-2$, $^+7$) represent the same point on the coordinate plane.

37. Every point on the number line except 0 has an opposite.

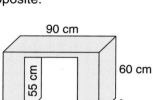

Solve.

38. Calculate the total surface area of this solid wood utility table.

39. During summer vacation Fernando saved $400. He budgeted his money as shown in the graph. How much money is budgeted for each item?

Clothes 25%
Books 25%
$12\frac{1}{2}\%$ Other
15% Savings
$12\frac{1}{2}\%$ Recreation
10% Stationery

40. Make a double bar graph to show money spent on space activities ($ in millions).

Date	Total Amount Spent	Amount Spent by NASA
1955	$ 75	$ 74
1957	150	76
1959	521	146
1961	1518	944
1963	4114	2552
1965	6782	4990

41. How many square meters of cloth were used to make 30 triangular banners each with a base of 62 cm and a height of 20 cm?

42. Which is the better buy, one 6-oz can at 89¢ or three 6-oz cans at 98¢?

43. An item regularly priced at $75 is on sale for 15% less. Find the cost of the item after a sales tax of 6% is added.

44. At a sale, Agnes received a discount of $40 on a coat marked $120. What was the rate of discount? What was the cost of the coat?

45. Out of the $42,000 Mr. Ackert earns each year, he invests $12\frac{1}{2}\%$. How much money does he invest?

46. $16\frac{2}{3}\%$ of each radio hour is spent in advertising. The rest is spent in programming. How many minutes are spent in programming?

47. A new road 3.6 km long was built in $4\frac{1}{2}$ weeks. About how much road was built per week?

48. Mrs. Morgan has $2\frac{1}{2}$ candy bars to give to 5 children. How much will each child receive?

Algebra and Rational Number Topics

In this chapter you will:

- Classify and evaluate algebraic expressions
- Add, subtract, multiply, and divide monomials
- Add and subtract polynomials
- Multiply and divide polynomials by monomials
- Compute with signed rational numbers
- Solve problems: more than one equation

Do you remember?

How to distribute a number or variable across a sum:

$$5(3 + 2) = (5 \times 3) + (5 \times 2)$$
$$a(b + c) = (a \times b) + (a \times c)$$

Be a Star Mathlete

You will be if you can open these mysteries of math.

Identify these numbers and then see if your friends can.

- I am the only *even* prime number.
- I am the first *perfect* number—the sum of my proper divisors.
- I am the smallest number other than one that is both a square and a cube.
- I am the only palindromic 3-digit cube.

Prepare other "Who am I's" to stump your friends.

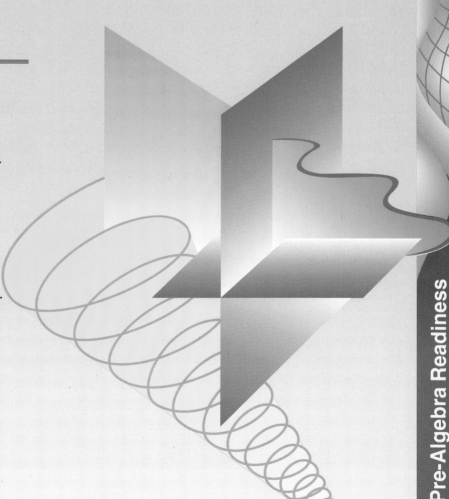

Pre-Algebra Readiness

Algebraic Expression	Definition	Examples
Monomial	A numeral (constant), a variable, or a product of a numeral and one or more variables	6, x, $4a$, b^3, $-3x^2y$
Polynomial	A monomial or the sum or difference of two or more monomials	$(-3x^2 + 8x)$, $(a^3 + a^2b + ab^2 + b^3)$
Binomial	A polynomial that is the sum or difference of *two* monomials	$(6 + x)$, $(4a + b^3)$, $(b^3 - 3x^2y)$
Trinomial	A polynomial that is the sum or difference of *three* monomials	$(x^2 + 2x + 3)$, $(a^2 - 5ab^2 + b^4)$

$$8x^4 \; = \quad \underset{\substack{\text{coefficient} \\ \text{of base}}}{8} \qquad \underset{\text{base}}{x} \qquad \overset{\text{exponent}}{4}$$

▶ **Exponent:** tells how many times a base is used as a factor.

$$8x^4 = 8(x \cdot x \cdot x \cdot x)$$

If no exponent is expressed, *one* is understood.

$$8x = 8(x) = 8x^1$$

▶ **Like terms:** terms that are the same except for their coefficients.

Like terms can be simplified:

$$6x^2y^3 + 8x^2y^3 = (6 + 8)x^2y^3 = 14x^2y^3$$

$6x^2y^3$ and $8x^2y^3$ are like terms.
$6x^2y^2$ and $6x^2y^3$ are unlike.

Classify each polynomial as a monomial, a binomial, or a trinomial.

1. $-2xy$
2. $a^3 + b^2$
3. 21
4. $-16x + y$
5. $x^2 - 8x + 16$
6. $-6x^2y$
7. $y^4 - 6y^2 + 9$
8. $3t$
9. $xy - x + 1$
10. st
11. $-3a + 5b$
12. $km + m^2 + 1$
13. $x^2 - 6$
14. -4
15. $x^2 + 2xy + y^2$
16. $1 - k$
17. $110x$
18. $\frac{1}{2}y^2 + y$
19. $6 + 4b + 2b^2$
20. $x^2y^2 + 17x - 6$

Name the coefficient, base, and exponent in each term.

21. $7x^2$

22. $2x^3y^2 + 8x^2y$

23. $6m^4 - 2m^3 + 3$

Identify which expressions have like terms. Simplify these.

24. $8c + 5c$

25. $8c^2 + 5c$

26. $19x^2y - 12x^2y$

27. $36b^3d^2 - 8b^3d^2$

28. $6(x^3 + x^3)$

29. $7y^3z^2 + 3y^2z^3$

The Degree of Polynomials

Degree of a monomial: the sum of the exponents of the variables in the monomial. A monomial that is a nonzero constant and has no variables has zero degree.

$4x^2 \longrightarrow$ Degree 2

$-\dfrac{1}{2}xy = -\dfrac{1}{2}x^1y^1 \longrightarrow$ Degree 2

$2a^3b = 2a^3b^1 \longrightarrow$ Degree 4

44 has no variable.
It is a nonzero constant. $\longrightarrow$ Degree 0

The degree of a polynomial is the *greatest* of the degrees of its *monomials.*

Polynomial	Degrees of Its Monomials	Degree of Polynomial
$4x^2 + 6x$	2, 1	2
$7a^2b^3 + 4ab + b^4$	5, 2, 4	5
$-2r^2 + rs^2 + 5s$	2, 3, 1	3
$17 + 2t$	0, 1	1

Give the degree of each monomial.

30. $-14x$

31. $3xy$

32. $4x^3y^2$

33. -7

34. $23a$

35. c^2d

36. $-16y^3$

37. $-b$

38. t^2

39. $9r^2s$

40. $5xyz$

41. -2

42. m^4

43. $7xy^2$

44. $14a^2b^2$

Give the degree of each polynomial.

45. $-13a^2b$

46. $2z + 7$

47. $4x^3y^2z$

48. -56

49. $17 + 2t$

50. $16x^2 - 9$

51. $m^4n - 3m^2n^2$

52. $8a^3 + 2ab + b^2$

53. $7b^2 - 2abc$

54. $xy - 7$

55. $a^3b^2 - ab^3$

56. $x^3y^2 + 5x^3 - 10y$

57. $t^3 - 9t^2 + 7$

58. $100s - 5$

59. -11

60. $r^4s + r^3s^3 + rs$

17-2 Evaluating Algebraic Expressions

To find the value of an algebraic expression:

- Substitute the assigned value for each variable.
- Perform the computation, following the order of operations.

Order of Operations
1. ( ) before [ ]
2. Exponents
3. "×" or "÷" left to right
4. "+" or "−" left to right

Evaluate.

$2x^2 - 6x + 5$ when $x = 3$	$9xy - y + 7$ when $x = 5$ and $y = -2$
$2(3)^2 - 6(3) + 5$	$9(5)(-2) - (-2) + 7$
$= 2(9) - 18 + 5$	$= 9(-10) - (-2) + 7$
$= 18 - 18 + 5$	$= -90 + 2 + 7$
$= 0 + 5 = 5$	$= -88 + 7 = -81$
$2x^2 - 6x + 5 = 5$ when $x = 3$	$9xy - y + 7 = -81$ when $x = 5$ and $y = -2$

Evaluate each expression when $a = 3$, $b = -2$, and $c = 5$.

1. $a^2 - b$
2. $2ab + c$
3. $-5a^2$
4. $c^2 - 4ab$
5. $3a^2 - b^2$
6. abc
7. $b^3 - 1$
8. $a^3 - 2b^2 + c$
9. $\frac{1}{2} a + \frac{1}{3} c$
10. $-a^2 b$
11. $50b + 20c$
12. $a^2 b^2 - c$
13. $-6a^2 b + ab$
14. $b^2 c^2$
15. $2a^2 + 8a - 20$
16. $10a + 30c$
17. $(bc)^2 - 3a$
18. $b(a + c)$
19. $ba + bc$
20. $\frac{1}{10} bc^2$

21. Find the value of $r^2 - 2rs + s^2$ when:
    - **a.** $r = 5$ and $s = 1$
    - **b.** $r = 4$ and $s = 2$
    - **c.** $r = -2$ and $s = 3$
    - **d.** $r = -3$ and $s = -1$
    - **e.** $r = 10$ and $s = -7$
    - **f.** $r = 2$ and $s = 5$

22. Examine your answers for parts **a** through **f** of exercise 21. To what special group of numbers do all these answers belong?

Evaluate each equation. Then write *True* or *False*.

23. $2x^2 + 3xy \overset{?}{=} 9$ if $x = -3$ and $y = 1$
24. $x - y \overset{?}{=} 8$ if $x = 10$ and $y = -2$
25. $3s^2 + 2t \overset{?}{=} 0$ if $s = 2$ and $t = -6$
26. $2a^2 - b \overset{?}{=} 32$ if $a = -3$ and $b = 4$
27. $r^2 \overset{?}{=} 4r$ if $r = 10$
28. $5xy - 3x \overset{?}{=} 50$ if $x = -20$ and $y = \frac{1}{10}$
29. $(rs)^2 \overset{?}{=} \frac{1}{8}$ if $r = -1$ and $s = \frac{1}{4}$
30. $k^2 + 7k \overset{?}{=} 98$ if $k = 7$

Ascending and Descending Order of Polynomials

A polynomial with one variable is written in **ascending order** when the degree of each term is **less than** the degree of the term that follows it.

$$4 + 9x + 11x^2 = 4 + 9x^1 + 11x^2$$

Nonzero constant: degree 0

$$0 < 1 < 2$$
Ascending order

$$2y + 7y^3 = 2y^1 + 7y^3$$

$$1 < 3$$
Ascending order

A polynomial with one variable is written in **descending order** when the degree of each term is **greater than** the degree of the term that follows it.

$$4a^3 - 9a^2 + a - 6 = 4a^3 - 9a^2 + a^1 - 6$$

$$3 > 2 > 1 > 0$$
Descending order

$$t^2 - t = t^2 - t^1$$

$$2 > 1$$
Descending order

A polynomial with more than one variable can be written in either ascending order or descending order for any *one* of its variables.

$$4x^2y + xy^2 - 6 = 4x^2y + x^1y^2 - 6 \qquad \text{Descending order of } x$$

$$8s^2 + 4st - t^2 = 8s^2 + 4st^1 - t^2 \qquad \text{Ascending order of } t$$

Arrange in ascending order.

31. $8x^3 - 2 + 6x$

32. $5 - k^2 + 11k + 2k^3$

33. $m^2 - 9 + m$

34. $12 + r^2 - 9r$

35. $100t + t^2 - 18$

36. $r^4 + 3r^2 + 2r - 8$

37. $6t^4 + t^2 - t^3$

38. $44 - 19x^3 + 22x^2$

Arrange in descending order.

39. $t + t^5 + 25$

40. $-x^3 + 7 + x + 2x^2$

41. $9 - r$

42. $s + 2s^2 + 7$

43. $16 - 20k + 11k^2$

44. $p^4 + p^3 + 8 - p$

45. $x + x^5 - x^2 - 2$

46. $m^9 - m^5 + 8m^7$

Arrange in descending order of x.

47. $5x^3y^2 - 18xy + 10x^2y + y^3$

49. $9 - 3x^2y + 10xy$

51. $xy^3 + x^2y + 7 + x^3$

48. $y^2 + 11x - 2x^2$

50. $7x^2 + 21 - 19xy$

52. $x^4y + x^2y^2 + x^3y^3 + 11x$

17-3 Addition and Subtraction of Monomials

> **To add or subtract monomials that have like terms:**
>
> Use the distributive property to simplify each algebraic expression by adding or subtracting the coefficients of like terms.

$9x + 11x =$ ___?___

$9x + 11x = (9 + 11)x$ — Distributive Property

$\qquad = 20x$

$9x + 11x = 20x$

$14b - 6b =$ ___?___

$14b - 6b = (14 - 6)b$ — Distributive Property

$\qquad = 8b$

$14b - 6b = 8b$

$10x^2y + 6x^2y =$ ___?___

$10x^2y + 6x^2y = (10 + 6)x^2y$ — Distributive Property

$\qquad = 16x^2y$

$10x^2y + 6x^2y = 16x^2y$

$-20cd - 11cd =$ ___?___

$-20cd - 11cd = (-20 - 11)cd$

$\qquad = -31cd$ — Distributive Property

$-20cd - 11cd = -31cd$

$-8ab^2 + 10ab^2 =$ ___?___

$-8ab^2 + 10ab^2 = (-8 + 10)ab^2$ — Distributive Property

$\qquad = 2ab^2$

$-8ab^2 + 10ab^2 = 2ab^2$

$7fg - 7fg =$ ___?___

$7fg - 7fg = (7 - 7)fg$ — Distributive Property

$\qquad = 0fg = 0$ — Zero Property

$7fg - 7fg = 0$

In exercises 1–6 identify which terms are like the first term.

1. $7a^2b$ **a.** $7a^2b^2$ **b.** $9a^2b$ **c.** $7a^2b$
2. $2m^2n^2$ **a.** $4m^2n^2$ **b.** $8m^2n^2$ **c.** $5mn^2$
3. abc **a.** $5abc$ **b.** $3abc$ **c.** $2abc^2$
4. $9k^2$ **a.** $11k^2m^2$ **b.** $5k$ **c.** k^2
5. $3xy$ **a.** xy^2 **b.** yx **c.** xy
6. stv^2 **a.** $7tsv^2$ **b.** stv^2 **c.** stv

Simplify by adding or subtracting like terms.

7. $6s + 17s$
8. $9t - 5t$
9. $4x + 28x$
10. $-6y + 11y$
11. $16k - (-3k)$
12. $44m + (-8m)$
13. $15st - 20st$
14. $-8ab - 17ab$
15. $-4z - 4z$
16. $-6k^2 - 15k^2$
17. $m^4 + 13m^4$
18. $-2m^2n + 7m^2n$
19. $20k + 8k - 6k$
20. $8x - 12x - x$
21. $t^2 - 9t^2 + 4t^2$
22. $yz - 8yz + 9yz$
23. $2b^3 - 5b^3 + 7b^3$
24. $r^2 + r^2 + r^2$
25. $15pq - 15pq$
26. $3ac^2 - 4ac^2 + ac^2$
27. $2d + 2d - 2d$

Simplifying Polynomials

A polynomial is in **simplest form** when all like terms have been combined.

To simplify a polynomial, rewrite it, if necessary, such that like terms are grouped together.

$$5x + 3y - 2x + 9y$$
$$= 5x - 2x + 3y + 9y$$
$$= (5 - 2)x + (3 + 9)y$$
$$= 3x + 12y$$

$$4r^2s + 9s^2 + 8r^2s - 6s^2$$
$$= 4r^2s + 8r^2s + 9s^2 - 6s^2$$
$$= (4 + 8)r^2s + (9 - 6)s^2$$
$$= 12r^2s + 3s^2$$

Polynomials often have unlike terms, which cannot be simplified.
DO NOT FORGET to put these terms back in the original polynomial after it has been simplified.

$$5s^2t + 8st^2 - 9s^2t + 11st^2 + 6$$
$$= (5 - 9)s^2t + (8 + 11)st^2 + 6$$
$$= -4s^2t + 19st^2 + 6$$

Do not forget the 6.

$$8x + 7x - 9 + 3x$$
$$= (8 + 7 + 3)x - 9$$
$$= 18x - 9$$

Do not forget the 9.

Simplify the following polynomials.

28. $9r + 6t + 5r - 2t$

29. $12x + 17y - 4x - 9y$

30. $15a - 6b + 10b - b$

31. $4a^2 + 11b - 9a^2 - 5b$

32. $-9 + 10b + 15 + 11b$

33. $-6k + 2m - 9m + k$

34. $25x^2 - 36 + 9 - 10x^2$

35. $r + 30s - 14r + 9s$

36. $18 - 2t + 4t - 9 + t$

37. $14c^2 + 7c - 7c^2 + 20c$

38. $16 + 11g - 12g + 6$

39. $m^2 - 3m + 8m^2 - 9m$

40. $5k - 11k + 9 + 8k$

41. $d^2 - 4 + 11d^2 + 30$

42. $12 + 12r - 20 - r$

43. $-14z + 8z + 4y + 6z$

44. $4x^2y + 8xy^2 - 2xy^2$

45. $9rs^2 + 15r^2s - 10r^2s$

Simplify the following polynomials. Arrange answers in descending order of x.

46. $x^2y + x^3 + 2x^2y + 5y^3$

47. $6xy + 9x^2 + 4xy + 8$

48. $14x^2 - 5x + 9x - 12x^2$

49. $3x^2y - 9xy^2 - 4xy^2 + 6x^2y$

50. $13x^2y - 4x^2y + 7x^3 - 2x^3$

51. $-xy + 7x^2y + 6 + 5xy$

52. $-12x^3y + 4x^2y^3 + 6x^3y + x^2y^3$

53. $x^3 - 6x + 7x^4 + 3x$

The Laws of Exponents

THE LAWS OF EXPONENTS

For Multiplication	For Division

For Multiplication

Add the exponents.

$$a^m \cdot a^n = a^{m+n}$$

$$x^2 x^3 = x^2 \cdot x^3$$

$$= (x \cdot x)(x \cdot x \cdot x) = x^5$$

OR

$$x^2 x^3 = x^{2+3} = x^5$$

For Division

Subtract the exponents.

$$\frac{a^m}{a^n} = a^{m-n}; \; a \neq 0$$

$$\frac{x^5}{x^3} = \frac{\cancel{x} \cdot \cancel{x} \cdot \cancel{x} \cdot x \cdot x}{\cancel{x} \cdot \cancel{x} \cdot \cancel{x}}$$

$$= x \cdot x = x^2$$

OR

$$\frac{x^5}{x^3} = x^{5-3} = x^2$$

To multiply monomials: $6m \cdot 5m^2 n = \underline{\;?\;}$

- Multiply the coefficients.

$$6m \cdot 5m^2 n = (6 \cdot 5) \, m \cdot m^2 n$$
$$= 30m \cdot m^2 n$$

- Use the *Law of Exponents for Multiplication.* (Add the exponents.)

$$30m \cdot m^2 n = 30 \, (m^1 \cdot m^2) n$$
$$= 30 m^{(1+2)} n$$
$$6m \cdot 5m^2 n = 30 m^3 n$$

$3x^2 \cdot 4x^3 = \underline{\;?\;}$

$$(3x^2)(4x^3) = (3 \cdot 4)(x^2)(x^3)$$
$$= 12 \, (x \cdot x)(x \cdot x \cdot x)$$
$$= 12x^5$$

OR $3x^2 \cdot 4x^3 = (3 \cdot 4)(x^{2+3})$
$$= 12x^5$$

$9a^2 b \cdot 3ab^3 = \underline{\;?\;}$

$$(9a^2 b)(3ab^3) = (9 \cdot 3)(a^2 b)(ab^3)$$
$$= 27 \, (a^2)(a)(b)(b^3)$$
$$= 27 \, (a \cdot a)(a)(b)(b \cdot b \cdot b)$$
$$= 27 a^3 b^4$$

OR $(9a^2 b)(3ab^3) = (9 \cdot 3)(a^2 \cdot a)(b \cdot b^3)$
$$= (9 \cdot 3) \, a^{2+1} \, b^{1+3}$$
$$= 27 a^3 b^4$$

To divide monomials: $-35c^2 d^5 \div 5cd^3 = \underline{\;?\;}$

- Divide the coefficients.
- Use the *Law of Exponents for Division.* (Subtract the exponents.)

$$\frac{-35 c^2 d^5}{5 cd^3} = \frac{-7 \, (\cancel{c} \cdot c)(\cancel{d} \cdot \cancel{d} \cdot \cancel{d} \cdot d \cdot d)}{(\cancel{c})(\cancel{d} \cdot \cancel{d} \cdot \cancel{d})} = -7cd^2$$

OR

$$\frac{-35 c^2 d^5}{5 c^1 d^3} = -7 c^{2-1} d^{5-3} = -7cd^2$$

Study this: $\dfrac{48 a^3 b^2}{8 a^2 b^1} = \underline{\;?\;} \longrightarrow \dfrac{48}{8} \, a^{3-2} b^{2-1} = 6a^1 b^1 = 6ab$

Multiply.

1. $8(2x)$

2. $9t \cdot 6t$

3. $-4y \cdot 12y$

4. $10r^2 \cdot 38r$

5. $5b^2 \cdot (-9b^4)$

6. $(-2m)(-16m^2)$

7. $(6ab^2)(-4b)$

8. $15x \cdot 3xy$

9. $-7cd \cdot 8c^3d^2$

10. $14st \cdot 4st$

11. $(-9a)(-11a^2c^2)$

12. $3m^2 \cdot 7m^5$

13. $25k^2 \cdot 4k^2$

14. $-rs \cdot 8r^2s$

15. $(6p^4q^2)(-2p^2q)$

16. $16xy \cdot 8y$

17. $(-30b)(-8bc)$

18. $(40x)(-20xy^3)$

19. $\frac{1}{2}m^4 \cdot 80m$

20. $\frac{1}{5}t^2 \cdot 50t^2$

21. $\frac{1}{3}m^2 \cdot \frac{1}{9}n$

22. $3x^2 \cdot 4x^3 \cdot 5y$

23. $2b \cdot 5b \cdot b^3$

24. $6y^2 \cdot 5xy \cdot 2x^3$

Divide.

25. $\dfrac{18ab^2}{2b}$

26. $\dfrac{-21x^2y}{3xy}$

27. $\dfrac{50x^3y^3}{-10xy}$

28. $\dfrac{-45r^2s}{-3r^2}$

29. $\dfrac{60d^4e^5}{-12de^4}$

30. $\dfrac{27b^3c^4}{9b^3c^4}$

31. $\dfrac{-33s^4t^6}{-11s^2t^4}$

32. $\dfrac{-71r^3t}{71r^2}$

33. $\dfrac{72g^2h}{6g}$

34. $\dfrac{63t^5v^5}{-9t^3v^4}$

35. $\dfrac{56w^4z^3}{8wz^3}$

36. $\dfrac{-13k^2p^2}{-13k^2p^2}$

37. $\dfrac{-21d^3e^2}{7d^3e}$

38. $\dfrac{4ab^2c^3}{8abc}$

39. $\dfrac{15x^4y^3x}{-5x^2y^3}$

40. $\dfrac{-36m^5n^2}{-9m^5n}$

Find the missing factor.

41. $3x^2 \cdot \underline{\ ?\ } = 42x^5$

42. $8a^2b \cdot \underline{\ ?\ } = 72a^4b$

43. $10xy \cdot \underline{\ ?\ } = 110x^4y$

44. $5s^2t^2 \cdot \underline{\ ?\ } = 30s^4t^3$

45. $4x \cdot \underline{\ ?\ } = 100xy^4$

46. $6k^2 \cdot \underline{\ ?\ } = -600k^2$

47. $2ab \cdot \underline{\ ?\ } = -6abc$

48. $-3m \cdot \underline{\ ?\ } = -6m^2n$

49. $7d^2 \cdot \underline{\ ?\ } = 70d^2e^2$

CHALLENGE   Find the area for these rectangles.

50.
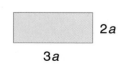
$2a$
$3a$

51.

$2xy$
x^2

52.

$4a^2c^2$
$20ac$

53.
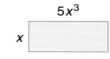
$5x^3$
x

17-5 Addition and Subtraction of Polynomials

To add polynomials: $(6x + 9) + (3x + 11) = \underline{\ ?\ }$

- Rewrite each polynomial with like terms grouped together.

$6x + 9 + 3x + 11$
$= (6x + 3x) + (9 + 11)$
　　Like terms　　　Like terms

- Simplify by adding the coefficients of the like terms and any constants.

$= (6 + 3)x + (9 + 11)$
$= 9x + 20$

$(5x^2 - 11x) + (7x^2 - 3x) = \underline{\ ?\ }$

$= (5x^2 + 7x^2) + (-11x - 3x)$
$= (5 + 7)x^2 + (-11 - 3)x$
$= \qquad 12x^2 + (-14)x$
$= 12x^2 - 14x$

$(2y^2 + 9y) + (-12y + 17) = \underline{\ ?\ }$

$= 2y^2 + (9y - 12y) + 17$
$= 2y^2 + (9 - 12)y + 17$
$= 2y^2 + \qquad (-3)y \quad + 17$
$= 2y^2 - 3y + 17$

Adding is sometimes easier if the like terms in each polynomial are arranged in a column.

$(9x^2 - 6x) + (11x^2 + 20x) = \underline{\ ?\ }$

$\quad 9x^2 - \ \ 6x$
$\underline{11x^2 + 20x}$
$20x^2 + 14x$ **Sum**

$(5y^3 - 9y^2) + (3y^2 + 7) = \underline{\ ?\ }$

$\quad 5y^3 - 9y^2$
$\underline{\qquad + 3y^2 + 7}$
$5y^3 - 6y^2 + 7$ **Sum**

To subtract polynomials: $(7a^2 + 4a) - (5a^2 + 9a) = \underline{\ ?\ }$

Add the opposite of the polynomial to be subtracted.

$(7a^2 + 4a) - (5a^2 + 9a)$

$(5a^2 + 9a) \xrightarrow{\text{opposite}} -5a^2 - 9a$

$= (7a^2 + 4a) + (-5a^2 - 9a)$ ← Adding the opposite of $(5a^2 + 9a)$
$= [7a^2 + (-5a^2)] + (4a - 9a)$
$= (7 + {}^-5)a^2 + (4 - 9)a$
$= 2a^2 + (-5)a = 2a^2 - 5a$ **Difference**

Subtracting is also sometimes easier if the like terms in each polynomial are arranged in a column.

$(-6cd + 11d) - (14cd - 5d) = \underline{\ ?\ }$

$\quad - 6cd + 11d$
$\underline{- (+14cd - \ \ 5d)}$ ← Change the signs in the subtrahend and add. →

$\quad - 6cd + 11d$
$\underline{-14cd + \ \ 5d}$
$-20cd + 16d$ **Difference**

Add.

1. $(14x + 9) + (27x - 15)$

2. $(-6t + 40) + (15t - 35)$

3. $(8k^2 - 14k) + (-12k^2 - 9k)$

4. $(26y^2 + z) + (15y^2 + 13z)$

5. $(3x - 11y) + (-x + 6y)$

6. $(-4m^2 - m) + (-6m^2 + 7m)$

7. $(31x^2 - 15x) + (-16x^2 - 9x)$

8. $(4cd + 11d) + (15cd - 9d)$

9. $(13xy - 9x) + (-2xy + 16x)$

10. $(b^2 - 4b) + (9b^2 - 12b)$

11. $(0.05e + 0.15f) + (0.62e + 0.33f)$

12. $(3.8a + 8.1b) + (5.9a - 3.6b)$

13. $\begin{array}{r} 77a^2 + 25a \\ 44a^2 + 75a \\ \hline \end{array}$

14. $\begin{array}{r} 10m^2 - 19m \\ 31m^2 - 24m \\ \hline \end{array}$

15. $\begin{array}{r} -43b - 29c \\ -58b - 37c \\ \hline \end{array}$

16. $\begin{array}{r} 35t^2 + 17 \\ 9t^2 - 39 \\ \hline \end{array}$

17. $\begin{array}{r} 82c - 10d \\ -48c + 9d \\ \hline \end{array}$

18. $\begin{array}{r} 16 - 96b \\ 4 - 25b \\ \hline \end{array}$

19. $\begin{array}{r} 4.8x + 9.2y \\ 3.9x - 9.2y \\ \hline \end{array}$

20. $\begin{array}{r} 0.55s + 0.41t \\ 0.77s - 0.40t \\ \hline \end{array}$

Subtract.

21. $(5x + 11y) - (2x + 7y)$

22. $(12a - 3b) - (15a + 10b)$

23. $(16c + b) - (5c + 12b)$

24. $(40a - 25b) - (16a - 5b)$

25. $(-12x + 17y) - (-3x + 26y)$

26. $(36e - 29f) - (19e + 28f)$

27. $(-2x + 8y) - (16x - 8y)$

28. $(72y^2 + 36y) - (20y^2 + 15y)$

29. $(-18cd + 7) - (14cd - 12)$

30. $(100k^2 - 50k) - (38k^2 - 72k)$

31. $(-18a^2 + 13a) - (10a^2 + a)$

32. $(0.82s - 1) - (0.24s - 1)$

33. $\begin{array}{r} 43rs - 25s^2 \\ 17rs - 12s^2 \\ \hline \end{array}$

34. $\begin{array}{r} -71s^2t + 11st \\ -15s^2t - 30st \\ \hline \end{array}$

35. $\begin{array}{r} 48k^2 + 19k \\ -15k^2 - 26k \\ \hline \end{array}$

36. $\begin{array}{r} -31ab - 17ab^2 \\ 27ab - 14ab^2 \\ \hline \end{array}$

37. $\begin{array}{r} 85t^2 - 27t \\ -76t^2 + 28t \\ \hline \end{array}$

38. $\begin{array}{r} 40x^2y + 54xy \\ 27x^2y + 38xy \\ \hline \end{array}$

39. $\begin{array}{r} 0.84k + 1.2k^2 \\ 0.16k - 0.9k^2 \\ \hline \end{array}$

40. $\begin{array}{r} 2.67a + 5.19b \\ 0.29a - 9.14b \\ \hline \end{array}$

41. $\begin{array}{r} 6.3a^2b + 0.9b \\ -0.7a^2b - 0.6b \\ \hline \end{array}$

Compute.

42. $(6a^2 + 10) + (-4a^2 - 17)$

43. $(5t - 6) - (17t + 16s)$

44. $(-42a^2 + 16a) - (-19a^2 + 28a)$

45. $(8p - 21p^2) + (17p + 18p^2)$

46. $(10x^3 + 8x^2) + (-15x^2 + 7)$

47. $(-23r + 19s) - (25s - 6)$

48. Write two binomials whose sum is:
    a. a trinomial   b. a monomial   c. 0

49. Write two binomials whose difference is:
    a. a trinomial   b. a monomial   c. 0

17-6 Multiplying Polynomials

To multiply polynomials: $c(2b + 3c) + c^2 = $ ___?___

- Use the distributive property.

$$c(2b + 3c) + c^2 = 2bc + 3c \cdot c + c^2$$
$$= 2bc + 3c^2 + c^2$$

- Group any like terms.

$$= 2bc + 4c^2$$

- Simplify the polynomial. Write in descending order, if possible.

$$c(2b + 3c) + c^2 = 4c^2 + 2bc \quad \text{Simplified}$$

Study these.

$a(b + c) = $ ___?___

$a(b + c) = ab + ac$

$c(2b + 3c) = $ ___?___

$c(2b + 3c) = 2bc + 3c \cdot c$
$$= 2bc + 3c^2$$

$4a^2(2a + 5) = $ ___?___

$4a^2(2a + 5) = (4a^2)(2a) + (4a^2)(5)$
$$= (4 \cdot 2)(a^2 \cdot a) + (4 \cdot 5)(a^2)$$
$$= 8a^3 + 20a^2$$

Use the distributive property to multiply.

1. $10(a + 2) = 10a + $ ___?___

2. $3(b + 7) = 3b + $ ___?___

3. $m(15 - m) = 15m - $ ___?___

4. $r(2 - 8r) = 2r - $ ___?___

5. $2d(7 + d) = $ ___?___ $+$ ___?___

6. $5y(8 + y) = $ ___?___ $+$ ___?___

7. $a^2(a + 5) = $ ___?___ $+$ ___?___

8. $x^2(x + y) = $ ___?___ $+$ ___?___

9. $6t(t - t) = $ ___?___ $-$ ___?___

10. $9z(z - 2z) = $ ___?___ $-$ ___?___

11. $7(a^2 + 2ab + b) = 7a^2 + $ ___?___ $+$ ___?___

12. $4(b^2 - 3b - 2) = 4b^2 - $ ___?___ $-$ ___?___

13. $v(v^2 - 5v + 6) = $ ___?___ $-$ ___?___ $+$ ___?___

14. $r^2(r^2 + r - 6) = $ ___?___ $+$ ___?___ $-$ ___?___

Multiply.

15. $3(a + b)$

16. $7(c - 2d)$

17. $-(4b - 5c)$

18. $-2c^2(-4e - 6f^2)$

19. $x^2(2x + 6y)$

20. $-b(-3b + 4c)$

21. $8rs(6r^2 - 4s^2)$

22. $-c(5c^2 + 6b)$

23. $-4x^2(6xy - 7y)$

24. $-2ab^2(7a - 3ab)$

25. $4r^2(8m^2n + 3n^2)$

26. $-xy^3(xy - x^2y^2)$

430

Find the product.

27. $m(4n - 3m) + 2mn$

28. $a^2(3a - 2b + 7) - a^2$

29. $ab(7 + 2b - 3a) - 6ab$

30. $2r(s^2 - 5rs + r) - 3r^2$

31. $4x^2(2x + 3xy - 5) + 8x^2$

32. $\frac{1}{2}c(2a + 4b - 6c) + ac$

33. $\frac{1}{3}d(6e - 18d + 3) - 2d^2$

34. $\frac{1}{5}y(10x - 20y + 25z) - 3y^2 + 2z$

35. $3st(s^2 - 2st - t^2) + 4s^2t^2$

36. $6xy(x^2 - y^2 + xy) + 7x^3y - 5xy^3$

37. $a^2b^2(10 - 2b + 3a) - 5a^2b^2$

38. $5z^3(x^3 - 2y^3 - 10) + 10y^3z^3 + 4$

39. $12(a^2 - 2ab + 1) + 3(a^2 - 2)$

40. $d(d + e + e^2) + 2d(e - e^2)$

41. $s(4 - st - t^2) + t(s + st - 1)$

42. $3k(k^2 - 2k + 4) + 6(k^3 - k^2 - 12k)$

43. $c^2(4c - 2cd - d) + 2(c^3 - c^2d)$

44. $5a(a + ab - b) + b(a^2 + 10ab)$

45. $\frac{1}{4}m(12n - 8m) + m^2(6 - 3mn)$

46. $\frac{2}{3}b(6a - 9b) + ab(4 - ab)$

Solve. (The first one is done.)

47.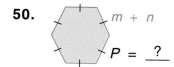

$P = \underline{\ ?\ }$
$P = 4s$
$P = 4(a + b)$
$P = 4a + 4b$

$a + b$

48.

m

$m + 5$

$P = \underline{\ ?\ }$

49.

x $x + y$

y $P = \underline{\ ?\ }$

50.

$m + n$

$P = \underline{\ ?\ }$

51.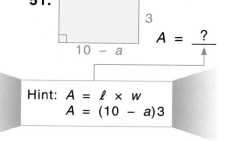

3

$10 - a$

$A = \underline{\ ?\ }$

Hint: $A = \ell \times w$
$A = (10 - a)3$

52.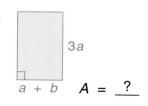

$3a$

$a + b$ $A = \underline{\ ?\ }$

53. The perimeter of a square equals $8m + 24$.
Write an algebraic expression for each side.

54. The perimeter of a regular pentagon is $15a - 35$. Find each side.

55. The radius of a circle equals $5 + n$. Find the circumference in π units.

17-7 Dividing Polynomials by Monomials

To divide a polynomial by a monomial: $\quad 12mn + 8n \div 4n = \underline{\ ?\ }$

- Divide *each* term of the polynomial by the monomial.

$$12mn + 8n \div 4n = \frac{12mn + 8n}{4n}$$

$$= \frac{12mn}{4n} + \frac{8n}{4n}$$

- Divide coefficients; then divide like variables.

$$= \left(\frac{12}{4}\right)m\left(\frac{n}{n}\right) + \left(\frac{8}{4}\right)\left(\frac{n}{n}\right)$$

$$= (3)\,m\,(1) \quad + \quad (2)(1)$$

$$= (3)(1)\,m \quad + \quad (2)(1)$$

$$= 3m + 2 \quad \text{Quotient}$$

The quotient must have the *same* number of terms as the dividend.

$\underbrace{12mn + 8n}_{\text{2 terms in dividend}} \div 4n = \underbrace{3m + 2}_{\text{Must be 2 terms in quotient}}$

Study these.

$27a^2b - 18ab^2 \div 9ab = \underline{\ ?\ }$

$$27a^2b - 18ab^2 \div 9ab = \frac{27a^2b - 18ab^2}{9ab} = \frac{27a^2b}{9ab} - \frac{18ab}{9ab}$$

$$= \frac{\overset{3}{27}a \cdot ab}{\underset{}{9} \quad ab} - \frac{\overset{2}{18}ab \cdot b}{9ab}$$

$$= 3a - 2b \quad \text{Quotient}$$

OR, $\dfrac{27a^2b}{9ab} - \dfrac{18ab^2}{9ab} = \left(\dfrac{27}{9}\right)(a^{2-1})(b^{1-1}) - \left(\dfrac{18}{9}\right)(a^{1-1})(b^{2-1}) = 3a^1b^0 - 2a^0b^1$

$$= 3a - 2b \quad \text{Quotient}$$

Division of polynomials can be expressed 3 ways:

I	II	III
$\dfrac{3a^2 + 9ab}{3a} = \underline{\ ?\ }$	$(3a^2 + 9ab) \div 3a = \underline{\ ?\ }$	$3a\overline{)3a^2 + 9ab}$ $\overset{?}{}$
$= \dfrac{3a^2}{3a} + \dfrac{9ab}{3a}$	$= (3a^2 \div 3a) + (9ab \div 3a)$	$= 3a\overline{)3a^2} + 3a\overline{)9ab}$
$= \left(\dfrac{3}{3}\right)a^{2-1} + \left(\dfrac{9}{3}\right)a^{1-1}b$	$= \left(\dfrac{3}{3}\right)\dfrac{a^2}{a} + \left(\dfrac{9}{3}\right)\left(\dfrac{a}{a}\right)b$	$= 3\overline{)3} \cdot a\overline{)a^2} + 3\overline{)9} \cdot a\overline{)ab}$
$= 1a + 3b$	$= 1a + 3b$	$= 1 \cdot a + 3 \cdot b$
$= a + 3b \quad \text{Quotient}$	$= a + 3b \quad \text{Quotient}$	$= a + 3b \quad \text{Quotient}$

Divide: $24s^2t + 6st + 8st^2$ by $2st = \underline{\ ?\ }$

$$= \frac{24s^2t + 6st + 8st^2}{2st} = \frac{24s^2t}{2st} + \frac{6st}{2st} + \frac{8st^2}{2st}$$

$$= 12s \quad + \quad 3 \quad + \quad 4t$$

Divide.

1. $(a + 3a) \div a$
2. $(x - 5x) \div x$
3. $(4m + mn) \div m$
4. $(7c + cd) \div c$
5. $(5y + 20) \div 5$
6. $(8z^2 - 4z) \div z$
7. $(18b^2 + 12b) \div 6b$
8. $(45d - 36d^2) \div 9d$
9. $(14r + 21r^2) \div 7r$
10. $(6xy - 8y^2) \div 2y$
11. $(10a^2b + 3ab) \div ab$
12. $(4xy^2 - 7xy) \div xy$
13. $\dfrac{18mn - 3n + 6n^2}{3n}$
14. $\dfrac{4b^2 - 6ab + 10b}{2b}$
15. $\dfrac{15ab - 3bc + 6bd}{3b}$
16. $\dfrac{16xy + 8xz - 20x}{4x}$
17. $\dfrac{15s^2t^2 - 5st^2 - 10st}{5st}$
18. $\dfrac{2a^3b^3c^3 - 5a^2b^2c^2 + 7abc}{abc}$
19. $\dfrac{15x^4 - 5x^3 - 10x^2}{5x^2}$
20. $\dfrac{8a^3b^3 - 4a^2b^2 - 2ab^2}{2ab^2}$
21. $\dfrac{12m^3n^3 + 6m^3n^2 - 18m^2n}{6mn}$
22. $\dfrac{3d^3e^2 + 5d^2e^3 - d^2e^2}{d^2e^2}$
23. $\dfrac{x^3y^3z^3 - 2x^2y^2z^2 + 7xyz^2}{xyz^2}$
24. $\dfrac{14a^5b^5 + 7a^4b^4 - 7a^3b^3}{7a^3b^2}$
25. $\dfrac{14k^6m^8 - 42k^4m^9 + 56k^2m^4}{14k^2m^4}$

Divide.

26. $a^2 \overline{)\, a^4 + 2a^3 + a^2}$
27. $c^2 \overline{)\, c^5 - 3c^4 + 2c^2}$
28. $xy \overline{)\, x^3y^3 + 2x^2y^2 - xy}$
29. $mn \overline{)\, m^4n^3 + m^3n^2 + m^2n^3}$
30. $5a^2 \overline{)\, 5a^4 + 10a^3 + 15a^3}$
31. $3b^3 \overline{)\, 3b^3 + 9b^3c^2 + 27b^3c^3}$

Find the missing factor. (The first one is done.)

32. $t^2 \cdot \underline{\ ?\ } = 3t^2 + 8t^2v \longrightarrow (3t^2 + 8t^2v) \div t^2 = 3 + 8v$
33. $x^3 \cdot \underline{\ ?\ } = 7x^3 + 9x^3y^3$
34. $2a \cdot \underline{\ ?\ } = 12a^2 + 18a^4$
35. $5y \cdot \underline{\ ?\ } = 10y^3 - 15y^5$
36. $7xy \cdot \underline{\ ?\ } = 7x^2y^2 + 14x^2y^3 + 21xy^4$
37. $3fg \cdot \underline{\ ?\ } = 9fg - 3f^2g^2 - 12f^2g^3$
38. $14k^2 \cdot \underline{\ ?\ } = 14k^6 - 42k^4 + 56k^2$
39. $abc \cdot \underline{\ ?\ } = 7abc - 21ab^2c + a^2b^3c^2$
40. $3x^2y \cdot \underline{\ ?\ } = 15x^3y^3 - 21x^2y^2 + 6x^2y$

Solve.

41. Find the length of a rectangle having a width of $2n$ units and an area of $6n + 4mn$ square units.

42. Find the width of a rectangle having a length of $10t$ units and an area of $50t^2 - 10t$ square units.

Addition and Subtraction of Signed Rational Numbers

Addition Rules			Subtraction Rules
+, +	$\xrightarrow{\text{Add}}$	+	• Change subtrahend to its opposite.
−, −	$\xrightarrow{\text{Add}}$	−	• Then use addition rules.
+, − or −, +	$\xrightarrow{\text{Subtract}}$	Use the sign of addend farther from zero.	

In adding and subtracting signed rational numbers follow the *same rules* as for adding and subtracting integers.

To add rational numbers having *like* signs: $^-10.5 + {}^-25.2 = \underline{\ ?\ }$

• Find their sum.

• Use the sign of the addends.

Sign of the addends $\longrightarrow$

$$
\begin{array}{r}
^-1\ 0.5 \\
+\ ^-2\ 5.2 \\
\hline
^-3\ 5.7 \quad \textbf{Sum}
\end{array}
$$

To add rational numbers having *unlike* signs: $^-35.7 + {}^+14.9 = \underline{\ ?\ }$

• Find their *difference.*

• Use sign of addend farther from zero.

Negative sign because $^-35.7$ is farther from zero on the number line than $^+14.9$ is. $\longrightarrow$

$$
\begin{array}{r}
^-3\ 5.7 \\
+\ ^+1\ 4.9 \\
\hline
^-2\ 0.8 \quad \textbf{Sum}
\end{array}
$$

To subtract rational numbers: $^-16.25 - {}^+10.3 = \underline{\ ?\ }$

• Add the *opposite* (inverse) of the subtrahend.

$$
\begin{array}{r}
^-1\ 6.2\ 5 \\
-\ ^+1\ 0.3 \\
\end{array}
$$

Opposite of $^+10.3$ = $^-10.3$

$$
\begin{array}{r}
^-1\ 6.2\ 5 \\
+\ ^-1\ 0.3 \\
\hline
^-2\ 6.5\ 5 \\
\end{array}
$$

Difference

Find opposite of $^+10.3 \longrightarrow {}^-10.3$
Then add.

Add.

1. $^-2.4 + {}^-3.1$

2. $^-6.8 + {}^+2.4$

3. $^+10.1 + {}^-8.1$

4. $^+11.02 + {}^+3.89$

5. $^-7.3 + {}^-7.4$

6. $^-4.5 + {}^+7.6$

7. $^-15 + {}^+10.6$

8. $^+15.9 + {}^-15.9$

9. $^-1.7 + {}^-13.5$

10. $^-2.8 + {}^+7.7$

11. $^-3.3 + {}^-8.8$

12. $^-6.13 + {}^+10$

Subtract.

13. $^+1.6 - ^-2.1$

14. $^-3.2 - ^+4.4$

15. $^+7.5 - ^+3.4$

16. $^-7.8 - ^-5.2$

17. $^+1.4 - ^+3.7$

18. $^-6.6 - ^-8.7$

19. $^+2.9 - ^+7$

20. $^-8.6 - ^-9$

21. $^+10.3 - ^-15.31$

22. $^-3.1 - ^-6.05$

23. $^+4.2 - ^-1.8$

24. $^+5.02 - ^+7.1$

25. $^-8.02 - ^+10.6$

26. $^-0.15 - ^-3.402$

27. $^+1.9 - ^+3.607$

Simplify.

28. $^+6.2 + (^+8.3 - ^+10.3)$

29. $^+7.3 - (^+8.4 - ^+2.2)$

30. $^+4.05 + (^-2.3 + ^-1.05)$

31. $^-20 - (^+5.1 - ^-5.4)$

32. $(^-8 - ^-2.6) + (^+4.2 - ^+7.1)$

33. $(^+6.4 + ^-3.9) + (^-2.4 - ^-11)$

34. $(^+6 + ^-4.81) - (^-2.1 + ^+3.7)$

35. $(^-11 - ^-1.05) - (^+2 - ^-6.15)$

Use < or > to compare.
(Hint: Use estimation. The first one is done.)

36. $^+13.81 + ^-21.75 \underline{\ ?\ } ^-10 \longrightarrow ^+14 + ^-22 \approx ^-8$ and $^-8 > ^-10$

37. $^+72.09 + ^+14.83 \underline{\ ?\ } ^+85$

38. $^-10.03 + ^+18.71 \underline{\ ?\ } ^+8$

39. $^-8.03 + ^-9.27 \underline{\ ?\ } ^-18$

40. $^-27.1 - ^-30.3 \underline{\ ?\ } ^-3$

41. $^+50.11 - ^+61.09 \underline{\ ?\ } ^-12$

42. $^-38.85 - ^-12.7 \underline{\ ?\ } ^-26$

43. $^+18.92 - ^-5.85 \underline{\ ?\ } ^+13$

44. $^-62.07 - ^-71.93 \underline{\ ?\ } ^-10$

Write an equation; then solve.

45. The temperature of chemical A was $^-10.5\,°C$. The scientists decreased its temperature by $25.2\,°C$ to cause it to freeze. What is the freezing point of chemical A?

46. A submarine dived 132.58 m to reach a resting depth of 700 meters below sea level. What was its original depth?

47. Josh deposited a loan of $500.00 in his account to cover bills totaling $2059.28. If his balance before the loan was $1875.50, what is his final balance?

17-9 Multiplication and Division of Signed Rational Numbers

In multiplying and dividing signed rational numbers follow the same rules as for multiplying and dividing integers.

Multiplication and Division Rules		
Signs	*Operations*	*Answers*
+, +	× or ÷ →	+
−, −	× or ÷ →	+
+, −	× or ÷ →	−
−, +	× or ÷ →	−

To multiply signed rational numbers:

- Find their products.

- If the factors have *like* signs, the product is *positive*. If the factors have *unlike* signs, the product is *negative*.

```
      ⁻3.1
    ×  ⁻0.2
    ⁺0.6 2
```

Like signs:
Product is positive.

```
         ⁻6.4
    ×   ⁺0.1 1
           6 4
         6 4
    ⁻0.7 0 4
```

Unlike signs:
Product is negative.

To divide signed rational numbers:

- Find their quotients.

- If *both* rational numbers have *like* signs, the quotient is *positive*. If the rational numbers have *unlike* signs, the quotient is *negative*.

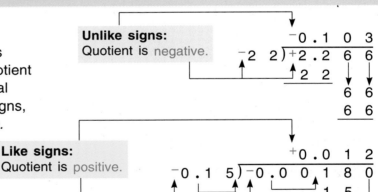

Unlike signs:
Quotient is negative.

```
                 ⁻0.1 0 3
      ⁻2 2)⁺2.2 6 6
            2 2
                6 6
                6 6
```

Like signs:
Quotient is positive.

```
                      ⁺0.0 1 2
      ⁻0.1 5)⁻0.0 0 1 8 0
                      1 5
                      3 0
                      3 0
```

Multiply each factor by ⁺0.2.

1. ⁺1.2   2. ⁻2.1   3. ⁻0.6   4. ⁺0.9   5. ⁻2.4

Multiply each factor by ⁻0.03.

6. ⁺1.1   7. ⁻0.4   8. ⁺2.1   9. ⁻3.2   10. ⁺10.3

Multiply each factor by itself.

11. ⁺1.2   12. ⁻0.6   13. ⁺2.7   14. ⁻3.1   15. ⁻0.14

Multiply.

16. $^-3.7$
     $\times\, ^-1.6$

17. $^+4.3$
     $\times\, ^-0.9$

18. $^-6.07$
     $\times\, ^-1.35$

19. $^-10.6$
     $\times\, ^+2.07$

20. $^+0.018$
     $\times\, ^-1.65$

21. $^-0.009$
     $\times\, ^+3.98$

22. $^+64.21$
     $\times\, ^-2.09$

23. $^-15.32$
     $\times\, ^-0.71$

Divide each dividend by $^+0.25$.

24. $^+16$ 25. $^-20$ 26. $^+4.8$ 27. $^-8.8$ 28. $^+12.4$

Divide each dividend by $^-0.5$.

29. $^+12$ 30. $^-24$ 31. $^-1.6$ 32. $^+2.2$ 33. $^-4.2$

Divide.

34. $^+11.55 \div\, ^+3.3$

35. $^-0.1224 \div\, ^+0.12$

36. $^-14.868 \div\, ^-2.124$

37. $^-0.16\,)\overline{\,^+9.28\,}$

38. $^-7\,)\overline{\,^-0.588\,}$

39. $^+0.5\,)\overline{\,^-47\,}$

40. $^-1.3\,)\overline{\,^+0.2626\,}$

41. $^+1.5\,)\overline{\,^-6\,}$

42. $^+1.8\,)\overline{\,^+0.054\,}$

43. $\dfrac{^+6.633}{^-0.11}$

44. $\dfrac{^-3.8}{^+2.5}$

45. $\dfrac{^-0.651}{^-3}$

Use the order of operations to simplify. (The first one is done.)

46. $^-4.5 - (^+2.4 +\, ^-3.1) = \underline{\ ?\ }$
     $^-4.5 -\quad\ \ (^-0.7)\quad =$
     $^-4.5 +\quad\ \ (^+0.7)\quad =\, ^-3.8$

47. $^+2.2 +\, ^-1.3 \times\, ^-5$

48. $^-8.6 -\, ^+2.4 -\, ^-0.3$

49. $^-0.18 + (^-6.4 -\, ^-11)$

50. $^-0.72 - (^+1.5 +\, ^-0.3)$

51. $^-0.4\,(^-6.5 -\, ^-1.5)$

52. $^+1.2\,(^-4.1 -\, ^-5.3)$

53. $^+10 + (^-8.2 -\, ^+1.4)$

54. $^-18 - (^+2.9 -\, ^-0.7)$

Solve.

55. Sara bought a dozen posters to decorate for a party. Each poster cost $2.59, and Sara had $30 in cash. How much more money was needed to pay the bill?

56. The initial temperature of a solution was $^-10°$C. If over a period of 5 minutes the temperature was raised 1.5 per minute, what was the final temperature of the solution?

STRATEGY
Problem Solving: Using More Than One Equation

Problem: Four seventh-grade students are comparing their special-edition baseball cards. How many special-edition baseball cards does each person have if:

I	Jed's cards	+	Tina's cards	+	Han's cards	= 124
II	Jed's cards	+	Margo's cards			= 134
III				2 ×	Han's cards	= 78
IV			Tina's cards	+	Han's cards	= 64

1 IMAGINE You want to trade baseball cards and need to know how many cards each person has.

2 NAME *Facts:* An equation for each clue

$$\begin{cases} \text{I} & J + T + H = 124 \\ \text{II} & J + M = 134 \\ \text{III} & 2H = 78 \\ \text{IV} & T + H = 64 \end{cases}$$

Question: How many cards does each person have in his or her collection?

3 THINK By combining the clues from two of the equations, the number of special-edition cards can be found.

4 COMPUTE Combine Clue I and Clue IV to solve for *J*.

$$\begin{array}{rl} \text{I} & J + T + H = 124 \\ \text{IV} & \underline{\quad T + H = 64} \\ & J \qquad\quad = 60 \end{array}$$

The difference between these equations equals the number Jed has. So, Jed has 60 cards.

Combine *J* = 60 with Clue II to solve for *M*.

$$\begin{array}{rl} \text{II} & J + M = 134 \\ & \underline{J \qquad = 60} \\ & M = 74 \end{array}$$

The difference between these equations equals the number Margo has. So, Margo has 74 cards.

Use Clue III to solve for *H*.

$$\text{III} \quad 2H = 78 \longrightarrow H = 39$$ So, Hans has 39 cards.

Combine *H* = 39 with Clue IV to solve for *T*.

$$\begin{array}{rl} \text{IV} & T + H = 64 \\ & \underline{\quad H = 39} \\ & T \quad = 25 \end{array}$$

The difference between these equations equals the number Tina has. So, Tina has 25 cards.

5 CHECK Place the values you have found into each clue. The first two are done.

I 60 + 25 + 39 = 124
   124 = 124

II 60 + 74 = 134
    134 = 134

Solve by using more than one equation.

1. The perimeter of a rectangle is 40 feet. The length is one foot less than twice the width. Find the length and width.

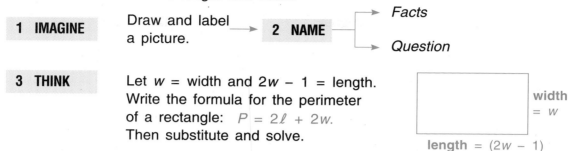

| 1 IMAGINE | Draw and label a picture. → | 2 NAME | → Facts |
| | | | Question |

3 THINK Let w = width and $2w - 1$ = length. Write the formula for the perimeter of a rectangle: $P = 2\ell + 2w$. Then substitute and solve.

width = w

length = $(2w - 1)$

4 COMPUTE ──→ **5 CHECK**

2. The perimeter of a rectangle is 26 m. Find the dimensions of the rectangle, given that the width is 3 m less than the length.

3. Use these clues to find the number of points scored by Shelly, Vera, and Cassandra in the basketball game. (Let S = Shelly, V = Vera, and C = Cassandra.)

$$S + V - C = 35$$
$$S + V = 53$$
$$S - C = 10$$

4. The perimeter of this parallelogram is 64 cm. If the length is 2 cm more than twice the width, what are its dimensions?

$(2w + 2)$

w

5. Only Doug, Ed, and Fran scored runs during the baseball game. Use the clues to find the total runs their team scored. (Let D = Doug, E = Ed, and F = Fran.)

$$D + E - F = 3$$
$$D - E = 2$$
$$F = 3$$

6. Use these clues to determine how much money Niki has saved in change. (Let Q = quarters, D = dimes, and N = nickels.)

$$N - D - Q = 2$$
$$D + Q = 11$$
$$3Q = 18$$

(Hint: Find how many there are of each coin; then convert to money.)

7. If the perimeter of this rectangle is 82 yards, what is its area?

$(5x + 1)$

$3x$

8. Find the weight (in lb) of each of Pia's three pets using these clues. (Let C = the weight of her cat, D = the weight of her dog, and S = the weight of her snake.)

$$D + C + S = 58$$
$$D - S = 40$$
$$5S = 10$$

Word Problems Involving Two Variables

Solve. (The first one is done.)

1. A bank held 8 more quarters than dimes. The value of the coins is \$3.05. Find how many of each coin there are.

   Let d = dimes and   1st equation: $q = d + 8$
   $\quad q$ = quarters   2nd equation: $25q + 10d = 305$

   Solve by substitution:
   $$25q \qquad\quad + 10d = 305$$
   $$25(d + 8) + 10d = 305$$
   $$25d + 200 + 10d = 305$$
   $$25d + 10d = 105$$
   $$35d = 105$$

   Replace d with 3.
   Solve for q.
   $$d = 3 \qquad \text{Number of dimes}$$

   $$q = d + 8$$
   $$q = 3 + 8$$
   $$q = 11 \qquad \text{Number of quarters}$$

2. The larger of two numbers is three times the smaller. Their sum is 68. Find both numbers.

3. A father is 5 times as old as his son. The difference in their ages is 44 years. How old is each?

4. The length of a rectangle is 8′ less than 3 times the width. The perimeter is 104′. Find both dimensions.

5. The second story of a building is 20′ less in height than the first. The two stories together are 64′ high. How high is each story?

USE THESE STRATEGIES:
Write an Equation
Use a Model/Drawing
Guess and Test
Hidden Information
Use a Formula

6. The length of a picnic table is 2′ less than twice its width. The perimeter is 26′. Find its length and width.

7. Separate 36 into two parts such that the first part is 28 less than three times the second. Find both parts.

8. The larger of two numbers is 3 more than the smaller. If 5 times the smaller is increased by 10 times the larger, the result is 165. Find both numbers.

9. A box of 18 coins in nickels and dimes was found in the attic. If the total value of the coins was \$1.45, how many of each coin were in the box?

Geometry Problems

10. In rectangle MNOP $\overline{MO}$ and $\overline{NP}$ are diagonals.
If $\overline{MO}$ is $(2x + 5)$ in. long and $\overline{NP}$ is $(3x - 1)$ in. long,
what is the length of $\overline{MO}$?

- Draw and label the figure. ⟶

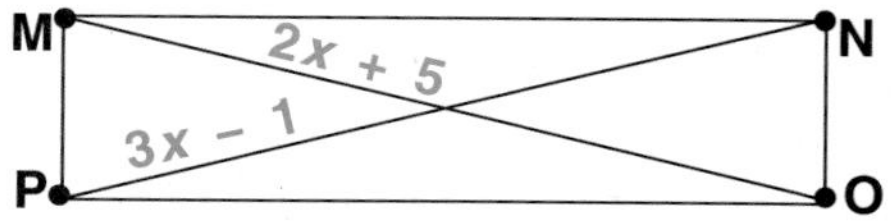

- Write and solve an equation. ⟶
 (The diagonals are congruent segments.)

$$2x + 5 = 3x - 1$$
$$2x - 2x + 5 = 3x - 2x - 1$$
$$1 + 5 = x - 1 + 1$$
$$6 = x$$

- Substitute 6 for x in $2x + 5$.
 $$\overline{MO} = 2x + 5$$
 $$\overline{MO} = 2(6) + 5$$
 $$\overline{MO} = 17 \text{ in.}$$

11.

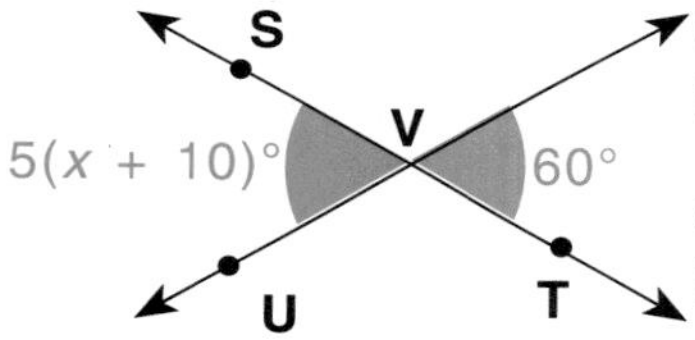

Given these intersecting lines:

a. Find x.
b. What is m $\angle SVU$?
c. What is m $\angle TVU$?

12.

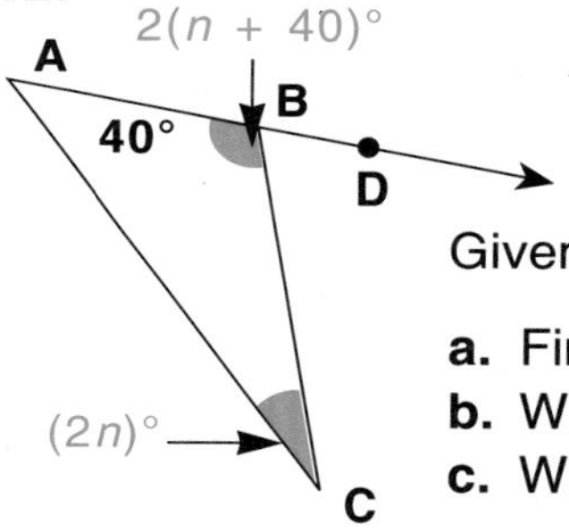

Given $\triangle ABC$:

a. Find n.
b. What is m $\angle ABC$?
c. What is m $\angle CBD$?

13.

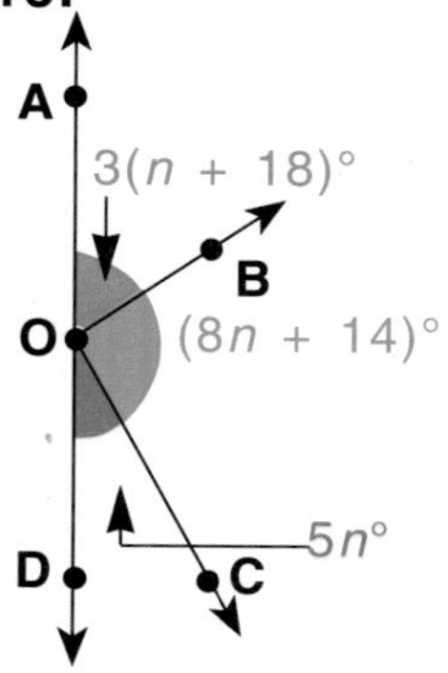

Given $\overline{AD}$ is a straight line:

a. Find n.
b. What is m $\angle AOB$?
c. What is m $\angle BOC$?

14.

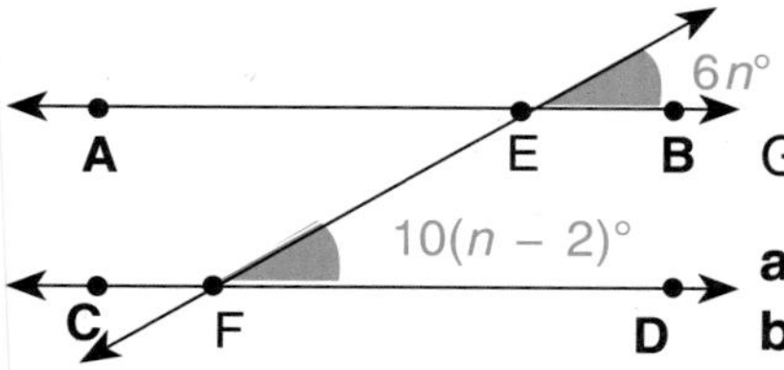

Given $\overline{AB} \parallel \overline{CD}$:

a. Find n.
b. What is m $\angle DFE$?
c. What is m $\angle CFE$?

15. The sum of two angles is 90 degrees. The smaller angle is 9 degrees more than half the larger angle. What is the measure of each angle?

16. Angle NPM and angle ERF are supplementary angles. Angle ERF is 15 degrees less than twice angle NPM. Find the measure of each angle.

17. The perimeter of parallelogram ABCE is 74 cm. The length of $\overline{AB}$ is 1 cm less than the length of $\overline{BC}$. Find the length of each side. (Hint: Draw the parallelogram.)

More Practice

Classify each polynomial as a monomial, a binomial, or a trinomial.

1. $-6x^2y$
2. r^2
3. $p^3z + sz^3 + pz$

4–6. Name the coefficient, base, and degree of each term in exercises 1–3.

Which expressions have like terms? Simplify these.

7. $25y^3x - 10y^3x$
8. $4p^2s^5 + 6p^2s^5$
9. $2r^3z^2 - 1r^3z$

Find the value of $h^3 - 4hp + p^2$ when:

10. $h = 4$ and $p = 2$
11. $h = -6$ and $p = 3$
12. $h = -2$ and $p = -6$

Arrange in ascending order of x.

13. $12x^2y - 7xy + 10x^3y^2$
14. $x^3 + xy^2 - 6 + x^2y$

Add or subtract these monomials.

15. $-5r^3m^2 + 9r^3m^2$
16. $4p - 9p - p$
17. $x^2 - 3x^2 + 5x^2$
18. $8mn - (2mn - 4mn)$
19. $(4a^2 + 3a^2) - (5a^2 + 2a^2)$
20. $(a^2 + 3a^2) - 7a^2$

Add.

21. $46b^2 + 35r$
$\ \underline{23b^2 + 55r}$

22. $13x^2 - 7x$
$\ \underline{18x^2 - 9x}$

23. $-27p - 32c$
$\ \underline{-29p - 38c}$

Subtract.

24. $-36a^2t + 26at$
$\ \underline{-18a^2t - 19at}$

25. $57y^2 - 37y$
$\ \underline{-49y^2 + 38y}$

26. $1.8z^2r + 0.7r$
$\ \underline{+0.9z^2r + 0.4r}$

Multiply or divide.

27. $-r(5r + 3x)$
28. $2dt(3d^2 - 2t^2)$
29. $-5z^2(3zt - 12t)$
30. $(2bt^2 - 5bt) \div bt$
31. $(12p^2 - 3p) \div p$
32. $(12br - 14r^2) \div 2r$

Add or subtract.

33. $^-2.6 + {}^-8.3$
34. $^-4.7 - {}^+2.9$
35. $^+14.6 - {}^-20.78$
36. $^-4.2 - {}^-1.03$
37. $^-6.01 - {}^-6.01$
38. $0 - {}^-2.75$

Multiply or divide.

39. $^-2.6 \cdot {}^-5.4$
40. $^+3.9 \cdot {}^-0.6$
41. $^+9.57 \div {}^+3.3$
42. $^-2.04 \div {}^-5.1$
43. $(^-6.2 - {}^+3.8) \div {}^-2.5$
44. $(^+3.5 - {}^+10.7) \div {}^+0.36$

More Practice

Simplify. Then arrange terms in descending order where appropriate.

45. $13m^2 - 8mn + 2mn - 11m^2$

46. $8 + 2d^2 - 3 - 4d + 6d^2$

47. $5a^2b - 2a^2b^2 + 4a^2b$

48. $11r - 2 + 7r^2 - 3r^2 - 15r$

49. $2c(8 - 4cd + 4c^2)$

50. $-5n(10 - n + mn) - 3n^2$

51. $(3x^2y)y + 2x^2y^3$

52. $10m^4n^2 - [(2m^2n)(3m^2n^2)]$

53. $(3s^2t)(2st)$

54. $\dfrac{(5x^2y^2)(8x^2y^2)}{4x^2y}$

55. $(3de^2 + 15e^3) - (6e^3 - 2de^2 + 1)$

56. $(-6x^2 + 4xy) - (4 + 3xy)$

57. $\dfrac{15h^2 + 21hq}{3h}$

58. $\dfrac{24a^3b^2 - 8a^2b^2 + 4ab}{-4ab}$

Simplify, using the order of operations.

59. $^-3.2 - (6.3 - 8.5)$

60. $^-293.15 \div (20.3 + {}^-91.8)$

61. $(^+9.2 - {}^+0.15) \times {}^-2.8 \div {}^+7$

62. $(^-3.6 + {}^-2.4) \div (^-8.16 - {}^-8.31)$

Solve.

63.
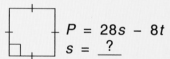
$P = \underline{\ ?\ }$

64.
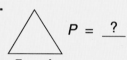
$P = \underline{\ ?\ }$

65.
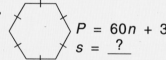
$P = \underline{\ ?\ }$

66.
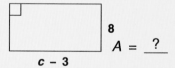
$P = 28s - 8t$
$s = \underline{\ ?\ }$

67.

$P = 15 + 27c$
$s = \underline{\ ?\ }$

68.
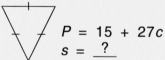
$P = 60n + 3$
$s = \underline{\ ?\ }$

69.
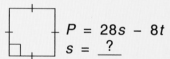
$A = \underline{\ ?\ }$

70. Suppose that the radius equals $a - 1$. What is the diameter?

71. The area of a rectangle is $14m + 7mn$ square feet. If its width is $7m$, what is its length?

72. The number of points scored in the first half was 10 points less than twice the number of points scored in the second half of the game. If the total points scored equal 29, how many points were scored in each half?

73. Ms. Ling used 25 coins to pay for a movie. How much did she spend if she had twice as many dimes as nickels and 9 more quarters than nickels?

74. One side of a triangle is 12 cm, another side is 3 cm less than the length of the third side. If the perimeter is 36 cm, find the length of the longest side.

More Practice

Solve. (The first one is done.)

75. The numerator of a certain fraction is 3 more than a number, and the denominator of the fraction is 2 more than twice the number. If the fraction in simplest form equals $\frac{5}{8}$, what is the fraction?

Let a = the number; then the numerator is ⟶ $a + 3$
and the denominator is ⟶ $2a + 2$

- Write the equivalent fractions. ⟶ $\frac{a + 3}{2a + 2} = \frac{5}{8}$

- Use the cross-product rule. ⟶ $8(a + 3) = 5(2a + 2)$

- Solve the equation.

$$8a + 24 = 10a + 10$$
$$8a - 8a + 24 = 10a - 8a + 10$$
$$-10 + 24 = 2a + 10 - 10$$
$$14 = 2a$$
$$7 = a$$

- Substitute $a = 7$ in the fraction.

$$a = 7 \longrightarrow \frac{a + 3}{2a + 2} = \frac{7 + 3}{14 + 2} = \frac{10}{16}$$

The fraction is $\frac{10}{16}$.

76. The sum of two numbers is 26. Their quotient is $\frac{4}{9}$. Find the numbers. (Hint: Let n = one number and $26 - n$ the other.)

77. When the same number, x, is added to both the numerator and the denominator of $\frac{5}{9}$, the resulting fraction equals $\frac{2}{3}$. What is the number?

78. The numerator of a fraction is 5 more than twice the denominator. If $\frac{7}{3}$ is an equivalent fraction, find the original fraction. (Hint: Let d = denominator.)

79. The area of a rectangle can be represented by the expression $4a^2 + 2ab$. If the length of the rectangle can be expressed as $2a$, what is the width?

80. The number of flowering trees in a park is 8 less than twice the number of evergreens. If the park contains 34 trees, how many of them are flowering and how many are evergreens?

81. Each of the two equal sides of an isosceles triangle is 4 cm less than twice the third side. If the perimeter is 52 cm, how long is each side?

Math Probe

ORIGINAL OPERATIONS

Mathematicians have used the basic operations of addition, subtraction, multiplication, and division for many centuries. The symbols for most of these operations are less than 800 years old.

Below is a new symbol for an "original operation." This original operation combines numbers in a special way.

Operation "DOUB-SUB":

$a \; ⍡ \; b$ means $2a - b$

Compute: $6 \; ⍡ \; 3$

$2a - b = 2 \times 6 - 3 = 12 - 3 = 9$

> ⍡ means "double the first number and subtract the second number."

To compute with "original operations":

- Substitute the values for the first and second numbers.
- Simplify, using the order of operation.

Here is another "original operation."

Operation "SUB-SQR":

$a \; ⊞ \; b$ means $(a - b)^2$

Compute: $3 \; ⊞ \; 5$

$(a - b)^2 = (3 - 5)^2 = (-2)^2 = 4$

Compute.

1. $7 \; ⍡ \; 4$
2. $5 \; ⍡ \; 2$
3. $8 \; ⍡ \; 5$
4. $2 \; ⍡ \; 4$
5. $6 \; ⊞ \; 5$
6. $4 \; ⊞ \; 2$
7. $9 \; ⊞ \; 5$
8. $3 \; ⊞ \; 8$

9. Explain what $⊞$ means.

10. Create a symbol for an operation that means "square the first number and add the second number."

11. Invent your own "original operation" and create a symbol for it. Make up four problems involving the operation, and then solve them.

445

Check Your Mastery

Classify each polynomial as a monomial, a binomial, or a trinomial. See pp. 420–421

1. dm

2. $y^2 + 4zy + b^2$

3. $3 + r$

Simplify like terms.

4. $14r^2z - 9r^2z$

5. $3t^3s^2 + 2t^2s^2$

6. $18x^2m^2 + 9x^2m^3$

What is the degree of each polynomial?

7. $-26x^2r$

8. -37

9. $\frac{1}{4}r^2 + r$

10. $y^3 + 5m^2y^2 + t^3$

Arrange in descending order of y. See pp. 422–425

11. $9t^2y - 2ty^3 + 6ty^2$

12. $xy^3 + xy^2 - 6xy^4$

Simplify the following polynomials.

13. $7b^3 + 16b - 8b^3 - 4b$

14. $15zt^2 + 6z^2r - 3z^2r$

Multiply or divide. See pp. 426–427

15. $-xz \cdot 3x^2z$

16. $\dfrac{12z^3b^3}{-4zb}$

17. $\dfrac{15k^4p^2}{-5k^2p}$

18. $3h \cdot 8h \cdot h^3$

Add or subtract. See pp. 428–429

19. $(15m^2 + n) + (2n + 13m^2)$

20. $(t^2 - 3t) + (4t^2 - 15t)$

21. $(-9sm + 5) - (18sm - 6)$

22. $(30r + 15r^2) - (46r^2 - 56r)$

Multiply or divide. See pp. 430–433

23. $s(6s^2 - 4s + 9) - s^2$

24. $3d(t^3 - 2dt + d) - 9d^2$

25. $(16a - 12a^2) \div 2a$

26. $(9z^2 - 3z) \div z$

Compute. See pp. 434–437

27. $^-5.1 + {}^-9.9$

28. $^-76.1 - {}^+49.1$

29. $^+6.83 - {}^-9.081$

30. $^-5.3 \cdot {}^-4.4$

31. $^+37.6 \cdot {}^-8.2$

32. $^+64.8 \div {}^+8.1$

Still More Practice

CHAPTER 1

Practice 1-1

Name the period of the underlined digits.

1a. 9,<u>2</u>76,410   **b.** <u>8</u>,092,583,021

2a. 7,063,<u>521</u>   **b.** 6,<u>024</u>,926,530

In what place is the underlined digit?

3a. 50,<u>8</u>16   **b.** 2<u>2</u>6,690,374

4a. 6.01<u>5</u>   **b.** 37.5<u>8</u>12

Write the standard numeral.

5a. $(5 \times 0.1) + (7 \times 0.001) + (3 \times 0.0001)$

  b. $(9 \times 100,000) + (9 \times 1000) + (6 \times 10) + (6 \times 1)$

6. Write 9,875,643,100 in expanded form.

7. Write thirty-two million, three hundred thousand as a standard numeral.

8. Write three hundred twenty-one thousand, three hundred four as a standard numeral.

9. Write the decimal 5 thousandths as a standard numeral.

10. Write six hundred and twenty-four hundredths as a standard numeral, and in expanded form.

Practice 1-2

Round to the nearest underlined digit:

1a. 9,<u>2</u>76,410   **b.** <u>8</u>,092,583

2a. 39.<u>0</u>91   **b.** 0.5<u>1</u>73

Compare. Write < or >.

3a. 0.021 _?_ 0.012   **b.** 0.064 _?_ 0.604

4a. 987,012 _?_ 987,102   **b.** 15.03 _?_ 15.036

Arrange from greatest to least:

5a. 61,807; 61,087; 61,708

  b. 2.047; 2.0471; 2.407; 2.0457

6. The amounts spent in a 5-day period were: $62,422; $62,421, $62,412; $62,142; $62,400

7. Mr. Sica won 572,108 votes for governor. Round this number to the nearest hundred thousand.

8. The highest four scores in a contest were: 26.8, 29.4, 26.49, and 26.76. Arrange the scores in order from greatest to least.

9. In the 100-meter dash, Dorian received 9.086 points and Starr received 9.608 points. Which runner won the greater number of points?

10. The average speed of an Indianapolis 500 winner was 163.529 mph. Round this average speed to the nearest whole number.

CHAPTER 2

Practice 2-1

1a. 27,825 +14,290   **b.** 84,627 +90,603

2a. 47 + 29 + 32 + 83   **b.** 46 + 897 + 2031

3a. 62,483 −19,876   **b.** 4004 −2067

4a. 94,612 − 8359   **b.** 80,046 − 62,915

5. The express train can carry 80,631 passengers per day. Today 51,972 people rode the train. How many more people could the train have serviced?

6. A plane traveled 4283 miles, 9602 miles, 596 miles, and 4727 miles. Estimate the total mileage to the nearest hundred miles.

7. If a stadium holds 15,273 seats and 6,953 are filled, how many seats remain?

8. Estimate the difference between 6300 and 2196.

9. Kim read 27 pages of her book on Monday, 90 pages on Tuesday, 47 on Wednesday, and 89 on Thursday. Estimate how many pages she has read.

10. By how much is the sum of 2639 and 1582 greater than the sum of 628 and 1597?

Practice 2-2

1a.
```
   4.63
 +8.29
```
b.
```
   24.63
 +89.76
```

2a.
```
   21.32
 +156.273
```
b.
```
   0.26
 +0.051
```

3a.
```
    1.4
    8.23
   91.4
 +  6.7
```
b.
```
    2.51
    3.62
    8.97
 +14.2
```

4a. 5.4 + 2.9 + 0.21 + 8   **b.** 3 + 8.7 + 9.1

5. Marie spent $146.29 on groceries and $38.76 for clothes. Estimate the total amount she spent.

6. Mario finished a race in 2.4 minutes, Louis in 1.78 minutes, and Tom in 3.05 minutes. Estimate their team total.

7. The phone bills for three months were $59.95, $18.25, and $10.27. Would a check of $88.47 be enough to cover the cost of the 3 bills?

8. Estimate the sum of 17.5 and the number which is twice 17.5.

9. Which sum is greater: 6.79 + 27.3 + 18.003 or 9.76 + 3.72 + 30.081?

10. Estimate this sum to the nearest dime: $100.61, $97.38, $17.96, $147.92, $349.82

Practice 2-3

1a.
```
   24.7
 −18.6
```
b.
```
   92.61
 −18.97
```

2a.
```
   27.9
 −14.82
```
b.
```
   183.7
 − 97.24
```

3a.
```
   92
 − 1.83
```
b.
```
   125
 − 27.62
```

4a. $100 − $63.75   **b.** $892 − $799.95

5a. 20.19 − 8.72   **b.** 0.39 − 0.039

6. Estimate the difference to the nearest whole number of 17.9 and 6.1.

7. How much wider is a 60-cm-wide door than one that is 52.9 cm wide?

8. Tim has $20.50. How much has he left after he buys a football for $12.85?

9. The annual rainfall in a certain city is 52.8 inches. How much greater is this than the rainfall in a city with 23.9 inches annually?

10. Emil spent $12.75 for a catcher's mitt, $3.25 for a baseball, and $3.99 for a bat. He then had $5.01 left. Estimate how much money he had at first.

CHAPTER 3

Practice 3-1

1a. 30 × 60   **b.** 15 × 100

2a. 3267 × 10   **b.** 90 × 70

3a. 2754 × 100   **b.** 5021 × 1000

4a.
```
    388
 × 141
```
b.
```
    823
 ×104
```

5a. 8753 × 307   **b.** 8924 × 347

Estimate; then compute:

6a. 41 × 935   **b.** 58 × 607

7. If 15 seats form a row, estimate how many seats there are in 29 rows.

8. A plane travels 960 km/h. How far did the plane travel in 12 hours?

9. Ride and Shine Car Wash services 203 cars each week. How many cars are washed in a year? (Hint: 1 year = 52 weeks)

10. Which is smaller: the product of 5824 and 473 or the product of 7513 and 283?

Practice 3-2

1a. $\begin{array}{r} 23 \\ \times 0.8 \\ \hline \end{array}$

b. $\begin{array}{r} 81 \\ \times 0.14 \\ \hline \end{array}$

2a. $\begin{array}{r} 89 \\ \times 0.02 \\ \hline \end{array}$

b. $\begin{array}{r} 147 \\ \times 0.09 \\ \hline \end{array}$

3a. $\begin{array}{r} 72.1 \\ \times 2.8 \\ \hline \end{array}$

b. $\begin{array}{r} 81.3 \\ \times 4.7 \\ \hline \end{array}$

Estimate; then compute:

4a. 7.25×0.57

b. 81.4×1.73

5a. 0.082×0.03

b. 0.55×0.08

6. If Maria has $10, can she buy 12 tennis balls at 80¢ each?

7. Estimate the cost of 11 markers if each one costs $0.89.

8. Tony can run 4.5 miles in 20 minutes. How far can he run in 2 hours?

9. If Mr. Kemp needs 6.8 yards of fabric to make one drape, estimate how many yards he will need for 21 drapes.

10. Rocco bought 1 dozen party favors at $1.25 each, 4 door prizes at $3.98 each, and 4.3 lb of lunch meat at $2.95 a pound. Estimate the total bill.

Practice 3-3

1a. $69 \div 10$

b. $108 \div 100$

2a. $350 \div 70$

b. $4800 \div 40$

3a. $16{,}371 \div 9$

b. $39{,}907 \div 7$

4a. $62\overline{)75{,}826}$

b. $97\overline{)88{,}852}$

Estimate; then compute:

5a. $210\overline{)120{,}540}$

b. $569\overline{)117{,}783}$

6. If 738 tickets to the school play must be distributed equally among 6 classrooms, estimate the number each room will receive.

7. B.&C. Shipping can transport 966 lb of cargo on each trip. If Mr. Morris wants to send 11,592 lb, how many trips will be needed?

8. Mrs. Juris must inspect every 24th item on the assembly line. If 1584 items pass by her work area each day, estimate how many items she checks each day.

9. There were 7001 visitors to the Mercury Space Display each day. Estimate the number of days it took to sell 279,013 tickets.

10. The quotient of 92,907 divided by 837 is how much less than the quotient of 37,557 divided by 117?

Practice 3-4

1a. $8\overline{)642.4}$

b. $54\overline{)55.62}$

2a. $23\overline{)0.828}$

b. $27\overline{)0.675}$

3a. $100\overline{)993}$

b. $1000\overline{)162.9}$

4a. $0.75\overline{)324}$

b. $0.014\overline{)0.427}$

5a. $3.2\overline{)166.4}$

b. $0.003\overline{)0.01896}$

6. A rope 8.4 meters long is cut into ten pieces. What is the length of each piece?

7. Linda collected $33.25 for selling tickets at $.95 each. How many tickets did she sell?

8. A stack of 10 books measured 41.4 cm high. How thick was each book?

9. If one page is 0.41 mm, how thick will 1000 pages be?

10. How many $.68 cans of soup can be bought for $16.32?

CHAPTER 4

Practice 4-1

Find the value of the expression.

If t = 7, 10, 17, 15, 26:

1a. $t + 12$ **b.** $t - 7$

2a. $t + t$ **b.** $\dfrac{15 - t}{2}$

If r = 10, 30, 4, 5, 0:

3a. $6r$ **b.** $10r$

4a. $\dfrac{r}{4}$ **b.** $\dfrac{r}{5}$

Write a mathematical expression for each of the following:

5a. 7 more than a number   **b.** a number decreased by 6

6a. 25 less than a number
b. the difference of a number and 6

7a. a number doubled
b. a number less 25

8a. the product of 13 and a number
b. the quotient of a number doubled and the number

9a. someone's age 8 years ago
b. someone's age 12 years from now

10a. the product when 7 and a number are multiplied by 2
b. the sum when twice a number is added to 6

Practice 4-2

Write a mathematical sentence for each of the following:

1a. half a number is less than 6.
b. three years from now she will be older than 18.

Solve for n and name the property.

2a. $6n = 6$ **b.** $(n + 4) + 6 = 7 + (4 + 6)$

3a. $7 + n = 7$ **b.** $n \times (15 + 4) = (3 \times 15) + (3 \times 4)$

Write a statement that will "undo":

4a. $r + 9 = 15$ **b.** $b - 5 = 46$

5a. $3d = 21$ **b.** $\dfrac{c}{8} = 20$

Solve the equation for the variable.

6a. $m + 7 = 15$ **b.** $r + 11 = 25$

7a. $x - 8 = 7$ **b.** $p - 10 = 3$

8a. $w - 41 = 15$ **b.** $93 = t + 18$

9. Marie paid $47.55 for a jacket after a reduction of $10. What was the price before the discount?

10. Beth Anne is 12 years younger than Marie who is 17. How old is Beth Anne?

Practice 4-3

Solve the equation for the variable.

1a. $10t = 100$ **b.** $5t = 125$

2a. $11t = 132$ **b.** $13d = 91$

3a. $\dfrac{a}{5} = 10$ **b.** $\dfrac{d}{9} = 20$

4a. $\dfrac{r}{4} = 15$ **b.** $\dfrac{b}{6} = 30$

5a. $2n + 1 = 7$ **b.** $\dfrac{n}{2} - 1 = 5$

Write an equation and solve.

6. Philip studies three times as long as his younger brother who studies 40 minutes. How long does Philip study?

7. A number divided by 12 is 8. What is the number?

8. The product of a number and 7 is 161. What is the number?

9. Mr. Croppe has 54 head of cattle. This is 18 times as many as Mr. Rowe has. How many cattle does Mr. Rowe have?

10. The present enrollment at Lewis School is 471. This is three times the enrollment 2 years ago. How many students attended the school 2 years ago?

Practice 4-4

Evaluate the expression.

1a. $2(6 - 3)$ **b.** $(20 + 4) \div 6$

2a. $(11 - 7) \div (3 - 1)$ **b.** $25 - (3 + 4) \times 2$

3a. $[(10 + 5) \div 3] \times 2$ **b.** $24 \div [10 - (2 + 3)]$

Use the given value to solve each formula.
$b = 6$ m $w = 3$ m $h = 5$ m $\ell = 10$ m

4a. $A = \ell w$ **b.** $P = 2(\ell + w)$

5a. $A = bh$ **b.** $A = \dfrac{bh}{2}$

6. What is the perimeter of a square 6 in. on a side? $(P = 4s)$

7. What is the area of a square 12 cm on a side? $(A = s \times s)$

8. How many miles does a car travel in 2 hours at a rate of 45 mph? $(D = rt)$

9. Find the width if the area is 55 ft^2 and the length is 11 ft. $(A = \ell \times w)$

10. How long will it take a car to travel 220 miles at a rate of 55 mph?

CHAPTER 5

Practice 5-1

Compare. Write $<$ or $>$.

1a. $^-8$ _?_ $^-4$ **b.** $^-15$ _?_ $^+12$

Order from least to greatest.

2a. $^-6, \, ^+5, \, ^+4, \, ^-8$ **b.** $^-2, \, ^+3, \, ^-17, \, ^-8, \, ^+1$

Order from greatest to least.

3a. $^+25, \, ^-4, \, 0, \, ^+18, \, ^-26$ **b.** $^+9, \, ^+4, \, ^-7, \, ^+6, \, ^-9$

Add.

4a. $^+7 + \, ^+4 = \underline{\,?\,}$ **b.** $^+9 + \, ^+8 = \underline{\,?\,}$

5a. $^-6 + \, ^-7 = \underline{\,?\,}$ **b.** $^-16 + \, ^-4 = \underline{\,?\,}$

6a. $^+3 + \, ^-8 = \underline{\,?\,}$ **b.** $^+12 + \, ^-5 = \underline{\,?\,}$

7. a. Positive 1 plus negative 1 is equal to _?_

   b. Negative 1 plus negative 1 is equal to _?_

8. Joan walked forward 15 steps, stopped, then walked forward 7 more steps. How many steps was she from the starting point?

9. Susan walked forward 20 steps then walked backward 15 steps. How many steps was she from the starting point?

10. Peter and Ellen started out on their bicycles from the same point. Peter rode 6 km to the right while Ellen rode 8 km to the left. Write an integer to show the distance and direction of each. Who rode farther? How much farther?

Practice 5-2

Write the opposite of each integer.

1a. $^+6$ **b.** $^-4$

2a. $^-24$ **b.** $^-(^+50)$

Subtract.

3a. $^+20 - \, ^+15 = \underline{\,?\,}$ **b.** $^+6 - \, ^+3 = \underline{\,?\,}$

4a. $^-6 - \, ^-6 = \underline{\,?\,}$ **b.** $^-9 - \, ^-2 = \underline{\,?\,}$

5a. $^+5 - \, ^-2 = \underline{\,?\,}$ **b.** $^+7 - \, ^-6 = \underline{\,?\,}$

6. a. $^-14 - \, ^+10 = \underline{\,?\,}$ **b.** $^-24 - \, ^+8 = \underline{\,?\,}$

7. a. $^-6 - \, ^-2 = \underline{\,?\,}$ **b.** $^-2 - \, ^-6 = \underline{\,?\,}$

8. a. $^+5.4 - \, ^+2.1 = \underline{\,?\,}$ **b.** $^+3.2 - \, ^+3.2 = \underline{\,?\,}$

9. a. Positive 1 minus negative 1 is equal to _?_

   b. Negative 1 minus positive 1 is equal to _?_

10. Joan walked forward 15 steps, stopped, then walked backward 7 steps. How many steps was she from the starting point?

CHAPTER 6

Practice 6-1

Use braces to list the members of:

1a. the set of multiples of 10.
b. the set of integers between $^-3$ and 0.

2a. the set of odd numbers.
b. the set of whole numbers less than 0.

$A = \{1, 2, 3, 4\}$ $B = \{2, 3\}$ $C = \{5\}$

3a. $A \cup B$ **b.** $A \cap B$

4a. $A \cup C$ **b.** $A \cap C$

$R = \{0, 1, 2, 3, 4, 5\}$ Find the solution set(s) for:

5a. $3n < 10$ **b.** $n + 2 \geq 5$

Write each product in exponent form.

6a. $10 \times 10 \times 10$ **b.** $10 \times 10 \times 10 \times 10$

Write as a decimal.

7a. 10^{-2} **b.** 10^{-4}

Write in expanded form using exponents.

8a. 703,210   **b.** 2,400,603

9a. 1.23   **b.** 2.004

Write the standard number.

10a. $(4 \times 10^6) + (2 \times 10^4) + (1 \times 10^0)$
b. $(6 \times 10^2) + (3 \times 10^1) + (7 \times 10^{-1})$

Practice 6-2

Which of these numbers are:

divisible by 2?

1a. 30,208   **b.** 16,293

divisible by 3?

2a. 1634   **b.** 1732

divisible by 4?

3a. 46,772   **b.** 308,222

divisible by 5?

4a. 4810   **b.** 79,425

divisible by 6?

5a. 2304   **b.** 4302

Choose the prime numbers from:

6a. (11, 12, 13, 14, 15)   **b.** (25, 26, 27, 28, 29)

Find Prime Factors.

7a. 81   **b.** 900

8a. 66   **b.** 40

Using exponents write the prime factorization of:

9a. 42   **b.** 63

10a. 120   **b.** 48

Practice 6-3

Write the set of factors for:

1a. 18   **b.** 80

2a. 75   **b.** 48

Write the first six multiples for:

3a. 9   **b.** 12

4a. 15   **b.** 30

Find the Greatest Common Factor.

5a. (8, 16)   **b.** (16, 36)

6a. (36, 45)   **b.** (100, 75)

Find the Least Common Multiple.

7a. (6, 24)   **b.** (5, 8)

8a. (24, 36)   **b.** (10, 15, 25)

9. What prime factor of 78 divides 297?

10. What prime factor of 165 is the GCF of 27 and 93?

Practice 6-4

Write in scientific notation.

1a. 37,000　　　　　**b.** 431,000

2a. 7,600,000　　　**b.** 25,800,000

3a. 2900　　　　　　**b.** 61,000,000,000

Write the standard numeral.

4a. 8.71×10^3　　　**b.** 2.25×10^5

5a. 1.4×10^7　　　**b.** 3×10^8

6a. 6.2×10^6　　　**b.** 9×10^9

7. Name two numbers between 100 and 110 that are divisible by both 2 and 3.

8. Name a number between 250 and 260 that is divisible by both 3 and 4.

9. A hand lens increased the thickness of a line 10^3 times. The line was 0.5 mm thick. How thick did the line appear?

10. A truck holds 10^4 times more pounds than when it is empty. The empty truck weighs 3 tons. How much does the loaded truck weigh?

CHAPTER 7

Practice 7-1

1a. $\dfrac{3}{4} = \dfrac{?}{8}$　　　**b.** $\dfrac{2}{10} = \dfrac{1}{?}$

2a. $\dfrac{6}{7} = \dfrac{?}{14}$　　　**b.** $\dfrac{5}{3} = \dfrac{?}{15}$

Express in lowest terms.

3a. $\dfrac{24}{36}$　　　**b.** $\dfrac{14}{18}$

4a. $\dfrac{15}{50}$　　　**b.** $\dfrac{44}{48}$

Write as a mixed number.

5a. $\dfrac{75}{9}$　　　**b.** $\dfrac{77}{4}$

Change to an improper fraction.

6a. $8\frac{2}{9}$　　　**b.** $29\frac{1}{6}$

Compare. Write $<$, $=$, $>$.

7a. $\dfrac{6}{8}\ \underline{\ ?\ }\ \dfrac{9}{12}$　　　**b.** $\dfrac{3}{10}\ \underline{\ ?\ }\ \dfrac{7}{25}$

Which is closer to 0?

8a. $\dfrac{3}{20}$ or $\dfrac{5}{6}$　　　**b.** $\dfrac{7}{9}$ or $\dfrac{1}{4}$

Which is closer to 1?

9a. $\dfrac{5}{8}$ or $\dfrac{1}{12}$　　　**b.** $\dfrac{7}{10}$ or $\dfrac{18}{20}$

Which is a little less than $\dfrac{1}{2}$?

10a. $\dfrac{5}{9}$ or $\dfrac{4}{10}$　　　**b.** $\dfrac{17}{20}$ or $\dfrac{2}{5}$

Practice 7-2

1a. $\dfrac{4}{9} + \dfrac{8}{9}$　　　**b.** $\dfrac{3}{11} + \dfrac{4}{11}$

2a. $\dfrac{8}{13} + \dfrac{1}{4}$　　　**b.** $\dfrac{5}{12} + \dfrac{2}{9}$

3a. $\dfrac{2}{3} - \dfrac{5}{8}$　　　**b.** $\dfrac{11}{12} - \dfrac{5}{24}$

4a. $5 - \dfrac{3}{8}$　　　**b.** $3 - \dfrac{3}{11}$

5a. $8 - 5\frac{2}{3}$　　　**b.** $12 - 7\frac{1}{8}$

6. Annette had 22 yd of gift wrap. After using $12\frac{2}{9}$ yd, how much was left?

7. What is the difference between $\dfrac{9}{10}$ and $\dfrac{4}{5}$?

8. What is the difference between 7 and the sum of $\dfrac{1}{8}$ and $\dfrac{5}{6}$?

9. Find the sum of $\dfrac{2}{5}$, $\dfrac{2}{3}$, and $\dfrac{1}{6}$.

10. Estimate the sum of $\dfrac{3}{4}$, $\dfrac{1}{2}$, $\dfrac{1}{4}$, $\dfrac{2}{3}$, and $\dfrac{1}{3}$.

Practice 7-3

1a. $8\frac{2}{5} + 9\frac{1}{5}$ **b.** $17\frac{5}{6} + 4\frac{5}{6}$

2a. $15\frac{8}{12} + 4\frac{4}{16}$ **b.** $1\frac{11}{30} + 2\frac{7}{12}$

3a. $2\frac{5}{7} - 1\frac{3}{7}$ **b.** $12\frac{3}{5} - 9\frac{1}{2}$

4a. $12\frac{3}{11} - 9\frac{7}{11}$ **b.** $13\frac{6}{13} - 10\frac{7}{13}$

5a. $46\frac{1}{5} - 37\frac{1}{4}$ **b.** $18\frac{3}{7} - 12\frac{3}{4}$

Round to the nearest whole number.

6a. $22\frac{3}{7}$ **b.** $44\frac{11}{16}$

7. Estimate the difference between $61\frac{1}{4}$ and $9\frac{7}{8}$.

8. Estimate the sum of $36\frac{1}{11}$ and $65\frac{6}{7}$.

9. To make a ribbon pillow Jenny needs $2\frac{3}{4}$ yd of blue ribbon, $1\frac{3}{5}$ yd of white, and $2\frac{1}{8}$ yd of red. How many yards of ribbon does she need?

10. Archie worked $3\frac{5}{6}$ hours on Friday, $2\frac{1}{3}$ hours on Saturday, and $1\frac{1}{4}$ hours on Sunday. Estimate his total work time for the three days.

CHAPTER 8

Practice 8-1

1a. $\frac{2}{6} \times \frac{9}{16}$ **b.** $\frac{4}{9} \times \frac{3}{7}$

2a. $\frac{6}{21} \times \frac{18}{30}$ **b.** $\frac{4}{7} \times \frac{49}{64}$

3a. $121 \times \frac{3}{11}$ **b.** $18 \times \frac{2}{27}$

4a. $2\frac{1}{4} \times 3\frac{1}{3} \times \frac{2}{5}$ **b.** $4\frac{1}{8} \times 4\frac{2}{7}$

5a. $6\frac{2}{9} \times \frac{3}{8} \times \frac{6}{7}$ **b.** $3\frac{1}{5} \times 2\frac{1}{4}$

6. How much does Greg earn at his after-school job if he works 5 days for $2\frac{3}{4}$ hr each and is paid $3.50 an hour?

7. One package of fruit snacks weighs $1\frac{1}{8}$ oz. How much will 12 packages weigh?

8. Molly sold $18\frac{1}{3}$ dozen cupcakes at the school sale. Rita sold $\frac{3}{4}$ as many. How many cupcakes did Rita sell?

9. Lucy bought $2\frac{1}{8}$ yd of fabric at $2.60 a yd and $1\frac{3}{4}$ yd of lace at $.96 a yd. What did she pay for her purchases?

10. Mrs. Scott bought a $14\frac{1}{2}$ lb turkey at $1.35 a pound. What change should she receive from $30.00?

Practice 8-2

1a. $12 \div \frac{6}{7}$ **b.** $81 \div \frac{9}{10}$

2a. $\frac{5}{6} \div 35$ **b.** $\frac{2}{7} \div 30$

3a. $\frac{4}{7} \div \frac{3}{14}$ **b.** $\frac{3}{8} \div \frac{15}{16}$

4a. $1\frac{5}{9} \div 2\frac{1}{3}$ **b.** $3\frac{1}{4} \div 4\frac{1}{3}$

5a. $6 \times \underline{\ ?\ } = 1$ **b.** $2\frac{1}{2} \times \underline{\ ?\ } = 1$

6a. $\dfrac{10}{\frac{2}{9}}$ **b.** $\dfrac{\frac{5}{8}}{\frac{3}{4}}$

7. How many $\frac{1}{4}$-lb patties can be made from $2\frac{1}{2}$ lb of hamburger?

8. A length of rubber tubing is $12\frac{1}{2}$ ft. How many $2\frac{1}{2}$ ft pieces can be cut from it? Will there be any tubing left?

9. How many times as long would it take to travel the same distance by train as by plane if the plane time was $2\frac{1}{2}$ hours and the train time was 20 hours?

10. If $\frac{5}{8}$ of Jan's money is $25.00, how much money has Jan?

Practice 8-3

1a. $\frac{3}{5}$ of 60 = __?__   **b.** $\frac{2}{3}$ of 99 = __?__

Change to a decimal.

2a. $1\frac{4}{5}$ **b.** $\frac{33}{8}$

3a. $\frac{5}{8}$ **b.** $4\frac{7}{8}$

Change to a fraction or mixed number in lowest terms.

4a. 0.6   **b.** 0.85

5a. 6.15   **b.** 0.004

6a. 1.25   **b.** 8.02

Change to a repeating decimal.

7a. $\frac{2}{9}$ **b.** $\frac{2}{7}$

8a. $\frac{7}{15}$ **b.** $\frac{7}{12}$

9. Leo has a batting average of .660. Mark had 7 hits in 9 times at bat. Who has the better average?

10. Rafael pitched 36 games. He won $\frac{5}{9}$ of them. How many games did he lose?

CHAPTER 9

Practice 9-1

1a. 8 cm = __?__ mm   **b.** 5 m = __?__ cm

2a. 4 km = __?__ m   **b.** 8.9 m = __?__ cm

3a. 4.83 kL = __?__ L   **b.** 0.18 L = __?__ cL

4a. 890 mL = __?__ L   **b.** 5 cg = __?__ mg

5a. 0.9 mg = __?__ g   **b.** 35.1 kg = __?__ g

6. What is the mass of 2 mL of water?

7. A train can travel 2100 m/min. How many km/min is this? How many km/hr?

8. Mr. Leiden used 50 liters of gasoline on a trip. What part of a kiloliter would he use on 10 trips of the same distance?

9. How much water would fit in a 6 cubic decimeter container?

10. A piece of wood 2 m long has a mass of 10 kg. What would be the mass of the same wood 20 cm long? 2 cm long?

Practice 9-2

Name and give the **GPE** of the more precise measurement.

1a. mg or g   **b.** in. or ft

2a. oz or lb   **b.** km or m

Copy and complete.

3a. 48 in. = __?__ ft   **b.** 15 ft = __?__ yd

4a. 32 oz = __?__ lb   **b.** $1\frac{1}{2}$ lb = __?__ oz

5a. 0.5 qt = __?__ pt   **b.** 5 qt 3 pt = __?__ gal

6. Which is greater: 18 qt or $4\frac{1}{2}$ gal?

7. How many pint containers can be filled from 36 gallons of milk?

8. A 40-yd piece of wire can be cut into how many pieces of 1-ft length?

9. Which weighs more: a 12-oz jar of peanut butter or a $\frac{3}{4}$-lb jar?

10. How many square yards of carpet are needed to cover a floor that measures 4 yd by 2 yd 2 ft?

Practice 9-3

1a. 1 ft 3 in. = _?_ in.   **b.** 2 ft 9 in. = _?_ in.

2a. 1 yd 2 ft = _?_ ft   **b.** 2 yd 3 in. = _?_ in.

3a. 3 gal 1 qt = _?_ qt   **b.** 3 lb 2 oz = _?_ oz

4a. 3 ft 2 in. × 2 = _?_   **b.** 2 gal 2 qt × 4 = _?_

5a. 5 yd 2 ft   **b.**   4 pt 1 c
       +1 yd 2 ft   −1 pt 3 c

6. At what temperature on the Celsius scale does water boil?

7. What is normal body temperature on the Celsius scale and on the Fahrenheit scale?

8. What is the combined weight of a pumpkin that weighs 9 lb 6 oz and one that weighs 12 lb 11 oz?

9. A recipe calls for 1 qt and 3 c of water. How much will be needed if you triple the recipe?

10. One can of carrots weighs 20 oz. How many lb will be in 5 cans?

CHAPTER 10

Practice 10-1

1. Identify each symbol:

    a. $\overleftrightarrow{AB}$ **b.** $\overline{AB}$ **c.** $\overrightarrow{AB}$

2. What kind of angle is:

    a. ∠ABC   **b.** ∠DEF   **c.** ∠GHJ

3. Tell the relation of:
    a. $\overline{AB}$ to $\overline{CD}$ **b.** ∠EJH to ∠GJF   **c.** $\overline{JK}$ to $\overline{LM}$

  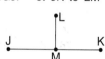

4. Draw a line segment 3 cm long. Using a compass and a straightedge, construct a congruent segment.

5. What is the measurement of the two congruent angles formed by bisecting an angle of:

    a. 90°   **b.** 18°   **c.** 180°   **d.** 36°

6. Copy the figures below. Use a compass and straightedge to construct congruent figures.

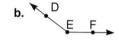

    a.   **b.**

7. Draw the lines of symmetry in:

    a. square   **b.** rhombus

Do the figures below have rotational symmetry of the degree shown?

8. 90°   **9.** 180°

10. 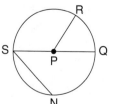 90°

CHAPTER 11

Practice 11-1

Name the quadrilaterals shown below.

1a.   **b.**

2a.    **b.**

Use circle *P* for exercises 3–6.

3. $\overline{SQ}$ is a _?_ .

4. $\overline{PQ}$ is a _?_ .

5. ∠SPR is a _?_ .

6. $\overline{SN}$ is a _?_ .

Answer true or false.

7a. All polygons have 4 sides.

    b. All triangles have 3 angles.

    c. All squares are quadrilaterals.

    d. All trapezoids are polygons.

    e. All quadrilaterals are squares.

Use circle *P* for exercises 8–10.

8. If m∠RPQ = 70°, then m$\overset{\frown}{RPQ}$ = _?_ .

9a. If SP = 6 in., then SQ = _?_ .

10. If m∠SPR = 110°, then m$\overset{\frown}{SR}$ = _?_ .

Practice 11-2

Find the perimeter of a figure with these
dimensions:

1 a.
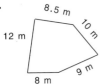

b.
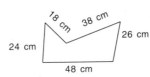

2. Rectangle
 a. 2 m by 11 m   **b.** 4 ft by $3\frac{1}{2}$ ft

3. Square
 a. 13 in.   **b.** $10\frac{3}{4}$ yd

4. Triangle with sides measuring:
 a. 7 cm; 2 cm; 6 cm **b.** 10 in.; 4 in.; 0.5 in.

Find the side of a square given a perimeter of:

5. a. 18 m   **b.** $20\frac{1}{2}$ ft

6. Find the width of each rectangle given the
 perimeter of 48 m and:
 a. $\ell = 20$ m   **b.** $\ell = 16$ m

7. The side of a square measures 40 cm. Find
 its perimeter in meters.

8. Fencing costs $5.75 per yard. Find the cost
 of enclosing a rectangular garden that is 2
 yards wide and 11 yards long.

9. Find the difference between the perimeter of a
 4 ft square and the perimeter of a rectangle
 42 in. wide and 48 in. long.

10. A square poster has a perimeter of 120.6 cm.
 What is the measure of each side?

Practice 11-3

Find the area of a figure with these dimensions.

1 a. Rectangle: 8 cm by 26 cm

 b. Square: 9 m

2 a. Parallelogram: $b = 14$ cm; $h = 10$ cm

 b. Triangle: $b = 21$ cm; $h = 9$ cm

3. Rectangle
 a. 15 ft by 2 ft   **b.** 3 m by 7.5 m

4.
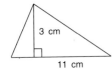

5.
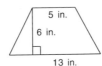

6. Find the area of a trapezoid whose bases are
 8 m and 12 m and height is 9 m.

7. How much area is enclosed by a
 parallelogram whose height is 3 cm and base
 is 12.5 cm?

8. How many square centimeters of cardboard
 are needed for each place card if each
 measures 6.5 cm by 3.6 cm?

9. Tell which is greater, and by how much: the
 area of a 13-dm square or the area of a
 trapezoid with bases 122 m and 125.6 m and
 a height of 14 m.

10. How many triangular wall hangings can be cut
 from 224 m^2 of canvas if each hanging has a
 height of 3.2 m and a base of 4 m?

Practice 11-4

Find the length of the figure given the area and
the width.

1. Rectangle
 a. 32 cm^2; 8 cm   **b.** 195.6 ft^2; 12 ft

2. Square
 a. 225 cm^2 **b.** 900 cm^2

Find the height of the figure given the area and
the base.

3. Triangle
 a. 60 yd^2; 20 yd   **b.** 192 cm^2; 16 cm

4. Parallelogram
 a. 168 ft^2; 16 ft   **b.** 113.4 m^2; 12 m

5. A triangle has an area of 30 m^2 and a height
 of 500 cm. What is the measure of its base?

6. Find the side of a square whose area is the
 same as that of a triangle having a base of
 112 cm and a height of 56 cm.

7. 360 cm^2 of cork covered a bulletin board
 24 cm long. How wide was the board?

8. A farmer marked off a section of his farm for
 soybeans. If this plot was in the shape of a
 parallelogram covering an area of 2112 m^2
 with a base of 96 m, what is the height?

9. What is the length of a piece of plastic that
 covers a table with an area of 0.88 m^2 and
 a width of 80 cm?

10. A triangular banner made from 460 cm^2 of felt
 has a height of 20 cm. How many centimeters
 are in the base?

Practice 11-5

Find the circumference and the area of the circles with these dimensions.

1a. radius: 9 cm   **b.** radius: 24 dm

2a. diameter: 24 m   **b.** diameter: 16 m

3a. radius: 14 cm   **b.** radius: 2.5 m

4a. diameter: 40 dm   **b.** diameter: 9 m

5a. radius: 12 m   **b.** radius: 10 mm

6. Find the circumference in terms of π of a circle whose radius is 2.75 ft.

7. Find the area in terms of π of a circle whose diameter is 14 m.

8. What is the circumference (in feet) of a circle whose diameter is 10.5 yd?

9. One circle has a radius of 10 cm. Another circle has a radius of 0.5 m. By how much do the areas of these circles differ?

10. What is the greatest number of circular disks, each with a radius of 5.2 cm, that can be printed on a 1-meter square board? (In each step, round to the nearest tenth.)

CHAPTER 12

Practice 12-1

Express each ratio as a fraction. Simplify.

1a. 2 : 3   **b.** 49 : 14

2a. 6 : 27   **b.** 5 to 9

3a. 16 : $\frac{4}{5}$ **b.** 10 to $2\frac{1}{2}$

Express each rate as a fraction. Simplify.

4. 5 miles in 10 minutes

5. 135 people and 3 buses

6. 3 pizzas for 12 people

7. 18 windows washed in 30 min

8. If Curtis can decorate 3 cakes in 18 minutes, how long does it take to decorate one cake?

9. One dozen oranges cost \$.72. How much does one orange cost?

10. Mrs. Percy spent \$19.20 for 12 gallons of gas. What was the cost per gallon?

Practice 12-2

1a. $\frac{2}{8} \overset{?}{=} \frac{3}{24}$ **b.** $\frac{4}{12} \overset{?}{=} \frac{5}{15}$

2a. $\frac{5}{18} \overset{?}{=} \frac{10}{36}$ **b.** $\frac{9}{12} \overset{?}{=} \frac{15}{20}$

3a. $\frac{5}{0.75} \overset{?}{=} \frac{3}{0.15}$ **b.** $\frac{7}{8} \overset{?}{=} \frac{1.75}{2}$

4a. $\frac{12}{x} = \frac{4.8}{6}$ **b.** $\frac{x}{7.2} = \frac{0.28}{2.4}$

5a. $\frac{x}{20} = \frac{50}{2.5}$ **b.** $\frac{7}{5.6} = \frac{8}{x}$

6. A picture 6.4 cm long and 3.2 cm wide is to be enlarged to have a width of 9.6 cm. Will the length be 19.2 cm?

7. If the interest on a loan is \$160 for $3\frac{1}{3}$ years, what will it be for $2\frac{1}{2}$ years?

8. What distance on a map represents 1500 km if 1.5 cm represents 300 km?

9. At the same time a tree 8 meters high casts a shadow of 6 meters long, how long will the shadow be of a man 2 meters tall?

10. Mr. Holt's motorboat travels at a rate of 100 km per $2\frac{1}{2}$ hours. At this rate, how far will it go in 7 hours?

CHAPTER 13

Practice 13-1

1a. $\frac{27}{100} = ?\%$ **b.** $\frac{17}{100} = ?\%$

2a. $\frac{1}{10} = ?\%$ **b.** $\frac{9}{50} = ?\%$

3a. $\frac{5}{8} = ?\%$ **b.** $\frac{6}{6} = ?\%$

4a. $0.25 = ?\%$ **b.** $0.08 = ?\%$

5a. $0.001 = ?\%$ **b.** $2.5 = ?\%$

6a. $0.3 = ?\%$ **b.** $1.12 = ?\%$

7. If $\frac{3}{4}$ of the flowers in the garden are blue, what percent are not blue?

8. Rocky has 9 books about space travel. He has finished 4 of them. What percent has he read?

9. Five out of every six junior high students wear braces. What percent do not?

10. Mr. Tapi sent out 700 questionnaires. 100 people responded. What percent did not respond?

Practice 13-2

Change to a decimal.

1a. 6.3%   **b.** 112%

2a. 2.75%   **b.** $37\frac{1}{2}\%$

3a. $6\frac{2}{3}\%$ **b.** $13\frac{1}{3}\%$

Change to a fraction.

4a. 12.6%   **b.** $333\frac{1}{3}\%$

5a. 9.9%   **b.** $4\frac{1}{6}\%$

Write <, =, or >.

6a. 5.08% __?__ $5\frac{1}{2}\%$

b. 1.2% __?__ $1\frac{1}{5}\%$

7a. 61% __?__ 0.610

b. 310% __?__ 0.31

8a. 1.75 __?__ $17\frac{1}{2}\%$

b. 0.085 __?__ $8\frac{1}{4}\%$

9a. $\frac{1}{2}\%$ __?__ 0.05

b. $\frac{1}{4}\%$ __?__ 0.025

10a. $11\frac{1}{9}\%$ __?__ $\frac{1}{11}$

b. $8\frac{1}{3}\%$ __?__ $\frac{1}{15}$

Practice 13-3

1a. Find 25% of 160.   **b.** Find 60% of 245.

2a. Find 50% of 420.   **b.** Find 75% of 80.

3a. Find $6\frac{1}{4}\%$ of 128.

b. Find $26\frac{2}{3}\%$ of 1215.

4a. Find $4\frac{1}{6}\%$ of 720.

b. Find $18\frac{3}{4}\%$ of 288.

5a. Find $37\frac{1}{2}\%$ of 200.

b. Find $66\frac{2}{3}\%$ of 270.

6a. Find $114\frac{2}{7}\%$ of 140.

b. Find 250% of 20.88.

7. 15% of a class of 60 students attend swimming classes. How many students attend?

8. $33\frac{1}{3}\%$ of Mary's books are mysteries. If there are 117 books in her collection, how many are mysteries?

9. Linda dieted and lost $16\frac{2}{3}\%$ of her total body weight. If she had weighed 162 lb, how many pounds did she lose?

10. At the grand opening of his store, Mr. Miller sold $37\frac{1}{2}\%$ of the 832 articles he had on sale. How many items did he sell on the opening day?

Practice 13-4

1a. 70% of what number is 21?

b. 75% of what number is 37.5?

2a. 25% of what number is 9?

b. 24% of what number is 15.6?

3a. 150% of what number is 90?

b. 130% of what number is 91?

4a. $66\frac{2}{3}$% of what number is 144?

b. $83\frac{1}{3}$% of what number is 360?

5a. $11\frac{1}{9}$% of what number is 60?

b. $16\frac{2}{3}$% of what number is 70?

6. The Davis family has traveled $12\frac{1}{2}$% of the total distance of their trip. If they have gone 360 km, how much further will they travel?

7. 14 or 30% of the offices in one building are carpeted. How many are in the building?

8. If 80% of the enrollment of a school is 416, what is the total enrollment?

9. If 70% of the height of a tree is 14 meters, how high is the tree?

10. Teresa's age is $8\frac{1}{3}$% of her grandmother's age. If Teresa is 5 years old, how old is her grandmother?

Practice 13-5

What percent:

1a. of 20 is 4?   **b.** of 16 is 2?

2a. of 40 is 10?   **b.** of 60 is 12?

3a. of 1 ft is 1 in.?   **b.** of 1 day is 1 hour?

4a. of 0.25 is 1.5?   **b.** of 1.24 is 0.03?

5a. of $\frac{5}{6}$ is $\frac{1}{12}$?   **b.** of 4.5 is 2.2?

6. Dr. Brundt spends 3 out of every 4 hours of working time making hospital visits. What percent of his day is spent in the hospital?

7. Out of 20 problems Beth solved 16 correctly. What percent did she have correct?

8. From a high school class of 280 students, 14 received awards for excellence in mathematics. What percent was this?

9. During one month, 12 out of 30 days had rainfall. On what percent of the days did it rain? On what percent did it not rain?

10. A basketball team won 5 of their 40 games. What percent did they win?

CHAPTER 14

Practice 14-1

Complete the chart.

	Discount	List Price	Rate	Selling Price
1a.	?	$460	24%	?
b.	?	$728	$62\frac{1}{2}$%	?
2a.	?	$225	15%	?
b.	?	$68.90	10%	?
3a.	$180	$540	?	?
b.	$370	$1110	?	?
4a.	?	$85.95	?	$51.57
b.	?	$16.08	?	$14.07
5a.	$.32	?	5%	?
b.	$96	?	32%	?

6. What is the amount saved if a discount of 30% is given on bedroom furniture listed at $2250?

7. At a sale Jean received a discount of $10 on a coat marked $80. What was the rate of discount?

8. The original price of a suitcase is $190. The price is reduced $24.70. What is the percent of reduction?

9. A copper teakettle is on sale for $16.00, which is a reduction of $33\frac{1}{3}$%. What is the regular price of the teakettle? (Find the discount first.)

10. Mr. Schmidt bought a piano listed at $3510 with a $33\frac{1}{3}$% discount. What was the sale price?

Practice 14-2

Round to the nearest cent where necessary.

Find 3% of:

1a. $18   **b.** $20

Find 2% of:

2a. $19.90   **b.** $38.64

Find 3.5% of:

3a. $30   **b.** $100

Given a rate of 4%, find the sales tax on:

4a. $38.60   **b.** $75.50

Given a rate of 5%, find the sales tax on:

5a. $126.24   **b.** $90.48

6. What is the total cost of an item marked $201.32 after a discount of 25% and a sales tax of 5%?

7. What sales tax must be paid on a set of table flatware marked $158.99 if the rate is 8%?

8. The marked price of a lamp is $19.95. What will be the total cost if there is a discount of 10% but a sales tax of 4% is added?

9. A sweater $53.50 is reduced 20%. What is the total cost including a sales tax of 6%?

10. Which item has a higher final cost: one listed at $42.88 plus a 6% sales tax or one listed at $54.98 plus a 4% sales tax?

Practice 14-3

1a. 5% of $600 = __?__   **b.** 3% of $80 = __?__

2a. $4\frac{1}{2}$% of $100 = __?__

  b. 2% of $4200 = __?__

Find the commission, given the selling price and the rate of commission.

3a. $1750, 4%   **b.** $1295, 3%

4a. $895, 7%   **b.** $600, 4.5%

5a. $1000, 5.5%   **b.** $6842, 5%

6. A salesman received a commission of 6% on all sales above $1000. What was his commission on a car sold for $12,000?

7. What would Mr. Haynes earn on a desk marked $198 at a rate of 4.5% commission?

8. Mrs. Clark earned 6.5% on an entertainment set marked $1264. What commission did she receive?

9. During a week a salesperson sold the following amounts: $479, $248, $189, $1210, $990. At a rate of 6%, what commission did she earn for the week?

10. One real estate agent sold property totaling $88,700 at a commission rate of 9%. Another sold property for $92,580 at a rate of 7.5%. Who earned more money? How much more?

Practice 14-4

What part of a year is:

1a. 9 months?   **b.** 3 months?

2a. 4 months?   **b.** 6 months?

Find the interest, given the principal, the rate, and the time.

3a. $550, 4%, 1 yr   **b.** $325, 5%, 2 yr

4a. $1200, 6%, $2\frac{1}{2}$ yr   **b.** $900, $12\frac{1}{2}$%, 3 yr

5a. $575, 6%, 1 yr 9 mo

  b. $600, 4%, 2 yr 3 mo

6. What is the interest on $725 at $6\frac{1}{4}$% for $2\frac{1}{2}$ years?

7. Maryann deposited a small legacy of $1500 in a bank that paid $6\frac{1}{2}$% interest. How much did her money earn after 3 years?

8. Find the amount received on $240 at $4\frac{1}{2}$% when the money was withdrawn at the end of 4 years.

9. Mr. Dolan borrowed $12,000 from a bank that charges 15% interest. What amount must he pay the bank at the end of the year?

10. What interest will be due on $5000 at 12% for 2 years 4 months?

CHAPTER 15

Practice 15-1

1a. 2 (14 ft × 8 ft) = <u> ? </u> ft²

b. 2 (20 yd × 16 yd) = <u> ? </u> yd²

Find the surface area of a rectangular prism, given the length (ℓ), width (*w*), and height (*h*).

2a. ℓ = 12 cm, *w* = 9 cm, *h* = 6 cm

b. ℓ = 24 cm, *w* = 22 cm, *h* = 10 cm

3a. ℓ = 7 ft, *w* = 9 ft, *h* = 6 ft

b. ℓ = 20 ft, *w* = 15 ft, *h* = 10 ft

Find the surface area of each cube, given an edge.

4a. 20 cm　　　**b.** 25 cm

5a. 7 m　　　**b.** 4 m

6. a. 12 ft　　　**b.** 20 ft

7. A rectangular prism is 35 cm long, 15 cm wide, and 3 cm high. Find the surface area.

8. How many square feet of wood paneling would be needed to cover the four walls of a playroom 14 ft long, 10 ft wide, and 9 ft high?

9. Nicole wants to cover a box with contact paper. How much will it cost at $2.78 a square meter if the box measures 20 cm by 15 cm by 30 cm?

10. How many square inches of lining would be used for the inside of a box, including the lid, if the box is 18 in. long, 15 in. wide, and $5\frac{1}{2}$ in. deep?

Practice 15-2

Find the circumference of the base of a cylinder, given the diameter (*d*).

1a. *d* = 24 cm　　　**b.** *d* = 5 yd

Find the area of the rectangular surface of a cylinder, given the diameter (*d*) and height (*h*).

2a. *d* = 8 cm　　　*h* = 4.2 cm

b. *d* = 9 ft　　　*h* = 18 ft

Find the area of the bases of a cylinder, given the diameter (*d*) or radius (*r*).

3a. *d* = 7 m　　　**b.** *r* = 10 ft

Find the total surface area of a cylinder, given the diameter or radius and the height.

4a. *r* = 3 yd　　　*h* = 14 yd

b. *d* = 16 in.　　　*h* = 20 in.

5a. *d* = 6 m　　　*h* = 5.2 m

b. *r* = 7 m　　　*h* = 3.6 m

6. How many square meters of tin would be used to make 30 soup cans each 7 cm in diameter and 9.2 cm high?

7. Mr. Slonim painted the inside (including the floor) of a circular swimming pool that has a diameter of 32 ft and a depth of 8 ft. How many square feet of surface did he paint?

8. Joan bought a cylindrical waste basket. What is the surface area of the basket if it is 10 in. in diameter and 15 in. high? (There is no lid.)

9. How many square meters of labels were used to cover 100 cans of fruit each measuring 14 cm in diameter and 18 cm in height? (Round to the nearest tenth.)

10. How many square yards of glass enclose a tourist information center 30 ft in diameter and 12 ft high?

Practice 15-3

Find the volume of a rectangular prism, given the length (ℓ), width (*w*), and height (*h*).

1a. ℓ = 28 cm, *w* = 25 cm, *h* = 10 cm

b. ℓ = 30 cm, *w* = 18 cm, *h* = 12 cm

2a. ℓ = 16 ft, *w* = 20 ft, *h* = 4 ft 3 in.

b. ℓ = 2 yd, *w* = 1 yd 2 ft, *h* = 8 ft

Find the volume of a cube, given an edge.

3a. 3 m　　　**b.** 20 cm

4a. 5 dm　　　**b.** 8 m

5a. 2 ft 4 in.　　　**b.** 15 yd

6. Find the volume of a sand box 1.3 m long, 0.8 m wide, and filled to a depth of 0.2 m.

7. How many liters of water are in an aquarium 50 cm by 42 cm by 30 cm? (1 liter = 1000 cm³)

8. Find the volume of a cube of ice 1 m 30 cm on each edge.

9. How many cubes 6 m on an edge are equivalent in volume to an 18-cm cube?

10. A closet is $5\frac{1}{2}$ ft long, 4 ft wide, and 8 ft high. How many cubic feet of air does the closet enclose?

Practice 15-4

Find the surface area of each prism.

1a.
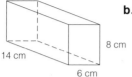
14 cm
8 cm
6 cm

b.
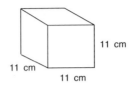
11 cm
11 cm
11 cm

Find the volume of each prism in exercise 1.

2a. rectangular prism   **b.** cube

Find the volume of each triangular prism.

3a.
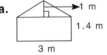
1 m
1.4 m
3 m

b.
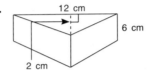
12 cm
6 cm
2 cm

Given the area of the base and height of the triangular prism, find the volume.

4. a. $B = 16$ in.^{2} $h = 4$ in. **b.** $B = 50$ m^2 $h = 5$ m

5. a. $B = 24$ cm^2 $h = 3$ cm **b.** $B = 1.8$ m^2 $h = 0.2$ m

6. A monument has a base 2.3 m long, 1.6 m wide, and 2 m high. How many cubic meters of concrete are needed for this structure?

7. How much sand will a box hold if it is 4 ft long, 3 ft 6 in. wide, and 1 ft 6 in. deep?

8. What is the difference in the volume of a cube 10 cm on each edge and a rectangular prism 3 cm × 5 cm × 6 cm?

9. A triangular prism has a base with an area of $4\frac{1}{2}$ ft^2 and a height of 4 ft. Find its volume.

10. The base of a triangular prism covers 15.4 cm^2. It is 9 cm high. What is its volume?

CHAPTER 16

Practice 16-1

A set of cards are numbered 0, 1, 2, 3, 4, 5. What is the probability of drawing:

1a. the number 3?   **b.** an even number?

2a. a prime number?   **b.** the number 8?

Find the range, mean, median, and mode of the following sets of data.

3. 9, 5, 7, 10, 6, 7

4. 80, 75, 90, 75, 85, 80

Find the value of each.

5a. 5!   **b.** 3! + 4!

6. How many different arrangements can be made of the digits 2, 4, 6, 8?

7. A box of crayons has 3 red, 5 green, and 1 yellow crayons. What is the probability of picking out a green crayon? a red crayon? a yellow crayon?

8. The temperature readings for a week were: 42°, 45°, 29°, 34°, 41°, 48°, 50°. Find the range, mean, median, and mode for this data.

9. From cards labeled A, B, and C choose a card, then flip a coin. Show all possible outcomes on a tree diagram.

10. How many different monograms can be made from the letters T, M, R?

Practice 16-2

1–2. The 7th and 8th grades sold tickets for the play for 10 days. Make a double bar graph to display their sales.

	1	2	3	4	5	6	7	8	9	10
7th	80	90	64	98	46	120	70	55	45	105
8th	72	86	86	105	52	85	70	75	40	60

3–4. Construct a double line graph to show this information:

Grade							
	K	1	2	3	4	5	6
Girls	13	35	31	25	41	21	19
Boys	10	29	37	22	19	33	18

Give the size of the central angle on a circle graph that represents:

5. a. 30%   **b.** $16\frac{2}{3}$%

6. a. 45%   **b.** 90%

7. a. 20%   **b.** 15%

8. Construct a circle graph to show:

Activity	Number of Hours Spent Per Day	Activity	Number of Hours Spent Per Day
Sleeping	7	Helping at Home	1
Eating	2	Studying	2
Playing	4	Other	1
School	7		

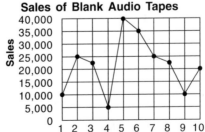

Sales of Blank Audio Tapes

9. On what day did sales reach the highest point?

10. How many more sales were made on day 7 than on day 4?

463

Practice 16-3

In which quadrant does each point lie?

1a. $(^+3, ^+4)$ **b.** $(^-3, ^-2)$

2a. $(^-1, ^-1)$ **b.** $(^+1, ^-3)$

3a. $(^+3, ^-3)$ **b.** $(^-2, ^+2)$

Write y in terms of x.

4a. $x - y = 7$ **b.** $x + y = 9$

5a. $4x - y = 8$ **b.** $3x + 5y = 15$

6a. $3x + 2y = 1$ **b.** $2x - 5y = 6$

Which points lie on $2x + y = 10$?

7a. $(0, 10)$ **b.** $(3, 5)$

8a. $(2, 6)$ **b.** $(^-1, 12)$

Copy and complete the table of values and write ordered pairs for each equation.

9.

x	$x + 4$	y	ordered pair
$^-2$			
$^-1$			
0			
$^+1$			
$^+2$			

10. Find the perimeter and area in square units for the figure named by the following coordinates.

$(^+4, ^+3), (^-2, ^+3), (^-2, ^-1), (^+4, ^-1)$

CHAPTER 17

Practice 17-1

Solve.

1a. $5x + x + 4 = 82$ **b.** $t + 4 = 15$

2a. $3y + y = 36$ **b.** $r - 9 = 16$

Simplify.

3a. $(b^8)(b^5)$ **b.** $(12c^4)(^-3c^7)$

4a. $(^-4h^3 + 3h)2h$ **b.** $(6b)(2b - 1)$

5a. $(3a^3b^3)^3$ **b.** $(2x)^2(3y)^2$

Solve.

6. Find 3 consecutive integers whose sum is 96.

7. Find 3 consecutive odd integers whose sum is 321.

8. Find 4 consecutive integers whose sum is 158.

9. Find 3 consecutive even integers whose sum is 84.

10. Find 5 consecutive integers whose sum is 95.

Practice 17-2

1a. $\dfrac{x^8}{x^2}$ **b.** $\dfrac{t^3}{t^3}$

2a. $\dfrac{^-c^8}{^-c^4}$ **b.** $\dfrac{y^4}{y^2}$

3a. $\dfrac{^-8c + 24d}{^-8}$ **b.** $\dfrac{12t + 12r}{6}$

4a. $\dfrac{8e^3 - 6e^7}{e^3}$ **b.** $\dfrac{^-4c + 20d}{4}$

5a. $\dfrac{5x^2 + 17x}{x}$ **b.** $\dfrac{9y^3 + 13y^4}{y}$

6. Show that the sum of 2 consecutive integers is always an odd number.

7. Show that the sum of 3 consecutive integers is always a multiple of 3.

8. Show that the sum of 4 consecutive integers is always divisible by 2 but never divisible by 4.

9. Are there 2 consecutive odd integers whose sum is 90?

10. Are there 3 consecutive integers whose sum is 80?

Brain Builders

TEST 1

1a. $2.6 + 0.3 + 16.7$ **b.** $67 + 6 + 7 + 9$

2a. From $\frac{1}{3}$ take $\frac{5}{18}$ **b.** Take $5\frac{5}{7}$ from $7\frac{13}{14}$

3a. 304×67 **b.** 4050×60

4a. $1672 \div 19$ **b.** $29{,}792 \div 49$

5a. $9\frac{11}{20} + 4\frac{7}{10}$ **b.** $1\frac{9}{10} + \frac{5}{12}$

6. A jacket reduced by $4.95 was sold for $50. Find the original price.

7. Steve bought some fabric for $10.80. If the price was $1.20 a yard, how many yards did he buy?

8. At $1.89 for a container of butter, how much do 5 containers cost?

9. After using her car for a year, Mrs. Whelen's odometer read 10,920 km. How many kilometers did she average a week?

10. Ruth had a bolt of ribbon 15 meters long. She used 7.75 m. How many meters did she have left?

TEST 2

1a. $234 + 23.4 + 2.34$ **b.** $0.3 + 0.04 + 0.125$

2a. $6.98 - 5.04$ **b.** Take 0.07 from 9.07

3a. 1704×246 **b.** 1800×207

4a. $56{,}268 \div 521$ **b.** $16{,}920 \div 423$

5a. $4\frac{1}{9} \div 2\frac{1}{3}$ **b.** $3\frac{5}{8} + 2\frac{5}{6}$

6. John is 1.42 m tall. His father is 2 m tall. How much taller is his father?

7. Kevin can swim a distance in 5.24 minutes. Bob swims it in 7.5 minutes. How much faster is Kevin?

8. A plane averages 125.5 km/h. How far does it go if it flies for 15.25 hours?

9. Ellen earns $23.92 a week selling papers. In how many weeks will she have earned $143.52?

10. At $27.50 a yard, how much will 37.5 yards of carpet cost?

TEST 3

1a. $7\frac{3}{5} + 5\frac{1}{2}$ **b.** $3\frac{7}{10} + 7\frac{3}{4} + 1\frac{4}{5}$

2a. From 43 take 0.43   **b.** Take $15\frac{2}{7}$ from $25\frac{1}{14}$

3a. $557.6 \div 680$ **b.** $76{,}188 \div 907$

4a. 2108×58 **b.** $\$8.90 \times 70$

5a. $60{,}716 \div 706$ **b.** $0.325 + 32.5 + 3.25$

6. If a worker earns $2500 a month, how much will she earn in $1\frac{1}{2}$ years?

7. At 3 for $27, how much will 15 books cost?

8. Joan bought a hairbrush for $2.75. How much change did she get if she paid for it with a $10 bill?

9. A pilot flew 140.9 km, 201.7 km, and 365.86 km. What was her total flying distance?

10. At $1.29 a dozen, how much will $7\frac{1}{2}$ dozen oranges cost?

TEST 4

1a. $3.5 + 3.76 + 0.049$ **b.** $6 - 0.4$

2a. $18 \div 0.24$ **b.** 0.04×0.03

3a. $3\frac{7}{8} + 8\frac{1}{3}$ **b.** $1 - \frac{3}{5}$

4a. $8\frac{2}{3} \times 1\frac{1}{2}$ **b.** $7\frac{1}{3} \div 8$

5a. $1.5 = \underline{\ ?\ }\%$ **b.** $0.125 = \frac{?}{?}$

6. Find the cost of 2.5 ft of wire at $.78 a foot.

7. A plane flew 2062.5 km in $8\frac{1}{4}$ hr. What was its average speed per hour?

8. 42 is $16\frac{2}{3}\%$ of what number?

9. If 3.75 m of fabric cost $13.20, what is the cost of 1 meter?

10. Tom had $16. He spent $2. What fractional part of $16 did he have left?

TEST 5

1a. 8000 − 674   **b.** 3050 × 920

2a. $6\frac{1}{4} \div 25$ **b.** Take 0.3 from 16

3a. 26,752 ÷ 76   **b.** $3\frac{2}{3} \times 3\frac{3}{11}$

4a. 108.5 ÷ 3.5   **b.** 4.5 × 0.048

5a. $12\frac{1}{6} - 8\frac{7}{8}$ **b.** $5\frac{4}{9} + 3\frac{3}{4}$

6. The speed of 360 km/h is how many kilometers per minute?

7. How many 6-cm pieces can be cut from 1.2 meters of ribbon?

8. Leo cut grass from 9:30 A.M. to 12:00 noon. He earned $9.25. How much was he paid per hour?

9. At $1.08 per dozen, what will 8 eggs cost?

10. Paul's job started at $5.50 per hour. If his wage was raised by $\frac{1}{10}$, what was his new rate of pay?

TEST 6

1a. 90,000 − 46,972   **b.** 8030 × 90

2a. 9.6 − 0.85   **b.** 43 + 0.065 + 25

3a. $6\frac{1}{4} + \frac{8}{9}$ **b.** $9\frac{1}{8} - 2\frac{1}{4}$

4a. $\dfrac{1\frac{2}{3}}{16\frac{2}{3}}$ **b.** $\dfrac{\frac{5}{12}}{\frac{5}{8}}$

5a. 36.34 ÷ 0.79   **b.** 22,464 ÷ 576

6. What was Jane's attendance percentage at school last month if she was absent 3 days out of 20?

7. A mathematics book that lists at $16.08 is discounted $16\frac{2}{3}\%$. What would 100 copies cost?

8. A discount of 15% brings a bill to $289. What was the original bill?

9. At the rate of 3 for 25¢, what is the cost of 7 dozen pencils?

10. Which sale offers a greater savings: a $16\frac{2}{3}\%$ reduction on a $30 item or 10% on a $40 item?

TEST 7

1a. 80,007 − 60,948   **b.** 70,400 × 608

2a. 2.9148 ÷ 8.4   **b.** 50,008 ÷ 658

3a. $1\frac{5}{8} + 3\frac{5}{6}$ **b.** $3\frac{3}{8} - 1\frac{7}{12}$

4a. 7 tenths + 7 hundredths   **b.** $(^-6 + {}^+3)^-2$

5a. $5\frac{1}{4} \div 6\frac{1}{4}$ **b.** $8\frac{5}{9} \times 3\frac{2}{11}$

6. What is the perimeter of a square field that contains 9 hectares?

7. An article sold for $11.70 brings a gain of $12\frac{1}{2}\%$. What was the cost of the article to the merchant?

8. If the gain is $6 and the selling price is $36, what is the rate of gain?

9. If the rail fare between 2 cities is $17.20 plus 15% tax, what is the total cost of the train ticket?

10. If the $9.89 you pay for a railway ticket includes a 15% federal tax, how much of the money does the railroad company receive? How much is tax?

TEST 8

1a. 80,004 − 79,658   **b.** 40,060 × 9008

2a. 0.004 ÷ 0.16   **b.** 4356 ÷ 968

3a. $6\frac{7}{12} + 9\frac{5}{8}$ **b.** $8\frac{1}{9} - 4\frac{7}{10}$

4a. 920 − 7.48   **b.** 25% of 0.12

5a. $17 \div 0.3\frac{2}{5}$ **b.** $5 = 6\frac{1}{4}\%$ of ___

6. The amount sold is $960. The rate of commission is $4\frac{1}{2}\%$. Find the proceeds.

7. The selling price is $1728. The rate of gain is $12\frac{1}{2}\%$. Find the cost.

8. The selling price of $960 reflects a discount of $16\frac{2}{3}\%$. Find the original price.

9. By selling for $60 less than cost, a merchant lost 15%. What was the cost?

10. If the loss is $4 and the cost is $36, what is the rate of loss?

TEST 9

1a. $^-3 + (^-2 - ^-12)$ b. $(^-2 - ^+5) - ^-8$

2a. $\dfrac{15}{\frac{3}{8}}$ b. $\dfrac{10\frac{1}{2}}{42}$

3a. $12{,}532 \div 208$ b. 25% of 0.2

4a. $18 = 4.5\%$ of $\underline{\ ?\ }$ b. $\frac{1}{2}\%$ of 0.5

5a. $3(0.08 + 0.2)$ b. $2.8 = \underline{\ ?\ }\%$ of 8

6. A cabinet was sold at a discount of $3\frac{1}{4}\%$ for \$619.20. Find the original price.

7. Donna borrowed \$7480 at a 6% rate of interest to start a small business. She paid back the loan at the end of $2\frac{1}{3}$ years. What amount did she pay in all?

8. Find the ratio of 20 cm to 1 meter.

9. George covers 80 cm in one step. How many steps must he take to cover 200 m?

10. An antique dealer bought a chair for \$124. If she is to make a profit of 20%, what must her selling price be?

TEST 10

1a. $^-10 - (^+3 - ^+11)$ b. $(^+9 - ^+12) - ^-4$

2a. $5\frac{1}{8} - 2\frac{3}{5}$ b. $7\frac{1}{9} \times 4\frac{1}{8}$

3a. $\frac{1}{4}(0.4 - 0.04)$ b. $12\frac{1}{2}\%$ of 0.64

4a. $4224 \div 704$ b. $76.8 \div 0.18$

5a. 0.5% of 35 b. $5 = \underline{\ ?\ }\%$ of 15

6. Find the cost of an article listed at \$45 but subject to a 25% discount.

7. Find the cost of 2.4 meters of ribbon at 80¢ a meter.

8. Mrs. Brady receives a salary of \$480 a week in addition to a 5% commission on all sales. What were her earnings for a 4-week period during which she sold a new car for \$12,000 and four used cars at \$3000 each?

9. John has \$14.75. How much more money does he need to buy a new baseball glove priced at \$25 less 10% discount?

10. If 150 gallons of fuel oil cost \$247.50, what will 250 gallons cost?

TEST 11

1a. $0.0035 = 3.5 \times 10^?$ b. $6.03 \times 10^4 = \underline{\ ?\ }$

2a. $5\frac{1}{4} - 2\frac{3}{8}$ b. $3\frac{4}{7} \times 2\frac{4}{5}$

3a. $\frac{1}{8}(0.8 - 0.08)$ b. $37\frac{1}{2}\%$ of 0.48

4a. $9090 \div 303$ b. $26.8 \div 0.36$

5a. 150% of 150 b. $4 = \frac{5}{12}$ of $\underline{\ ?\ }$

6. If 21 pieces are $8\frac{1}{3}\%$ of all the pieces in a puzzle, how many pieces are in the puzzle?

7. If Mr. Jones paid \$66.88 for 6.4 meters of curtain material, what was the cost per meter?

8. Mr. Tallon receives a salary of \$380 a week and a commission of 5% on sales over \$4000. Last week his sales amounted to \$8500. What were his total earnings?

9. What is the total amount collected when a commission of 4% brings an agent \$72?

10. Which is a better salary and by how much: \$32,000 a year or \$620 a week for 45 weeks?

TEST 12

1a. $^-24 \div (^-8 - ^-4)$ b. $^+6 \times (^-5 + ^+9)$

2a. $0.425 = \frac{?}{?}$ b. $\frac{5}{12} = \underline{\ ?\ }\%$

3a. $0.08 = \underline{\ ?\ }\%$ of 8 b. \$19.50 is $12\frac{1}{2}\%$ of $\underline{\ ?\ }$

4a. 125% of $10 = \underline{\ ?\ }$ b. $3\frac{1}{2}\%$ of $\underline{\ ?\ } = 14$

5a. $15.6 \div 0.015$ b. $4\frac{1}{2} \div \frac{1}{8}$

6. If a plane can go 240 km in $1\frac{1}{2}$ hr, how far can it go in $3\frac{1}{2}$ hrs ?

7. A grocer lost 4% by selling peas at 48¢ a can. How much did the grocer pay for each can?

8. Eggs costing 85¢ per dozen sold at a gain of $3\frac{1}{4}\%$. Find the gain on 24 doz.

9. The discount on a bill is \$15. If the rate of discount is $33\frac{1}{3}\%$, what was the marked price?

10. Peter has driven 340 km. If this is $66\frac{2}{3}\%$ of the entire journey, how many more km must he drive?

1a. $80{,}020 - 49 \div 7$ **b.** $907{,}040 \times 408$

2a. $0.4\frac{1}{6} = \frac{?}{?}$ **b.** $^-20 + {}^+4 \div {}^-2$

3a. $5\frac{4}{7} + 3\frac{3}{4}$ **b.** $6\frac{5}{6} + 8\frac{7}{18}$

4a. $1\frac{1}{4} = \underline{\ ?\ }\%$ **b.** 0.4% of $17\underline{\ ?\ }$

5a. $35 = 41\frac{2}{3}\%$ of $\underline{\ ?\ }$ **b.** 18 is $\underline{\ ?\ }\%$ of 54

6. By selling a boat for $2820, a gain of 20% over the original price was made. What did the boat cost originally?

7. A 28.3 kg weight is made of an alloy consisting of 2 parts copper and 3 parts lead. How much copper and how much lead does it contain?

8. At $3.80 an hour, how much will Mark earn for $38\frac{1}{4}$ hours of work?

9. The regular price of a bracelet is $48.50. What is the total cost if, in addition, there is a federal tax of 30% and a city tax of 1%?

10. Mr. Dunn's bank balance is $2204.95. He deposits $190.60 and then issues checks for $95.17 and $146.25. Find his new balance.

1a. 4500 g $= \underline{\ ?\ }$ kg   **b.** $340\% = \frac{?}{?}$

2a. $590 \div 100$ **b.** 1% of 350

3a. $80{,}207 - 56 \div 7$ **b.** $90{,}200 \times 680 \div 10$

4a. $0.2\,(0.8 - 0.08)$ **b.** 6 is $\underline{\ ?\ }\%$ of 24

5a. $21 = 87\frac{1}{2}\%$ of $\underline{\ ?\ }$ **b.** 0.5% of 20

6. If 1.2 kg of coffee will serve 160 people, how many kg are needed to serve 300 people?

7. Find the cost of 5 m of wire at 98¢ per m.

8. Dick worked $9\frac{1}{2}$ hr Monday, $8\frac{3}{4}$ hr Tuesday, and $6\frac{1}{8}$ hr Wednesday. If he is paid $4.40 per hour, how much did he earn in these three days?

9. A car was sold for $9280.30. This was 15% less than the original price. What was the original price?

10. If 3 km of the distance between two cities is 0.4%, how far apart are they?

1a. $0.8\frac{1}{3} = \frac{?}{?}$ **b.** $\frac{1}{16} = 0.\underline{\ ?\ }$

2a. $1 \div 0.08$ **b.** $0.3\,(0.6 - 0.04)$

3a. $7\frac{1}{8} - 4\frac{5}{12}$ **b.** $6\frac{4}{7} + 8\frac{2}{3}$

4a. 16 is 0.4% of $\underline{\ ?\ }$ **b.** 8 is $\underline{\ ?\ }\%$ of 200

5a. $4.5 = \underline{\ ?\ }\%$ **b.** 3.6×18

6. If 4.5 meters of cotton cost $12.60, what will 6 meters cost?

7. Felice borrows $960 at 5% interest. In 2 yr 6 mo what is the total amount she will pay?

8. An agent sold 100 bags of corn at $2.50 a bag, 100 sacks of potatoes at $1.90 a sack, and 400 kg of coffee at $1.80 a kg. What commission did he receive at 3%?

9. Find the average temperature from these readings: 36°, 32°, 47°, 45°, 50°, and 42°.

10. Find the cost of 6.5 liters of milk at 68¢ a liter.

1a. $0.6\frac{1}{4} = \frac{?}{?}$ **b.** $\frac{1}{32} = 0.\underline{\ ?\ }$

2a. $0.6 \div 0.004$ **b.** $261 \div 348$

3a. 120 dL $= \underline{\ ?\ }$ L   **b.** 78 cm $= \underline{\ ?\ }$ m

4a. $3\frac{2}{9} - 2\frac{5}{6}$ **b.** $0.8\,(0.6 - 0.04)$

5a. 36 is $37\frac{1}{2}\%$ of $\underline{\ ?\ }$ **b.** 36 is $\underline{\ ?\ }\%$ of 4.5

6. Judy's starting salary was $28,800 a year. If she received a raise amounting to $120 a month, by what percent did her salary increase?

7. What is the sale price if the original price is $6.40 and the discount rate is $6\frac{1}{4}\%$?

8. The sale price of a TV set is $270. If the discount is 20%, what was the original price?

9. Find the area of a circular table top 42 cm in diameter. (Use $\pi = \frac{22}{7}$.)

10. Ben had 10.5 meters of rope. He gave 3.25 m to one friend and 3.125 m to another. How much did he have left?

1a. $0.3\frac{1}{3} = \frac{?}{?}$ **b.** $\frac{1}{40} = 0.\underline{\ ?\ }$

2a. $9\frac{1}{6} - 4\frac{4}{9}$ **b.** $7\frac{5}{8} + 3\frac{4}{5}$

3a. $13.5 \div 0.045$ **b.** $3.5L = \underline{\ ?\ }\ dL$

4a. $3\frac{1}{3} = \underline{\ ?\ }\ \%$ **b.** $2\frac{1}{2}\%$ of $\underline{\ ?\ } = 75$

5a. 6 is $\underline{\ ?\ }\%$ of 99 **b.** 250% of 10

6. If a plumber charged $55.20 for 6 hours work, what would he charge for $4\frac{1}{4}$ hours work?

7. A commission agent sent a scrap dealer $249.40 as proceeds for a sale of $320. Find the rate of commission.

8. A cloth dealer bought fabric at $4.80 a yard subject to a discount of 25%. She sold it at $4.40 a yard. What was her percent of gain?

9. If 16.4 kilograms of chicken cost $44.28, what is the cost per kilogram?

10. Find the interest on $800 for 3 years at 4.5%.

1a. $\frac{1}{2} + \frac{1}{8} + \frac{1}{4} + \frac{1}{5}$ **b.** $6 \div 0.096$

2a. 4.65×0.027 **b.** $(^{+}12 - {}^{-}3) \times {}^{-}2$

3a. $3.098 = \underline{\ ?\ }\%$ **b.** 3.7% of 134

4a. 75.6 is 36% of $\underline{\ ?\ }$ **b.** 72 is $\underline{\ ?\ }\%$ of 75

5a. $2(18 + 0.5)$ **b.** $2[3 + (4 - 0)]$

6. If the $178,500 that was paid for a house includes an agent's commission of 2%, what did the house cost?

7. If Fred's wages were $4.20 an hour after receiving an increase of 20%, what were his wages before the increase?

8. Find the amount due on a loan of $480 at $4\frac{1}{2}\%$ for 3 months.

9. The floor plan for a kitchen in a new home is 1.2 cm long and 2.2 cm wide. What are the actual dimensions if the scale is 1 cm = 3 m?

10. Rosemary paid $2.80 for 0.84 meter of fabric and Kathleen paid $2.75 for 0.8 meter. Who paid more per meter? How much more?

1a. $\frac{3}{4} + \frac{5}{8} + \frac{5}{6} + \frac{1}{16}$ **b.** $\frac{3}{4} - \frac{3}{8}$

2a. 60.75×0.246 **b.** $565.2 \div 6.28$

3a. 7.35 is $\underline{\ ?\ }\%$ of 98 **b.** $0.085 = \underline{\ ?\ }\%$

4a. 125% of 40 **b.** 88.2 is 45% of $\underline{\ ?\ }$

5a. $3.45 \div 10^2$ **b.** 3.45×10^2

6. A stationer bought pencils at 10¢ each and sold 12 doz for $24.76. What was the total gain?

7. Find the amount due on $400 for 2 yr 2 mo at 4.5%.

8. Nan weighs 40.9 kg; Mary, 41.8 kg; Jane, 45.9 kg; and Ann, 45 kg. What is their average weight?

9. Find the perimeter of an equilateral triangular flower bed 3.4 m on a side.

10. A driver averaging 90 km/hr left one city for another 210 km away. Allowing for a 1-hour stop for lunch, how many hours elapsed before he reached his destination?

1a. $3\frac{1}{5} - 1\frac{5}{6}$ **b.** $8\frac{5}{9} + 6\frac{5}{12}$

2a. $0.002 + 0.2 + 0.02$ **b.** $10 - 0.9604$

3a. $2.9484 \div 0.054$ **b.** 0.175×0.03

4a. 65 is $\underline{\ ?\ }\%$ of 125 **b.** 52.8 is 96% of $\underline{\ ?\ }$

5a. 5.2% of 624 **b.** $1.1 = \underline{\ ?\ }\%$

6. How many 1.5-kg boxes can be filled from 3 cartons of hard candy each containing 60 kg?

7. Mrs. Darby left $60,000 to be divided among her three children in proportion to their ages. Juan was 24 years old; Carlos, 20 years; and Sena, 16 years. How much did each receive?

8. Gloves bought for $36 per $\frac{1}{2}$ dozen pair are sold for $5.50 a pair. What is the rate of loss computed on the cost?

9. One factor of 80 is $\frac{2}{5}$. What is the other factor?

10. If a scale on a map shows 1 cm = 26 km, what is the distance between 2 cities 4.5 cm apart on that map?

TEST 21

1a. $0.6\frac{2}{3} = \frac{?}{?}$ **b.** $\frac{5}{16} = \underline{\ ?\ }\%$

2a. $3\frac{1}{2} = \underline{\ ?\ }\%$ **b.** 65 is $37\frac{1}{2}\%$ of $\underline{\ ?\ }$

3a. $5 \times 24 \div 10^2$ **b.** $6 \times 10^3 \times 470$

4a. $1\frac{3}{4} = \underline{\ ?\ }\%$ **b.** 6.6 is $\underline{\ ?\ }\%$ of 12

5a. $\frac{3}{2}$ is $133\frac{1}{3}\%$ of $\underline{\ ?\ }$ **b.** 1.5% of 18

6. How much would a $\frac{1}{2}$-minute TV commercial cost at $1500 a second?

7. A grain dealer paid $1650 for wheat. If a 20% shipping charge is added, what is the total cost?

8. How many pieces 0.5 m long can be cut from 2.5 m of ribbon?

9. A farmer paid $249 for a piece of machinery that had been discounted 17%. What was the original price for the piece of machinery?

10. If 4 m of plastic cost $6.20, what would 2.5 m cost?

TEST 22

1a. $2n - 1 = {}^-9$ **b.** $3n + 4 = {}^-14$

2a. 84 is 120% of $\underline{\ ?\ }$ **b.** $83\frac{1}{3}\%$ of 420

3a. $9\frac{1}{4} - 3\frac{5}{6}$ **b.** $2.5\ m = \underline{\ ?\ }\ dm$

4a. $\dfrac{6\frac{1}{4}}{10}$ **b.** $8\frac{4}{9} + 3\frac{3}{4}$

5a. $2\frac{1}{5} = \underline{\ ?\ }\%$ **b.** $\dfrac{2\frac{2}{3}}{10} - \frac{1}{5}$

6. What is the surface area of a cylinder 12 cm in diameter and 16 cm high?

7. A portable sewing machine was sold for $150. If this was a reduction of 26%, what was the marked price? (Round to the nearest cent.)

8. If 145 g of salted peanuts cost 80¢, what will 1189 g cost?

9. If Mr. Kane suffered a $2\frac{1}{2}\%$ loss by selling property for $23,459.28, how much had he paid for it?

10. Mr. Myles had a debt of $12,000. He has paid off $87\frac{1}{2}\%$ of it. How much does he still owe?

TEST 23

1a. $6\frac{3}{8} + 13\frac{1}{2}$ **b.** ${}^-3 \times {}^-4 + {}^+10$

2a. ${}^-3n = {}^-21$ **b.** $n + 5 = {}^-10$

3a. $0.5(0.6 - 0.04)$ **b.** $0.06 + 0.5 + 0.003$

4a. $(6 - 2)! \div 3!$ **b.** $(2 + 4)! \div 3!$

5a. $0.9[0.7 + (0.3 \times 6)]$ **b.** 1.9 is $2\frac{1}{2}\%$ of $\underline{\ ?\ }$

6. What is Matt's percent of absence for a month in which he attended school 18 out of 20 days?

7. Mrs. Vitulli saved $4.25 by buying a blanket at a reduction of $12\frac{1}{2}\%$. Find the original price.

8. The proceeds from a sale are $6133.20. What was the amount sold if the commission rate was 5%?

9. Mr. Carlin was receiving $560 a week. His salary was increased 15% but later reduced 5%. What was his annual salary after these adjustments?

10. Mrs. Mellon receives a fixed salary of $540 a week and a commission of 5% on all sales. If last month's sales amounted to $3000, what were her earnings?

TEST 24

1a. $3.4 \div 10^2$ **b.** $14.7 - 7.35$

2a. $\frac{3}{4}\left(1\frac{1}{9} \div \frac{5}{8}\right)$ **b.** $\left(6\frac{3}{8} - 4\frac{7}{12}\right) \div \frac{1}{2}$

3a. $8 \times 10,000 - 68$ **b.** ${}^-16 + {}^-25 \div {}^+5$

4a. $\dfrac{\frac{5}{6}}{1\frac{1}{5}}$ **b.** $\frac{1}{2}$ of $8 = \underline{\ ?\ }\%$ of 12

5a. 30% of 5 m $= \underline{\ ?\ }$ cm   **b.** 144 is 25% of $\underline{\ ?\ }$

6. If on a map 7.5 cm represents 90 m, what does 15.5 cm represent?

7. Michael paid $42 for one new tire and one used tire for his bicycle. The new tire was $6 more than the used one. What did the used tire cost?

8. Our team won 45 games and lost 15. What was the team's winning percentage?

9. A swimming pool 200 m by 100 m has around it a walk 1.5-m wide. How many square meters are in the walk?

10. Find the perimeter of a rectangle with an area of 124.7 square meters and an altitude of 8.6 meters.

Drill and Mental Math

1

DRILL

1. Add 5 to: 19, 48, 77, 26, 35, 58
2. Subtract 5 from: 14, 71, 32, 54, 63, 90
3. Multiply by 5: 9, 0, 12, 8, 6, 20, 7, 11, 4
4. Find $\frac{1}{5}$ of: 60, 30, 40, 55, 35, 45, 25, 20
5. Round to the nearest thousand: 24,123   986   7645

MENTAL

1. How many boxes of 1000 can you fill if you have 2052 + 3910 items?
2. How many boxes of 1000 can you fill if you have 396 + 210 + 415 items?
3. Which is greater: 456 + 324 + 316 or 1000?
4. Find 1498 rounded to the nearest hundred.
5. Find 976,123 rounded to the nearest million.
6. Carlos has $10. Does he have enough money to buy a game for $3.98 and a mitt for $5.75?
7. One factor of 45 is 5. What is the other factor?
8. At $9 a yard, how many yards of fabric can be bought for $63?
9. At $8 a yard, find the cost of 5 yards of lace.
10. At $14.75 each, how much will 4 blouses cost?

2

DRILL

1. Add 7 to: 73, 42, 65, 76, 41, 64, 33
2. Subtract 7 from: 73, 42, 65, 76, 41, 64
3. Multiply by 7: 9, 6, 8, 10, 12, 5, 11, 0
4. Divide by 7: 9, 11, 16, 19, 22, 27, 30
5. From 9636 subtract: 1000, 4000, 9000

MENTAL

1. Which is greater: 1096 − 324 or 700?
2. Find the difference to the nearest thousand: 5962 − 2052.
3. Find the difference to the nearest hundred: 1021 − 396.
4. Find the difference to the nearest thousand: 2654 − 200.
5. Find the difference to the nearest hundred: 2654 − 200.
6. Bruce had $10. He spent $5.23. How much money did he have left?
7. Peter has 74¢, which is 7¢ less than the amount Ellen has. How much money does Ellen have?
8. At 55 km/h, how far can you go in 7 hours?
9. If a plane travels 800 miles in 4 hours, what is its speed per hour?
10. From 3 hundred take 4 tens.

3

DRILL

1. Add 9 to: 54, 27, 36, 43, 28, 62, 51
2. Subtract 9 from: 54, 27, 36, 43, 28, 62
3. Multiply by 9: 4, 8, 5, 12, 6, 9, 7, 11, 3
4. Find $\frac{1}{9}$ of: 72, 81, 54, 45, 27, 63, 99, 36
5. Multiply: 3 × 8, 3 × 80, 3 × 800, 3 × 8000, 3 × 8,000,000, 3 × 8,000,000,000

MENTAL

1. What is the perimeter of a square that measures 42 m on each side?
2. There are 30 boxes of candy. If each box holds 200 pieces, how many pieces of candy are there?
3. Would 9 boxes at $18 a box cost more than $200?
4. If n = 9, what number is named by: (24 − 6) ÷ n?
5. What number is 9 less than 34?
6. What number is 10 times 60,000?
7. What number is 20 times 2323?
8. At 6¢ each, how many items can be bought for 54¢?
9. An airplane traveled 110 miles in $\frac{1}{3}$ hour. What was its speed per hour?
10. Compare. Write <, =, or >.
    9 × 8 _?_ 70   4 × 12 _?_ 50
    24 + 26 _?_ 50   87 − 29 _?_ 50

DRILL

1. Add 8 to: 24, 35, 43, 26, 78, 67, 49, 32
2. From 6000 subtract: 6000, 600, 60, 6, 66, 666
3. Multiply by 8: 6, 0, 7, 10, 12, 4, 8, 9, 5
4. Find $\frac{1}{8}$ of: 32, 64, 40, 72, 56, 24, 48, 96
5. Multiply: 9×8, 9×80, 90×80, 90×800, 90×8000, 900×800

MENTAL

1. The factors are 40 and 80. What is the product?
2. If a car travels 50 miles per hour, how far will it travel in 8 hours?
3. At $2.60 each, what will 7 items cost?
4. Which is greater: 49×210 or 10,500?

5. Round to the nearest 10 to estimate: There were 19 events and 28 athletes in each event. How many athletes competed?
6. Round to the nearest 100 to estimate: $2795 \div 375$.
7. Round to the nearest 1000 to estimate: There were 3150 men, 2725 women, and 5225 children at the meet. How many people were there?
8. Estimate to the nearest 100: There were 12,515 programs. If 11,415 were sold, how many were left?
9. What number is 4 more than 8×6?
10. Tim has 56 coins in his collection. Donna has $\frac{1}{8}$ of that number. How many coins does Donna have?

DRILL

1. Tell the equivalent fraction that has a denominator of 10: $\frac{1}{2}$, $\frac{2}{5}$, $\frac{5}{5}$, $\frac{4}{5}$, $\frac{3}{5}$
2. Tell the equivalent fraction that has a denominator of 100: $\frac{3}{10}$, $\frac{1}{2}$, $\frac{1}{4}$, $\frac{2}{5}$, $\frac{9}{20}$, $\frac{4}{25}$
3. Tell the equivalent fraction that has a denominator of 1000: $\frac{1}{100}$, $\frac{1}{10}$, $\frac{1}{2}$, $\frac{1}{50}$, $\frac{1}{8}$
4. Express as hundredths: 0.5, 0.1, 0.7, 0.9, 0.4
5. Tell the correct relation ($<$, $=$, $>$):
   0.2 _?_ 0.3   0.3 _?_ 0.03
   0.006 _?_ 0.02   0.125 _?_ 0.12
   0.009 _?_ 0.001   0.32 _?_ 0.4

MENTAL

1. Write 0.5 as a fraction in simplest form.
2. Which is the approximate thickness of a sheet of paper? 3 in. or 0.3 in. or 0.03 in.
3. Order from smallest to greatest: 0.1, 0.03, 0.001, 0.031.
4. Order from greatest to smallest: 0.009, 0.9, 0.912, 0.091, 0.91.
5. How much greater than 0.8 is 1?
6. How much greater than 0.02 is 0.06?
7. What number added to 0.04 is equal to 1?
8. Which is greater: $0.04 + 0.9$ or 1?
9. Harry had $20. He spent $\frac{4}{5}$ of it. How much money did he have left?
10. How much greater than 950 is 1500?

DRILL

1. Name the opposite integer for each: $^{+}5$, $^{+}8$, $^{+}11$, $^{-}18$, $^{-}4$, $^{-}7$
2. Read the integers in order from smallest to greatest: $^{+}4$, $^{+}2$, $^{+}3$, $^{+}1$, 0
3. Tell what integer is one more than: $^{+}8$, $^{+}16$, $^{-}6$, $^{-}11$, $^{-}5$
4. Tell what integer is one less than: 0, $^{-}3$, $^{+}14$, $^{-}4$, $^{-}10$, $^{-}8$
5. Evaluate $a + b$ when $a = {}^{+}2$ and b is: 0, $^{+}3$, $^{-}3$, $^{-}2$, $^{-}5$

MENTAL

1. Name the integer that represents the opposite of negative two.
2. On a horizontal number line, what integer is 1 unit to the left of negative two?

3. On a horizontal number line, what integer is 1 unit to the right of negative two?
4. On a horizontal number line, what integer is 7 units to the right of $^{-}7$?
5. What integer is suggested by a temperature reading of 5 degrees below zero?
6. On a horizontal number line, what integer is 3 units to the left of zero?
7. What operation is suggested by moving on a number line from $^{-}5$ to $^{-}1$?
8. Use $<$ or $>$ to compare: $^{-}3$ _?_ $^{-}4$
9. If the base (or length) is 5 meters and the height (or width) is 6 meters, what is the area of a parallelogram? ($A = bh$)
10. Evaluate $\frac{ab}{2}$ when $a = 3$ and $b = 7$.

7

DRILL

1. Subtract from 2: 0.3, 0.5, 0.8, 0.4, 0.6, 0.2, 0.1, 0.9, 1.3, 1.6
2. Tell the correct relation (< or >): $^-4$? $^-2$ $^+4$? $^+2$ $^+3$? $^-3$ 0 ? $^-3$
3. Tell what integer is 2 greater than: 0, $^+4$, $^-7$, $^-5$, $^-3$
4. Read the integers in order from greatest to smallest: $^+3$, $^-1$, 0, $^-4$, $^+4$
5. Solve for n: $^-4 + n = 0$; $^-4 + 0 = n$; $^-4 - n = ^-4$

MENTAL

1. On a horizontal number line, what integer is 6 units to the left of zero?
2. The temperature was 7°C. By midnight it had dropped 9°. Was the midnight temperature above or below zero?
3. The temperature was $^-3$°C. By noon it had risen 7°. Was the noon temperature above or below zero?
4. What integer is suggested by a temperature of 15 degrees below zero?
5. Pedro spent $5 of his earnings and saved the rest. If he saved $45, how much had he earned?
6. What is $\frac{1}{3}$ of $2.70?
7. If 40 armchairs cost $7840, how much would one cost: $960 or $196 or $121?
8. If $15 is $\frac{3}{8}$ of Stephen's money, how much money does he have?
9. At $1.10 a dozen what will $1\frac{1}{2}$ dozen eggs cost?
10. When $n = 9$, what number is named by $4n$?

8

DRILL

1. Tell what number is named: $3 + 4 + 6 + 2$ $3000 + 4000 + 6000 + 2000$
2. On a horizontal number line, what integer is: 2 units left of zero; 2 units right of zero; 4 units left of zero
3. Tell what must be added to each in order to reach zero: $^-16$, $^-9$, $^+3$, $^+15$
4. What integer is 2 less than: 0, $^-2$, $^+1$, $^-10$?
5. Describe the number-line action:
   $^+4 - ^+7 = $? $^-3 - ^+5 = $?
   $^-4 + ^+2 = $? $0 - ^-2 = $?

MENTAL

1. Write the number sentence suggested by a number-line move from $^-6$ to zero.
2. Write the number sentence suggested by a number-line move from $^-5$ to $^+3$.
3. Use <, =, or > to compare: $^+7 - ^+9$? $^-7 + ^+9$.
4. Use <, =, or > to compare: $^-3 + ^+6$? $^-6 + ^+3$.
5. Amy earned $4.50 a day after school. How much did she earn in 4 days?
6. If a dozen items cost $19.20, how much does one item cost?
7. What part of an hour is 20 minutes?
8. Mr. Thorsland bought a car for $9435. He was allowed $1500 trade-in for his old car. How much cash did he need to pay for his new car?
9. Evaluate $3(a + b)$ when $a = ^-4$ and $b = ^+6$.
10. Estimate the sum to the nearest whole number: $2\frac{7}{8} + \frac{3}{4} + 2\frac{1}{8}$

9

DRILL

1. If $n = 9$, what number is named by: $16 + n$ $25 - n$ $n(4 + 7)$ $189 \div n$
2. On a horizontal number line, how many units to the right or left of zero is each integer: $^-1$, $^+1$, $^-6$, $^-3$, $^-16$, $^+10$
3. Read the integers in order from smallest to greatest: $^-5$, $^-6$, 0, $^-1$, $^-3$
4. Tell the correct relation (< or >):
   $^-5$? $^-4$ $^+8$? 0 $^+8$? $^+9$
   $^+1$? $^-1$
5. Describe the number-line action. Tell where you would start and the number of units left or right you would move: $^+2 + ^+4$
   $^+4 - ^+5$ $^-3 - ^+2$ $^-3 + ^+4$

MENTAL

1. What operation is sugested by moving on a number line from $^+2$ to $^-1$?
2. What operation is suggested by moving on a number line from $^+4$ to $^-4$?
3. The temperature was $^-5$°C. It dropped 10°. What was the new temperature?
4. The temperature was $^-8$°C. It fell 20°. What was the new temperature?
5. How much money must be added to $1.90 to make $2.50?
6. Ned earned $94.50 in 3 days. This is an average of how much money per day?
7. When $\frac{2}{3}$ of a number is 12, what is the number?
8. What number is $^-6$ less than 4?
9. If $\frac{5}{6}$ of a number is 10, what is $\frac{2}{3}$ of the same number?
10. Find the difference to the nearest thousand: $47,246 - 18,964$.

473

10 **DRILL**

1. Tell the correct relation ($<$ or $>$):
   $^+6$ _?_ $^-7$ $^-3$ _?_ $^-2$ $^-5$ _?_ $^-9$
   $^+4$ _?_ $^-4$

2. Add: $^+8 + {}^+7$ $^-7 + {}^+6$ $^+5 + {}^-5$
   $^+4 + 0$ $^+4 + {}^-3$

3. Complete: $^-4 - {}^-2 = {}^-4 +$ _?_
   $^-6 - {}^-7 = {}^-6 +$ _?_
   $^-5 - {}^-5 = {}^-5 +$ _?_

4. Subtract $^-3$ from: $^+4$, $^+5$, $^+7$, $^+3$, $^+2$

5. Multiply by $^-3$: 0, $^+1$, $^-2$, $^+4$, $^-5$, $^-6$

MENTAL

1. Danny mixed 4.75 g and 1.25 g. How many grams in all?

2. How much less than 10 kg is 3.2 kg?
3. Evaluate $a \div b$ when $a = {}^+12$ and $b = {}^-3$.
4. What is the largest whole number that makes this true: $0.25 + 2.9 >$ _?_ ?
5. Write the numeral: four million four.
6. One factor of 15 is $^-5$. What is the other factor?
7. On a horizontal number line, what integer is 6 units to the left of zero?
8. One morning the temperature registered 3°C and then dropped 6 degrees. What was the later temperature reading?
9. Evaluate $^-24 \div d$ when $d = {}^-6$.
10. Use $<$, $=$, or $>$ to compare:
    $^+36 \div {}^+4$ _?_ $^-18 \div {}^-2$.

11 **DRILL**

1. Subtract: $^+3 - {}^+6$, $^+2 - {}^+5$,
   $^+4 - {}^+5$, $^+5 - {}^+9$, $^-7 - {}^+1$, $^-6 - {}^-8$

2. Multiply: $^+2 \times {}^+4$, $^-6 \times {}^-3$, $^-8 \times {}^+5$,
   $^+6 \times {}^-9$, $^-7 \times {}^-3$, $^+10 \times {}^+4$, $^-9 \times {}^-9$

3. Tell the correct relation ($<$ or $>$):
   $^-5$ _?_ $^-6$, $^+9$ _?_ $^-7$, 0 _?_ $^+6$,
   $^+4$ _?_ $^+8$

4. Subtract $^-10$ from: $^+2$, $^+6$, $^-4$, $^-15$, $^+7$

5. Divide by $^-3$: $^+12$, $^-9$, $^-3$, $^+6$, $^-15$

MENTAL

1. Ten gallons of regular gasoline cost $12.99. Write the price of 1 gallon, using a fraction to express part of a penny.
2. Ten gallons of premium gasoline cost

$15.49. Write the price of 1 gallon, using a fraction to express part of a penny.
3. Write the difference, placing a decimal point in the correct position:
   $25 - 0.76 = 2\ 4\ 2\ 4$.
4. Write the sum, placing a decimal point in the correct position:
   $0.42 + 2.4 + 7 + 0.125 = 9\ 9\ 4\ 5$.
5. 90 is $\frac{3}{4}$ of what number?
6. After Tim travels 0.75 of his journey, he still has 15 miles to go. What is the total distance he will travel?
7. If 30 items cost $4563, how much is 1 item: $1521 or $152.10 or $15.21?
8. $5346 + 700 + 60 +$ _?_ $= 5346 + 764$
9. What is $^-3 + {}^-2 \times {}^-5$?
10. What is 40% of 15?

12 **DRILL**

1. Tell which fraction is greater:
   $\frac{7}{9}$ or $\frac{5}{9}$, $\frac{11}{25}$ or $\frac{7}{25}$, $\frac{9}{100}$ or $\frac{90}{100}$

2. Tell which fraction is smaller:
   $\frac{7}{3}$ or $\frac{3}{7}$, $\frac{4}{9}$ or $\frac{9}{4}$, $\frac{3}{2}$ or $\frac{2}{3}$

3. $\frac{1}{2} = \frac{?}{4} = \frac{?}{16} = \frac{?}{8} = \frac{?}{14} = \frac{?}{18} = \frac{?}{10} = \frac{?}{12}$

4. $\frac{1}{3} = \frac{?}{12} = \frac{?}{18} = \frac{?}{24} = \frac{?}{30} = \frac{?}{21} = \frac{?}{27} = \frac{?}{15}$

5. Let $\ell = 5$ and $w = 4$. Evaluate: $\ell + w$, $2\ell + 2w$, ℓw, $2(\ell + w)$.

MENTAL

1. Order from least to greatest:
   $\frac{1}{2}$, $\frac{1}{10}$, $\frac{1}{3}$, $\frac{1}{5}$, $\frac{1}{100}$

2. Tim ran $\frac{3}{4}$ of the way around the track. Molly ran $\frac{7}{8}$ of the way around the track. Who ran farther?
3. Which of these is greatest:
   $\frac{9}{10}$, $\frac{9}{9}$, $\frac{9}{100}$, $\frac{90}{100}$, $\frac{900}{1000}$?
4. Which of these is smallest:
   $\frac{1}{2}$, $\frac{50}{100}$, $\frac{5}{10}$, or $\frac{2}{6}$?
5. Cal ran $\frac{6}{8}$ of the way around the track. Juan ran $\frac{3}{4}$ of the way around the track. Who ran farther?
6. What fractional part of a yard is 9 inches?
7. At the rate of 60 miles an hour, how far will you travel in 30 minutes?
8. At $7 a yard, what part of a yard of fabric can you buy for 70¢?
9. Write an equivalent fraction for $\frac{2}{3}$.
10. Megan had $16. She spent $12 of it. What fractional part of the money did she spend?

13 DRILL

1. Add 11 to: 29, 54, 73, 65, 36, 28, 47, 92
2. Subtract 11 from: 32, 45, 63, 24, 56, 41
3. Multiply by 7 and add 11: 9, 6, 10, 8, 7, 5
4. Change to whole numbers:
$\frac{0}{6}, \frac{6}{6}, \frac{48}{6}, \frac{0}{10}, \frac{8}{7}, \frac{21}{7}, \frac{45}{5}$
5. Change to mixed form:
$\frac{14}{4}, \frac{25}{4}, \frac{9}{4}, \frac{21}{4}, \frac{13}{4}, \frac{35}{4}, \frac{19}{4}$

MENTAL

1. What fraction added to $\frac{5}{8}$ is equal to 1?
2. How much greater than $\frac{7}{8}$ is 1?

3. What fraction represents $\frac{1}{5}$ more than the original whole?
4. What fraction represents $\frac{1}{3}$ less than the original whole?
5. What must be taken from $1\frac{1}{2}$ feet to leave 10 inches?
6. The difference between two numbers is 6. If the greater number is 15, what is the other number?
7. How many days are there in 3 weeks and 5 days?
8. Is $\frac{3}{4}$ yd of ribbon longer than $\frac{5}{8}$ yd?
9. At a speed of 250 miles per hour, how long will it take to travel 750 miles?
10. If Esther earns $5 an hour, how long will it take her to earn $45?

14 DRILL

1. Tell the correct relation (= or ≠):
   $56 \div 7$ __?__ 8 $34 - 5$ __?__ 30
   $56 + 7$ __?__ 60 9×7 __?__ 70
2. Change to improper fractions:
   $7\frac{1}{3}, 4\frac{2}{3}, 9\frac{1}{4}, 6\frac{3}{4}$
3. Add: $\frac{1}{3} + \frac{2}{3}, \frac{4}{5} + \frac{1}{5}, \frac{2}{9} + \frac{7}{9}, \frac{5}{8} + \frac{3}{8}$
4. Subtract: $\frac{3}{5} - \frac{1}{5}, \frac{1}{2} - \frac{1}{2}, \frac{3}{8} - \frac{3}{8},$
   $\frac{5}{6} - \frac{1}{6}$
5. Subtract from 1: $\frac{1}{2}, \frac{3}{4}, \frac{5}{8}, \frac{7}{10}, \frac{4}{5}, \frac{7}{16},$
   $\frac{3}{100}$

MENTAL

1. $\frac{3}{8} + \frac{3}{4}$ is equal to how many eighths?

2. How much greater than 1 is $\frac{3}{4} + \frac{1}{4} + \frac{3}{4}$?
3. How much less than 5 is $6 - 1\frac{2}{5}$?
4. What fraction subtracted from 2 equals $1\frac{1}{3}$?
5. One factor of 24 is 8. What is the other factor?
6. Find the cost of 50 cards at $30 per hundred.
7. If a train travels 90 miles per hour, how far will it travel in 12 hours?
8. Mt. Whitney is 14,484 ft high. If you climbed to a point 200 ft below the top, how high would you have climbed?
9. Is $\frac{22}{3}$ greater or less than 7?
10. From a box of 1000 envelopes Ralph used 684. How many were left?

15 DRILL

1. Tell which are prime numbers:
   11, 18, 23, 63, 61, 7
2. Tell the greatest common factor for each pair:
   24 and 8   12 and 18
   36 and 24   20 and 32
3. Simplify: $\frac{7}{14}, \frac{2}{4}, \frac{3}{9}, \frac{12}{18}, \frac{7}{35}, \frac{21}{28}, \frac{15}{25}$
4. Multiply: $\frac{1}{3} \times \frac{5}{6}, \frac{3}{5} \times \frac{1}{4}, \frac{2}{7} \times \frac{2}{3}, \frac{3}{4} \times \frac{3}{8}$
5. Change to mixed form:
   $\frac{65}{9}, \frac{76}{9}, \frac{47}{9}, \frac{58}{9}, \frac{40}{9}, \frac{28}{9}, \frac{89}{9}$

MENTAL

1. What is 7 more than 3×30?
2. Write the prime factors of 24.

3. Joel has $1\frac{1}{4}$ pounds of raisins. Can he put all the raisins into 11 boxes so that each box contains exactly $\frac{1}{8}$ pound?
4. If $\frac{3}{4}$ of the cost of a dress is $28, what is the cost of the dress?
5. A small jet travels at a speed of 300 miles per hour. How far will it go in $2\frac{1}{3}$ hr?
6. At $4.50 per ticket, how many tickets can be bought for $690?
7. Ned earns $8.10 a day working after school. How much does Ned earn in 5 days?
8. Find the difference to the nearest thousand: 42,125 - 19,625.

16

DRILL

1. Multiply by 6: 7, 9, 5, 4, 3, 8, 10, 6
2. Find $\frac{1}{6}$ of: 36, 6, 12, 18, 54, 48, 24, 30
3. Multiply $\frac{2}{3} \times 6$, $\frac{4}{5} \times 5$, $\frac{3}{6} \times 12$, $7 \times \frac{5}{7}$
4. Tell the reciprocal of: $\frac{2}{3}$, $\frac{1}{6}$, $\frac{5}{9}$, $\frac{3}{4}$, $\frac{7}{10}$, $\frac{4}{7}$
5. Complete: $12 \div 2 = 12 \times \underline{\ ?\ }$
   $\frac{1}{2} \div \frac{2}{9} = \frac{1}{2} \times \underline{\ ?\ }$ $2 \div \frac{3}{5} = 2 \times \underline{\ ?\ }$

MENTAL

1. Dividing a number by 9 is equivalent to multiplying it by $\underline{\ ?\ }$.
2. Dividing a number by $2\frac{1}{2}$ is the same as multiplying it by $\underline{\ ?\ }$.

3. How many hamburgers can be made from $2\frac{1}{2}$ lbs of meat if each weighs $\frac{1}{4}$ lb?
4. Which is greater: $\frac{7}{8} \div \frac{1}{8}$ or $\frac{7}{8} \times \frac{1}{8}$?
5. The difference between two numbers is 8. If the smaller number is 9, what is the greater number?
6. At a speed of 200 miles per hour, how long will it take to travel 500 miles?
7. Which is greater: 28×695 or $21,000$?
8. There are 31 classes with 39 pupils in each class. Estimate to the nearest hundred the total number of pupils.
9. Rosa read $\frac{1}{2}$ of her book on one day and $\frac{1}{4}$ the second day. What part of her book is still to be read?
10. In a box of 18 pens $\frac{1}{6}$ of them are green. How many pens are not green?

17

DRILL

1. Tell whether or not the fraction is greater than $\frac{1}{2}$: $\frac{3}{4}$, $\frac{1}{5}$, $\frac{7}{8}$, $\frac{11}{15}$, $\frac{2}{3}$
2. Round to the nearest whole number: $4\frac{3}{4}$, $1\frac{1}{5}$, $6\frac{7}{8}$, $24\frac{11}{15}$, $16\frac{2}{3}$
3. Tell whether or not the sum is greater than 1: $\frac{1}{4} + \frac{3}{4}$, $\frac{5}{9} + \frac{5}{9}$, $\frac{6}{8} + \frac{2}{8}$, $\frac{4}{5} + \frac{2}{5}$
4. Change to improper fractions: $3\frac{3}{7}$, $4\frac{2}{7}$, $2\frac{4}{7}$, $5\frac{5}{7}$, $6\frac{6}{7}$, $7\frac{1}{7}$
5. Change to mixed form: $\frac{15}{7}$, $\frac{29}{7}$, $\frac{32}{7}$, $\frac{40}{7}$, $\frac{57}{7}$

MENTAL

1. How much less than 1 is $\frac{1}{3} + \frac{4}{9}$?
2. Use $<$, $=$, or $>$ to compare: $3\frac{3}{4} + \frac{1}{8} \ \underline{\ ?\ }\ 4$.

3. Use $<$, $=$, or $>$ to compare:
   $15\frac{1}{2} - 6\frac{7}{8} \ \underline{\ ?\ }\ 9$.
4. Which would you rather have: $3\frac{1}{2}$ lb of candy or $\frac{5}{4}$ lb?
5. Find the product by renaming one factor:
   $3\frac{3}{4} \times \frac{8}{9} = \left(3 \times \frac{8}{9}\right) + \left(\underline{\ ?\ } \times \frac{8}{9}\right) = \underline{\ ?\ }$
6. Which is equal to $2\frac{7}{10} \times 4\frac{1}{3}$: $11\frac{7}{10}$ or $8\frac{7}{30}$?
7. If $\frac{3}{4}$ of a pound costs $1.80, what is the price per pound?
8. At the rate of $2\frac{1}{2}$ miles an hour, how long will it take to walk 5 miles?
9. If $33.24 is to be divided equally among 3 people, how much money should each person receive?
10. If Rebecca earns $42.40 a day, how much will she earn in 4 days?

18

DRILL

1. Multiply by 40: 10, 1000, 100, 20, 40
2. Read the numeral, then rename as factors: 10^3, 4^2, 5^4, 10^6, 3^5, 4^4
3. Give the simplest number name for each: 10^3, $10 \times 10 \times 10$, 10^2, 10×10^2
4. Give the simplest number name for each: 2^2, 3^2, 9^2, 4^1, 2^3, 1^3, 5^0
5. Tell the correct relation ($<$, $=$, $>$):
   $10^4 \ \underline{\ ?\ }\ 10^4$ $3^2 \ \underline{\ ?\ }\ 2^3$ $2^4 \ \underline{\ ?\ }\ 4^2$
   $10^0 \ \underline{\ ?\ }\ 6^1$

MENTAL

1. Write as a numeral with a base and an exponent:
   $10 \times 10 \times 10 \times 10 \times 10 \times 10 \times 10$.
2. Write the number named by: $3 \times 3 \times 3$.

3. Write the standard numeral: $(4 \times 10^4) + (2 \times 10^3) + (5 \times 10^2) + (3 \times 10^1)$.
4. Use exponent form and expand: 63,575.
5. At 40 km/h, how long will it take to go 200 km?
6. Express 3.01×10^6 as a standard numeral.
7. Vincente spent $6, which was $\frac{2}{3}$ of his money. How much money did Vincente have at first?
8. There are 24,500 people at the game. If 7000 people are in the grandstands, how many are not in the grandstands?
9. How many times greater than 6 is 72?
10. 30 boxes each contained 24 cards. How many cards were there in all?

19 DRILL

1. Tell what number is named: 4×10^2
   5×10^1 3×10^0 $6 \times \frac{1}{10^2}$ $2 \times \frac{1}{10^2}$

2. Tell what number is named: $4^2, 9^2, 7^2, 5^2$

3. Tell the value of the 3 in: 0.003, 0.03, 0.3

4. Name an equivalent fraction that has a denominator of 10 or 100: $\frac{1}{4}, \frac{2}{5}, \frac{3}{25}, \frac{1}{5}, \frac{3}{4}$

5. Tell an equivalent fraction that has a denominator of 100 or 1000: $\frac{1}{250}, \frac{1}{8}, \frac{2}{50}$

MENTAL

1. Name 0.75 as a fraction.
2. Find $\frac{1}{8}$ as a percent.

3. Name $\frac{4}{25}$ as a decimal.

4. Guess the thickness of sewing thread: 0.25 cm or 0.025 cm or 2.5 cm.

5. Express 0.7 as thousandths.

6. Use <, =, or > to compare:
   $0.12 + 0.02 + 0.012$ __?__ 0.2.

7. At 30¢ a yard, how many yards can be bought for 90¢?

8. If a plane travels 320 miles in one hour, how far will it go in $\frac{3}{4}$ hour?

9. If a plane travels 150 miles in $\frac{3}{4}$ hour, how far will it go in one hour?

10. If $15 is $\frac{3}{8}$ of Stephen's money, how much money does he have?

---

20 DRILL

1. What fractional part of 50 is:
   20, 40, 10, 30, 25, 35, 45, 15, 5

2. Express as decimals:
   $\frac{1}{2}, \frac{1}{4}, \frac{3}{4}, \frac{3}{10}, \frac{7}{10}, \frac{9}{10}, \frac{1}{5}, \frac{2}{5}, \frac{3}{5}, \frac{4}{5}, \frac{1}{10}$

3. Read each numeral, changing the fractional part to its equivalent decimal:
   $0.5\frac{1}{2}, 0.5\frac{3}{4}, 0.07\frac{3}{10}, 0.16\frac{1}{2}, 0.3\frac{1}{4}$

4. Express as fractions in simplest form:
   0.5, 0.25, 0.3, 0.4, 0.75, 0.80, 0.125, 0.375

5. Round to the nearest whole number:
   7.5, 8.25, 6.19, 5.763, 4.199

MENTAL

1. The principal bought 40 new books for the library. 12 of them are biographies. What fractional part is this?

2. Write the fractional equivalent for 0.

3. Write the quotient, placing a decimal point in the correct position:
   $378.02 \div 41 = 9\,2\,2$.

4. A pamphlet of 30 sheets of paper is 0.12 inch thick. About how thick is one sheet?

5. 90 is $\frac{3}{4}$ of what number?

6. After Mr. Boyle travels $\frac{5}{8}$ of his journey, he still has another 15 miles to go. What is the total distance he will travel?

7. If 30 items cost $4563, how much is 1 item?

8. $\frac{7}{8}$ of a number is 14. What is the number?

9. Is $\frac{3}{5}$ more or less than 0.7?

10. If $n = 7$ what number is named by $\frac{18 + n}{5}$?

---

21 DRILL

1. Simplify: $\frac{6}{24}, \frac{18}{45}, \frac{24}{60}, \frac{12}{36}, \frac{36}{42}, \frac{54}{60},$
   $\frac{18}{72}, \frac{21}{24}$

2. $\frac{5}{8} = \frac{?}{48} = \frac{?}{64} = \frac{?}{40} = \frac{?}{56} = \frac{?}{80} = \frac{?}{32} = \frac{?}{72} = \frac{?}{24}$

3. Multiply 8 by: $\frac{1}{4}, \frac{3}{4}, \frac{1}{2}, \frac{1}{8}, \frac{3}{8}, \frac{5}{8}$

4. Express as fractions: 0.05, 0.04, 0.06, 0.6, 0.125

5. Express as decimals: $\frac{1}{3}, \frac{2}{3}, \frac{1}{6}, \frac{5}{6}, \frac{1}{8}, \frac{3}{8}$

MENTAL

1. If 6.5 kg of rice is put into 100 boxes, how much rice will there be in each box?

2. Round 345.5 to the nearest hundred.

3. Write in numerals: eighteen and three thousandths.

4. Wendy's watch gains 2.1 minutes in a week. What is the average number of minutes it gains per day?

5. Caroline's mother works 8 hours a day. What part of a day does she work?

6. Melanie earned $11.75. She spent $7.25. Now she needs $1.35 more to buy a tape. How much is the tape?

7. The total weight of candy bars in a box is 60 ounces. If each candy bar weighs $3\frac{3}{4}$ ounces, how many candy bars are in the box: 30 or 15 or 16?

8. At $6.40 a yard, what is the cost of $1\frac{3}{4}$ yd?

9. If $\frac{5}{8}$ of Gail's height is 80 cm, how tall is Gail?

10. If $n = 9$, find the value of $2(17 - n)$.

22 DRILL

1. Read each ratio: 4 to 5, $\frac{4}{5}$, 4:5, 12:13, 13:12, $\frac{1}{4}$, 4 to 3, 3 to 5

2. What ratios are equivalent to 2:5? 4: ? , 6: ? , 8: ? , 10: ? , 14: ? , 18: ?

3. Tell the equivalent ratio: 3:4 = 6: ? 2:3 = 4: ? 5:6 = ? :30 7:8 = ? :24

4. Express each rate in simplest form: 2 dimes:1 dollar, 2 months:1 year, 220 miles:4 hours

5. Tell the correct relation (= or ≠): 2:3 ? 5:9 2:4 ? 5:7 7:5 ? 14:10

MENTAL

1. If the ratio of the number of red cars to the number of blue cars is 2:5, do you know the number of red cars?

2. What might be the number of red cars and the number of blue cars?

3. Which ratios are not equivalent to 5:8? $\frac{5}{8}$, 5 to 8, $\frac{10}{16}$, 15:27, 25 to 40

4. If x:12 is equivalent to 6:18, which is greater, x or 12?

5. Which ratio is equivalent to 5:4? 20:25 or 25:20

6. A car traveled 100 miles in 2 hours. What was its rate of speed?

7. If 215 cards were to be put into boxes of 30 cards each, how many cards would be left over?

8. Donna had $\frac{3}{4}$ yard of blue ribbon and $1\frac{1}{8}$ yards of red ribbon. How much more red ribbon did she have?

23 DRILL

1. Multiply by 3: $\frac{1}{9}$, $\frac{1}{3}$, $\frac{1}{6}$, $\frac{2}{9}$, $\frac{2}{3}$, $\frac{1}{15}$

2. Tell the equivalent ratios: $\frac{3}{4} = \frac{6}{?} = \frac{12}{?} = \frac{15}{?} = \frac{21}{?} = \frac{30}{?}$

3. Name three ratios equivalent to 3:2.

4. Express each rate in simplest form: 100 miles in 2 hours, 5 pencils for 40¢ 40 gallons in 4 seconds, 150 meters in 3 seconds, 9 apples for 75¢

5. Using 3 pens for $1.00 as the rate, tell the price of: 6 pens, 9 pens, 12 pens, 15 pens, 21 pens

MENTAL

1. There are 13 boys and 14 girls in a class. What is the ratio of boys to girls?

2. A team won 8 games and lost 4. What is the simplest ratio of wins to losses?

3. A piece of plywood is 30 cm wide and 90 cm long. What is the simplest ratio of length to width?

4. A salt solution contains 2 grams of salt to every 3 grams of water. What is the ratio of water to salt?

5. Write two numbers greater than 5 and less than 11 that are divisible by 2.

6. Stan spent $2\frac{1}{2}$ hours on a project. How many minutes was that?

7. How many 0.4L glasses can you fill if you have 12 liters of lemonade?

8. Write the Arabic numeral for MCC.

9. Write the decimal equivalent for $\frac{8}{20}$.

10. Is $\frac{12}{36}$ more or less than $\frac{2}{3}$?

24 DRILL

1. Tell the equivalent fraction that has a denominator of 100: $\frac{7}{50}$, $\frac{1}{2}$, $\frac{11}{20}$, $\frac{4}{25}$, $\frac{7}{10}$, $\frac{1}{4}$, $\frac{3}{4}$, $\frac{11}{10}$

2. Read each percent: 2%, 52%, 65%, 17%, 100%, 23%

3. Tell 100% of: 7, 10, 100, 50, 6, 18, 25

4. Write as percents: $\frac{5}{100}$, $\frac{16}{100}$, $\frac{75}{100}$, $\frac{90}{100}$, $\frac{50}{100}$, $\frac{2}{100}$

5. Express as percents: 0.25, 0.75, 0.15, 0.05

MENTAL

1. Write 35% as a fraction.

2. What percent expresses a 5% decrease in the original whole?

3. Earl receives 60% of the profits and Lukas receives 40%. How much more of the profits does Earl receive?

4. Beatrice gave correct answers to 80 out of 100 questions. What percent of her answers were correct?

5. Complete: 9:10 = ? :100 = ? .

6. Diane gave correct answers to 90% of the test questions. If there were 10 questions, how many were correct?

7. Emil saved $15 out of $20. What percent of his money did he save?

8. Rosalie won 3 out of 4 games. What percent of the games did she win?

9. Which is greater: 5.22 ÷ 3 or 1?

10. What is the decimal equivalent of 35%?

25

DRILL

1. Tell the simplest form of the ratio of 8 to: 16, 24, 48, 32, 56, 40, 72
2. How much less than 1 is: 5%, 25%, 4%, 3%, 15%
3. How much greater than 1 is: 125% 101% 110% 112% 175%
4. Express as percents: 1, 1.5, 0.15, 0.015, 2, 2.3, 0.23, 0.023
5. Express as decimals: 72%, 7.2%, 720%, 5.5%, 55%, 550%

MENTAL

1. What is 110% of 50?
2. Laura misspelled 10% of 40 words. How many words did she misspell?
3. What is 90% of 40?

4. About 60% of air pollution is caused by manufacturing. What percent is caused by other factors?
5. If 24 out of 30 students went on a field trip, what is the ratio of the number of students who did *not* go to the number of students who went?
6. It takes Jason 10 minutes to walk a mile. How far can he go in an hour?
7. Irene's watch gains 26.6 seconds in one week. About how many seconds does her watch gain in a day?
8. What is the smallest number represented by four digits that are all alike and even?
9. Sugar is on sale at 3 kg for $2.00. How much will 6 kg of sugar cost?
10. Which of these are equivalent: 2, 2.00, 0.002, 0.20, 2.000?

26

DRILL

1. Round to the nearest whole number: 37.1, 37.3, 37.8, 92.36, 37.08, 16.67
2. Express as fractions in simplest form: $12\frac{1}{2}$%, $37\frac{1}{2}$%, $62\frac{1}{2}$%, $87\frac{1}{2}$%, 25%, 75%, 20%, 40%, 80%
3. Find 25% of: 32, 8, 16, 64, 24, 56, 48
4. Find $12\frac{1}{2}$% of: 80, 40, 96, 16, 320, 560
5. Find 20% of: 5, 25, 60, 20, 45, 30, 40

MENTAL

1. What is 30% of 300?
2. If 80% of the homes in a town are heated by oil, what percent are *not* heated by oil?

3. About 42% of the cars in the world belong to people in the U.S. What percent belong to people living elsewhere?
4. Find 50% of 250.
5. To travel 15 miles in 20 minutes, how many miles per hour must you be going?
6. Use <, =, or > to compare: 30% of 60 ___?___ 50% of 20.
7. Edna wants a radio that costs $90. If she has $75, how much more money does she need?
8. How many swatches of fabric 0.03 m long can be cut from a 6 m length?
9. Which of these are equivalent: 0.175, 1.75, $0.17\frac{1}{4}$, $1\frac{3}{4}$?
10. What does the bar over the 6 in $0.\overline{6}$ mean?

27

DRILL

1. Give two ratios equivalent to: $\frac{2}{7}$, $\frac{3}{8}$, $\frac{1}{4}$, $\frac{5}{6}$
2. Tell if the expression is true: $\frac{3}{8} = \frac{24}{64}$, $\frac{1}{5} = \frac{4}{20}$, $\frac{1}{6} = \frac{5}{25}$, $\frac{4}{8} = \frac{10}{19}$, $\frac{5}{9} = \frac{45}{81}$
3. What percent of 80 is: 10, 20, 40, 60
4. Express as decimals: 110%, 130%, 150%, 125%, 190%, 200%, 240%
5. Find 150% of: 10, 20, 50, 60, 80, 18, 24

MENTAL

1. What percent expresses a 10% increase in the original whole?
2. Use <, =, or > to compare: 1.4 ___?___ 140%.

3. If 200% of the cost of a dress is $64, what is the cost?
4. If 10% of the amount sold is $50, what is the amount of sales?
5. It is 15 minutes past 4:50 P.M. What time is it?
6. A snail moved 14 cm in $3\frac{1}{2}$ minutes. How many cm did it move in 1 minute?
7. Complete the proportion: 3:7 = ___?___ :21
8. There are 4678 men, 3415 women, and 2998 children at the county fair. Estimate to the nearest thousand the total number of people at the fair.
9. At $1.20 a yard what will 0.4 yard of fabric cost?
10. In a bouquet of 1 dozen carnations, 8 were pink. What percent of the bouquet was pink?

479

28

DRILL

1. Multiply by 1000: 4, 14, 0.6, 0.4, 0.004
2. Express as fractions in simplest form: 50%, 5%, 25%, 20%, 10%, 75%, 30%, 4%, $12\frac{1}{2}$%, 40%, 2%
3. Express as decimals: $\frac{1}{5}$, $\frac{2}{5}$, $\frac{1}{8}$, $\frac{5}{8}$, $\frac{1}{3}$, $\frac{1}{9}$
4. Express as percents: 1.25, 1.4, 1.8, 1.1, 1.5, 1.6, 1.75, 0.3, 0.4, 0.2, 0.25, 0.025
5. What percent of 20 is: 2, 4, 5, 10, 15, 18, 20

MENTAL

1. A book listed at $8 is sold for $6. Find the rate of reduction.

2. A discount of 5% brings a bill down to $19. Find the amount of the original bill.
3. Find 50% of $800.
4. By selling for $100 above cost, a dealer makes $12\frac{1}{2}$ % profit. Find the cost.
5. A school team won 15 out of 20 games. Express this as a ratio.
6. How many packages weighing 0.07 kg can be made from 49 kg of raisins?
7. How much rope would you buy if you needed 8 pieces, each $1\frac{1}{2}$ yards long?
8. Find the sum: 2125 + 144 + 75 + 6.
9. What fractional part of 60 is 40?
10. From 100% subtract 15%.

29

DRILL

1. Find $\frac{1}{2}$ of: 20, 40, 50, 18, 16, 84, 140
2. Multiply by 6: $\frac{1}{3}$, $\frac{1}{6}$, $\frac{1}{12}$, $\frac{1}{2}$, $1\frac{1}{6}$, $2\frac{1}{6}$, $3\frac{1}{6}$
3. Express as percents: 6, 6.5, 0.65, 0.065, 4, 4.5, 0.45, 0.045
4. What percent of 10 is: 5, 2, 3, 7, 8, 4, 9
5. Find $12\frac{1}{2}$ % of: 48, 64, 24, 72, 32, 16, 96, 56

MENTAL

1. $112\frac{1}{2}$ % of a number is the same as what fraction times the number?
2. 300 is 150% of what number?
3. What percent of 50 is 75?

4. Gina wanted to leave a 20% tip for the waiter. The meal cost $24.90. What did Gina leave for a tip?
5. What fractional part of 16 is 12?
6. Mrs. Jacobs cut $\frac{3}{8}$ of a yard of ribbon from a roll that contained 8 yards. How much ribbon was left on the roll?
7. Complete the proportion: 4:7 = 6: _?_ .
8. 20% of the students in a school went to the baseball game. If 450 students attended the game, how many students are there in the school?
9. What does the bar over the 3 in $0.8\overline{3}$ mean?
10. At a rate of 80 words per minute, how long would it take to type 480 words?

DRILL

1. Multiply by 10: 8, 17, 4, 3.6, 7.5, 1.56, 19.3, 4.15
2. Multiply by 100: 7, 12, 0.02, 0.006, 0.005, 0.27, 0.657, 0.3
3. Express in meters: 4 km, 14 km, 7 km, 0.5 km, 3.5 km
4. Express in kilometers: 5000 m, 500 m, 5500 m, 250 m, 5250 m, 765 m
5. What decimal part of a kilometer is: 500 m, 250 m, 100 m, 750 m, 10 m

MENTAL

1. Use <, =, or > to compare: 325 m + 575 m _?_ 1 km.
2. If a truck travels at 90 km/h, how far does it travel in one minute?

3. Edwin walked 0.2 km in 5 minutes. How long will it take him to walk 1 km?
4. A bicycle path is 750 m long. How much less than 1 km is that?
5. At a class party there were 10 doughnuts for every 3 students. How many doughnuts were there if there were 24 students in the class?
6. 90 is what percent of 30?
7. Find the difference to the nearest million: 36,925,125 − 9,123,995.
8. Sally ran 100 meters in 10 seconds. How many meters did she run in one second?
9. Write the decimal equivalent for $\frac{5}{6}$.
10. Subtract $\frac{3}{8}$ from $\frac{1}{2}$.

DRILL

1. Tell the correct relation ($<$, $=$, or $>$):

   $\frac{1}{5}$? $\frac{3}{10}$ $\frac{5}{7}$? $\frac{10}{14}$

   $\frac{2}{3}$? $\frac{1}{6}$ $\frac{12}{15}$? $\frac{4}{5}$

2. Give two equivalent ratios for: $\frac{10}{15}$, $\frac{18}{20}$, $\frac{7}{8}$

3. Find 20% of: 150, 40, 60, 250, 55, 100

4. Tell the unit of measure represented by:
   L, kL, mL, cm^3, dm^3, m^3

5. Express in liters: 1000 mL, 1 dm^3,
   4000 mL, 5 dm^3, 14 000 mL

MENTAL

1. Dana mixed 500 mL of lemon juice,
   500 mL of water, and 100 mL of syrup.
   How many liters of lemonade did she
   make?

2. A pitcher contained 1.4 liters of juice. If
   400 mL were used at breakfast, how many
   liters of juice were left?

3. Gene drank 0.3 liter of milk. How many
   milliliters did he drink?

4. At $1.25 a liter, how much will 3L cost?

5. 12 is 120% of what number?

6. Find the value of x. $x:32 = 3:8$

7. At a cake sale the class received $6.75 for
   cookies, $9.65 for pies, and $4.30 for
   pieces of cake. If the class paid $5.00 for
   expenses, how much money was left?

8. A 3 foot 6 inch board is cut evenly into
   7 pieces. How long is each piece?

9. 1020 cards were packaged 50 to a pack.
   How many packs were there and how many
   loose cards?

10. If $\frac{3}{4}$ of a number is 18, what is the
    number?

32

DRILL

1. Tell the number named by: 6^2, 1^2, 4^2,
   2^2, 7^2, 5^2, 10^2, 8^2, 3^2

2. Tell what number can be multiplied by itself
   to give: 1, 9, 49, 36, 25, 16, 64

3. Find $\frac{3}{4}$ of: 8, 12, 16, 24, 36, 48, 40, 72

4. At 90¢ a yard, find the cost of: 6 ft,
   4 ft, 8 ft, 9 ft, 5 ft

5. Round to the nearest inch: $7\frac{2}{3}$ in.,
   $10\frac{1}{8}$ in., $4\frac{5}{6}$ in., $39\frac{3}{16}$ in., $9\frac{3}{4}$ in.

MENTAL

1. What number multiplied by itself has 81 as
   the product?

2. The perimeter of a square is 12 feet. Find
   the length of one side in inches.

3. Which is greater: 4 yd 27 in. or 13 ft?

4. A man's suit is priced at $190. The tax is
   5%. How much tax must be paid?

5. Larry's lunches cost $4.95. If he wants to
   leave a 15% tip, how much money should
   he leave?

6. Out of 150 tickets, 20% were not sold.
   How many tickets were not sold?

7. After a radio had been reduced 10%, the
   net price was $63. What was the original
   price?

8. If 20 out of 50 people in a room are
   children, what percent of the people are
   children?

9. A room is 25 ft long. Express the length in
   yards and feet.

10. If $n = 0.7$, find the value of $n - 0.7$.

33

DRILL

1. Multiply by 100: 34, 3.4, 0.34, 25, 2.5,
   0.25, 0.025, 43, 4.3, 0.43, 0.043

2. Multiply by 10^2: 45, 59, 0.04, 0.05,
   0.24, 0.65, 0.004, 0.006

3. Multiply by 1000: 23, 2.3, 0.23,
   0.023, 125, 12.5, 1.25, 0.125

4. Tell whether to multiply by 10, 100, or
   1000 to get the smallest whole-number
   product: 0.1, 0.01, 0.001, 0.323,
   0.43, 1.04

5. Divide by 100: 200, 16, 12, 3.5, 24.6

MENTAL

1. Right triangles have two acute angles. What
   kind of angle is the third angle?

2. A triangle has an altitude of 20 m and a
   base of 13.7 m. Find its area.

3. Into what two shapes does a diagonal
   divide a rectangle?

4. Find the perimeter of an equilateral triangle
   whose sides are 30 cm.

5. What number is named by 1.6×10^3?

6. A rectangle 5 inches wide has an area of
   60 square inches. What is its length?

7. What percent of 1 yd^2 is 1 ft^2?

8. Which is the better buy:
   5 oz for 39¢ or 1 lb for $1.20?

9. Round 0.2939 to the nearest hundredth.

10. A table is 1 m long and 60 cm wide. What
    is the ratio of the width to the length?

34 . DRILL

1. Tell the number named by: 3^2, 9^2, 12^2, 11^2, 6^2, 8^2, 20^2, 10^2, 7^2
2. Express in meters to the nearest 0.1 m: 60 cm, 456 cm, 142 cm, 35.5 cm, 289 cm
3. Express in cm: 42 mm, 8 dm, 1.4 m
4. Express in milliliters: 1 L, 0.5 L, 0.25 L, 0.58 L, 32 L, 0.6 L
5. Tell the simplest ratio of 4 centimeters to: 8 cm, 12 cm, 20 cm, 100 cm, 36 cm, 50 cm, 60 cm, 24 cm

MENTAL

1. In $C = \pi d$ what does d stand for?
2. What does the symbol π stand for?

3. What is the radius of a bicycle wheel whose diameter is 26 inches?
4. If the diameter of a circle is 20 cm, what is the circumference?
5. The thickness of a washer is 2.8 cm. What is the thickness of 5 washers?
6. What is $\frac{3}{4}$ of $2.40?
7. 10% of what amount equals $20?
8. Carmen eats $4\frac{3}{4}$ ounces of dried fruit each day. About how much dried fruit does she eat in 5 days: $9\frac{3}{4}$ oz or $20\frac{3}{4}$ oz or $23\frac{3}{4}$ oz?
9. What is the area of a square measuring 15 cm on a side?
10. If one side measures 9 centimeters, what is the perimeter of an equilateral triangle?

35 DRILL

1. What percent of 6 is: 2, 1, 4, 3, 6, 9, 12
2. Divide by 100: 1, 3, 4, 13, 12, 14, 16
3. What is the ratio of red balloons to yellow balloons if there are 4 red and: 8 yellow, 4 yellow, 2 yellow, 1 yellow, 6 yellow
4. Using the data in example 3, what is the ratio of yellow to red balloons?
5. Give two ratios each for: 6 boys and 4 girls, 10 batters and 7 hits, 5 cats and 8 dogs

MENTAL

1. Using these ranked scores, give the frequency of a score of 4: 12, 8, 7, 4, 4, 4, 3, 3, 2, 2.
2. Using the data in example 1, find the mode.

3. Using the data in example 1, find the median, or midscore.
4. Using the data in example 1, find the mean, or average, score.
5. Find the sum by adding and subtracting: 34,605 + 46,798 = (34,605 + _?_) $^-$3,202 = _?_ .
6. What is the measure of an angle on a circle graph that equals 10%?
7. At $4 per 100, how much does 1 cost?
8. If $\frac{3}{8}$ of Bill's age is 6 years, how old is he?
9. A wall plaque that is 10 cm long has a perimeter of 30 cm. How wide is the plaque?
10. Each side of a hexagon measures 1.3 m. What is its perimeter?

36 DRILL

1. To the nearest penny, find 5% of: 60¢, 20¢, 40¢, 45¢, 95¢
2. Read each probability: $\frac{2}{3}$, $\frac{4}{5}$, $\frac{2}{7}$, $\frac{8}{9}$, $\frac{1}{4}$
3. In what quadrant is each: ($^+$4, $^+$1), ($^-$3, $^-$8), ($^-$1, $^+$1), ($^+$1, $^-$2)
4. The numbers 1, 2, 3, 4 are on a spinner. Tell the probability of spinning: 1, 2, 3, 4, an odd number, an even number, a number other than 1, 2 or 4.
5. Tell about the marbles if the probability of picking a blue marble is: $\frac{1}{2}$, $\frac{2}{5}$, $\frac{1}{4}$, $\frac{3}{4}$

MENTAL

1. What decimal part of 20 is 1?
2. Use <, =, or > to compare: $\frac{1}{2} \times \frac{1}{2}$ _?_ $\frac{1}{2}$.

3. You have 3 cans of peaches, 2 cans of peas, and 2 cans of pineapple. If the labels are removed, what is the probability of picking a can of peaches?
4. Using the data in example 3, what is the probability of picking a can of peas?
5. Using example 3, what is the probability of picking either peas or peaches?
6. Using example 3, what is the probability of picking a can other than peas?
7. Use <, =, or > to compare: $\frac{1}{2} \div \frac{1}{2}$ _?_ $\frac{1}{2}$.
8. How many fabric combinations can be made if there are 3 colors and 4 patterns?
9. Out of 800 employees, 480 are women. What is the ratio of women to the total?
10. The temperature was 5°C. If it dropped 15 degrees, what was the new temperature?

482

37 DRILL

1. What fractional part of an hour is:
   20 min, 40 min, 10 min, 5 min,
   15 min, 12 min
2. What is the greatest possible remainder
   when dividing by: 3, 5, 6, 8, 11, 15,
   20, 100
3. Find the dividend if the divisor was 10 and
   the quotient was: 3 R7, 2 R5, 6 R9,
   8 R2, 5 R4
4. Multiply by $^-2$: $^+3$, $^-7$, 0, $^+4$, $^-10$, $^+12$
5. Divide by $^-5$: $^-15$, $^-40$, $^+10$, $^+35$, $^-60$, 0

MENTAL

1. What is the volume of a cube 5 cm on
   each edge?
2. How many cm are in 26 m?

3. Evaluate $2 \cdot (3n - 5)$ if $n = 10$?
4. What is 50% of 2 hrs and 10 min?
5. If a leaky faucet loses 28 gal of water in
   24 hrs, how much water is lost in a week?
6. One-ninth of 27 flags were torn. How many
   flags are not torn?
7. The difference between two numbers is 10.
   If the smaller number is $^-10$, what is the
   greater number?
8. The side of both a square and a regular
   octagon is 10 cm. How many times larger is
   the perimeter of the octagon than that of
   the rectangle?
9. 30% of Jon's day is spent in bed. How long
   is that?
10. The opposite of double a number is $^+10$.
    What is the number?

38 DRILL

1. Add 10^2 to: 320, 65, 450, 1800,
   2700, 16
2. Divide by $\frac{1}{3}$: 7, 10, 1, $\frac{1}{3}$, $\frac{1}{6}$, $\frac{2}{3}$, $\frac{5}{6}$
3. Name the reciprocal of: 2, $\frac{1}{2}$, 4, $\frac{3}{4}$, 0.3,
   0.7, 0.11, 0.9, 1.5, 3.5
4. Which numbers are divisible by 3? 216,
   117, 405, 112, 98, 75, 705, 408
5. Solve for n if half of n equals: $^+7$, $^+10$,
   $^+4$, $^+5$, $^-10$, $^-3$, $^-14$, $^-12$, $^+6$

MENTAL

1. $3 \div \left(2 + \frac{1}{3}\right)$
2. Which is smaller: the sum of 0.5 and 1.2
   or the product of 0.5 and 1.2?

3. $\dfrac{3}{1 + \frac{1}{8}}$
4. A regular pentagon has a perimeter of
   20.5 m. Find each side.
5. How much did Sally pay for dog food if
   $\frac{1}{3}$ off list price saved her 63¢?
6. Which is greater 1.2×10^2 or $1.2 + 10^2$?
7. The area of a rectangle is $1\frac{1}{3}$ times the
   area of the square. If the rectangle's area is
   12 ft^2, what is the area of the square?
8. Dividing a number by 2.5 is the same as
   multiplying it by __?__ .
9. Solve for a: $\frac{a}{2} = 3\frac{1}{4}$
10. Dividing a number by $3\frac{1}{3}$ is the same as
    multiplying it by __?__ .

39 DRILL

1. What number is named by: 3×10^3,
   5×10^5, 2×10^4, 6.1×10^4,
   8.2×10^3, 2.5×10^4, 4.7×10^5
2. Express in scientific notation: 13,000;
   2700; 450,000; 6,000,000; 108,000;
   9,810,000
3. Divide by $^+10$ then add $^-2$ to: $^+100$,
   $^-100$, $^+50$, $^-50$, $^+40$, $^-40$, $^+200$,
   $^-200$, $^+70$
4. Which fractions are a little less than 1:
   $\frac{1}{9}$, $\frac{4}{5}$, $\frac{7}{15}$, $\frac{9}{10}$, $\frac{45}{47}$, $\frac{3}{50}$, $\frac{6}{11}$, $\frac{25}{28}$
5. Multiply by 3: 6, $6\frac{1}{3}$, 8, $8\frac{1}{3}$, 5, $5\frac{2}{3}$, 2,
   $2\frac{2}{3}$, $7\frac{1}{3}$, $4\frac{2}{3}$, $9\frac{1}{3}$, 10

MENTAL

1. Add: $(2.0 \times 10^2) + (5.0 \times 10^4)$.
2. Subtract: $(6.5 \times 10^3) - (5 \times 10^2)$.
3. Which is closer to 0.995: $\frac{7}{9}$ or $\frac{10}{11}$?
4. Simplify: $4 + 3.5 - 6$.
5. Which is greater: 0.14 or $\frac{1}{7}$?
6. Multiply: (2) (3) (4) (0.5) (0.25) (0.1)
7. The length of a rectangle is double
   the width. The length is 10 in. What is
   the area?
8. $7 : n = 5 : 12$
9. $\dfrac{2}{1 + \frac{1}{2}}$
10. Which is greater: (6.03×10^0) or
    $0.0603 \div 0.001$?

40

DRILL

1. Which are divisible by 9: 16, 207, 558, 106, 185, 387, 7083
2. Find the GCF of 24 and: 4, 12, 8, 48, 16, 32, 27, 30, 64
3. Take 10^3 from: 2000, 1000, 3500, 64,000, 10,000
4. Multiply by 5: 1, $1\frac{2}{5}$, 2, $2\frac{1}{5}$, $2\frac{3}{5}$, 3, $3\frac{4}{5}$
5. Which fractions are close to 0: $\frac{4}{9}$, $\frac{11}{90}$, $\frac{7}{8}$, $\frac{3}{80}$, $\frac{18}{97}$, $\frac{16}{31}$, $\frac{24}{49}$, $\frac{4}{35}$

MENTAL

1. Which number is divisible by 2 and 5: 8, 15, or 30?
2. One factor of 1.4 is 70. What is the other factor?

3. Which number is divisible by 4 and 6: 32, 44, or 60?
4. How far will a spaceship travel in an hour if it travels $\frac{1}{12}$ mile per second?
5. What is the perimeter of a paper measuring $8\frac{1}{2}''$ × 11"?
6. What is the area of a poster $14\frac{1}{4}$ in. by 8 in.?
7. Let $a(b + c) = 27$. Find b if $a = 13$ and $c = 5$.
8. Let $\frac{m - n}{s} = 5$. Find m if $s = 4$ and $n = 6$.
9. Estimate the difference: $16\frac{1}{2} - 1\frac{3}{41}$.
10. Estimate the sum: $4\frac{2}{27} + \frac{3}{89}$.

41

DRILL

1. What is the measurement of each congruent angle formed by bisecting an angle of: 20°, 32°, 90°, 42°, 180°, 25°
2. From 1000 take: 10^1, 10^2, 10^3, 10^0, 2^2, 3^2, 5^2
3. Find the LCM of 36 and: 4, 9, 12, 10, 5, 15, 6, 2, 120, 45
4. Solve for n if one third n equals: 5, 10, 4, 1, 3, 20, 12, 6, 11, 15
5. Solve: $3^2 - 2^2$, $5^2 - 4^2$, $7^2 - 6^2$, $9^2 - 8^2$, $11^2 - 10^2$, $6^2 - 5^2$

MENTAL

1. Which is greater: $\sqrt{1\frac{7}{9}}$ or $(1.1)^2$?

2. $\dfrac{3^4 \times 5^2}{3^2 \times 5}$
3. Choose the best estimate for $48\frac{7}{9}$ % of 203: 1000, 100, 450.
4. 11 out of 42 is about: 50%, 25%, 20%
5. Which is greater: 10^2 or 4 × 5?
6. Which is less: 12^2 or 4^2?
7. $\frac{n}{3} = {}^-7$ $n = $ ___?___
8. Compare. Use <, =, or >. $20^2 - 19^2$ ___?___ $25^2 - 24^2$
9. The diagonal of a square forms two congruent isosceles right triangles. What is the measure of each congruent angle?
10. ∠ ABC was formed by bisecting ∠ ABD. Find m ∠ ABC if m ∠ ABD equals 72°.

42

DRILL

1. From 10 take: 2, 1, $\frac{1}{7}$, $\frac{2}{3}$, $\frac{3}{5}$, $\frac{7}{8}$, $1\frac{1}{5}$, $\frac{2}{7}$, $3\frac{3}{8}$
2. What is the ratio of a nickel to: a dime, a quarter, a nickel, a penny, a dollar
3. Add $3m$ to: $2m$, $6m$, $10m$, $12m$, $3m$, m
4. Subtract $2d$ from: $5d$, $3d$, $2d$, $11d$, $20d$
5. Multiply by 3.14: 10, 2, 4, 3, $\frac{1}{2}$, 4, 100

MENTAL

1. What percent of the diameter of a circle is its radius?
2. Compare. Use <, =, or >. 3.14 ___?___ $\frac{22}{7}$

3. A circle has a circumference of 8π cm. Find its diameter.
4. A circle has an area of 25π ft². Find its diameter.
5. Which is the better option: renting 2 videos for $7.75 or renting 3 videos for $11?
6. Solve for d: $5d - 3d = 12$.
7. The circumference of a circle is 62.8 cm. What is the diameter? (Use 3.14 for π.)
8. Solve for c: $2c + 3c = 15$.
9. What is the area of a circle having a radius of 20 in.? (Use 3.14 for π.)
10. What is the area of a semicircle having a diameter of 6 cm? (Express answer in π units.)

GLOSSARY

The glossary lists in alphabetical order significant and recurring mathematical terms that appear throughout the text. It is intended not so much as a memorization device but as a quick and simple reference tool for the students.

A

absolute value   The distance of a number from zero on a number line. $|3| = 3; |-3| = 3$.

acute angle   An angle measuring less than 90°.

addend   Any one of a set of numbers to be added.

angle   Two rays with a common endpoint, called the vertex.

area   The number of square units a region contains.

associative property of addition   The grouping of the addends does not change the sum. For all numbers a, b, and c, $a + (b + c) = (a + b) + c$.

associative property of multiplication   The grouping of the factors does not change the product. For all numbers a, b, and c, $a \times (b \times c) = (a \times b) \times c$.

B

BASIC   A simple computer-programming language.

bisect   To divide a segment or an angle into two congruent parts.

C

cancellation   The dividing of both the numerator and the denominator of a fraction by any common factors before multiplying.

Celsius scale   The scale used to measure temperature in the metric system in which 0° is the freezing point of water and 100° is the boiling point.

centimeter   A unit of length in the metric system equal to 0.01 meter.

circle   A closed plane figure all of whose points are equidistant from a point within called the center.

circumference   The distance around, or the perimeter of, a circle.

commission   Money earned equal to a percent of the selling price of items sold.

common denominator   A multiple of the denominators of two or more fractions.

common factor   The set of common factors of 8 and 20 is {1,2,4} because each number is a factor of 8 and 20.

common multiple   The set of common multiples of 3 and 4 is {0,12,24, ...} because each number is a multiple of 3 and 4.

commutative property of addition   The order of the addends does not change the sum. For all numbers a and b, $a + b = b + a$.

commutative property of multiplication   The order of the factors does not change the product. For all numbers a and b, $a \times b = b \times a$.

complementary angles   Two angles the sum of whose measure is 90°.

complex fraction   A fraction having a fraction in the numerator, the denominator, or both.

composite number   A whole number greater than 1 that has more than two factors.

cone   A solid figure having one circular base and a curved surface.

congruent figures   Figures having the same size and shape.

coordinate plane   A grid divided into four quadrants used to locate points by naming ordered pairs.

coordinates   Ordered pair of numbers used to locate a point on a grid.

corresponding parts   Matching sides or angles of a figure.

cube   A rectangular prism with six congruent faces.

customary system of measurement   The system based on foot, pound, quart, and Fahrenheit measures.

cylinder   A solid figure having two parallel, congruent, circular bases and a curved surface.

D

data   Numbers that give information.

database   A collection of computer data arranged in files and used for more than one purpose.

decimal   A numeral that includes a decimal point, which separates the ones from the tenths place.

decimeter   A unit of length in the metric system equal to 0.1 meter.

dekameter   A unit of length in the metric system equal to 10 meters.

denominator   The denominator of a fraction names the total number of congruent or equal parts.

diameter   A segment passing through the center of a circle with both endpoints on the circle.

digits   The mathematical symbols used to express a standard numeral: 0, 1, 2, 3, 4, 5, 6, 7, 8, 9.

discount   A reduction on the regular, or list, price of an item.

distributive property of multiplication over addition   The product of a factor times a sum can be written as the sum of the two products. For all numbers *a, b,* and *c, a* × (*b* + *c*) = (*a* × *b*) + (*a* × *c*).

dividend   The number to be divided.

divisible   A number (*n*) is divisible by another number (*a*) if there is no remainder when *n* is divided by *a.*

divisor   The number by which the dividend is divided.

E

edge   The line segment formed by the intersection of two faces of a polyhedron.

empty set   A set having no elements (0 or { }).

equally likely outcomes   The chance is the same of getting any one of the desired outcomes.

equation   A mathematical sentence expressing equality.

equilateral triangle   A triangle with three congruent sides.

equivalent fractions   Fractions that name the same number: $\frac{2}{3}$ and $\frac{4}{6}$ are equivalent fractions.

estimate   To round one or more numbers in an operation to determine an approximate answer.

expanded numeral   A numeral expressed in terms of the place value of each digit:
324 means (3 × 100) + (2 × 10) + (4 × 1)
= 300   +   20   +   4

exponent   A numeral that tells how many times the base is to be used as a factor. In 6^3 the exponent is 3 and the base is 6.

expression   An open mathematical phrase or sentence containing one or more variables. The value of the expression depends upon the value given the variable.

F

factor   One of two or more numbers that are multiplied to form a product.

fraction   Part of a region, an object, or a set; any number $\frac{a}{b}$ where *b* ≠ *0.*

frequency distribution   A chart that records the number of times an event or response occurs.

G

gram   The basic unit of mass in the metric system.

graph   A point or a collection of points on a line or a coordinate plane.

greatest common factor (GCF)   The greatest factor that is common to two or more numbers.

H

hectometer   A unit of length in the metric system equal to 100 meters.

hexagon   A polygon with six sides.

I

identity element of addition   Zero is the identity element in addition because adding zero to a number does not change its value. For any number *a, a* + 0 = *a.*

identity element of multiplication   One is the identity element in multiplication because multiplying a number by 1 does not change its value. For any number *a, a* × 1 = *a.*

improper fraction   A fraction having its numerator equal to or greater than its denominator.

inequality   A mathematical sentence using an inequality symbol. The symbols used are <, >, ≠.

integers   Numbers that are either positive or negative and 0.

interest   The amount paid by the borrower for the use of the principal for a stated period of time.

intersection of sets   The intersection of two sets is the set of all the elements common to both sets.

inverse relationships   Opposite operations: addition and subtraction are inverse relationships; multiplication and division are inverse relationships.

isosceles triangle   A triangle having two opposite sides and two opposite angles congruent.

K

kilogram   A unit of mass in the metric system equal to 1000 grams.

kiloliter   A unit of capacity in the metric system equal to 1000 liters.

kilometer   A unit of length in the metric system equal to 1000 meters.

L

LOGO   A programming language used to create simple graphics.

least common denominator (LCD)   The least common multiple of the denominators of two or more fractions.

least common multiple (LCM)   The least number, other than 0, that is a common multiple of two or more numbers.

line   A set of points in order extending indefinitely in opposite directions.

line of symmetry   A line that divides a figure into two congruent parts.

liter (L)   The basic unit of capacity in the metric system.

M

mean   The average of a set of numbers.

median   The middle number in a set of numbers arranged in order. If there is an even number of entries, the median is the average of the two numbers in the middle.

meter The basic unit of length in the metric system.

metric system The system of measurement based on meter, gram, liter, and Celsius measures.

metric ton A unit of mass in the metric system equal to 1 000 000 grams.

milligram A unit of mass in the metric system equal to 0.001 gram.

milliliter A unit of capacity in the metric system equal to 0.001 liter.

millimeter A unit of length in the metric system equal to 0.001 meter.

mixed number A number having a whole number part and a fraction part.

mode The number that appears most frequently in a set of numbers.

multiple The product of a given number and any whole number. Some multiples of 3 are 0, 3, 6, 9, 12, ...

N

numerator The numerator of a fraction names the number of parts being considered.

O

obtuse angle An angle measuring between 90° and 180°.

octagon A polygon with eight sides.

ordered pair A pair of numbers that locate a point in a coordinate plane. The first number tells how far to move right or left on the x-axis. The second number tells how far to move up or down on the y-axis. (4,6) and (–3,5) are ordered pairs.

order of operations The order in which mathematical operations must be done when more than one operation is involved.

origin The point (0,0) in the coordinate plane where the x-axis and the y-axis intersect.

P

parallel lines Lines in a plane that never intersect.

parallelogram A quadrilateral with two pairs of parallel sides.

pentagon A polygon with five sides.

percent The ratio or comparison of a number to one hundred.

perimeter The measure of the distance around a figure.

period The name given to every group of three places in a numeral. The periods are called ones, thousands, millions, ...

perpendicular lines Lines in a plane that intersect to form right angles.

pi The ratio of the circumference to the diameter of any circle. The symbol for this constant ratio is π. (π = 3.14 or $\frac{22}{7}$)

pictograph A graph using a picture to represent a given quantity.

pi units A way of expressing circumference or area in terms of π.

place value The value of a digit depending upon its position or place in a standard numeral. In 843, the 4 is in the tens place and means 4 tens or 40.

point A location or position usually named by a capital letter of the alphabet.

polygon A simple closed figure with sides that are line segments.

polyhedron A space figure having polygons as faces.

power of a number The result of using a number as a factor a given number of times. An exponent is used to express the power. 10^3 = 10 × 10 × 10, or 1000

prime factorization Expressing a composite number as the product of two or more prime numbers.

prime number A whole number greater than 1 that has only two factors, itself and 1.

principal The amount of money borrowed from a bank.

prism A polyhedron having one pair of parallel faces for which the prism is named. The other faces are polygons.

probability A branch of mathematics that analyzes the chance that a given outcome will occur. The probability of an event is expressed as the ratio of a given outcome to the total number of outcomes possible.

product The result of multiplying two or more factors.

proportion An equation stating that two ratios are equal.

pyramid A polyhedron having one base for which the pyramid is named. The other faces are triangular and meet in a common vertex.

Q

quadrant One of four sections into which the coordinate plane is divided.

quadrilateral Any four-sided polygon.

quotient The answer that results from the division of the dividend by the divisor.

R

radius A segment from the center of a circle to a point on the circle.

range The difference between the greatest and the least number in a set of numbers.

ratio A comparison of two numbers by division.

ray   Part of a line with one endpoint.

reciprocal   The product of a number and its reciprocal is 1. 4 and $\frac{1}{4}$ are reciprocals since $4 \times \frac{1}{4} = 1$.

rectangle   A quadrilateral with all angles and all sides congruent.

rectangular prism   A prism with six rectangular faces.

rectangular pyramid   A pyramid with a rectangular base.

regular polygon   A polygon having all sides and all angles congruent.

regular price   The original, or list, price of an item; the price before a discount has been given.

repeating decimal   A decimal in which the last digit or group of digits of the quotient repeats.

replacement set   The set of numbers from which the variable can be replaced.

rhombus   A parallelogram with all sides congruent and opposite angles congruent.

right angle   An angle measuring exactly 90°.

rotation symmetry   Symmetry about a point; the figure appears the same after rotating through a specific number of degrees.

S

sale price   The sale price is the difference between the list price and the discount.

sales tax   The amount added to the marked price of an item and collected as tax.

scale   The ratio of a pictured measure to the actual measure.

scalene triangle   A triangle with no sides congruent.

scientific notation   The expression of large numbers as the product of a number from 0 through 9 and a power of 10.

segment   A part of a line with two end points.

set   A collection of elements having something in common.

similar figures   Figures having the same shape but varying in size. The corresponding sides of similar figures are in proportion.

simple interest   The amount paid by a bank to a depositor only on the principal for a stated period of time.

solution set   The set of numbers that make an equality or an inequality true.

sphere   A curved space figure having all points equidistant from the center.

square   A rectangle with four sides of the same measure.

square root   A number which when multiplied by itself gives the original number (radicand).

standard numeral   The name given to a number as it is written or read.

straight angle   An angle measuring 180°.

subset   A set with elements that belong to another set.

supplementary angles   Two angles the sum of whose measures is 180°.

surface area   The sum of the areas of all the faces of a solid figure.

T

terminating decimal   A decimal that has no remainders. 0.05 is a terminating decimal.

trapezoid   A quadrilateral with one pair of parallel sides.

tree diagram   A diagram showing all possible outcomes of an event or of more than one event.

triangle   A polygon with three sides.

triangular prism   A prism having two parallel triangular faces.

triangular pyramid   A pyramid with a triangular base.

U

union of sets   A set made up of the combination of all the members of two or more sets.

V

variable   A letter of the alphabet that stands for a number value in a mathematical expression or equation.

vertex   The common endpoint of two rays in an angle, two line segments in a polygon, or three or more edges in a polyhedron.

volume   The number of cubic units of space a figure contains.

W

whole number   Any of the numbers 0, 1, 2, 3 . . . .

X

x-axis   The horizontal number line in a coordinate plane.

Y

y-axis   The vertical number line in a coordinate plane.

Index

Answers to Selected Odd-Numbered Exercises

1 Numbers and Numeration

Page 2 **1.** thousands **3.** ones **5.** millions **7.** twenty-seven thousand, two hundred fifty **9.** four million, eight hundred fifty-seven thousand, nine hundred fourteen **11.** one hundred one billion, one hundred one thousand **13.** 405,003,020 **15.** 12,096,053,101 **17.** 142,000,200,026

Page 3 **1.** tens; ninety **3.** hundreds; zero **5.** hundred thousand; three hundred thousand **7.** ten million; eighty million **9.** 1000 **11.** No, they are both 40,000 **13.** 1000 **15.** 10 **17.** 10,000 **19.** 100,000 **21.** 1

Page 4 **1.** b **3.** b **5.** one tenth

Page 5 **7.** one thousand **9.** twenty-four and six tenths **11.** ninety-five thousandths **13.** eight ten thousandths **15.** three hundred sixty-four hundred thousandths **17.** sixty-eight and seventy-five ten thousandths **19.** thirty-nine and eight hundredths **21.** 406.95 **23.** 27.0536 **25.** 0.000095

Page 6 **1.** 371,324 **3.** 9,007,300 **5.** (3 × 10,000) + (7 × 1000) + (6 × 100) + (8 × 10) + (5 × 1) **7.** (4 × 100,000) + (4 × 10,000) + (4 × 1000) + (4 × 10) + (4 × 1) **9.** (6 × 1,000,000) + (4 × 100,000) + (2 × 10,000) + (5 × 10) + (8 × 1) **11.** (5 × 10,000) + (5 × 1) **13.** (2 × 100,000,000) + (9 × 1,000,000) + (1 × 100,000) + (9 × 1000) **15.** (7 × 100,000,000) + (8 × 10,000,000) + (6 × 100) + (4 × 10) + (5 × 1)

Page 7 **1.** tenths; seven tenths **3.** ten thousandths; eight ten thousandths **5.** ten thousandths; nine ten thousandths **7.** 60.008053 **9.** (2 × 0.1) + (1 × 0.01) + (1 × 0.001) + (5 × 0.0001) **11.** (4 × 10) + (1 × 1) + (7 × 0.1) **13.** (3 × 1) + (3 × 0.01) + (7 × 0.001) **15.** (6 × 10) + (3 × 1) + (4 × 0.1) + (9 × 0.0001) **17.** hundredths **19.** ones

Page 8 **1.** > **3.** < **5.** < **7.** < **9.** >

Page 9 **1.** 14,638; 14,683; 14,688; 14,863 **3.** 2.345; 2.435; 2.445; 2.453 **5.** 0.091; 0.109; 0.190; 0.9 **7.** c

Page 11 **1.** 7000 **3.** 90,000 **5.** 3,000,000 **7.** 900,000 **9.** 200 **11.** 300 **13.** 35,000 **15.** 12,000 **17.** 1,000,000 **19.** 3,000,000 **21.** 800,000,000 **23.** 300,000,000 **25.** b **27.** c **29.** 22,000 **31.** 550,000 **33.** 39,000 **35.** 20,000,000

Page 13 **1.** 0.3 **3.** 0.1 **5.** 7.1 **7.** 0.51 **9.** 8.05 **11.** 1 **13.** 7 **15.** 63 **17.** ones; 2 **19.** 20 days per month **21.** 21 petals **23.** d **25.** $18; $1.20; $.62

2 Addition and Subtraction of Whole Numbers and Decimals

Page 24 **1.** 753 **3.** 920 **5.** 912 **7.** 13,357 **9.** 115,739 **11.** 4076 **13.** 5100 **15.** 10,055 **17.** 1,078,264 **19.** 8,713,304 **21.** $394,292 **23.** $1,150,614 **25.** $178,014

Page 25 **29.** 46 **31.** 6163 **33.** 499,284 **35.** $5,076,473

Page 26 **1.** 78 **3.** 331 **5.** 3459 **7.** 3579 **9.** 728,577 **11.** 15,164 **13.** 32,359 **15.** $112,568 **17.** $26,879 **19.** $4,388,001 **21.** 1080

Page 27 **23.** b **25.** d **27.** 5785 **29.** 12,199 **31.** 947 **33.** 52,851 **35.** 37,146 **37.** 50,113 **39.** 1011 **41.** 234,568 **43.** 47,965 **45.** $320,868 **47.** $809,414 **49.** $2

Page 28 **1.** 50,000 + 30,000 **3.** 50,000 + 28,709 **5.** < **7.** < **9.** IV; 111,432; 111,753 **11.** IV; 97,462; 97,566 **13.** IV; 946,107; 946,040

Page 29 **1.** I; 80,000 − 40,000 **3.** IV; 76,453 −

39,000 **5.** < **7.** > **9.** IV; 19,643; 19,376 **11.** III; 67,000; 66,961 **13.** IV; 204,079; 204,390

Page 30 **1.** 19 **3.** 48 **5.** 21 **7.** 39 **9.** 53 **11.** 49 **13.** 54 **15.** 346 **17.** 127 **19.** 195 **21.** 559 **23.** 3307 **25.** 858 **27.** 3000

Page 31 **1.** 0.963 **3.** 26.02 **5.** 8.394 **7.** $12.31 **9.** $402.31 **11.** 91.633 **13.** 6.78 **15.** 164.821 **17.** 44.401 **19.** 56.2501 **21.** d

Page 32 **1.** 31.48 **3.** 63.161 **5.** 50.511 **7.** $93.84 **9.** $278.90 **11.** 41.58 **13.** 34.38 **15.** 737.073 **17.** 0.779 **19.** 0.597 **21.** $30.79 **23.** 7.9388 **25.** 5.4624 **27.** 140.432 **29.** 0.034 **31.** 0.2827

Page 33 **33.** 3.334 **35.** 3.19 **37.** 21.75

Page 34 **1.** 735 **3.** 1313 **5.** 160 **7.** 151.0 **9.** 7

Page 35 **11.** <; nearest ten **13.** <; nearest one **15.** <; nearest tenth **17.** 240; 236 **19.** 440; 441

3 Multiplication and Division of Whole Numbers and Decimals

Page 46 **1.** 5600 **3.** 164,200 **5.** 200; 20,000 **7.** 32,700; 3,270,000 **9.** <

Page 47 **1.** 700 **3.** 7,000 **5.** 2,400,000 **7.** 40,000; 60,000 **9.** 140,000; 180,000 **11.** < **13.** < **15.** >

Page 48 **1.** 4800; 5128 **3.** 12,000; 12,280 **5.** 10,000; 9162 **7.** 12,000; 11,058 **9.** 300,000; 307,173 **11.** 300,000; 265,736 **13.** 4,000,000; 3,572,889 **15.** 1,050,000; 1,051,447 **17.** 3,200,000; 3,273,028

Page 49 **19.** 40,000; 42,528 **21.** 40,000; 41,335 **23.** 60,000; 61,725 **25.** 180,000; 171,988 **27.** 28,200,000; 29,955,918 **29.** 12,000,000; 12,135,752 **31.** 6,400,000 **33.** 350,000; 361,012 **35.** 120,000 **37.** 400,000; 392,030 **39.** 4,500,000; 4,508,460 **41.** 28,147,504 **43.** 77,764,360 **45.** 66,888,080 **47.** 54,000; 52,510 **49.** 100,000; 118,404 **51.** 19,035; 18,800 **53.** 44,805; 43,500

Page 50 **1.** 23.8 **3.** 12.256 **5.** 0.0136 **7.** 54.366 **9.** 8.445 **11.** 3737.241 **13.** 0.0336 **15.** 0.005886 **17.** 0.91476 **19.** 1115.4641

Page 51 **1.** < **3.** > **5.** < **7.** 7.475 **9.** 117.3153 **11.** 0.0816 **13.** 0.0056 **15.** 0.0051 **17.** a **19.** b

Page 52 **3.** 360 ÷ 4; 90 **5.** 240 ÷ 6; 40 **7.** 685 ÷ 5; 137 **9.** 6300 ÷ 7; 900 **11.** 930 ÷ 3; 310 **13.** 15,000 ÷ 3; 5000

Page 53 **1.** c **3.** b **5.** 100 **7.** 100 **9.** 300 **11.** 150

Page 54 **1.** 9000 **3.** 8000 **5.** 20,000 **7.** 3000 **9.** 7000 **11.** 514 **13.** 9415 R4 **15.** a

Page 55 **1.** b **3.** b **5.** 20; 21 **7.** 320; 311 **9.** 200; 214 **11.** 50; 55 **13.** 900; 883 **15.** 900; 863 **17.** 1000; 968 **19.** 800; 726

Page 56 **1.** 3; 1 digit **3.** 120; 3 digits **5.** 53 R11 **7.** 33 **9.** 20; 21 **11.** 25; 22 **13.** 45; 41 **15.** 75; 82 **17.** 100; 96 **19.** 50; 44 **21.** 60; 63 R20 **23.** 30; 29 R4

Page 57 **1.** 1; 18<19 **3.** 0; 35>7 **5.** 3000; 3053 **7.** 3000; 3208 **9.** 2000; 2030 **11.** 1000; 809 R100

Page 58 **1.** partial dividend, 5, cannot be divided by 24

Page 59 **3.** so the division can be completed **5.** = **7.** < **9.** 0.49 **11.** 2.02 **13.** 0.592 **15.** 0.0143 **17.** 0.0375 **19.** 5.0805 **21.** 9.03 **27.** 40.02 **29.** 4.56

Page 61 **1.** b **3.** b **5.** 66.71 **7.** 2.08 **9.** 0.2233 **11.** 2.56 **13.** 0.00651 **15.** 0.305 **17.** 10 **19.** 100 **21.** 1000 **23.** 100 **25.** 1000 **27.** 105 **29.** 1.818 **31.** 0.758 **33.** 72; 136; 106.1 **35.** 25,000; 7200;

13,600; 2090; 10,610

1. 56>1; 56>16 **3.** 6; 70 **5.** 42,000; 7000 **7.** 420; 70 **9.** 4.2; 0.7
Page 63 **11.** a **13.** b **15.** 800.0 **17.** 5.76 **19.** 0.109 **21.** 3; 3.5 **23.** 2; 1.75 **25.** 400; 438 **27.** 4; 4.15 **29.** 54.33 **31.** 3.61 **33.** 166.67 **35.** 3.17 **37.** 0.01

4 Expressions, Equations, and Inequalities

Page 74 **1.** 13 **3.** 27 **5.** 6.06 **7.** 18 **9.** 18.18 **11.** 24 **13.** 127 **15.** 149.9 **17.** 21 **19.** 28 **23.** 0 **25.** 3
Page 75 **29.** 100 **31.** 3 **33.** 8 **35.** 3 **37.** 5 **39.** 0 **41.** 5 **43.** F; subtraction is not commutative **45.** F; reciprocals are not = when $c \ne d$ **47.** b **49.** d
Page 76 **5.** $x - 6$ **7.** $2a + 5$ **9.** $6 + 2y$ **11.** $2z + 3$
Page 77 **15.** $\frac{5 + x}{10}$ **17.** $\frac{x + 12}{10}$ **29.** $d - 2$ **31.** $\frac{c}{3}$ **33.** $c + \$1.12$
Page 78 **3.** $2 off means $2 subtracted from the regular price
Page 79 **5.** $2n + 1 > 3$; inequality **7.** $3n - 6 = 20$; equation **9.** $\frac{n}{4} < 2$; inequality **11.** $\frac{n}{6} < 2$; inequality **13.** $\frac{n}{3} < 20$; inequality **15.** > **17.** = **19.** = **21.** >
Page 80 **1.** 4; Commutative **3.** 0; Identity **5.** 8; Distributive **7.** 7; Distributive **9.** 8; Associative
Page 81 **1.** 5 **3.** 56 **5.** 4 **7.** $b = 8 + 4$ **9.** $d = 45 \div 5$ **11.** $t = 18 + 10$ **13.** $f = 18 - 9$ **15.** $k = 48 \div 16$
Page 83 **1.** 2 **3.** 23 **5.** 7 **7.** 113 **9.** 7 **11.** 17 **13.** 5 **15.** 44 **17.** 12 **19.** 8 **21.** 11 **23.** 47 **25.** a; c **27.** a; b **29.** a; b
Page 84 **1.** 7 **3.** 27 **5.** 20 **7.** 21 **9.** 150 **11.** 147
Page 85 **13.** b **15.** a **17.** c **19.** d **21.** a; c **23.** a; b **25.** b; c **27.** 2
Page 86 **1.** 2 **3.** 9 **5.** 23 **7.** 15 **9.** 1 **11.** 5 **13.** 22 **15.** 5 **17.** 7 **19.** 3 **21.** 4
Page 87 **23.** b **25.** + 12 **27.** + 18 **29.** + 99 **31.** $10 \div (2 \times 5) + 1 - 0 \times 2 \div (2 + 4 - 1) \times 6 = 2$; $(10 \div 2) \times (5 + 1) - 0 \times (2 \div 2) + (4 - 1) \times 6 = 48$ **33.** $12 - [(4 \times 6) \div 2] = 0$ **35.** $[7 + (3 \times 3)] \div 8 = 2$ **37.** $[(2 + 7) \times 3] + 3 = 30$ **39.** $6 + [(24 - 4) \div (4 + 6)] = 8$ **41.** $[(6 \div 3) + 2] \times 5 \div 5 + (7 \times 2) = 18$
Page 88 **1.** 30 ft^2 **3.** 8 m^3 **5.** 25 in.^{2} **7.** 15 in. **9.** 20 ft **11.** 24 m^2
Page 89 **13.** 3 m **15.** 75 mph **17.** 6 m

5 Integers

Page 100 **1.** H **3.** D **5.** E **7.** $^+7$ **9.** $^+5$ **11.** $^+8$ **13.** $^-3$; 3 **15.** $^+2$; 2 **17.** $^-2$; 2 **19.** $^-100$; 100 **21.** $^-486$; 486 **23.** $^+1025$; 1025
Page 101 **25.** $^-\$18$ **27.** 0 **29.** positive **31.** $^-17$ **33.** $^+7$ m, $^-7$ m **35.** $^-90$ m, $^+90$ m **37.** $^+2$ floors, $^-2$ floors **39.** $^-6$ points, $^+6$ points **41.** $^+10$, $^-10$ **43.** $^+\$4$, $^-\$4$ **45.** $^-10$ days, $^+10$ days **47.** $^-15$ steps, $^+15$ steps
Page 102 **1.** < **3.** > **5.** > **7.** < **9.** < **11.** < **13.** > **15.** > **17.** < **19.** > **21.** > **23.** >
Page 103 **29.** $^-6$, $^+1$, $^+5$, $^+7$ **31.** $^-8$, $^-2$, 0, $^+7$ **33.** $^-6$, 3, 0, $^+3$, $^+5$ **35.** $^+6$, 0, $^-5$, $^-11$ **37.** $^-6$, $^-9$, $^-16$, $^-19$ **39.** $^+40$, $^+32$, $^-18$, $^-20$
Page 104 **1.** $^-9$ **3.** $^-20$ **5.** $^+12$ **7.** $^+13$ **9.** $^+7$ **11.** $^-3$ **13.** $^+11$ **15.** $^-11$
Page 105 **17.** $^-6$ **19.** $^-11$ **21.** $^-7$ **23.** $^-6$ **25.** $^-12$ **27.** $^+17$ **29.** $^+10$ **31.** $^-18$ **33.** $^+63$ **35.** $^-25$ **37.** $^-1$ **39.** $^+2$ **41.** $^+1$ **43.** $^-4$ **45.** $^-5$ **47.** $^+7$ **49.** 0 **51.** $^+6$ **53.** $^-1$ **55.** $^-12$ **57.** $^-2$

59. $^-3$ **61.** $^-3$ **63.** $^+3$ **65.** $^-10$ **67.** $^-14$ **69.** $^-2$ **71.** $^-50$
Page 106 **1.** $^-14$ **3.** $^+16$ **5.** $^+9$ **7.** $^-2$ **9.** $^+11$ **11.** $^-3$ **13.** $^-15$ **15.** $^+10$ **17.** $^+4$ **19.** $^+2$ **21.** $^-6$ **23.** $^+14$ **25.** 0 **27.** $^+11$ **29.** $^-6$ **31.** $^-14$ **33.** $^-7$ **35.** $^+14$ **37.** $^+9$ **39.** $^+4$ **41.** $^-15$ **43.** $^+20$ **45.** $^-2$ **47.** $^-10$
Page 107 **49.** $^+1$ **51.** 0 **53.** $^+1$ **55.** $^-4$ **57.** $^-5$ **59.** $^-23$ **61.** $^+13$ **63.** $^+8$ **65.** 0 **67.** $^+24$ **69.** $^-1$ **71.** $^+48$ **73.** $^+100$ **75.** $^-36$ **77.** $^+11$ **79.** $^-1$ **81.** $^+15$ **83.** $^+1$ **85.** $^+6$ **87.** $^-8$ **89.** $^-3$ **91.** $^+24$ **93.** $^-21$ **95.** $^-15$ **97.** $^+10$ **99.** $^-12$ **101.** $^-2$ **103.** $^+10$
Page 108 **1.** $^+27$ **3.** $^+20$ **5.** $^+30$ **7.** $^+24$ **9.** $^-18$ **11.** $^-72$ **13.** $^-63$ **15.** $^+25$ **17.** $^-30$ **19.** $^-39$ **21.** $^-20$ **23.** 0 **25.** $^-4$ **27.** $^+8$ **29.** $^-40$ **31.** $^+36$ **33.** $^+42$ **35.** $^-24$ **37.** $^-20$ **39.** $^+54$ **41.** $^-9$ **43.** $^-24$ **45.** $^-20$ **47.** $^-55$ **49.** $^-8$ **51.** $^-27$ **53.** $^-63$ **55.** $^-33$ **57.** $^-54$ **59.** $^-3$
Page 109 **61.** $^+16$ **63.** $^-28$ **65.** $^+54$ **67.** $^-2$ **69.** $^-16$ **71.** $^-22$ **73.** $^-30$ **75.** $^+63$ **77.** $^+8$ **79.** 0 **81.** $^-1$ **83.** $^-88$ **95.** = **97.** < **99.** > **101.** = **103.** =
Page 110 **1.** $^+3$ **3.** $^+9$ **5.** $^-11$ **7.** $^-7$ **9.** $^+6$ **11.** $^+6$ **13.** $^-9$ **15.** $^-7$ **17.** $^-1$ **19.** 0 **21.** $^+8$ **23.** $^-2$ **25.** $^+6$ **27.** $^-4$ **29.** $^-4$ **31.** $^+66$ **33.** $^-48$ **35.** $^-7$ **37.** $^-25$ **39.** $^+96$
Page 111 **41.** $^-8$, $^+3$, $^+11$, $^-2$, $^+1$ **43.** $^-5$, $^-12$, $^+8$, $^+3$, $^+1$ **45.** T **47.** $^+32$, $^-64$, $^+128$ **49.** $^+625$; $^-3125$; $^+15,625$

6 Number Theory

Page 122 **1.** chemistry containers **3.** 4 wheel vehicles **5.** meals **7.** US coins
Page 123 **9.** $E = \{2, 4, 6, 8, 10, 12\}$ **11.** $W = \{26, 27, 28, 29, 30, 31, 32\}$ **13.** $F = \{40, 45, 50\}$ **15.** no **17.** no **19.** no **21.** T **23.** F **25.** T **27.** F **29.** set of 7 days of the week **31.** set of 12 months of the year
Page 124 **1.** $P = \{1, 3, 4, 5, 6\}$ **3.** $A \cup B = \{0, 2, 3, 4, 9\}$ **5.** $M \cup N = \{a, b, c, d, e, g\}$ **7.** $S = \{2\}$ **9.** $W \cap R = \{8, 12\}$
Page 125 **11.** $R \cup S = \{2, 3, 4, 5, 6, 7, 8, 9\}$; $R \cap S = \{2, 7\}$ **13.** $L \cup M = \{0, 1, 2\}$; $L \cap M = \{1\}$ **15.** $T \cup W = \{1, 2, 3, 4, 5, 6, 8, 9\}$; $T \cap W = \{\}$ **17.** $\{1, 2, 3, 4, 5, 6, 7, 8, \ldots\}$ **19.** $\{\}$ **21.** $\cap$ **23.** $\cup$ **25.** $\cup$
Page 126 **1.** 0, 1, 2, 3, 4, 5; 0, 1, 2, 3, 4, 5, 6, 7, 8, 9
Page 127 **3.** $S = \{5, 6, 7, 8, \ldots\}$ **5.** $S = \{7, 8, 9\}$ **7.** $S = \{5, 10, 15, \ldots\}$ **9.** $S = \{3, 6, 9, 12\}$ **11.** $S = \{\} = \phi$ **13.** $S = \{3, 4, 5\}$ **15.** nonzero multiples of 10 **17.** even whole numbers between 6 and 14 **19.** nonzero multiples of 3 **21.** nonzero multiples of 4 **23.** odd whole numbers between 9 and 21 **25.** $n \ne 98$
Page 128 **1.** 1,000,000 **3.** $10 \times 10 \times 10 \times 10$; 10,000 **5.** 10×10; 100 **7.** 1 **9.** 10 **11.** 100,000,000 **13.** 10^5 **15.** 10^4 **17.** 10^{12} **19.** 10^7 **21.** 10^3 **23.** 10^6 **25.** 10^5 **27.** 10^7 **29.** 10^0
Page 129 **31.** 4 **33.** 1 **35.** it has 1 less 0 **37.** 0.001 **39.** $10 \times 10 \times 10 \times 10 \times 10$; $\frac{1}{100,000}$; 0.00001
Page 130 **1.** 10^4; 10^3; 10^1; 10^0 **3.** 10^6; 10^5; 10^3; 10^1 **5.** 10^7; 10^5; 10^3 **7.** $6 \times 10^3 + 8 \times 10^2 + 7 \times 10^1$ **9.** $3 \times 10^4 + 7 \times 10^2 + 4 \times 10^1 + 1 \times 10^0$ **11.** $3 \times 10^5 + 9 \times 10^4 + 5 \times 10^3 + 7 \times 10^0$ **13.** $8 \times 10^6 + 1 \times 10^4 + 1 \times 10^2 + 7 \times 10^1 + 6 \times 10^0$ **15.** 1

× 10^5 + 6 × 10^4 + 4 × 10^3 **17.** 1 × 10^7 + 1 × 10^5 + 1 × 10^3 + 1 × 10^2 + 1 × 10^0 **19.** 40,630
21. 200,080 **23.** 61,000,400 **25.** 5,005,005
27. 90,060,300
Page 131 29. c **31.** d **33.** < **35.** = **37.** 10^1; 10^{-1}; 10^{-4} **39.** 10^{-1}; 10^{-3}; 10^{-5} **41.** > **43.** < **45.** <
Page 132 1. 2, 4 **3.** no **5.** 2, 4 **7.** no **9.** no
11. 2, 4 **13.** no **15.** 2, 4 **17.** 2 **19.** 2, 4 **21.** no; all even numbers are divisible by 2, but not all even numbers are divisible by 4
Page 133 23. 3 **25.** 3, 9 **27.** no **29.** no **31.** 3, 9 **33.** no **35.** 3, 9 **37.** 3, 9 **39.** 3, 9 **41.** no
43. no; 234; 3627; 9081; 27,072; 8631; 36,882; 26,451 are; 78; 501; 384; 345; 7440; 17,184 are not **45.** no
47. 5 **49.** no **51.** no **53.** no **55.** 5, 10 **57.** no
59. 5, 10 **61.** 5 **63.** no **65.** no; 6480; 6080; 110; 8450; 290; 41,390; 52,770 are; 295; 7135; 6955; 16,275; 3005 are not
Page 134 3. 4, 6, 8, 9, 10, 12, 14, 15, 16, 18 **5.** 1; by definition **7.** P **9.** C **11.** P **13.** C **15.** C
17. C **19.** C **21.** P **23.** C **25.** P
Page 135 1. 2^4 **3.** $2^4 × 3$ **5.** $2 × 3^2$ **7.** 3^4
9. $2^2 × 3^2$ **11.** 5^2 **13.** $2^3 × 11$ **15.** $2 × 5$
17. $5 × 11$ **19.** $2^3 × 7$ **21.** $2^2 × 5^2$ **23.** $2^2 × 3^2 × 5$ **25.** $2 × 3 × 5^2$ **27.** $2^3 × 5^2$ **29.** $2 × 3^3$
31. $2 × 3 × 13$ **33.** $2^5 × 3$ **35.** $2^2 × 3 × 7$ **37.** $2^2 × 5 = 20$ **39.** $2^2 × 5^2 = 100$ **41.** $2^2 × 3 × 7 = 84$
43. $2^2 × 3 × 5^2 = 300$ **45.** $5^3 = 125$ **47.** $2 × 3 × 5^2 = 150$ **49.** they are squares of 5, 10, 7, 9, and 6; 5^2, 10^2, 7^2, 9^2, 6^2
Page 136 1. {1, 2, 3, 6} **3.** {1, 3, 9, 27}
5. {1, 2, 3, 4, 6, 9, 12, 18, 36} **7.** {1, 2, 5, 10}
9. {1, 13} **11.** {1, 2, 3, 5, 6, 10, 15, 30} **13.** {1, 2, 4, 7, 14, 28} **15.** {1, 2, 7, 14} **17.** {1, 2, 4, 8}
19. {1, 3, 9} **21.** {1, 11} **23.** {1, 2, 7, 14} **25.** 8, 13, 9, 9,11, 25, 14, 3 **27.** 1 **29.** 7 **31.** 16 **33.** 7
35. no; when both are relatively prime or prime, one is a (trivial) common factor
Page 137 37. 1 **39.** 1 **41.** 2 **43.** 5 **45.** 1
47. 1 **49.** I, III **51.** III **53.** II **55.** 1, yes **57.** 7
59. 1, yes **61.** 1, yes **63.** 4
Page 138 1. {4, 8, 12, 16, 20, 24} **3.** {8, 16, 24, 32, 40, 48} **5.** {19, 38, 57, 76, 95, 114} **7.** {5, 10, 15, 20, 25, 30} **9.** {15, 30, 45, 60, 75, 90} **11.** {3, 6, 9, 12, 15, 18} **13.** {20, 40, 60, 80, 100, 120}
15. {16, 32, 48, 64, 80, 96} **17.** {210, 220, 230, 240, 250, 260, 270, 280, 290, 300, 310, 320, 330, 340, 350, 360, 370, 380, 390} **19.** 10 **21.** 28 **23.** 18
25. 30 **27.** 30 **29.** 45 **31.** 6 **33.** 8 **35.** 20
37. 14 **39.** 56 **41.** 24 **43.** 35 **45.** 20 **47.** 36
49. 72 **51.** 20 **53.** 24
Page 139 55. 12 **57.** 56 **59.** 9 **61.** 28 **63.** 33
65. 77 **67.** II **69.** b; there are no common multiples < the product of the numbers **71.** $\frac{11}{24}$ **73.** $\frac{19}{90}$ **75.** $\frac{1}{20}$
77. $\frac{3}{28}$
Page 140 1. 10^6 **3.** 10^5 **5.** 10^4 **7.** 10^5 **9.** 10^8
11. 9.4 **13.** 1.4 **15.** 4.36 **17.** 8.302 **19.** 2.012
Page 141 21. $5.31 × 10^5$ **23.** $6.1 × 10^6$
25. $9.14 × 10^5$ **27.** $4.004 × 10^7$ **29.** $5.29 × 10^7$
31. $2.105 × 10^9$ **33.** $3.0001 × 10^3$ **35.** $1.0101 × 10^9$ **37.** 37,000 **39.** 2930 **41.** 70,000,000
43. 541,000 **45.** 1,010,000 **47.** 66,000,000
49. 10,000,000,000 **51.** 20,100,000 **53.** $4.5 × 10^4$
55. $8 × 10^9$ **57.** $4.4 × 10^9$ **59.** $9.29 × 10^7$

7 Fractions: Addition and Subtraction
Page 152 1. 2 **3.** 1 **5.** 2 **7.** 6; 3 **9.** $1\frac{1}{4}$ **11.** $1\frac{1}{2}$
13. no; it could be = to one

Page 153 15. 10 **17.** 3 **19.** 8 **21.** 0 **23.** $\frac{1}{1}, \frac{2}{2}, \frac{3}{3}, \frac{5}{5}$, etc. **25.** 5 **27.** 4 **29.** 6 **31.** 25 **33.** 27
35. 49 **37.** $\frac{15}{24}, \frac{20}{32}, \frac{25}{40}$ **39.** $\frac{10}{8}, \frac{15}{12}, \frac{20}{16}$
Page 154 1. $\frac{1}{2}$ **3.** $\frac{3}{8}$ **5.** $\frac{7}{9}$ **7.** $\frac{4}{5}$ **9.** $\frac{2}{3}$ **11.** $\frac{8}{9}$
13. $\frac{9}{11}$ **15.** $\frac{9}{14}$ **17.** 1 **19.** yes; one is GCF of a fraction in simplest form
Page 155 1. no **3.** yes **5.** no **7.** yes **9.** no
11. $4\frac{2}{3}$ **13.** $3\frac{3}{4}$ **15.** 5 **17.** $1\frac{1}{19}$ **19.** 9 **21.** $7\frac{1}{3}$
23. 2 **25.** $4\frac{1}{2}$ **27.** $5\frac{5}{9}$ **29.** 1 **31.** $\frac{10}{5}, \frac{2}{1}, \frac{4}{2}, \frac{6}{3}$, etc.
33. $\frac{8}{7}, \frac{9}{7}, \frac{10}{7}$, etc. to $\frac{13}{7}$
Page 156 1. 15 **3.** 144 **5.** 30 **7.** $\frac{9}{6}, \frac{7}{6}$ **9.** $\frac{15}{20}, \frac{12}{20}$
11. < **13.** < **15.** > **17.** > **19.** < **21.** > **23.** >
25. > **27.** > **29.** >
Page 157 31. $\frac{5}{8}, \frac{3}{4}, \frac{4}{5}$ **33.** $\frac{1}{2}, \frac{2}{3}, \frac{3}{4}, \frac{5}{6}$
35. $\frac{1}{4}, \frac{3}{10}, \frac{2}{5}, \frac{1}{2}$
Page 158 1. yes **3.** no **5.** yes **7.** yes **9.** no
11. yes **13.** no **15.** yes **17.** yes **19.** yes **21.** yes
23. no
Page 159 25. $\frac{7}{8}$ **27.** $\frac{26}{27}$ **29.** $\frac{15}{16}$ **31.** $\frac{11}{24}$ **33.** $\frac{19}{39}$
35. $\frac{35}{71}$ **37.** $\frac{15}{28}$ **39.** $\frac{10}{19}$ **41.** $\frac{44}{87}$ **47.** $\frac{8}{35}, \frac{21}{40}, \frac{19}{22}$
49. $\frac{3}{26}, \frac{24}{29}, \frac{11}{20}$ **51.** $\frac{8}{71}, \frac{24}{49}, \frac{30}{59}$
Page 160 1. $\frac{2}{3}$ **3.** $1\frac{2}{5}$ **5.** 1 **7.** 2 **9.** 12 **11.** 21
13. 10 **15.** 24 **17.** 40
Page 161 19. $\frac{8}{9}$ **21.** $\frac{11}{12}$ **23.** $\frac{25}{48}$ **25.** $1\frac{5}{12}$ **27.** $\frac{17}{18}$
29. $\frac{23}{24}$ **31.** $1\frac{7}{40}$ **33.** $\frac{31}{36}$ **35.** $1\frac{1}{14}$ **37.** $\frac{17}{18}$
Page 162 1. $\frac{2}{3}$ **3.** $\frac{2}{15}$ **5.** $\frac{3}{7}$ **7.** 3 **9.** 3 **11.** 6
13. 5 **15.** 3
Page 163 17. $\frac{8}{33}$ **19.** $\frac{1}{21}$ **21.** $\frac{11}{18}$ **23.** $\frac{5}{9}$ **25.** $\frac{5}{24}$
27. $\frac{19}{30}$ **29.** $\frac{3}{35}$ **31.** $\frac{1}{22}$
Page 164 1. 17 **3.** 17 **5.** $\frac{3}{2}$ **7.** $\frac{19}{12}$ **9.** $\frac{5}{3}$ **11.** $\frac{11}{2}$
13. $\frac{127}{12}$ **15.** $\frac{51}{2}$ **17.** = **19.** > **21.** = **23.** >
25. > **27.** <
Page 165 1. 7 **3.** 16 **5.** $24\frac{2}{2}$ **7.** $3\frac{7}{8}$ **9.** $7\frac{1}{6}$
11. $6\frac{1}{5}$ **13.** $4\frac{3}{4}$ **15.** $15\frac{5}{11}$ **17.** $11\frac{3}{4}$ **19.** $13\frac{1}{3}$
21. $2\frac{6}{7}$ **23.** $32\frac{8}{19}$ **25.** $45\frac{1}{9}$
Page 166 1. 4 **3.** 8 **5.** 4 **7.** 17 **9.** 2
11. 28 **13.** 11 **15.** 4 **17.** 3
1. $13\frac{5}{7}$ **3.** $8\frac{2}{3}$ **5.** $18\frac{1}{4}$ **7.** $75\frac{14}{15}$
Page 167
9. $132\frac{19}{44}$ **11.** $25\frac{7}{8}$ **13.** $70\frac{19}{21}$ **15.** $45\frac{1}{18}$
Page 168 1. $2\frac{2}{3}$ **3.** $7\frac{3}{20}$ **5.** $16\frac{5}{8}$ **7.** 8 **9.** 3
11. 8; 18 **13.** 9; 33 **15.** $5\frac{1}{2}$ **17.** $8\frac{3}{4}$ **19.** $10\frac{11}{28}$
Page 169 21. a **23.** $3\frac{1}{2}$; $3\frac{1}{6}$ **25.** $8\frac{1}{2}$; $8\frac{5}{18}$ **27.** 1; $1\frac{11}{15}$ **29.** 5; $4\frac{31}{42}$ **31.** $\frac{1}{2}$; $\frac{7}{10}$ **33.** 5; $5\frac{1}{14}$ **35.** 2; $1\frac{34}{35}$
37. $2\frac{1}{2}$; $2\frac{7}{36}$ **39.** 3; $3\frac{4}{15}$ **41.** = **43.** > **45.** <

8 More Fractions and Decimals
Page 180 1. $\frac{9}{25}$ **3.** $\frac{1}{12}$ **5.** $\frac{4}{7}$ **7.** $2\frac{1}{10}$ **9.** $3\frac{2}{3}$ **11.** $\frac{7}{20}$
13. 5, 7, 8 (has 2), 9 (has 2), 10, 12
Page 181 1. 4 **3.** 12 **5.** $4\frac{2}{7}$ **7.** 8 **9.** 8 **11.** $6\frac{1}{4}$
13. < **15.** > **17.** = **19.** $1\frac{1}{2}$ **21.** $2\frac{2}{3}$ **23.** 1
25. 1 **27.** $22\frac{2}{5}$
Page 182 1. 56 **3.** 24 **5.** $3\frac{9}{16}$ **7.** $2\frac{5}{6}$ **9.** $5\frac{2}{7}$
11. $1\frac{7}{8}$ **13.** $23\frac{1}{3}$ **15.** 20 **17.** 11 **19.** 15 **21.** 54
23. 5 **25.** 9 **27.** 12 **29.** zero property **31.** Identity
Page 183 33. 60 **35.** 44 **37.** 92 **39.** 155 **41.** 70

43. 80 45. > 47. < 49. > 51. > 53. >
Page 184 1. 6 3. 3 5. 4 7. 1 9. $\frac{25}{72}$ 11. 18
13. $\frac{5}{8}$ 15. $\frac{2}{25}$ 17. $1\frac{1}{3}$ 19. $29\frac{1}{4}$ 21. 42 23. 12
25. 10 27. 20 29. $1\frac{1}{3}$ 31. $1\frac{1}{2}$
Page 185 33. $15\frac{7}{12}$ 35. $58\frac{3}{16}$ 37. $3\frac{16}{23}$ 39. $1\frac{4}{5}$
41. $14\frac{1}{2}$ 43. $42\frac{3}{7}$ 45. $3\frac{5}{41}$ 47. $3\frac{6}{7}$ 49. 54
51. $73\frac{1}{3}$ 53. T; cancelling on left gives same product
55. T; factors on left > ones on right 57. F; $\frac{3}{11} < 1$ and
$\frac{3}{4} < 1$ 59. 42; 35 61. 44; 36 63. $\frac{10}{16}$; $\frac{1}{2}$
65. $\frac{2}{10}$; $\frac{1}{16}$ 67. $\frac{12}{20}$; $\frac{3}{4}$ 69. $\frac{6}{24}$; $\frac{5}{8}$
Page 186 1. 2 3. $\frac{11}{6}$ 5. 4 7. $\frac{3}{10}$ 9. $\frac{9}{5}$ 11. $\frac{1}{16}$
13. $\frac{5}{12}$ 15. $\frac{4}{15}$ 17. $\frac{10}{19}$ 19. $\frac{1}{64}$ 21. $\frac{19}{2}$ 23. 4 25. 1
or $\frac{7}{7}$ 27. $\frac{10}{3}$ 29. $\frac{7}{6}$ 31. $\frac{1}{12}$ 33. 16 35. $\frac{2}{7}$ 37. $\frac{2}{17}$
39. $\frac{8}{49}$ 41. $\frac{3}{31}$ 43. $\frac{2}{17}$ 45. $\frac{3}{37}$ 47. No 49. No
Page 187 1. 6; 18 3. 10; 100 5. $3\frac{1}{3}$ 7. 24
9. $68\frac{4}{7}$
Page 188 1. $\frac{3}{2}$; $\frac{7}{6}$ 3. $\frac{1}{15}$; $\frac{1}{24}$ 5. $3\frac{3}{4}$ 7. $4\frac{4}{5}$ 9. $1\frac{1}{11}$
11. $\frac{2}{27}$ 13. $\frac{7}{36}$ 15. = 17. = 19. < 21. >
Page 189 1. 35 3. 9 5. 3 7. $\frac{1}{10}$ 9. $\frac{1}{15}$ 11. 12
13. $\frac{2}{7}$ 15. 1 17. $\frac{5}{7}$ 19. $\frac{3}{40}$ 21. $\frac{6}{7}$ 23. $21\frac{5}{7}$
Page 190 1. $1\frac{3}{4}$ 3. $1\frac{2}{3}$ 5. $2\frac{1}{4}$ 7. $1\frac{5}{18}$ 9. $1\frac{1}{2}$
11. $2\frac{17}{18}$ 13. $3\frac{1}{8}$ 15. $1\frac{1}{12}$ 17. $2\frac{2}{3}$ 19. $2\frac{25}{28}$ 21. =
23. < 25. > 27. > 29. >
Page 191 31. 2 33. 2 35. 2 37. 4 39. $\frac{1}{2}$ 41. 2
43. $\frac{4}{5}$ 45. $1\frac{1}{4}$ 47. 33 49. 24 51. 15
Page 192 1. 9 3. 20 5. 60 7. 16 9. 40 11. 27
13. 14 15. 150
Page 193 17. 256 19. 400 21. 36 23. 92
25. 150 27. 120 29. 119 31. 84 33. 128
35. 160 37. 171 39. 245
Page 194 1. 0.3 3. 0.12 5. 0.5625 7. 0.63
9. 0.75 11. 0.1875 13. 0.25 15. 0.625
Page 195 17. 6.2 19. 1.25 21. 8.5 23. 8.8
25. 2.25 27. < 29. < 31. < 33. = 35. >
37. < 39. > 43. 0.75 45. 0.4 49. 0.375
51. 0.875
Page 196 1. 4545 3. 5555 5. 0000 7. $0.\overline{15}$
9. $2.67\overline{87}$ 11. 2.678
Page 197 13. 0.3125 15. $0.\overline{45}$ 17. $0.\overline{65}$
19. 0.875 21. 0.38 23. 0.175 25. $0.\overline{285714}$
27. $0.41\overline{6}$ 29. 0.03125 31. > 33. > 35. =
37. < 39. = 41. <
Page 198 1. $\frac{3}{5}$ 3. $\frac{1}{20}$ 5. $\frac{3}{4}$ 7. $\frac{9}{200}$ 9. $\frac{1}{2}$ 11. $\frac{89}{1000}$
13. $\frac{11}{25}$ 15. $8\frac{3}{16}$ 17. $3\frac{1}{5}$ 19. $5\frac{11}{20}$ 21. $8\frac{3}{200}$ 23. $4\frac{1}{16}$
25. $3\frac{7}{8}$ 27. $7\frac{7}{80}$ 29. $1\frac{13}{2000}$
Page 199 35. > 37. > 39. > 41. > 43. <
45. > 47. 1.250 49. 1.250

9 Measurement
Page 212 1. cm 3. km 5. mm 7. mm 9. b
11. a
Page 213 21. 200 mm
Page 214 1. 0.006; 0.06; 6000 3. 0.035; 35; 350;
3500; 35 000 5. 0.2 7. 0.9 9. 1 875 000 11. 500
13. 0.8 15. 0.03546
Page 215 17. 2 19. 0.222 21. 6300 23. 1000
25. 8.3 27. 4 789 000 000 29. 93 000 31. >
33. > 35. < 37. > 39. <
Page 216 1. mL 3. mL 5. kL 7. L 9. L or kL

11. 2000 13. 6100 15. 2.483
Page 217 1. kg 3. g or mg 5. t 7. mg 9. g
11. kg 15. 24 000 17. 160 000 19. 55.5
21. 0.000001 23. < 25. = 27. > 29. < 31. >
Page 218 1. 2 3. 15; 15 5. 27 L; 27 kg
7. 12; 12 9. 13 cm^3; 13 mL 11. 11 cm^3; 11 g
Page 219 1. mm, 0.5 mm 3. oz, 0.5 oz 5. mg,
0.5 mg 7. 12 cm, 0.5 cm 9. 5 ft, 0.5 ft 11. 9.1 m,
0.5 m 13. 6210 m, 0.5 m 15. 24 900 g, 0.5 g
Page 220 1. 36 3. 48 5. 40 7. 2; 6 9. 39; 1
11. 16,340 13. 53 15. 74; 14 17. a 19. b
Page 221 27. 1 yd 2 ft 29. 2 mi 2 yd 1 ft
31. 6 ft 6 in. 33. 54 ft 3 in. 35. 21 yd 2 ft 4 in.
37. 9 in. 39. 55 ft
Page 222 1. tsp 3. tsp 5. gal or qt 7. 4 9. 16
11. 10 13. 18 15. 6 17. 6; 3 19. 7 gal 21. 2 gal
1 qt 1 pt 23. 27 pt 25. a
Page 223 1. oz 3. T 5. T 7. lb 9. oz 11. 6000
13. 104 15. 2.75 17. 5.5 19. 10 lb 8 oz 21. 1 lb
13 oz 23. 41 T 1506 lb 4 oz 25. =
Page 224 3. R 5. U 7. R 9. U 11. R
Page 225 17. a 19. a

10 Plane Geometry
Page 236 1. no 3. no 5. $\overrightarrow{EF}, \overrightarrow{EG}, \overrightarrow{EJ}$ 7. ∠ GEF
9. one
Page 237 13. a 15. c 17. ∠'s UMT, TMS
19. ∠ QMT 21. ∠ QMW 23. ∠'s AOB, COD;
∠'s AOC, BOD
Page 238 1. a 3. r 5. r 7. a 9. a 11. r
Page 239 13. 90° 15. 50° 17. 90° 19. 110°
21. 180° 35. 25°; 115° 37. 86°; 176° 39. 29°; 119°
41. 66°; 156° 43. 71°; 161° 45. 45°; 135°
Page 246 1. yes 3. no 5. yes
Page 247 1. 90°,180° 3. none 5. none
7. 120° 9. 180° 11. H, I, O, S, X, N

11 Polygons and Circles
Page 258 1. 3 3. 8 5. 10 7. 6 9. 3; 3
11. pentagon; 5 13. heptagon; 7 15. 9; 9
Page 259 17. ABCL, DEGL, GHIJ 19. DFGL
21. GHM, DEF, DGL, FGD, CKL, ABJ 23. DLG, DFG,
AJB 25. 85°; acute 27. 60°; acute 29. 105°;
obtuse 31. 45°; right
Page 260 1. $\overline{VY}, \overline{TY}, \overline{XY}$ 3. $\overline{VT}$ 5. 6 cm
Page 261 13. ∠ JNM, ∠ JNK, ∠ KNL, ∠ LNM
15. ∠ JNM, ∠ MNL; ∠ JNM, ∠ JNK; ∠ JNK, ∠ KNL;
∠ KNL, ∠ LNM 17. $\overline{KM}$
Page 262 1. 74 cm 3. 155 dm 5. 14 m 7. 320 yd;
2 (100 + 60) 9. 84 cm 11. 10.8 km; 4(2.7)
Page 264 1. 6 cm 3. 6.2 m 5. 4 cm 7. $\frac{9}{16}$ ft
9. 17.3 cm 11. 4 dm 13. 2.1 mm 15. 44 mm
Page 265 1. ∠ Q 3. ∠ X 5. $\overline{XY}$ 7. $\overline{RQ}$ 9. ∠ J
11. ∠ K 13. $\overline{GJ}$ 15. $\overline{DA}$ 17. $\overline{KH}$
Page 266 1. 35 m^2 3. 26.25 mm^2; 10.5(2.5)
5. 63 ft^2 or 7 yd^2; 9(7) or $3\left(\frac{7}{3}\right)$ 7. 360 in.^{2} or
$2\frac{1}{2}$ ft^2; 20(18) or $1\frac{2}{3}\left(1\frac{1}{2}\right)$ 9. 144 cm^2; 12(12)
11. 4.41 cm^2; 2.1(2.1) 13. 0.09 m^2; 0.3(0.3)
Page 267 15. 1024 yd^2 17. 41 m^2 19. 48 m^2
Page 268 1. 24 in.^{2} 3. 12.5 m^2 5. 40 m^2
7. 20.88 cm^2 9. 9600 cm^2 11. 15 in.^{2} 13. c
Page 269 1. 525 cm^2 3. 30 m^2 5. 18 yd^2
7. 2250 cm^2
Page 270 1. 48 cm^2 3. 32 m^2 5. 50 m^2 7. 385 m^2
9. 50.6 cm^2 11. 18 in.^{2}
Page 271 13. 16 cm^2 15. 24 ft^2 17. 153 mm^2
Page 272 1. 7 m 3. 3 m 5. 10 mm 7. 12 km

9. 8 ft 11. 15 yd 13. 4.2 in. 15. 4.5 dm
Page 273 17. 7 cm 19. $2\frac{2}{3}$ ft 21. 10 cm

23. 35 mm
Page 274 1. 12.56 cm 3. 50.24 m 5. 348.54 ft
7. 5.024 cm 9. 314 ft 11. 62.8 mm 13. 251.2 mm
15. 21.98 m 17. 47.1 mm 19. 28.888 in.
Page 275 21. 88 in. 23. 660 ft 25. 110 mm
27. 22 ft 29. 176 km 31. 24π mm 33. 30π m
35. $7\frac{1}{2}\pi$ ft 37. 400π yd 39. 0.3π cm
Page 276 1. 3.1 in.² 3. 314 m² 5. 12.6 km²
7. 19.6 m² 9. 616 mm² 11. 1386 cm² 13. $21\frac{21}{32}$ ft²
15. 346.5 mm²
Page 277 17. 17 mm; 907.46 mm² 19. 15 mm;
706.5 mm² 21. 4π m² 23. 9π in.²

12 Ratio and Proportion
Page 288 1. 9 to 9; 9 : 9; $\frac{9}{9}$ 3. 9 to 3; 9 : 3; $\frac{9}{3}$
5. 9 to 2; 9 : 2; $\frac{9}{2}$
Page 289 11. 4 : 3 13. 5 : 2 15. $\frac{1}{\$2}$ 17. $\frac{1}{5}$
19. $\frac{50}{1}$ 21. $\frac{5}{1}$ 23. $\frac{55}{1}$ 27. $\frac{8}{5}$ 29. $\frac{25}{3}$ 31. $\frac{5}{4}$ 33. $\frac{5}{2}$
Page 291 3. $\frac{4}{5}=\frac{8}{10}=\frac{16}{20}=\frac{32}{40}=\frac{64}{80}=\frac{96}{120}$ 5. $\frac{4}{5}$
7. $\frac{4}{1}$ 9. $\frac{6}{7}$ 11. $\frac{4}{3}$ 13. $\frac{5}{8}$ 15. $\frac{3}{2}$ 25. $\frac{9}{1}$ 27. $\frac{3}{4}$
29. $\frac{1}{4}$ 31. $\frac{1}{2}$ 33. b 35. a 37. $\frac{1}{33¢}$ 39. $\frac{3}{1}$ 41. $\frac{5}{1}$
Page 292 1. $\frac{30}{10}=\frac{15}{5}$ 3. $\frac{5}{250}=\frac{2}{100}$, $\frac{5}{2}=\frac{250}{100}$
Page 293 5. 192 = 192; yes 7. 120 ≠ 240; no
9. 54 ≠ 36; no 11. 9 ≠ 729; no 13. 180 = 180; yes
15. 25 ≠ 125; no
Page 294 1. 100; 10 3. 49; 1 5. 8t; 5.25
7. 3d = 105; 35 9. 21 11. 64
Page 295 13. a 15. c 17. 18 19. 3 21. 5 23. 6
25. 28 27. 6 29. 9 31. 9 33. 27 35. 60.5
37. 8 39. 3 41. 5.18
Page 296 1. 9 ft 3. 12 ft 5. 8 ft × 3 ft
Page 297 9. 0.7 cm × 1.1 cm 11. 2.4 cm × 3 cm
Page 298 1. 224 mi 3. 112 mi 5. 175 mi 7. 182 mi
Page 299 13. 175 km 19. 2.7 km 21. 11.1 km
Page 300 1. $\frac{6}{7.2}=\frac{6.5}{7.8}$; 46.8 = 46.8 3. $\frac{FG}{LK}=\frac{GH}{KJ}=\frac{HF}{JL}$,
$\frac{0.8}{2.4}=\frac{0.4}{1.2}=\frac{0.5}{1.5}=\frac{1}{3}$, yes
Page 301 5. 3 cm
Page 302 1. 8.4 m
Page 303 3. 40 m

13 Percent
Page 314 3. 0.05 5. 41% 7. 98% 9. 5%
11. 25% 13. 5% 15. 29% 17. 37.5% 19. 1.5%
21. 1.1%
Page 315 23. 35 : 100; $\frac{7}{20}$; 0.35; 35% 25. 12 : 100;
$\frac{3}{25}$; 0.12; 12% 27. 37 : 100; $\frac{37}{100}$; 0.37; 37%
29. 88 : 100; $\frac{22}{25}$; 0.88; 88% 31. 65 : 100;
$\frac{13}{20}$; 0.65; 65% 33. 80% 35. 20% 37. 8%
39. $87\frac{1}{2}$% 41. 2% 43. 15% 45. 90% 47. 36%
49. 28% 51. 10%
Page 316 1. 0.32 3. 0.77 5. 0.85 7. 0.07
9. 0.04 11. 0.06 13. 0.009 15. $\frac{9}{20}$ 17. $\frac{1}{5}$ 19. $\frac{21}{100}$
21. $\frac{7}{100}$ 23. $\frac{11}{20}$ 25. $\frac{41}{50}$ 27. $\frac{3}{100}$ 29. $\frac{1}{3}$ 31. $\frac{1}{16}$
Page 317 1. 3 3. 2.48 5. 3.13 7. 2.08 9. 1.02
11. 5.18 13. 2.89 15. $\frac{52}{25}$ 17. 3 19. $\frac{9}{5}$ 21. $\frac{29}{25}$
23. $\frac{9}{4}$ 25. $\frac{31}{10}$ 27. 400% 29. 120% 31. 158%
33. 108% 35. 130% 37. 350% 39. 125%
41. 500%
Page 318 1. 125% 3. 0.3% 5. 333.3% 7. 1.3%

9. 30% 11. 0.5% 13. 0.9% 15. 837.5%
17. 110% 19. 335% 21. 450% 23. 4%
25. 37.5% 27. $11\frac{1}{9}$% 29. $14\frac{2}{7}$% 31. $83\frac{1}{3}$%
33. 4% 35. $18\frac{2}{11}$% 37. $71\frac{3}{7}$% 39. $3\frac{1}{3}$%
41. 2.5% 43. $6\frac{2}{3}$% 45. 18.75% 47. $28\frac{1}{8}$%
49. 87.5% 51. $46\frac{2}{3}$%
Page 319 1. 0.809; $\frac{809}{1000}$ 3. 0.0375; $\frac{3}{80}$ 5. 0.003;
$\frac{3}{1000}$ 7. 0.181; $\frac{181}{1000}$ 9. 0.0075; $\frac{3}{400}$ 11. 0.125; $\frac{1}{8}$
13. 0.206; $\frac{103}{500}$ 15. 0.089; $\frac{89}{1000}$ 17. 0.0005; $\frac{1}{2000}$
19. 0.048; $\frac{6}{125}$ 21. $0.09\frac{1}{3}$; $\frac{7}{75}$ 23. 0.08125; $\frac{13}{160}$
Page 320 1. because equivalent fractions have the same percent equivalents 3. 10%, 20%, 30%,..., 90%
5. $9\frac{1}{11}$%, $18\frac{2}{11}$%, $27\frac{3}{11}$%, ..., $90\frac{9}{11}$% 7. = 9. <
11. = 13. < 15. < 17. >
Page 321 19. $\frac{3}{25}$; 12% 21. $\frac{8}{25}$; 32% 23. >
25. < 27. = 29. > 31. <
Page 322 1. 43 3. 410 5. 192 7. 30 9. 48
11. 110 13. 36 15. 3.2 17. $4.50 19. $5.17
Page 323 21. 26.4 23. 1 25. 24 27. 13.5 29. 9
31. 7.5 33. 325; 331.5 35. 8; 14 37. 60; 63.63
39. 12; 12.312 41. 27; 26.7 43. 90; 88.5
45. 4; 4.4 47. 2.8; 2.96
Page 324 1. 20% 3. 75% 5. 37.5% 7. $14\frac{2}{7}$%
9. 50% 11. 50% 13. 2.5% 15. $33\frac{1}{3}$% 17. 25%
19. $33\frac{1}{3}$%
Page 325 21. 25% 23. $33\frac{1}{3}$% 25. 2% 27. 20%
29. 12 31. 0.012 33. 6.48 35. 54 37. 450
39. 72 41. 10,800 43. 14,400
Page 326 1. 400 3. 60 5. 31.25 7. 72
Page 327 9. 40 11. 80 13. 108 15. 200
17. 15 19. 20 21. 200 23. 21

14 Consumer Mathematics
Page 338 1. $120; $240 3. $.90; $5.10
5. $7.50; $7.50
Page 339 7. $627; 5% 9. $5.44; 15%
11. $773.50; 9% 13. $19.47; $6.49
Page 340 1. $22.50, $472.50 3. $1.60, $41.60
5. $32.41, $958.41
Page 341 7. $.05, $.80 9. $1.23, $21.63
11. $1.23, $21.63
Page 342 1. $58.50 3. $20 5. $35.40 7. $1800
9. $26.95 11. $250
Page 343 13. $228 15. 4% 17. 9%
Page 344 1. $25, $525 3. $34, $459 5. $\frac{7}{12}$ 7. $\frac{2}{3}$
9. $3\frac{1}{2}$ 11. $2\frac{3}{4}$ 13. $2\frac{1}{6}$
Page 345 15. $6 17. $99 19. $169.40
21. $1418.63 23. $533 25. $725, $5725
27. $13.60, $213.60 29. $57, $437
Page 346 1. $60, $810 3. $300, $2800 5. $190,
$3990 7. $153.90, $1773.90 9. $285, $1235
11. $72, $1272 13. $31.88, $406.88
Page 347 15. $20 17. $27 19. $450
Page 348 1. 30 for $4.45 3. 2 liters for 79¢
5. 1 pint for 99¢ 7. 145 count for $1.10 9. 8 6-oz
cans for $1
Page 349 11. 1 48-oz bottle for $7 13. 6 pairs for
$6.99 15. $3.84 17. $6 19. $6.40
Page 350 1. Fifty-six and $\frac{85}{100}$ 3. Thirty-three and $\frac{xx}{100}$
Page 351 11. 7-6-92 13. Town Telephone Co.
15. 223.52; 207.57; 28.16; 159.72
Page 352 1. $120 3. $560 5. $160 7. $560
11. $\frac{30}{200}=\frac{3}{20}$; 15% 13. $\frac{12}{200}=\frac{3}{50}$; 6%

Page 353 **15.** $350,000 **17.** $1,500,000 **19.** $750
21. $500 **23.** $375

15 Surface Area and Volume
Page 364 **9.** 6, 12 **11.** 10,15
Page 365 **15.** 5 **17.** Base is not a polygon.
19. hemisphere **21.** rectangular pyramid **23.** cone
Page 366 **1.** 3, 6; 1, 5; 2, 4 **3.** 54 in.^{2} **5.** 99 in.^{2}
7. 198 in.^{2}
Page 367 **9.** 252 cm^2 **11.** 237 mm^2 **13.** 33.2 m^2
15. 69 in.^{2} **17.** 1536 in.^{2} **19.** 253.5 m^2
21. $541\frac{1}{2}$ yd^2
Page 368 **1.** 3; 3, 4, 5 **3.** 1, 2 **5.** 12 cm^2
7. 66 cm^2
Page 369 **9.** 304 in.^{2} **11.** 116 in.^{2} **13.** 93 ft^2
Page 370 **1.** 150.72 ft^2 **3.** 5495 m^2
Page 371 **5.** 477.28 in.^{2} **7.** 339.12 cm^2 **9.** 20π ft^2
11. 196π yd^2 **13.** 810π in.^{2}
Page 372 **1.** 630 in.^{3} **3.** 1386 cm^3 **5.** 27 in.^{3}
7. 857.375 ft^3 **9.** 2197 in.^{3}
Page 373 **1.** 88.8 cm^3 **3.** 96 in.^{3} **5.** 10,800 ft^3
7. 13.5 yd^3
Page 374 **1.** 100.48 cm^3 **3.** 18.84 yd^3 **5.** 4.71 cm^3
Page 375 **11.** 700π mm^3 **13.** 1764π cm^3
15. 150π ft^3 **17.** 500π cm^3

16 Probability, Statistics, and Graphing
Page 386 **1.** 3 **3.** 8 **5.** no
Page 387 **7.** 3 **9.** red **11.** 4 **13.** P **15.** yes
17. no
Page 388 **1.** $\frac{1}{4}$ **3.** 0 **5.** 1 **7.** 0 **9.** $\frac{1}{2}$ **11.** 1
Page 389 **13.** $\frac{2}{7}$ **15.** $\frac{5}{7}$ **17.** 0 **19.** $\frac{5}{7}$ **21.** 1 **23.** $\frac{1}{2}$
25. $\frac{1}{2}$ **27.** $\frac{1}{6}$ **29.** $\frac{1}{6}$ **31.** $\frac{1}{3}$ **33.** $\frac{1}{3}$ **35.** 0 **37.** $\frac{1}{4}$
39. $\frac{1}{2}$
Page 390 **1.** $\frac{1}{4}$ **3.** $\frac{3}{16}$ **5.** $\frac{1}{6}$ **7.** $\frac{1}{4}$
Page 391 **9.** (1, H) (2, H) (3, H) (1, T) (2, T) (3, T)
11. $\frac{1}{3}$ **13.** $\frac{1}{6}$ **15.** $\frac{1}{4}$ **17.** $\frac{1}{2}$ **19.** $\frac{1}{18}$ **21.** $\frac{1}{9}$ **23.** $\frac{1}{3}$
25. $\frac{2}{3}$
Page 392 **1.** 120 **3.** 720 **5.** 2 **7.** 1 **9.** 25
11. 5038
Page 393 **15.** $\frac{3}{28}$ **17.** $\frac{3}{5}$ **19.** $\frac{1}{2}$ **21.** T **23.** F **25.** T
Page 394 **1.** line **3.** circle
Page 400 **1.** 36° **3.** 144° **5.** 216° **7.** 45° **9.** 240°
Page 401 **11.** 40% = 144°; 25% = 90°; 10% = 36°;
15% = 54°; 5% = 18°
Page 402 **1.** 86; 85; 85; 20 **3.** $24; $25; $26; $15
5. 1006; 1003.5; none; 1017
Page 404 **1.** ($^-$1, $^-$2) **3.** ($^-$4, $^+$3) **5.** ($^-$4, $^-$3)
7. ($^+$3, $^-$1) **9.** F **11.** A and E
Page 405 **17.** ($^-$2, $^-$4) **19.** (0, 0)
Page 406 **1.** $^+$1, $^+$1, ($^-$1, $^+$1); $^+$2, $^+$2, (0, $^+$2);
$^+$3, $^+$3, ($^+$1, $^+$3); $^+$4, $^+$4, ($^+$2, $^+$4)

17 Algebra and Rational Number Topics
Page 420 **1.** monomial **3.** monomial **5.** trinomial
7. trinomial **9.** trinomial **11.** binomial **13.** binomial
15. trinomial **17.** monomial **19.** trinomial
Page 421 **21.** 7, x, 2 **23.** 6, m, 4; $^-$2, m, 3
25. none **27.** 28b^3d^2 **29.** none **31.** 2 **33.** 0 **35.** 3
37. 1 **39.** 3 **41.** 0 **43.** 3 **45.** 3 **47.** 6 **49.** 1

51. 5 **53.** 3 **55.** 5 **57.** 3 **59.** 0
Page 422 **1.** 11 **3.** $^-$45 **5.** 23 **7.** $^-$9 **9.** $3\frac{1}{6}$ **11.** 0
13. 102 **15.** 22 **17.** 81 **19.** $^-$16 **21. a.** 16; **b.** 4;
c. 25; **d.** 4; **e.** 289; **f.** 9 **23.** T **25.** T **27.** F **29.** F
Page 423 **31.** $^-$2 + 6x + 8x^3 **33.** $^-$9 + m + m^2
35. $^-$18 + 100t + t^2 **37.** t^2 - t^3 + 6t^4 **39.** t^5 + t +
25 **41.** $-r$ + 9 **43.** 11k^2 - 20k + 16 **45.** x^5 - x^2 +
x - 2 **47.** 5x^3y^2 + 10x^2y - 18xy + y^3 **49.** $-3x^2y$ +
10xy + 9 **51.** x^3 + x^2y + xy^3 + 7
Page 424 **1.** b, c **3.** a, b **5.** b, c **7.** 23s **9.** 32x
11. 19k **13.** $^-$5st **15.** $^-$8z **17.** 14m^4 **19.** 22k
21. $^-$4t^2 **23.** 4b^3 **25.** 0 **27.** 2d
Page 425 **29.** 8x + 8y **31.** 6b - 5a^2
33. $^-$5k - 7m **35.** 39s - 13r **37.** 27c + 7c^2
39. 9m^2 - 12m **41.** 12d^2 + 26 **43.** 4y
45. 5r^2s + 9rs^2 **47.** 9x^2 + 10xy + 8
49. 9x^2y - 13xy^2 **51.** 7x^2y + 4xy + 6
53. 7x^4 + x^3 - 3x
Page 427 **1.** 16x **3.** $^-$48y^2 **5.** $^-$45b^6 **7.** $^-$24ab^3
9. $^-$56c^4d^3 **11.** 99a^3c^2 **13.** 100k^4 **15.** $^-$12p^6q^3
17. 240b^2c **19.** 40m^5 **21.** $\frac{1}{27}m^2n$ **23.** 10b^5
25. 9ab **27.** $^-$5x^2y^2 **29.** $^-$5d^3e **31.** 3s^2t^2 **33.** 12gh
35. 7w^3 **37.** $^-$3e **39.** $^-$3x^3 **41.** 14x^3 **43.** 11x^3
45. 25y^4 **47.** $^-$3c **49.** 10e^2
Page 429 **1.** 41x - 6 **3.** $^-$4k^2 - 23k **5.** 2x - 5y
7. 15x^2 - 24x **9.** 11xy + 7x **11.** 0.67e + 0.48f
13. 121a^2 + 100a **15.** $^-$101b - 66c **17.** 34c - d
19. 8.7x **21.** 3x + 4y **23.** 11c - 11b **25.** $^-$9x - 9y
27. $^-$18x + 16y **29.** $^-$32cd + 19 **31.** $^-$28a^2 + 12a
33. 26rs - 13s^2 **35.** 63k^2 + 45k **37.** 161t^2 - 55t
39. 0.68k + 2.1k^2 **41.** 7a^2b + 1.5b **43.** $^-$12t - 16s - 6
45. 25p - 3p^2 **47.** $^-$23r - 6s + 6
Page 430 **1.** 20 **3.** m^2 **5.** 14d + 2d^2
7. a^3 + 5a^2 **9.** 6t^2 - 6t^2 or 0 **11.** 14ab + 7b
13. v^3 - 5v^2 + 6v **15.** 3a + 3b **17.** $^-$4b + 5c
19. 2x^3 + 6x^2y **21.** 48r^3s - 32rs^3 **23.** $^-$24x^3y +
28x^2y **25.** 32r^2m^2n + 12r^2n^2
Page 431 **27.** 6mn - 3m^2 **29.** ab + 2ab^2 - 3a^2b
31. 8x^3 + 12x^3y - 12x^2 **33.** d + 2de - 8d^2
35. 3s^3t - 2s^2t^2 - 3st^3 **37.** 3a^3b^2 + 5a^2b^2 - 2a^2b^3
39. 15a^2 - 24ab + 6 **41.** $^-s^2t$ + st + 4s - t
43. 6c^3 - 2c^3d - 3c^2d **45.** $^-$3m^3n + 4m^2 + 3mn
Page 433 **1.** 4 **3.** 4 + n **5.** y + 4
7. 3b + 2 **9.** 2 + 3r **11.** 10a + 3 **13.** 6m - 1 + 2n
15. 5a - c + 2d **17.** 3st - t - 2 **19.** 3x^2 - x - 2
21. 2m^2n^2 + m^2n - 3m **23.** x^2y^2z - 2xy + 7
25. k^4m^4 - 3k^2m^5 + 4 **27.** c^3 - 3c^2 + 2 **29.** m^3n^2
+ m^2n + mn^2 **31.** 1 + 3c^2 + 9c^3 **33.** 7 + 9y^3
35. 2y^2 - 3y^4 **37.** 3 - fg - 4fg^2 **39.** 7 - 21b + ab^2c
Page 434 **1.** $^-$5.5 **3.** $^+$2.0 **5.** $^-$14.7 **7.** $^-$4.4
9. $^-$15.2 **11.** $^-$12.1
Page 435 **13.** $^+$3.7 **15.** $^+$4.1 **17.** $^-$2.3 **19.** $^-$4.1
21. $^+$25.61 **23.** $^+$6 **25.** $^-$18.62 **27.** $^-$1.707
29. $^+$1.1 **31.** $^-$30.5 **33.** 11.1 **35.** $^-$18.1 **37.** >
39. > **41.** > **43.** >
Page 436 **1.** $^+$0.24 **3.** $^-$0.12 **5.** $^-$0.48 **7.** $^+$0.012
9. $^+$0.096 **11.** $^+$1.44 **13.** $^+$7.29 **15.** $^+$0.0196
Page 437 **17.** $^-$3.87 **19.** $^-$21.942 **21.** $^-$0.03582
23. $^+$10.8772 **25.** $^-$80 **27.** $^-$35.2 **29.** $^-$24
31. $^+$3.2 **33.** $^+$8.4 **35.** $^-$1.02 **37.** $^-$58 **39.** $^-$94
41. $^-$4 **43.** $^-$60.3 **45.** $^+$0.217 **47.** $^+$8.7 **49.** $^+$4.42
51. $^+$2 **53.** $^+$0.4

Tables for Measures

Length

1 millimeter (mm) = 0.001 meter (m)

1 centimeter (cm) = 0.01 meter

1 decimeter (dm) = 0.1 meter

1 dekameter (dam) = 10 meters

1 hectometer (hm) = 100 meters

1 kilometer (km) = 1000 meters

Mass

1 milligram (mg) = 0.001 gram (g)

1 kilogram (kg) = 1000 grams

1 metric ton (t) = 1000 kilograms

Capacity

1 milliliter (mL) = 0.001 liter (L)

1 kiloliter (kL) = 1000 liters

Temperature

0° Celsius (C) .... Water freezes.

100° Celsius (C) .... Water boils.

Customary Units

Length

1 foot (ft) = 12 inches (in.)

1 yard (yd) = 36 inches

1 yard (yd) = 3 feet

1 mile (mi) = 5280 feet

1 mile (mi) = 1760 yards

Capacity

3 teaspoons (tsp) = 1 tablespoon (tbsp)

1 cup (c) = 8 fluid ounces (fl oz)

1 pint (pt) = 2 cups

1 quart (qt) = 2 pints

1 quart (qt) = 4 cups

1 gallon (gal) = 4 quarts

Weight

1 pound (lb) = 16 ounces (oz)

1 ton (T) = 2000 pounds

Temperature

32° Fahrenheit (F) ... Water freezes.

212° Fahrenheit (F) ... Water boils.

Percent Table

$25\% = \dfrac{1}{4}$

$50\% = \dfrac{1}{2}$

$75\% = \dfrac{3}{4}$

$10\% = \dfrac{1}{10}$

$20\% = \dfrac{1}{5}$

$30\% = \dfrac{3}{10}$

$40\% = \dfrac{2}{5}$

$60\% = \dfrac{3}{5}$

$15\% = \dfrac{3}{20}$

$6\dfrac{1}{4}\% = \dfrac{1}{16}$

$8\dfrac{1}{3}\% = \dfrac{1}{12}$

$70\% = \dfrac{7}{10}$

$80\% = \dfrac{4}{5}$

$90\% = \dfrac{9}{10}$

$12\dfrac{1}{2}\% = \dfrac{1}{8}$

$37\dfrac{1}{2}\% = \dfrac{3}{8}$

$62\dfrac{1}{2}\% = \dfrac{5}{8}$

$87\dfrac{1}{2}\% = \dfrac{7}{8}$

$33\dfrac{1}{3}\% = \dfrac{1}{3}$

$3\dfrac{1}{3}\% = \dfrac{1}{30}$

$6\dfrac{2}{3}\% = \dfrac{1}{15}$

$8\% = \dfrac{2}{25}$

$66\dfrac{2}{3}\% = \dfrac{2}{3}$

$16\dfrac{2}{3}\% = \dfrac{1}{6}$

$83\dfrac{1}{3}\% = \dfrac{5}{6}$

$9\dfrac{1}{11}\% = \dfrac{1}{11}$

$11\dfrac{1}{9}\% = \dfrac{1}{9}$

$14\dfrac{2}{7}\% = \dfrac{1}{7}$

$12\% = \dfrac{3}{25}$

$5\% = \dfrac{1}{20}$

$1\% = \dfrac{1}{100}$

$2\% = \dfrac{1}{50}$

$4\% = \dfrac{1}{25}$

$\dfrac{1}{2}\% = \dfrac{1}{200}$